Partial Differential Equations

Analytical Solution Techniques

The Wadsworth & Brooks/Cole Mathematics Series

M. Adams, V. Guillemin, *Measure Theory and Probability*

W. Beckner, A. Calderón, R. Fefferman, P. Jones, *Conference on Harmonic Analysis in Honor of Antoni Zygmund*

G. Chartrand, L. Lesniak, *Graphs & Digraphs, Second Edition*

J. Cochran, *Applied Mathematics: Principles, Techniques, and Applications*

W. Derrick, *Complex Analysis and Applications, Second Edition*

J. Dieudonné, *History of Algebraic Geometry*

R. Dudley, *Real Analysis and Probability*

R. Durrett, *Brownian Motion and Martingales in Analysis*

R. Epstein, W. Carnielli, *Computability: Computable Functions, Logic, and the Foundations of Mathematics*

S. Fisher, *Complex Variables*

A. Garsia, *Topics in Almost Everywhere Convergence*

P. Garrett, *Holomorphic Hilbert Modular Forms*

J. Kevorkian, *Partial Differential Equations: Analytical Solution Techniques*

R. McKenzie, G. McNulty, W. Taylor, *Algebras, Lattices, Varieties, Volume I*

E. Mendelson, *Introduction to Mathematical Logic*

R. Salem, *Algebraic Numbers and Fourier Analysis* and L. Carleson, *Selected Problems on Exceptional Sets*

R. Stanley, *Enumerative Combinatorics, Volume I*

J. Strikwerda, *Finite Difference Schemes and Partial Differential Equations*

K. Stromberg, *An Introduction to Classical Real Analysis*

Partial Differential Equations
Analytical Solution Techniques

J. Kevorkian
University of Washington

Wadsworth & Brooks/Cole Advanced Books & Software
Pacific Grove, California

Wadsworth & Brooks/Cole Advanced Books & Software

© 1990 by Wadsworth, Inc., Belmont, California 94002. All rights reserved. No part of this book may be reproduced, stored in a retrieval system, or transcribed, in any form or by any means—electronic, mechanical, photocopying, recording, or otherwise—without the prior written permission of the publisher, Brooks/Cole Publishing Company, Pacific Grove, California 93950, a division of Wadsworth, Inc.

Printed in the United States of America

10   9   8   7   6   5   4   3   2   1

Library of Congress Cataloging-in-Publication Data

Kevorkian, J.
    Partial differential equations: analytical solution techniques/J. Kevorkian.
        p.   cm.
    Includes bibliographical references.
    ISBN 0-534-12216-7
    1. Differential equations, Partial.   I. Title.
QA377.K48   1989
515'.353—dc20
                                            89-22096
                                          CIP

Sponsoring Editor: *John Kimmel*
Marketing Representative: *Karen Buttles*
Editorial Assistant: *Mary Ann Zuzow*
Production Editor: *Penelope Sky*
Manuscript Editor: *Linda L. Thompson*
Interior Design: *Sharon L. Kinghan*
Cover Design: *Flora Pomeroy*
Art Coordinator: *Lisa Torri*
Interior Illustration: *John Foster*
Typesetting: *Asco Trade Typesetting Ltd., Hong Kong*
Cover Printing: *Phoenix Color Corporation, New York*
Printing and Binding: *Arcata Graphics/Fairfield, Pennsylvania*

Preface

This is a text for a two-semester or three-quarter sequence of courses in partial differential equations. It is assumed that the student has a good background in vector calculus and ordinary differential equations and has been introduced to such elementary aspects of partial differential equations as separation of variables, Fourier series, and eigenfunction expansions. Some familiarity is also assumed with the application of complex variable techniques, including conformal mapping, integration in the complex plane, and the use of integral transforms.

Linear theory is developed in the first half of the book and quasilinear and nonlinear problems are covered in the second half, but the material is presented in a manner that allows flexibility in selecting and ordering topics. For example, it is possible to start with the scalar first-order equation in Chapter 5, to include or delete the nonlinear equation in Chapter 6, and then to move on to the second-order equations, selecting and omitting topics as dictated by the course. At the University of Washington, the material in Chapters 1–4 is covered during the third quarter of a three-quarter sequence that is part of the required program for first-year graduate students in Applied Mathematics. We offer the material in Chapters 5–8 to more advanced students in a two-quarter sequence.

The primary purpose of this book is to analyze the formulation and solution of representative problems that arise in the physical sciences and engineering and are modeled by partial differential equations. To achieve this goal, all the basic physical principles of a given subject are first discussed in detail and then incorporated into the analysis. Although proofs are often omitted, the underlying mathematical concepts are carefully explained. The emphasis throughout is on deriving explicit analytical results, rather than on the abstract properties of solutions. Whenever a new idea is introduced, it is illustrated with an example from an appropriate area of application. There is a selection of additional problems at the end of each chapter, ranging in difficulty from straightforward extensions of the textual material to rather challenging departures, testing the student's skill at application. In addition, there are selected review exercises for

Chapters 1 and 2 that illustrate some of the prerequisite ideas that are not explicit in the text.

The numerical solution of partial differential equations is a vast topic requiring a separate volume; here, the emphasis is on analytical techniques. Numerical methods of solution are mentioned only in connection with particular examples and, more generally, to illustrate the solution of hyperbolic problems in terms of characteristic variables. Certain analytical techniques covered in specialized texts have also been left out. The notable omissions concern the asymptotic expansion of solutions obtained by integral transforms, integral equation methods, the Weiner-Hopf method, and inverse scattering theory.

Acknowledgments

I dedicate this book with admiration and gratitude to Paco A. Lagerstrom and Julian D. Cole, who got me started in this field. I thank my wife, Seta, without whose patience and support this book would have remained just a plan. I appreciate the help of Radhakrishnan Srinivasan and David L. Bosley, who read the manuscript and made valuable comments and corrections. I applaud the skill and efficiency demonstrated by Lilly Harper, who transformed barely legible handwriting into flawless copy throughout the manuscript.

J. Kevorkian

Contents

C H A P T E R   3

The Wave Equation   **117**

Partial Differential Equations
Analytical Solution Techniques

CHAPTER 1

The Diffusion Equation

In this chapter, we study the one-dimensional diffusion equation

$$u_t - u_{xx} = 0 \tag{1}$$

which describes a number of physical models, such as the conduction of heat in a solid or the spread of a contaminant in a stationary medium.

We shall use (1) as a model to illustrate various solution techniques that will be useful in our study of other types of partial differential equations. To begin with, it is important to have a physical understanding of how (1) arises, and we consider the simple model of heat conduction in a solid.

1.1   Heat Conduction

Consider a thin rod of some heat-conducting material with variable density $\rho(x)\,(\mathrm{g/cm^3})$ (for example, a copper-silver alloy with a variable copper/silver ratio along the rod). Let $A(x)\,(\mathrm{cm^2})$ denote the cross-sectional area and assume that the surface of the rod is perfectly insulated so that no heat is lost or gained through this surface. (See Figure 1.1.) Thus, the problem is one-dimensional in the sense that all material properties depend on the distance x along the rod. We assume that at each spatial position x and time t, there is one temperature θ that does not depend on the transverse coordinates y or z. Let x_1 and x_2 be two arbitrary fixed points on the axis.

In the basic law of conservation of heat energy for the rod segment $x_1 \leq x \leq x_2$, the rate of change of heat inside this segment is equal to the net flow of heat through the two boundaries at x_1 and x_2. Consider an infinitesimal section of length dx in the interval $x_1 \leq x \leq x_2$. Using elementary physics, we have dQ, the heat content in this section, proportional to the mass and the temperature:

$$dQ \equiv c(\rho A\ dx)\theta \tag{2}$$

where the constant of proportionality c is the specific heat in $(\mathrm{cal/g°C})$. Thus, the

1

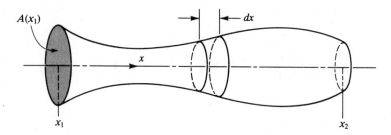

Figure 1.1

total heat content in the interval $x_1 \le x \le x_2$ is*

$$Q(t) \equiv \int_{x_1}^{x_2} c(x)\rho(x)A(x)\theta(x,t)\,dx \tag{3}$$

Next, we invoke Fourier's law for heat conduction, which states that the rate of heat flowing *into* a body through a small surface element on its boundary is proportional to the area of that element and to the *outward* normal derivative of the temperature at that location. The constant of proportionality here is $k \sim$ (cal/cm s °C), the *thermal conductivity*. Note that this sign convention implies the intuitively obvious fact that the direction of heat flow between two neighboring points is toward the relatively cooler point. For example, if the temperature increases as a boundary point is approached from inside a body, then the outward normal derivative of the temperature is positive, and this correctly implies that heat flows into the body.

For the present one-dimensional example, the net inflow of heat through the boundaries x_1 and x_2 is

$$R(t) \equiv A(x_2)k(x_2)\frac{\partial\theta}{\partial x}(x_2,t) - A(x_1)k(x_1)\frac{\partial\theta}{\partial x}(x_1,t)$$

The conservation of heat then implies

$$\frac{dQ}{dt} = R(t) \tag{4}$$

or

$$\frac{d}{dt}\int_{x_1}^{x_2} c(x)\rho(x)A(x)\theta(x,t)\,dx = A(x_2)k(x_2)\frac{\partial\theta}{\partial x}(x_2,t) - A(x_1)k(x_1)\frac{\partial\theta}{\partial x}(x_1,t) \tag{5}$$

Equation (5) is a typical integral *conservation law*, which has general applicability. For example, (5) remains true if material properties have a discontinuity at a given point $x = \xi$ inside the interval, as would be the case if we had a perfect thermal bond between two rods of different materials. We shall encounter other

---

* In this text we shall often use the notation $\equiv$ instead of $=$ when it is important to indicate that a new quantity is being defined, as in (2) and (3). As a special case of this notation, the statement $f(x,y) \equiv 0$ indicates that the function f of x and y vanishes identically; that is, it equals zero for all x and y by definition.

examples of such conservation laws later on in the book and shall study how discontinuities propagate in detail in Chapter 5.

For smooth material properties—that is, if c, ρ, A and k are continuous with continuous first derivatives—the solution $\theta(x, t)$ is also continuous with continuous first partial derivatives $\partial\theta/\partial x$ and $\partial\theta/\partial t$, and we may rewrite (5) in the following form after we express the right-hand side as the integral of a derivative:

$$\int_{x_1}^{x_2} \left\{ c(x)\rho(x)A(x)\frac{\partial\theta}{\partial t}(x, t) - \frac{\partial}{\partial x}\left[A(x)k(x)\frac{\partial\theta}{\partial x}(x, t) \right] \right\} dx = 0 \tag{6}$$

Since (6) is true for any x_1 and x_2, it follows that the integrand must vanish:

$$c(x)\rho(x)A(x)\frac{\partial\theta}{\partial t} - \frac{\partial}{\partial x}\left[A(x)k(x)\frac{\partial\theta}{\partial x} \right] = 0 \tag{7}$$

For constant material properties, this reduces to

$$\frac{\partial\theta}{\partial t} - \kappa^2 \frac{\partial^2\theta}{\partial x^2} = 0 \tag{8}$$

where $\kappa^2 = k/c\rho$ (cm^2/s) is the *thermal diffusivity*. The dimensionless form (1) follows from (8) when characteristic constants with dimensions of temperature, length, and time are used to define nondimensional variables.

For example, if we wish to solve (8) for a rod of length L, which is initally at a constant temperature θ_0 and has one end, $x = L$, held at $\theta = \theta_0$ while the other end, $x = 0$, has a prescribed temperature history $\theta(0, t) = \theta_0 f(t/T)$, we set

$$u \equiv \frac{\theta}{\theta_0}, \qquad x^* \equiv \frac{x}{L}, \qquad t^* \equiv \frac{t\kappa^2}{L^2}$$

and obtain the following dimensionless formulation:

$$\frac{\partial u}{\partial t^*} - \frac{\partial^2 u}{\partial x^{*2}} = 0, \qquad 0 \le x^* \le 1$$

$$u(x^*, 0) = 1$$

$$u(0, t^*) = f(\lambda t^*), \qquad t^* > 0$$

$$u(1, t^*) = 1$$

where λ is the dimensionless parameter $L^2/\kappa^2 T$.

The corresponding derivation for three-dimensional heat conduction follows from similar steps. If a solid occupies the domain G with surface S and outward unit normal $\mathbf{n}$, as shown in Figure 1.2, the total heat content of the solid is given by

$$Q(t) \equiv \iiint_G c\rho\theta \, dV \tag{9}$$

where dV is the volume element; for instance, $dV = dx \, dy \, dz$ in Cartesian variables. The net *inflow* of heat through the boundary S is

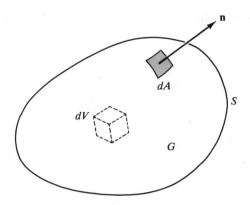

Figure 1.2

$$R(t) \equiv \iint_S k \operatorname{grad} \theta \cdot \mathbf{n} \, dA \tag{10}$$

Therefore, the integral conservation law of heat energy becomes

$$\frac{d}{dt} \iiint_G c\rho\theta \, dV = \iint_S k \operatorname{grad} \theta \cdot \mathbf{n} \, dA \tag{11}$$

where c, ρ, and k may depend on position.

For a medium where c, ρ, and k are smooth, we apply Gauss' theorem to express the right-hand side of (11) as the volume integral*

$$R(t) = \iiint_G \operatorname{div}(k \operatorname{grad} \theta) \, dV \tag{12}$$

and since the boundaries of G are fixed in space, we may rewrite the left-hand side of (11) as

$$\frac{dQ}{dt} = \iiint_G c\rho\theta_t \, dV \tag{13}$$

Therefore, assuming continuity of the integrands in (12) and (13), the three-dimensional version of (8) is

$$c\rho\theta_t - \operatorname{div}(k \operatorname{grad} \theta) = 0 \tag{14a}$$

For constant k, this reduces to

$$\theta_t - \kappa^2 \Delta\theta = 0 \tag{14b}$$

where $\kappa^2 = k^2/c\rho$ and Δ is the Laplacian operator $\Delta \equiv \operatorname{div} \operatorname{grad}$, given by

$$\Delta \equiv \frac{\partial^2}{\partial x^2} + \frac{\partial^2}{\partial y^2} + \frac{\partial^2}{\partial z^2} \tag{15}$$

in Cartesian coordinates.

---

*Thus, in writing $R(t)$ in the form given in (6), we have used the "one-dimensional version" of Gauss' theorem relating the definite integral of a derivative to values of the derivative at the endpoints.

1.2 Fundamental Solution

The fundamental solution of (1) gives the temperature over the infinite domain $-\infty < x < \infty$ and $t > 0$ as a consequence of an instantaneous unit source of heat at $x = 0$. Assume that we introduce this unit source at time $t = 0$ at the origin $x = 0$. Thus, we wish to solve

$$u_t - u_{xx} = \delta(x)\delta(t), \qquad -\infty < x < \infty; 0 \le t \tag{16}$$

with initial condition

$$u(x, 0^-) = 0 \tag{17}$$

Here δ is the Dirac delta function, and the notation $u(x, 0^-)$ indicates

$$u(x, 0^-) \equiv \lim_{t \uparrow 0} u(x, t) \tag{18a}$$

In the future, we shall also have occasion to use the notation

$$u(x, 0^+) = \lim_{t \downarrow 0} u(x, t) \tag{18b}$$

For a concise and systematic discussion of the delta function interpreted as a "symbolic function," see Chapter 3 of [1].

There is no loss of generality in taking the initial temperature equal to zero in (17); any constant value u_0 can be used and then reduced to (17) by simply considering a new dependent variable $u - u_0$. This is a consequence of the absence of nondifferentiated terms in (16).

Also, since the left-hand side of (16) does not involve x or t, we can translate the solution just obtained to determine the result for a source located at $x = \xi$ and turned on at $t = \tau$. More precisely, if $u = F(x, t)$ is the solution of (16)–(17), then the solution for

$$u_t - u_{xx} = \delta(x - \xi)\delta(t - \tau) \tag{19}$$

$$u(x, \tau^-) = 0 \tag{20}$$

is

$$u = F(x - \xi, t - \tau) \tag{21}$$

Returning to (16)–(17), we now illustrate various solution techniques for calculating the fundamental solution F.

1.2.1 Similarity (Invariance)

In this very useful approach, we ask under what scalings of the dependent and independent variables the system (16) and (17) is invariant. If such scalings exist, we can reduce (16) to an ordinary differential equation in terms of a "similarity" variable using arguments that go as follows.

Assume that we have found the solution of (16)–(17) in the form $u = F(x, t)$. Is it possible to use this result to obtain a second solution $u = G(x, t)$ by setting $\bar{x} = \beta x$ and $\bar{t} = \gamma t$ and defining G by

$$G(x, t) \equiv \alpha F(\beta x, \gamma t) \tag{22}$$

for positive constants α, β, and γ?

We compute

$$G_t = \alpha\gamma F_{\bar{t}} \quad \text{and} \quad G_{xx} = \alpha\beta^2 F_{\bar{x}\bar{x}}$$

and use the fact that for any constant c, we may use

$$\delta(cx) \to \frac{1}{|c|}\delta(x)$$

If $G(x, t)$ is to be a solution of (16)–(17), we must have

$$G_t - G_{xx} = \delta(x)\delta(t), \qquad G(x, 0^-) = 0$$

Expressing G_t and G_{xx} in terms of $F_{\bar{t}}$ and $F_{\bar{x}\bar{x}}$ and using $\delta(x)\delta(t) = \delta(\bar{x}/\beta)\delta(\bar{t}/\gamma) = \beta\gamma\delta(\bar{x})\delta(\bar{t})$ in the preceding equation gives:

$$\alpha\gamma F_{\bar{t}} - \alpha\beta^2 F_{\bar{x}\bar{x}} = \beta\gamma\delta(\bar{x})\delta(\bar{t}), \qquad \alpha F(\bar{x}, 0^-) = 0$$

or

$$F_{\bar{t}} - \left(\frac{\beta^2}{\gamma}\right)F_{\bar{x}\bar{x}} = \left(\frac{\beta}{\alpha}\right)\delta(\bar{x})\delta(\bar{t}), \qquad F(\bar{x}, 0^-) = 0$$

But we know that $F(\bar{x}, \bar{t})$ must satisfy (16)–(17) in terms of the $\bar{x}, \bar{t}$ variables. Therefore, $G(x, t)$, as defined by (22), can be a solution only if $\beta^2/\gamma = 1$ and $\beta/\alpha = 1$; that is, if $\beta = \alpha$ and $\gamma = \alpha^2$. Thus, (22) must be in the form

$$G(x, t) = \alpha F(\alpha x, \alpha^2 t) \tag{23}$$

Have we discovered a new solution of (16)–(17)? Of course not; the solution for this problem is unique, as is physically obvious, and can be proved. Therefore, (23) is just a statement of the *similarity structure* of the solution F, and it must read

$$\alpha F(\alpha x, \alpha^2 t) = F(x, t) \tag{24}$$

That is to say, if we replace x by αx and t by $\alpha^2 t$ in F and then multiply the result by α (for any $\alpha > 0$), *the resulting expression is identical to* $F(x, t)$. This property implies that $F(x, t)$ must be in the form

$$F(x, t) = \frac{1}{\sqrt{t}}f\left(\frac{x}{\sqrt{t}}\right), \quad \text{or} \quad \frac{1}{\sqrt{t}}g\left(\frac{x^2}{t}\right), \quad \text{or} \quad \frac{1}{x}h\left(\frac{x}{\sqrt{t}}\right), \dots$$

for certain functions $f, g, h, \dots$ of the indicated arguments.

Any one of an infinite number of possibilities that satisfy the similarity condition (24) may be used. Each choice will reduce (16) to an ordinary differential equation, which, when solved, will give the *same* result for F. Let us pick the form

$$F(x, t) = \frac{1}{\sqrt{t}}f(\zeta), \quad \zeta \equiv \frac{x}{\sqrt{t}}$$

We compute

$$F_x = \frac{1}{t} f'; \qquad F_{xx} = \frac{1}{t^{3/2}} f''; \qquad F_t = -\frac{1}{2t^{3/2}} f - \frac{x}{2t^2} f'$$

where $' \equiv d/d\zeta$.

Since the delta function on the right-hand side of (16) is identically equal to zero for $t > 0$, we need to solve only the homogeneous diffusion equation for $t > 0$. However, the initial condition $u(x, 0^-) = 0$ in (17) does not remain valid for $t = 0^+$. (If it did, the result would be the trivial solution $u(x, t) \equiv 0$.) The effect of the delta function on the right-hand side is to generate impulsively a nonzero value for $u(x, 0^+)$ (see (31)), which is the appropriate initial condition to be used in solving the homogeneous equation (16) for $t > 0$.

Consider now the homogeneous version of (16). Using the results we computed for F and its derivatives gives

$$-\frac{1}{2t^{3/2}} f - \frac{x}{2t^2} f' - \frac{1}{t^{3/2}} f'' = 0$$

which is the linear second-order ordinary differential equation

$$f'' + \frac{\zeta}{2} f' + \frac{1}{2} f = 0 \tag{25}$$

with the independent variable ζ.

Integrating once gives $f' + (\zeta/2)f = A = $ constant, and the solution of this is

$$f = A e^{-\zeta^2/4} \int^{\zeta} e^{s^2/4} \, ds + B e^{-\zeta^2/4}, \qquad B = \text{constant}$$

The constants A and B are determined by considering the total heat content $H(t)$ in the bar. In terms of our dimensionless units, the total heat is just the integral of the temperature:

$$H(t) \equiv \int_{-\infty}^{\infty} F(x, t) \, dx \tag{26}$$

$$= \frac{A}{\sqrt{t}} \int_{-\infty}^{\infty} g\left(\frac{x}{\sqrt{t}}\right) dx + \frac{B}{\sqrt{t}} \int_{-\infty}^{\infty} e^{-x^2/4t} \, dx$$

where

$$g(\zeta) = e^{-\zeta^2/4} \int^{\zeta} e^{s^2/4} \, ds = e^{-\zeta^2/4} \int^{\zeta^2/4} e^{\sigma} \sigma^{-1/2} \, d\sigma$$

Integrating the second expression for g by parts shows that

$$g(\zeta) = \frac{2}{|\zeta|} + 0(\zeta^{-3}) \quad \text{as} \quad |\zeta| \to \infty$$

Therefore, $(1/\sqrt{t}) \int_{-\infty}^{\infty} g \, dx$ in (26) is unbounded. Since the total heat must be finite, we set $A = 0$ and have

$$F = \frac{B}{\sqrt{t}} e^{-x^2/4t}, \qquad t > 0 \tag{27}$$

The idea now is to pick B in order to satisfy the inhomogeneous problem for $t = 0^+$. If we differentiate (26) with respect to t and use (16), we find

$$\frac{dH}{dt} = \int_{-\infty}^{\infty} F_t(x, t) \, dx = \int_{-\infty}^{\infty} [F_{xx}(x, t) + \delta(x)\delta(t)] \, dx$$

so that

$$\frac{dH}{dt} = F_x(\infty, t) - F_x(-\infty, t) + \delta(t) = \delta(t)$$

because the temperature gradient at $\pm\infty$ due to a unit heat source must be zero. Therefore, $H(t)$ is the Heaviside function—$H(t) = 1$ if $t > 0$, $H(t) = 0$ if $t < 0$—and for $t > 0$, (26) is just

$$1 = \int_{-\infty}^{\infty} \frac{B}{\sqrt{t}} e^{-x^2/4t} \, dx \tag{28}$$

Thus, as a result of switching on a unit source of heat at the origin, the total heat content in the rod does not vary with time, and this constant can be set equal to unity under an appropriate nondimensionalization.

We can rewrite (28) as

$$1 = 2B \int_{-\infty}^{\infty} \frac{e^{-x^2/4t}}{\sqrt{4t}} \, dx = 2B \int_{-\infty}^{\infty} e^{-\xi^2} \, d\xi = 2B\sqrt{\pi}$$

or

$$B = \frac{1}{2\sqrt{\pi}}$$

and the fundamental solution is

$$F(x, t) = \frac{1}{\sqrt{4\pi t}} e^{-x^2/4t} \tag{29}$$

More generally, the solution of (19)–(20) is

$$F(x - \xi, t - \tau) = \frac{1}{\sqrt{4\pi(t - \tau)}} e^{-(x-\xi)^2/4(t-\tau)} \tag{30}$$

It is important to note that this technique is not restircted to linear problems. For example, a classical use of similarity arguments is provided by the boundary-layer equations for viscous incompressible flow over an infinite wedge (or the special case of a semi-infinite flat plate if the wedge angle is zero). See Section B.14 of [2]. Here, the nonlinear partial differential equation for the flow stream function is reduced to a third-order nonlinear ordinary differential equation.

A crucial requirement for the applicability of similarity arguments is that both the governing equations *and* initial and/or boundary conditions be reducible to similarity form. In the preceding example, this was trivially true for the given initial condition $F = 0$ as this also immediately implied $G = 0$. In Problems 1, 2, and 5, these ideas are illustrated for sample linear and nonlinear problems.

For further reading on similarity methods, see [3] and [4].

1.2.2 Qualitative Behavior; Diffusion

Figure 1.3 shows three temperature profiles taken at three successive times $0 < t_1 < t_2 < t_3$. In each case, the area under the curve is, according to (28), equal to unity. As t gets smaller and smaller, the contribution to this area becomes more and more concentrated at the origin. This is just one of the many possible characterizations of the delta function (for instance, see page 319 of [5]), and we may write

$$F(x, 0^+) = \delta(x) \tag{31}$$

or

$$\delta(x) \overset{*}{=} \lim_{t \downarrow 0} \frac{1}{\sqrt{4\pi t}} e^{-x^2/4t} \tag{32}$$

where * indicates that the limit does not exist in the *strict sense*. Equation (31) also follows by integrating (16) with respect to t from $t = 0^-$ to $t = 0^+$ and noting that $\int_{0^-}^{0^+} u_{xx}\, dt = 0$.

The fundamental solution can be used to give a precise definition of *diffusion*. First, notice that if we regard the source at $x = 0$ as a disturbance introduced at time $t = 0$, the "signal speed" due to this disturbance is *infinite* because for any positive t, no matter how small, the value of u is nonzero for all x. Thus, the entire rod instantly "feels" the effect of the source. Of course, a real temperature gauge would fail to detect the very weak disturbance at large distances. Thus, the idea of a signal speed is not very useful in this case, and we

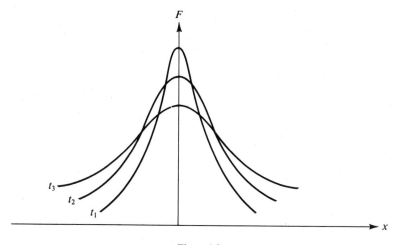

Figure 1.3

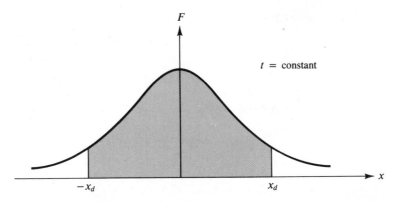

F

$t = \text{constant}$

$-x_d$ x_d x

Figure 1.4

would like to have a better characterization of how the rod "heats up" for $t > 0$. Suppose we ask, instead, where a given fraction of the *total heat* in the rod is to be found at any specified time. We know that at $t = 0^+$, all the heat is concentrated at the origin. For any $t > 0$, the heat is nonuniformly distributed over the entire rod with the maximum temperature at the origin, as shown in Figure 1.4.

Suppose that d is a fixed constant with $0 < d < 1$. At some time $t > 0$, the temperature distribution is the even function of x given by (29) and sketched in Figure 1.4. The shaded area represents the fraction d of the total area (which equals unity). Thus, as t increases, so does x_d. The question is, how does x_d depend on t? It follows from (29) and symmetry that

$$d = \frac{2}{\sqrt{4\pi t}} \int_0^{x_d} e^{-\sigma^2/4t} \, d\sigma$$

or, changing variables, that

$$d = \frac{2}{\sqrt{\pi}} \int_0^{x_d/2\sqrt{t}} e^{-\eta^2} \, d\eta \equiv \phi\left(\frac{x_d}{2\sqrt{t}}\right)$$

This implies that some function ϕ of $x_d/2\sqrt{t}$ is constant as t increases. Therefore, $x_d \sim \sqrt{t}$, and we say that heat *diffuses* in x at a rate proportional to $\sqrt{t}$.

1.2.3 Laplace Transforms

A more direct method for solving (16)–(17) is to take the Laplace transform with respect to t. We multiply (16) by e^{-st} and integrate with respect to t from $t = 0^-$ to $t = \infty$.

Denote the Laplace transform of a function $u(x, t)$ by $U(x, s)$; that is,

$$U(x, s) = \int_{0^-}^{\infty} e^{-st} u(x, t) \, dt \tag{33a}$$

Since $u(x, 0^-) = 0$, (16) gives

$$sU - U_{xx} = \delta(x) \tag{33b}$$

Therefore, U is in the form:

$$U(x, s) = \begin{cases} A(s)e^{-\sqrt{s}x} + B(s)e^{\sqrt{s}x}, & x > 0 \\ C(s)e^{-\sqrt{s}x} + D(s)e^{\sqrt{s}x}, & x < 0 \end{cases}$$

Clearly, we must set $B(s) = 0$ and $C(s) = 0$ to ensure that the solution remains bounded as $|x| \to \infty$. We conclude from (33b) that

$$U(0^+, s) - U(0^-, s) = 0$$

and, by integrating (33b) with respect to x from $x = 0^-$ to $x = 0^+$, that

$$U_x(0^+, s) - U_x(0^-, s) = -1$$

Therefore, $A = D = 1/2\sqrt{s}$ and

$$U(x, s) = \frac{1}{2\sqrt{s}} e^{-\sqrt{s}|x|}$$

It is a simple exercise in complex variables to show that the inversion formula for Laplace transforms; that is,

$$u(x, t) = \frac{1}{2\pi i} \int_\Gamma e^{st} U(x, s) \, ds, \qquad \Gamma = \text{Bromwich path} \tag{34}$$

gives (29). (See Review Problem 2.) The Bromwich path Γ is a vertical line taken to the right of all singularities of U in the s-plane.

1.2.4. Fourier Transforms

A third alternative is to use Fourier transforms with respect to x. Recall the definition for the Fourier transform $\bar{u}(k, t)$ of $u(x, t)$ (for example, Equations (7-7) and (7-8) in [5]):

$$\bar{u}(k, t) \equiv \frac{1}{\sqrt{2\pi}} \int_{-\infty}^{\infty} e^{ikx} u(x, t) \, dx \tag{35}$$

with the inversion formula

$$u(x, t) = \frac{1}{\sqrt{2\pi}} \int_{-\infty}^{\infty} e^{-ikx} \bar{u}(k, t) \, dk \tag{36}$$

We multiply (16) by $e^{ikx}/\sqrt{2\pi}$, integrate the result from $-\infty$ to ∞ with respect to x, and obtain

$$\bar{u}_t + k^2 \bar{u} = \frac{\delta(t)}{\sqrt{2\pi}} \tag{37}$$

and $\bar{u}(k, 0^-) = 0$ from (17).

Thus, for $t > 0$, $\bar{u}$ obeys the homogeneous version of (37) with $\bar{u}(k, 0^+) = 1/\sqrt{2\pi}$. The solution is therefore given by

$$\bar{u}(k,t) = \frac{e^{-k^2 t}}{\sqrt{2\pi}} \tag{38}$$

It is again left as an exercise to show that the inversion formula,

$$u(x,t) = \frac{1}{2\pi}\int_{-\infty}^{\infty} e^{-ikx - k^2 t}\, dk \tag{39}$$

gives (29). (See Review Problem 2.)

1.3   Initial-Value Problem (Cauchy Problem) on the Infinite Domain; Superposition

The general initial-value problem for the inhomogenous diffusion equation (1) on the infinite interval is:

$$u_t - u_{xx} = p(x,t), \qquad -\infty < x < \infty, t \geq 0$$
$$u(x,0^+) = f(x) \tag{40}$$

where p and f are arbitrarily prescribed functions. Here, p represents a dimensionless heat-source distribution that equals zero if $t < 0$ and f, an initial temperature distribution. Thus, for $t < 0$, $u \equiv 0$. At $t = 0$, the source distribution $p(x,t)$ is suddenly turned on everywhere and continues to be prescribed for all x and $t > 0$. Simultaneously, at $t = 0^+$, the initial temperature in the rod is prescribed as $f(x)$.

Because of linearity, the solution of (40) can be expressed as the sum of the following two problems:

1. $u_t - u_{xx} = p(x,t), \qquad -\infty < x < \infty, t \geq 0$
   $u(x,0^-) = 0$
2. $u_t - u_{xx} = 0, \qquad -\infty < x < \infty, t \geq 0$
   $u(x,0^+) = f(x)$

We now show that knowing the fundamental solution $F(x - \xi, t - \tau)$ allows us to write the solution of the first problem immediately in terms of a "superposition integral." To derive this superposition integral, we consider the solution of the first problem arising from the contribution of p coming from a small neighborhood of the fixed point $x = \xi, t = \tau$, with p set equal to zero everywhere outside this neighborhood. Let D denote the small neighborhood $\xi \leq x \leq \xi + \Delta\xi, \tau \leq t \leq \tau + \Delta\tau$, over which we may regard the value of p as the constant $p(\xi, \tau)$.

If $\tilde{p}$ denotes the incremental contribution to p from D, we have the following expression defining $\tilde{p}$:

$$\tilde{p} \equiv p(\xi,\tau)[H(t - \tau) - H(t - \tau - \Delta\tau)][H(x - \xi) - H(x - \xi - \Delta\xi)]$$

where H is the Heaviside function and the bracketed expressions ensure that $\tilde{p}$ vanishes outside D. We now multiply and divide this expression for $\tilde{p}$ by $\Delta\tau\Delta\xi$ and observe that since $dH/ds = \delta(s)$, the first bracketed expression divided by $\Delta\tau$ represents $\delta(t - \tau)$, whereas the second bracketed expression divided by $\Delta\xi$

represents $\delta(x - \xi)$. Therefore, in the limit as $\Delta\tau \to 0$, $\Delta\xi \to 0$, we have

$$\tilde{p} = p(\xi, \tau)\delta(t - \tau)\delta(x - \xi)\,d\tau\,d\xi$$

Since the solution of the diffusion equation with right-hand side $\delta(t - \tau)\delta(x - \xi)$ is the fundamental solution $F(x - \xi, t - \tau)$ defined in (30), linearity implies that the solution due to the right-hand side $\tilde{p}$ is just

$$\tilde{u} = p(\xi, \tau)F(x - \xi, t - \tau)\,d\tau\,d\xi$$

Linearity also implies that we may superpose the $\tilde{u}$ contributions arising from each of the infinitesimal domains D that cover the half-space $-\infty < x < \infty$, $0 \le t < \infty$, and this leads to the desired superpositon integral

$$\begin{aligned}
u(x, t) &= \int_{\xi=-\infty}^{\infty} \int_{\tau=0^-}^{t} F(x - \xi, t - \tau)p(\xi, \tau)\,d\tau\,d\xi \\
&= \int_{\xi=-\infty}^{\infty} \int_{\tau=0^-}^{t} \frac{p(\xi, \tau)}{\sqrt{4\pi(t - \tau)}} e^{-(x-\xi)^2/4(t-\tau)}\,d\tau\,d\xi
\end{aligned} \tag{41}$$

In addition to this formal derivation, it is easy to verify explicitly that (41) solves the first problem; this is left as an exercise (Problem 3).

To solve the second problem, we note that it is equivalent to

$$u_t - u_{xx} = \delta(t)f(x), \qquad -\infty < x < \infty, t \ge 0 \tag{42a}$$

$$u(x, 0^-) = 0 \tag{42b}$$

as can be verified by noting that integrating the inhomogeneous diffusion equation (42a) with respect to t from $t = 0^-$ to $t = 0^+$ gives $u(x, 0^+) = f(x)$. Since the right-hand side of (42a) vanishes when $t > 0$, the second problem and (42) are equivalent. To solve (42), we set $p(\xi, \tau)$ in (41) equal to $\delta(\tau)f(\xi)$ and obtain

$$u(x, t) = \frac{1}{\sqrt{4\pi t}} \int_{\xi=-\infty}^{\infty} f(\xi)e^{-(x-\xi)^2/4t}\,d\xi \tag{43}$$

Therefore, the solution of (40) is the sum of the solutions (41) and (43).
Note that

$$u(x, 0^+) = \lim_{t\to 0^+} \int_{-\infty}^{\infty} f(\xi)\frac{e^{-(x-\xi)^2/4t}}{\sqrt{4\pi t}}\,d\xi$$

and according to (32), this is just

$$u(x, 0^+) = \int_{-\infty}^{\infty} f(\xi)\delta(x - \xi)\,d\xi = f(x)$$

which is the correct initial condition.

We can also verify that the initial condition is satisfied by the following alternate approach that does not involve use of the delta function. We write (43) as the sum of three integrals over the intervals $(-\infty, x - \varepsilon)$, $(x - \varepsilon, x + \varepsilon)$ and $(x + \varepsilon, \infty)$, where ε is an arbitrarily small, fixed positive number. As $t \to 0^+$, the integrals tend to zero except over the interval $(x - \varepsilon, x + \varepsilon)$. Thus,

$$u(x, 0^+) = \lim_{t \to 0^+} \int_{x-\varepsilon}^{x+\varepsilon} f(\xi) \frac{e^{-(x-\xi)^2/4t}}{\sqrt{4\pi t}} \, d\xi$$

Changing variables of integration, we have

$$u(x, 0^+) = \lim_{t \to 0^+} \frac{1}{\sqrt{\pi}} \int_{-\varepsilon/\sqrt{4t}}^{\varepsilon/\sqrt{4t}} f(x + \sigma\sqrt{4t}) e^{-\sigma^2} \, d\sigma$$

$$= \frac{f(x)}{\sqrt{\pi}} \int_{-\infty}^{\infty} e^{-\sigma^2} \, d\sigma = f(x)$$

1.4   Initial- and Boundary-Value Problems on the Semi-Infinite Domain; Green's Functions

In studying the diffusion equation over the semi-infinite interval with a prescribed boundary condition at $x = 0$, it is useful first to consider the solution that results from a unit source somewhere in the domain and subject to a homogeneous (zero) boundary condition at the origin. This solution will be denoted as Green's function of the first kind, G_1, or second kind, G_2, depending on whether the boundary condition at $x = 0$ is $u = 0$ or $u_x = 0$.

1.4.1 Green's Function of the First Kind

Consider first the case where $u = 0$ at the origin; that is, we seek the solution for

$$u_t - u_{xx} = \delta(t)\delta(x - \xi) \tag{44a}$$

on $0 \le x \le \infty$, with ξ equal to a positive constant, and impose the boundary condition

$$u(0, t) = 0, \qquad t > 0 \tag{44b}$$

and initial condition:

$$u(x, 0^-) = 0 \tag{44c}$$

(Unless stated otherwise, we shall take the boundary condition for u at $x = \infty$ to be the same as the limit as $x \to \infty$ of the initial value. Thus, in the present case, we have $u(\infty, t) = u(\infty, 0) = 0$.)

Thus, we have introduced a concentrated unit source of heat at $x = \xi$ and $t = 0$. (Note that we can derive the solution for the case where (44a) involves $\delta(t - \tau)$ by replacing t everywhere in the solution by $t - \tau$.) The rod is initially at zero temperature, and its left end is maintained at zero temperature for all time, for example, by attaching this end to an infinite solid of zero temperature.

The only difference between this problem and the one discussed in Section 1.1 is the fact that we require u to vanish at $x = 0$ and $x \to \infty$ instead of $x \to \pm\infty$. Thus, Green's function is the response to a source with a homogeneous boundary condition imposed at a finite point.

An intuitively appealing procedure invokes symmetry relative to the origin to construct the solution once the fundamental solution is known. (This is often called the *method of images*.)

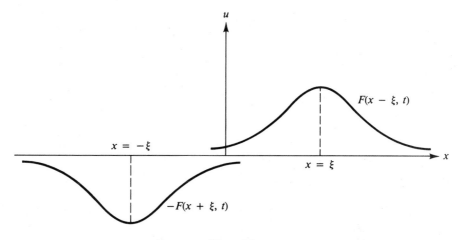

Figure 1.5

Consider the temperature that results in the *infinite* domain if we turn on a positive source of unit strength at $x = \xi$ and $t = 0$ and *simultaneously* turn on a negative source of unit strength at $x = -\xi$, the image point.

At any time $t > 0$, the temperature in the rod will be the sum of the two temperatures: $F(x - \xi, t)$ and $-F(x + \xi, t)$, corresponding to the positive and negative sources, respectively. These individual temperature profiles at some $t > 0$ are sketched in Figure 1.5. In particular, the combined temperature will always vanish at $x = 0$ for $t > 0$, by symmetry. Morever, since the image source is located at $x = -\xi$, outside the domain of interest, the combined temperature satisfies (44a). Therefore, the solution of (44) is Green's function:

$$G_1(x, \xi, t) \equiv F(x - \xi, t) - F(x + \xi, t) \tag{45}$$

where F is defined by (29).

More generally, the solution of

$$u_t - u_{xx} = \delta(x - \xi)\delta(t - \tau), \qquad \xi > 0, \tau > 0$$

with initial condition $u(x, \tau^-) = 0$ and boundary condition $u(0, t) = 0$ for $t > \tau$ and x on the semi-infinite interval $0 \leq x < \infty$ is Green's function of the first kind for the semi-infinite domain and has the form:

$$G_1(x, \xi, t - \tau) = \frac{1}{\sqrt{4\pi(t - \tau)}} [e^{-(x-\xi)^2/4(t-\tau)} - e^{-(x+\xi)^2/4(t-\tau)}] \tag{46}$$

1.4.2 Homogeneous Boundary-Value Problems

Consider the following *inhomogeneous* diffusion equation with *homogeneous* boundary condition:

$$u_t - u_{xx} = p(x, t), \qquad 0 \leq x, 0 \leq t \tag{47a}$$

$$u(x, 0^-) = 0 \tag{47b}$$

$$u(0, t) = 0, \qquad t > 0 \tag{47c}$$

Since Green's function (46) solves (47a) with $p = \delta(x - \xi)\delta(t - \tau)$ and the zero boundary condition (47c), then superposition of this Green's function, scaled in accordance with the given $p(x, t)$, will maintain the boundary condition and lead, as in the case of (41), to the integral representation

$$u(x, t) = \int_{0^-}^{t} d\tau \int_{0}^{\infty} [p(\xi, \tau) G_1(x, \xi, t - \tau)] d\xi \qquad (48)$$

It is important to bear in mind that *Green's function and the desired solution of* (47) *must both satisfy a zero boundary condition at the origin* in order for the super-position idea and the result (48) to make sense. For example, if $G_1(0, \xi, t - \tau) \neq 0$, then (48) does not satisfy (47c). Conversely, if we wish to solve the problem (47) with the right-hand side of (47c) replaced by some prescribed function $g(t)$, the representation (48) fails, since it automatically has $u(0, t) = 0$. We shall see in Section 1.4.3 next that this case is easily handled once the problem is transformed to one with a zero boundary condition at the origin.

Consider now the case where the initial condition (47b) is prescribed arbitrarily. Since the homogeneous problem

$$u_t - u_{xx} = 0, \qquad 0 \leq x, 0 \leq t \qquad (49a)$$

with nonzero initial condition

$$u(x, 0^+) = f(x) \qquad (49b)$$

and homogeneous boundary condition

$$u(0, t) = 0, \qquad t > 0 \qquad (49c)$$

is equivalent to

$$u_t - u_{xx} = \delta(t) f(x) \qquad (50)$$

with $u(x, 0^-) = 0$ and $u(0, t) = 0$, we can express the solution of (49) using the result (48) with $p = \delta(\tau) f(\xi)$; that is,

$$u(x, t) = \int_{0}^{\infty} \left[\int_{0^-}^{t} \delta(\tau) f(\xi) G_1(x, \xi, t - \tau) d\tau \right] d\xi = \int_{0}^{\infty} f(\xi) G_1(x, \xi, t) d\xi \qquad (51)$$

For the special case where $f(\xi) = c$, a constant, (51) gives

$$u(x, t) = c \left[\int_{0}^{\infty} \frac{e^{-(x-\xi)^2/4t}}{\sqrt{4\pi t}} d\xi - \int_{0}^{\infty} \frac{e^{-(x+\xi)^2/4t}}{\sqrt{4\pi t}} d\xi \right] \qquad (52)$$

Changing the variables of integration results in

$$u(x, t) = c \left\{ -\int_{x/2\sqrt{t}}^{0} \frac{e^{-\eta^2} d\eta}{\sqrt{\pi}} + \int_{0}^{\infty} \frac{e^{-\eta^2} d\eta}{\sqrt{\pi}} - \int_{0}^{\infty} \frac{e^{-\eta^2} d\eta}{\sqrt{\pi}} + \int_{0}^{x/2\sqrt{t}} \frac{e^{-\eta^2} d\eta}{\sqrt{\pi}} \right\}$$

$$= \frac{2c}{\sqrt{\pi}} \int_{0}^{x/2\sqrt{t}} e^{-\eta^2} d\eta = c \operatorname{erf}\left(\frac{x}{2\sqrt{t}} \right) \qquad (53)$$

where we have defined the error function in the usual way:

$$\text{erf}(y) \equiv \frac{2}{\sqrt{\pi}} \int_0^y e^{-\eta^2} d\eta \tag{54}$$

In particular, $\text{erf}(-y) = -\text{erf}(y)$, $\text{erf}(0) = 0$, and $\text{erf}(\infty) = -\text{erf}(-\infty) = 1$; this function is sketched in Figure 1.6.

Thus, the qualitative behavior of the solution (53) has u rising rapidly from its zero boundary value to the asymptotic value $u = c$. Temperature profiles at various times are sketched in Figure 1.7.

Notice that $\lim_{\substack{t \to 0^+ \\ x > 0}} u(x, t) = c$, in agreement with (49b), and that

$\lim_{\substack{x \to 0^+ \\ t > 0}} u(x, t) = 0$, in agreement with (49c). In particular, $u(0^+, 0^+)$ is undefined, as is to be expected from (49b) and (49c).

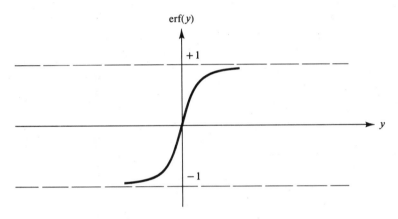

Figure 1.6

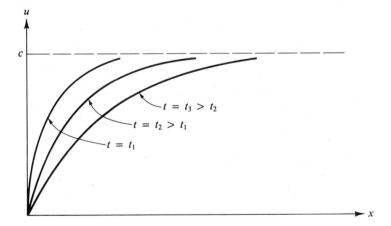

Figure 1.7

1.4.3 Inhomogeneous Boundary Condition $u = g(t)$

As pointed out in Section 1.4.2, the crucial requirement for applying super-position is that the boundary conditions are homogeneous. Does this mean that we cannot use Green's functions to solve an inhomogeneous boundary-value problem? We shall show next that if it is possible to transform the problem to one with homogeneous boundary conditions (as is often the case), a solution derived by superposition of Green's functions can still be used.

Consider the inhomogeneous boundary value problem

$$u_t - u_{xx} = 0, \qquad 0 \le x < \infty, 0 \le t < \infty \tag{55a}$$

with zero initial condition

$$u(x, 0^+) = 0 \tag{55b}$$

and a prescribed boundary condition at $x = 0$:

$$u(0, t) = g(t), \qquad t > 0 \tag{55c}$$

Again, in view of (55b), it is understood that $u(\infty, t) = 0$.

The idea is to transform $u(x, t)$ to a new dependent variable $w(x, t)$, which obeys a homogeneous boundary condition at the origin. Clearly, the simple choice

$$w(x, t) \equiv u(x, t) - g(t) \tag{56}$$

works, since w obeys the inhomogeneous equation

$$w_t - w_{xx} = -\dot{g}(t), \qquad t > 0 \tag{57}$$

with constant initial condition

$$w(x, 0^+) = -g(0^+) \tag{58a}$$

and zero boundary condition

$$w(0, t) = 0, \qquad t > 0 \tag{58b}$$

Equivalently, we have:

$$w_t - w_{xx} = -\dot{g}(t) - g(0^+)\delta(t) \tag{59a}$$

$$w(x, 0^-) = 0 \tag{59b}$$

$$w(0, t) = 0, \qquad t > 0 \tag{59c}$$

System (59) is a special case of (47), with $p(x, t) = -\dot{g}(t) - g(0^+)\delta(t)$. Writing out the solution (48) for this case gives

$$w(x, t) = \int_{0^+}^t \int_0^\infty \frac{-\dot{g}(\tau)}{2\sqrt{\pi(t - \tau)}} \left[e^{-(x-\xi)^2/4(t-\tau)} - e^{-(x+\xi)^2/4(t-\tau)} \right] d\xi \, d\tau$$

$$- \int_0^\infty \frac{g(0^+)}{2\sqrt{\pi t}} \left[e^{-(x-\xi)^2/4t} - e^{-(x+\xi)^2/4t} \right] d\xi \tag{60}$$

Solution (60) involves the two integrals

$$I = \frac{1}{\sqrt{\pi}} \int_0^\infty \frac{e^{-(x-\xi)^2/4(t-\tau)}}{[4(t-\tau)]^{1/2}} \, d\xi \tag{61a}$$

and

$$K = \frac{1}{\sqrt{\pi}} \int_0^\infty \frac{e^{-(x+\xi)^2/4(t-\tau)}}{[4(t-\tau)]^{1/2}} \, d\xi \tag{61b}$$

In preparation for evaluating I, we set the exponent in the integrand equal to $-\eta^2$, where η is a new variable of integration. We must be careful to take into account the fact that this exponent vanishes at the point $\xi = x > 0$, which is inside the interval of integration. Thus, we first split (61a) into two integrals over $0 \le \xi \le x$ and $x \le \xi < \infty$; then we change variable $\xi \to \eta$ by setting $(x - \xi)/[4(t - \tau)]^{1/2} = \eta$, $d\xi = -[4(t - \tau)]^{1/2} \, d\eta$ to obtain

$$I = \frac{1}{\sqrt{\pi}} \left[\int_{x/\sqrt{4(t-\tau)}}^0 e^{-\eta^2}(-d\eta) + \int_0^{-\infty} e^{-\eta^2}(-d\eta) \right]$$

$$= \frac{1}{\sqrt{\pi}} \left[\int_0^{x/\sqrt{4(t-\tau)}} e^{-\eta^2} \, d\eta + \int_0^\infty e^{-\eta^2} \, d\eta \right]$$

It then follows from (54) that

$$I = \frac{1}{2} \operatorname{erf}\left(\frac{x}{\sqrt{4(t - \tau)}}\right) + \frac{1}{2} \tag{61c}$$

Since $(x + \xi)$ does not vanish for $x > 0$ if $0 \le \xi < \infty$, we evaluate K directly by setting $(x + \xi)/[4(t - \tau)]^{1/2} = \eta$ to obtain

$$K = \frac{1}{\sqrt{\pi}} \left[\int_{x/\sqrt{4(t-\tau)}}^\infty e^{-\eta^2} \, d\eta \right] = \frac{1}{2} \operatorname{erfc}\left(\frac{x}{\sqrt{4(t - \tau)}}\right), \tag{61d}$$

where erfc denotes the complementary error function, which is defined by

$$\operatorname{erfc}(y) \equiv \frac{2}{\sqrt{\pi}} \int_y^\infty e^{-\eta^2} \, d\eta = 1 - \operatorname{erf}(y) \tag{62}$$

Thus, I may also be written as

$$I = 1 - \frac{1}{2} \operatorname{erfc}\left(\frac{x}{\sqrt{4(t - \tau)}}\right) \tag{63}$$

and (60) becomes

$$w(x, t) = \int_{0^+}^t \dot{g}(\tau) \operatorname{erfc}\left(\frac{x}{\sqrt{4(t - \tau)}}\right) d\tau + g(0^+) \operatorname{erfc}\left(\frac{x}{\sqrt{4t}}\right) - g(t)$$

$$= u(x, t) - g(t) \tag{64a}$$

In particular, if $g(t) = d = $ constant, we have $u(x. t) = d \operatorname{erfc}(x/\sqrt{4t})$. Here again, as for (53), $u(0^+, 0^+)$ is undefined. However, $\lim_{t \to 0^+ \atop x > 0} u(x, t) = 0$, in agreement with (55b), and $\lim_{x \to 0^+ \atop t > 0} u(x, t) = d$, in agreement with (55c).

Integrating the first term by parts in (64a) gives the alternate form

$$u(x, t) = w(x, t) + g(t) = \frac{x}{\sqrt{4\pi}} \int_0^t \frac{g(\tau) e^{-x^2/4(t-\tau)}}{[t - \tau]^{3/2}} \, d\tau$$

$$= \frac{x}{\sqrt{4\pi}} \int_0^t \frac{g(t - \tau) e^{-x^2/4\tau}}{\tau^{3/2}} \, d\tau \tag{64b}$$

In Problem 6, the solution (64b) is obtained using Laplace transforms or as the solution of a related integral equation. Problem 7 explores the application of the preceding ideas to the case of discontinuous material properties. Problem 8 concerns the effect of moving boundaries.

Next, we consider problems on the semi-infinite domain subject to the homogeneous boundary condition $u_x = 0$ at $x = 0$ and see how Green's function may also be used to solve the problem where u_x is specified at $x = 0$.

1.4.4 Green's Function of the Second Kind

We can also use a symmetry argument to solve

$$u_t - u_{xx} = \delta(x - \xi)\delta(t) \tag{65a}$$

on $0 \le x < \infty$, with $\xi > 0$ subject to the boundary condition

$$u_x(0, t) = 0, \qquad t > 0 \tag{65b}$$

and initial condition

$$u(x, 0^-) = 0 \tag{65c}$$

Here again, we assume that as $x \to \infty$, u tends to the value it has at infinity initially.

We might interpret the solution of (65) as the temperature in a semi-infinite rod in response to a unit source of heat at $x = \xi$, $t = 0$ for the case where the rod is insulated (that is, there is no heat flow) at the left end.

In order to ensure that condition (65b) holds for all $t > 0$ at the origin, we need to introduce an *image*, or *reflected*, source of unit *positive* strength at the image point $x = -\xi$. The figure corresponding to Figure 1.5 now has the two bell-shaped profiles above the x-axis and centered around the points $x = \pm \xi$. Therefore, the slope of the combined profile vanishes at $x = 0$, since the contributions to u_x from the source at $x = \xi$ and $x = -\xi$ cancel out exactly for all $t > 0$.

Thus, the solution of (65) is

$$G_2(x, \xi, t) \equiv F(x - \xi, t) + F(x + \xi, t) \tag{66}$$

where F is defined by (29).

More generally, if the source is turned on at $t = \tau > 0$, we have

$$G_2(x, \xi, t - \tau) = \frac{1}{\sqrt{4\pi(t - \tau)}}[e^{-(x-\xi)^2/4(t-\tau)} + e^{-(x+\xi)^2/4(t-\tau)}] \tag{67}$$

1.4.5 Homogeneous Boundary-Value Problems

As in Section 1.4.2 we can use superposition to express the solution of

$$u_t - u_{xx} = p(x, t), \qquad 0 \le x, 0 \le t \tag{68a}$$

$$u(x, 0^-) = 0, \tag{68b}$$

$$u_x(0, t) = 0, \qquad t > 0 \tag{68c}$$

in the form

$$u(x, t) = \int_0^t d\tau \int_0^\infty [p(\xi, \tau)G_2(x, \xi, t - \tau)]\, d\xi \tag{69}$$

Also, we can accommodate a nonzero initial condition

$$u(x, 0^+) = f(x)$$

by adding to (69) the contribution

$$u(x, t) = \int_0^\infty f(\xi)G_2(x, \xi, t)\, d\xi \tag{70}$$

For the case $f(\xi) = c = $ constant, it is easily seen by changing the sign of the second term in (52) that (70) reduces to $u = c$, as expected.

1.4.6 Inhomogeneous Boundary Condition

To solve the problem

$$u_t - u_{xx} = 0, \qquad 0 \le x < \infty, 0 \le t \le \infty \tag{71a}$$

$$u(x, 0^+) = 0 \tag{71b}$$

$$u_x(0, t) = h(t), \qquad t > 0 \tag{71c}$$

we introduce the *homogenizing* transformation

$$w(x, t) \equiv u(x, t) - xh(t) \tag{72}$$

It then follows that if u solves (71), w solves

$$w_t - w_{xx} = -x\dot{h}(t) \tag{73a}$$

$$w(x, 0^+) = -xh(0^+) \tag{73b}$$

$$w_x(0, t) = 0 \tag{73c}$$

Using the results in (69) and (70), we have

$$u(x, t) - xh(t) = -\int_0^t d\tau \int_0^\infty \xi\dot{h}(\tau)G_2(x, \xi, t - \tau)\, d\tau - h(0^+)\int_0^\infty \xi G_2(x, \xi, t)\, d\xi \tag{74a}$$

This can be simplified to the form:

$$u(x,t) = -\frac{1}{\sqrt{\pi}} \int_0^t h(\tau)(t-\tau)^{-1/2} e^{-x^2/4(t-\tau)} \tag{74b}$$

In Problem 9, you are asked to show this result and to reconcile it with the result obtained by Laplace transforms.

1.4.7 The General Boundary-Value Problem

The general boundary-value problem over the semi-infinite domain is

$$u_t - u_{xx} = p(x,t) \tag{75a}$$

$$u(x,0^+) = f(x) \tag{75b}$$

$$a(t)u(0,t) + b(t)u_x(0,t) = c(t) \tag{75c}$$

where we have the most general linear boundary condition (75c) at the left end with arbitrarily prescribed nonvanishing functions a, b, and c. In our previous discussion, we have solved the two special cases $a = 0$ or $b = 0$.

A Green's function approach is not feasible if a, b, and c are all nonzero, and in general we must use Laplace transforms (see Problem 10).

1.5   Initial- and Boundary-Value Problems on the Finite Domain; Green's Functions

The next step in our development involves problems on the finite domain, which may be taken as the unit interval $0 \le x \le 1$ with no loss of generality (that is, we choose the length L of the domain as the scale to normalize (8)). As in Section 1.4, we distinguish problems that have $u = 0$ or $u_x = 0$ at either end. Thus, we need to study *four* different Green's functions, and we start with the simplest case.

1.5.1 Green's Function of the First Kind

We refer to the solution satisfying the boundary condition $u = 0$ at both ends as Green's function of the first kind, G_1. More precisely, denote the solution of

$$u_t - u_{xx} = \delta(x - \xi)\delta(t - \tau), \quad 0 \le x \le 1, \tau \le t \tag{76a}$$

$$u(x,\tau^-) = 0 \tag{76b}$$

$$u(0,t) = u(1,t) = 0, \quad t > 0 \tag{76c}$$

as Green's function $G_1(x,\xi,t-\tau)$. Here, ξ and τ are constants with $0 < \xi < 1$, $0 < \tau$.

Let us construct G_1 using symmetry arguments in terms of appropriate fundamental solutions.

Consider the "primary" source $\delta(x - \xi)\delta(t - \tau)$ sketched as ↑ at the point $x = \xi$, $0 < \xi < 1$, on the unit interval in Figure 1.8.

In order to cancel the contribution of the primary source at the left boundary $x = 0$, we need to introduce a reflected (or mirror) source of negative

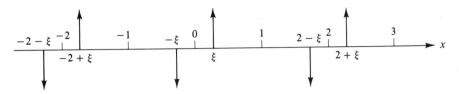

Figure 1.8

unit strength (sketched as $\downarrow$) at the image point $x = -\xi$. This image source must also be turned on at $t = \tau$. Similarly, to take care of the boundary contribution of the primary source at $x = 1$, we introduce another image source at $x = 2 - \xi$, also turned on at $t = \tau$. But now, the image source at $x = -\xi$ contributes to the boundary value at $x = 1$, and the image source at $x = 2 - \xi$ contributes to the boundary value at $x = 0$. To take care of the first, we introduce the $\uparrow$ unit source at $x = 2 + \xi$, to take care of the second, we introduce the $\uparrow$ unit source at $-2 + \xi$, and so on. The pattern that emerges has positive unit sources at $x = 2n + \xi$, $n = 0, \pm 1, \pm 2, \dots$, and negative unit sources are at $x = 2n - \xi, n = 0, \pm 1, \pm 2, \dots$. Using the expression defined by (30) for the response to each of the preceding sources, we obtain a representation for Green's function G_1 in the following series from:

$$G_1(x, \xi, t - \tau) \equiv \sum_{n=-\infty}^{\infty} \{F[x - (2n + \xi), t - \tau] - F[x - (2n - \xi), t - \tau]\}$$

(77)

where (compare with (30))

$$F(\alpha, \beta) \equiv \frac{1}{\sqrt{4\pi\beta}} e^{-\alpha^2/4\beta}$$

Green's function G_1 has the interesting symmetry property

$$G_1(x, \xi, t - \tau) \equiv G_1(\xi, x, t - \tau)$$

(78)

To demonstrate this symmetry property, we note that the right-hand side of (78) is by definition given by

$$G_1(\xi, x, t - \tau) = \sum_{n=-\infty}^{\infty} [F(\xi - 2n - x, t - \tau) - F(\xi - 2n + x, t - \tau)]$$

Since F is an even function of its first argument, we can rewrite the first term in the summation as $F(-\xi + 2n + x, t - \tau)$. Furthermore, since the summation ranges over $-\infty < n < \infty$, the infinite sum of these terms remain the same if we replace n by $-n$. Therefore, we may write

$$G_1(\xi, x, t - \tau) = \sum_{n=-\infty}^{\infty} [F(-\xi - 2n + x, t - \tau) - F(\xi - 2n + x, t - \tau)]$$

which is just $G_1(x, \xi, t - \tau)$.

In terms of heat conduction, the result (78) is intuitively obvious and physically consistent. Suppose we consider a homogeneous conductor with its two endpoints maintained at the same temperature, here normalized to be zero. Fix any two distinct locations x and ξ on the conductor and carry out the following two experiments. In the first experiment we turn on a unit source of heat at time τ at the point ξ and measure the temperature at the point x and time $t > \tau$. This gives the result $G_1(x, \xi, t - \tau)$ for the measured temperature. In the second experiment, we reverse the locations of the source and observer without changing the values of τ or t and find that the temperature at ξ, given by $G_1(\xi, x, t - \tau)$, is the same as that measured in the first experiment.

Using G_1 and superposition, we can now solve the inhomogeneous problem (compare with (47))

$$u_t - u_{xx} = p(x, t), \qquad 0 \le x \le 1, 0 \le t \tag{79a}$$

with zero initial condition

$$u(x, 0^-) = 0 \tag{79b}$$

and zero boundary conditions at both ends,

$$u(0, t) = u(1, t) = 0 \quad \text{for} \quad t > 0 \tag{79c}$$

in the form

$$u(x, t) = \int_0^t d\tau \int_0^1 [p(\xi, \tau) G_1(x, \xi, t - \tau)] \, d\xi \tag{80}$$

Similarly, as in (49)–(50), we solve the problem with $p(x, t) = 0$ and nonzero initial condition

$$u(x, 0^+) = f(x) \tag{81}$$

instead of (79b), in the form

$$u(x, t) = \int_0^1 f(\xi) G_1(x, \xi, t) \, d\xi \tag{82}$$

Green's functions for the remaining three homogeneous boundary-value problems are listed in Problem 11.

1.5.2 Connection with Separation of Variables

You may be wondering how the result in (82) is related to the solution we obtain by the more conventional separation of variables approach that is usually discussed in a first course in partial differential equations. We explore this question next.

To solve

$$u_t - u_{xx} = 0 \tag{83a}$$

$$u(x, 0^+) = f(x) \tag{83b}$$

$$u(0, t) = u(1, t) = 0 \qquad t > 0 \tag{83c}$$

we assume u in "separated" form:

$$u(x,t) = X(x)T(t) \tag{84}$$

Substituting (84) into (83a) gives

$$X\dot{T} - X''T = 0, \quad \text{or} \quad \frac{\dot{T}}{T} = \frac{X''}{X} \tag{85}$$

where the dot indicates d/dt and the double prime indicates d^2/dx^2. The second part of (85) can only hold if it equals a constant, and we quickly convince ourselves that this constant must be negative, say $-\lambda^2$. (Why?)

So, we obtain the *eigenvalue problem*

$$X'' + \lambda^2 X = 0, \qquad X(0) = X(1) = 0 \tag{86}$$

associated with (83). The solution is the *eigenfunction*

$$X_n = b_n \sin \lambda_n x, \qquad \lambda_n = n\pi$$

where b_n is arbitrary and n is an integer. Thus, the solution of (83) in a series of eigenfunctions is just the *Fourier sine series*:

$$u(x,t) = \sum_{n=1}^{\infty} B_n(t) \sin n\pi x \tag{87}$$

Substituting (87) into (83a), or using $\dot{T}_n + \lambda_n^2 T_n = 0$, gives $B_n = c_n e^{-n^2\pi^2 t}$, where $c_n = \text{constant}$.

To determine the c_n, we impose the initial condition (83b) and make use of orthogonality to obtain

$$c_n = 2 \int_0^1 f(\xi) \sin n\pi\xi \, d\xi \tag{88}$$

Thus, the solution of (83) may be written in series form as

$$u(x,t) = \sum_{n=1}^{\infty} \left[2 \int_0^1 f(\xi) \sin n\pi\xi \, d\xi \right] e^{-n^2\pi^2 t} \sin n\pi x$$

If we interchange summation and integration (a step that is nearly never questioned in a course in applied mathematics!), we find

$$u(x,t) = \int_0^1 f(\xi) H(x, \xi, t) \, d\xi \tag{89a}$$

where

$$H(x, \xi, t) \equiv 2 \sum_{n=1}^{\infty} (\sin n\pi\xi) e^{-n^2\pi^2 t} \sin n\pi x \tag{89b}$$

Comparing (89) with (82) gives a disconcerting indication that these two results do not agree. In fact, in order for the two results to agree, we must be able to show that $G_1 = H$. This is indeed the case and is a consequence of a certain identity for the "theta" function. For example, see page 75 of [6]. It is instructive to work out this identity in detail next.

We may use trigonometric identities to rewrite H in the form

$$H(x, \xi, t) = \frac{1}{2} \sum_{n=-\infty}^{\infty} e^{-n^2 \pi^2 t} \cos n\pi(x - \xi) - \frac{1}{2} \sum_{n=-\infty}^{\infty} e^{-n^2 \pi^2 t} \cos n\pi(x + \xi) \quad (90)$$

Now, the expression for G_1 in (77) agrees with (90) if we can show that

$$\sum_{n=-\infty}^{\infty} F(x + \xi - 2n, t) = \frac{1}{2} \sum_{n=-\infty}^{\infty} e^{-n^2 \pi^2 t} \cos n\pi(x + \xi) \quad (91a)$$

and

$$\sum_{n=-\infty}^{\infty} F(x - \xi - 2n, t) = \frac{1}{2} \sum_{n=-\infty}^{\infty} e^{-n^2 \pi^2 t} \cos n\pi(x - \xi) \quad (91b)$$

These two conditions are equivalent and reduce to one condition:

$$\frac{1}{\sqrt{\eta}} \sum_{n=-\infty}^{\infty} e^{-\pi(z-n)^2/\eta} = \sum_{n=-\infty}^{\infty} e^{-n^2 \pi \eta} \cos 2n\pi z \quad (92)$$

if we write $(x + \xi)$ or $(x - \xi)$ as $2z$, set $\eta = \pi t$, and use expression (78) defining F. Denote $\sqrt{\eta}$ times the expression on the left-hand side of (92) by ϕ; that is,

$$\phi(\eta, z) \equiv \sum_{n=-\infty}^{\infty} e^{-\pi(z-n)^2/\eta} \quad (93)$$

Clearly, ϕ is an even function of z (that is, $\phi(\eta, -z) = \phi(\eta, z)$). Also, it is periodic in z with unit period: $\phi(\eta, z + 1) = \phi(\eta, z)$. Therefore, we may expand ϕ in a Fourier cosine series:

$$\phi(\eta, z) = \sum_{v=-\infty}^{\infty} \alpha_v(\eta) \cos 2\pi v z \quad (94a)$$

where

$$\alpha_v(\eta) = \int_0^1 \sum_{n=-\infty}^{\infty} e^{-\pi(\zeta-n)^2/\eta} \cos 2\pi v \zeta \, d\zeta \quad (94b)$$

Interchanging integration and summation in (94b) gives

$$\alpha_v(\eta) = \sum_{n=-\infty}^{\infty} \int_0^1 e^{-\pi(\zeta-n)^2/\eta} \cos 2\pi v \zeta \, d\zeta \quad (95)$$

Now change variables and let $s = n - \zeta$. Therefore,

$$\alpha_v(\eta) = \sum_{n=-\infty}^{\infty} \int_{n-1}^{n} e^{-\pi s^2/\eta} \cos 2\pi v s \, ds$$

$$= \int_{-\infty}^{\infty} e^{-\pi s^2/\eta} \cos 2\pi v s \, ds = \sqrt{\eta} e^{-\pi v^2 \eta} \quad (96)$$

Thus, we have proven the identity

$$\sum_{n=-\infty}^{\infty} e^{-\pi(z-n)^2/\eta} = \sum_{n=-\infty}^{\infty} \sqrt{\eta} e^{-\pi n^2 \eta} \cos 2\pi n z$$

which is (92) when we divide by $\sqrt{\eta}$.

In conclusion, the series representation for G_1 converges to the same result as the series for H, even though these series *do not agree term by term*. This latter observation means that if we truncate the series for G_1, the resulting approximation will be valid in a different sense than the approximation obtained by truncating the Fourier series H. Let us pursue this idea further, as it will provide a useful characterization of the two approaches we have used.

Consider first what happens if we truncate the series (77) at $n = N$ for G_1. Clearly, we are neglecting all the heat sources located at distances greater than $2N + \xi$ on the positive axis and greater than $2N - \xi$ on the negative axis. For short times, the response due to these sources is very small over the unit interval (because we are ignoring only the weak exponential tails of the corresponding F functions). Thus, the Green's function representation (82), when G_1 is truncated for some $n = N$, *should be valid for short times*. In particular, the boundary conditions at $x = 0$ and $x = 1$ are only approximately satisfied with the truncated series, and this approximation deteriorates as t gets large. On the other hand, if we truncate the Fourier series representation (89), the boundary conditions are *exactly* satisfied for all times, but the initial condition will be described only approximately. Thus, the truncated series (89) should provide a *good approximation for t large*. A more careful analysis of the convergence properties of the G_1 and H series confirms the above intuitive conclusions.

We reiterate that both expressions converge to the same solution if the infinite series are summed. We shall see in Chapter 3 in examples for the wave equation that this property of Green's functions versus eigenfunction expansions is also true there. It is a useful result, as we are able to have an approximation involving a finite number of terms for both t small and t large.

1.5.3 Connection with Laplace Transform Solution

A third approach for solving the problem in (83) is to use Laplace transforms with respect to t. For simplicity, consider the special case $f = 1$. Using the notation of (33), we obtain the following problem for $U(x, s)$, the Laplace transform of $u(x, t)$:

$$U_{xx} - sU = -1, \qquad U(0, s) = U(1, s) = 0$$

The solution is, therefore,

$$U(x, s) = \frac{1}{s(e^{\sqrt{s}} - e^{-\sqrt{s}})} [e^{\sqrt{s}} - e^{-\sqrt{s}} + (e^{-\sqrt{s}} - 1)e^{\sqrt{s}x} - (e^{\sqrt{s}} - 1)e^{-\sqrt{s}x}] \tag{97}$$

The solution for $u(x, t)$ is then given by the inversion integral (34); that is,

$$u(x, t) = \frac{1}{2\pi i} \int_\Gamma e^{st} U(x, s) \, ds \tag{98}$$

Note the branch point at $s = 0$ and $s = \infty$. Since we must choose the branch of $\sqrt{s}$ that is positive when s is along the positive real axis, it is convenient to cut the s-plane along the negative real axis.

Expression (98) cannot be evaluated in terms of a finite number of elementary functions. One standard approximation for a Laplace transform inversion is the "large s" approximation, which consists of expanding (97) in series form for s large and then integrating the result, term by term, in (98). As discussed in texts in complex variables (for example, see page 279 of [5]), this gives an approximation for $u(x, t)$ valid for t *small*.

To see this, just change variables in (98), setting $s = \sigma/t$ and $ds = d\sigma/t$ and consider the limit $|s| \to \infty$, $|\sigma|$ *fixed*. Clearly, this implies that we need to take $t \to 0$, and, in effect, the substitution σ/t for s in $U(x, s)$ accomplishes this.

If we expand the denominator of (97) and rearrange the numerator, we obtain

$$U(x, s) = \frac{1}{s} + \frac{e^{-\sqrt{s}}}{s}(1 + e^{-2\sqrt{s}} + e^{-4\sqrt{s}} + \cdots)(e^{-\sqrt{s}(1-x)} - e^{\sqrt{s}(1-x)} - e^{\sqrt{s}x} + e^{-\sqrt{s}x})$$

Taking the product, we find that U equals the particular solution $1/s$ plus four series in the form:

$$U(x, s) = \frac{1}{s} + \frac{1}{s}[e^{-\sqrt{s}(2-x)} + e^{-\sqrt{s}(4-x)} + e^{-\sqrt{s}(6-x)} + \cdots]$$

$$- \frac{1}{s}[e^{-\sqrt{s}x} + e^{-\sqrt{s}(2+x)} + e^{-\sqrt{s}(4+x)} + \cdots]$$

$$- \frac{1}{s}[e^{-\sqrt{s}(1-x)} + e^{-\sqrt{s}(3-x)} + e^{-\sqrt{s}(5-x)} + \cdots]$$

$$+ \frac{1}{s}[e^{-\sqrt{s}(1+x)} + e^{-\sqrt{s}(3+x)} + e^{-\sqrt{s}(5+x)} + \cdots]$$

These series can be rearranged in the form

$$U(x, s) = \frac{1}{s} + \frac{1}{s} \sum_{n=1}^{\infty} (-1)^n e^{-\sqrt{s}(n-x)} - \sum_{n=0}^{\infty} (-1)^n e^{-\sqrt{s}(n+x)} \tag{99}$$

Using (98) or tables of Laplace transforms, we find that the transform of

$$f(t) = \mathrm{erfc}\left(\frac{\lambda}{2\sqrt{t}}\right)$$

with $t > 0$ and λ real is

$$F(s) = \frac{1}{s} e^{-\lambda\sqrt{s}}$$

Therefore, the termwise inversion of (99) gives the series

$$u(x, t) = 1 + \sum_{n=1}^{\infty} (-1)^n \mathrm{erfc}\left(\frac{n - x}{2\sqrt{t}}\right) - \sum_{n=0}^{\infty} (-1)^n \mathrm{erfc}\left(\frac{n + x}{2\sqrt{t}}\right) \tag{100}$$

It is left as an exercise (Problem 12) to show that this series is the same as the one resulting from the Green's function representation (82) when we take $f = 1$ and integrate the series for G_1 term by term. This gives a confirmation of our earlier intuitive argument that the truncated Green's function representation of the solution is valid for t small.

At any rate, the *exact* expressions (82), (89a) with $f = 1$, and (98) define the same function $u(x, t)$. The advantage of (82) and (89a) over (98) is that these are in terms of real quadratures, whereas (98) is a complex integral.

1.5.4 Uniqueness of Solutions

In this section, we show that solutions of the initial- and boundary-value problem for the diffusion equation are unique. We consider solutions of

$$u_t - u_{xx} = 0, \qquad 0 \leq x \leq 1, 0 \leq t < \infty \tag{101a}$$

with initial condition

$$u(x, 0^+) = f(x) \tag{101b}$$

and one of the following four boundary conditions:

$$u(0, t) = g(t), \qquad u(1, t) = h(t) \tag{102a}$$

$$u(0, t) = g(t), \qquad u_x(1, t) = h(t) \tag{102b}$$

$$u_x(0, t) = g(t), \qquad u(1, t) = h(t) \tag{102c}$$

$$u_x(0, t) = g(t), \qquad u_x(1, t) = h(t) \tag{102d}$$

Here g and h are arbitrarily prescribed in each case.

In preparation for this proof, we first derive an integral identity for solutions of (101a). Multiply (101a) by $u(x, t)$ and integrate the result with respect to x on the unit interval to obtain:

$$\int_0^1 uu_t \, dx = \int_0^1 uu_{xx} \, dx$$

Since the interval is independent of t, we may write the left-hand side of this expression as $(d/dt) \int_0^1 (u^2/2) \, dx$, and integrating the right-hand side by parts, we obtain

$$\frac{1}{2} \frac{d}{dt} \int_0^1 u^2(x, t) \, dx = uu_x \Big|_0^1 - \int_0^1 u_x^2(x, t) \, dx \tag{103}$$

Identity (103) is true for any solution of (101a). Suppose that u_1 and u_2 are two solutions of (101a), each of which satisfies the initial condition (101b) and one of the four pairs of boundary conditions (102). If we denote the difference by $u_1 - u_2 \equiv v(x, t)$, then $v(x, t)$ satisfies the problem:

$$v_t - v_{xx} = 0$$

$$v(x, 0) = 0$$

$$vv_x = 0 \qquad \text{at } x = 0 \quad \text{and} \quad x = 1$$

Therefore, identity (103) for v becomes

$$\frac{1}{2} \frac{d}{dt} \int_0^1 v^2(x, t) \, dx = -\int_0^1 v_x^2(x, t) \, dx \leq 0$$

Or, if we let

$$I(t) \equiv \frac{1}{2} \int_0^1 v^2(x, t) \, dx \geq 0$$

and

$$G(t) \equiv -\int_0^1 v_x^2(x, t) \, dx \leq 0$$

we have

$$\frac{dI}{dt} = G(t), \qquad \text{that is,} \qquad I(t) - I(0) = \int_0^t G(\tau) \, d\tau \leq 0$$

Thus, $I(t) - I(0) \leq 0$. But $I(0) = 0$; hence, $I(t) \leq 0$. According to its definition, $I(t) \geq 0$. So, we must have $I(t) = 0$, and the integral of a nonnegative quantity, such as v^2, can vanish only if $v(x, t) = 0$.

Thus, we have proven that $u_1(x, t) = u_2(x, t)$.

1.5.5 Inhomogeneous Boundary Conditions

As discussed in Section 1.4, we can transform a homogeneous equation with inhomogeneous boundary conditions to an inhomogeneous equation with homogeneous boundary conditions. To illustrate the idea, consider equation (101a) with initial condition (101b) and boundary conditions (102a).

To homogenize the boundary conditions, assume a transformation of dependent variable $u \to w$ in the linear form

$$u(x, t) \equiv w(x, t) + \alpha(t)x + \beta(t) \tag{104}$$

with as yet unspecified functions α and β of the time, to be chosen so that the boundary conditions for the resulting problem for w are homogeneous.

Using (104), we compute

$$u_t = w_t + \dot{\alpha}x + \dot{\beta}$$

$$u_x = w_x + \alpha, \qquad u_{xx} = w_{xx}$$

Therefore,

$$u_t - u_{xx} = w_t + \dot{\alpha}x + \dot{\beta} - w_{xx} = 0$$

that is, w obeys the inhomogeneous problem

$$w_t - w_{xx} = -\dot{\alpha}x - \dot{\beta}$$

In order to have $w(0, t) = 0$, we find from (104) that we must set $\beta(t) = g(t)$. Similarly, in order to have $w(1, t) = 0$, we must set $\alpha(t) = h(t) - g(t)$. Thus, the transformation relation is

$$u(x, t) \equiv w(x, t) + x[h(t) - g(t)] + g(t) \tag{105}$$

and w obeys the inhomogeneous equation

$$w_t - w_{xx} = [\dot{g}(t) - \dot{h}(t)]x - \dot{g}(t) \equiv p(x, t) \tag{106a}$$

subject to the initial condition

$$w(x,0) = f(x) - x[h(0^+) - g(0^+)] + g(0^+) \equiv F(x) \tag{106b}$$

and homogeneous boundary conditions

$$w(0,t) = w(1,t) = 0 \tag{106c}$$

The solution of problem (106) is just the sum of the solutions (80) and (82) with $f = F$; that is,

$$w(x,t) = \int_0^t d\tau \int_0^1 [p(\xi,\tau)G_1(x,\xi,t-\tau)\,d\xi + \int_0^1 F(\xi)G_1(x,\xi,t)\,d\xi \tag{107}$$

Having found $w(x,t)$, we obtain $u(x,t)$ from (105). Note that the form (106) is also appropriate for a solution using Fourier series, as homogenous boundary conditions are also crucial in being able to superpose eigensolutions. Problem 13 concerns the solution for the case (102b).

1.5.6 Higher-Dimensional Problems

The diffusion equation in two or more space dimensions is (compare with (14b))

$$u_t - \Delta u = 0 \tag{108}$$

where Δ is the Laplacian operator.

The fundamental solution of (108) is easily calculated because the response to a unit source must have spherical symmetry (or axial symmetry in two dimensions). Thus, we need consider only the homogeneous equation

$$u_t - u_{rr} - \frac{n-1}{r}u_r = 0 \tag{109}$$

associated with the spherically symmetric Laplacian, where n is the dimension of the space and r is the radial distance. The solution of (109) can be derived by similarity (or transforms) and is left as an exercise (Problem 14).

Symmetry arguments can also be used to derive Green's functions for certain "simple" domains. The ideas are identical to those used in studying Laplace's equation itself, so we defer this discussion to Chapter 2. Actually, if the time is eliminated from (108), using Laplace transforms the result is

$$\Delta U - sU = -u(x,y,z,0) \tag{110}$$

which is an inhomogeneous Helmholtz equation (to be considered in Chapter 2) in the given geometry.

1.6   Burgers' Equation

In 1948, Burgers proposed the following nonlinear diffusion equation as a mathematical model for turbulence:

$$u_t + uu_x - \varepsilon u_{xx} = 0 \tag{111}$$

In (111), ε is a positive parameter that measures the dissipative term. Although we can derive (111) as a limiting form of the x-component of the momentum equation for viscous flows (under certain assumptions), it *does not model turbu-lence* in any significant manner. Nevertheless, (111) is a fundamental "evolution equation," which arises in a number of unrelated applications where *viscous* and *nonlinear effects*, represented by εu_{xx} and uu_x, respectively, are equally important. For example, see Chapter 4 of [7]. Also, see [8] for a discussion of how (111) arises in applications.

In 1950 and 1951, E. Hopf and J. D. Cole independently showed (see the references cited in [7]) that (111) may be transformed to the linear diffusion equation (1) of this chapter. We now derive this transformation and discuss how it may be used to study initial- and boundary-value problems. Although the parameter ε may be transformed out of (111) with an appropriate scaling of the x and t variables, it is more instructive to retain it in the solution because we can then study how the results behave as $\varepsilon \to 0$. This is a "singular perturbation" problem, which is discussed in Sections 5.3.6 and 8.3.5. The reader is also referred to Chapter 4 of [7] and to Section 4.1.3 of [9] for more details.

1.6.1 The Cole-Hopf Transformation

This transformation of dependent variable $u \to v$ is defined by

$$u \equiv -2\varepsilon \frac{v_x}{v} \tag{112}$$

We then calculate

$$u_t = -2\varepsilon \frac{v_{xt}}{v} + 2\varepsilon \frac{v_x v_t}{v^2}$$

$$u_x = -2\varepsilon \frac{v_{xx}}{v} + 2\varepsilon \frac{v_x^2}{v^2}$$

$$u_{xx} = -2\varepsilon \frac{v_{xxx}}{v} + 6\varepsilon \frac{v_x v_{xx}}{v^2} - \frac{4\varepsilon v_x^3}{v^3}$$

Substituting these expressions into (111) gives

$$\frac{v_x}{v}(\varepsilon v_{xx} - v_t) - (\varepsilon v_{xx} - v_t)_x = 0 \tag{113}$$

Thus, *any* solution $v(x, t)$ of (113), when used in (112), gives an expression $u(x, t)$, which satisfies (111).

In particular, if v satisfies the diffusion equation

$$\varepsilon v_{xx} - v_t = 0$$

it also solves (113) trivially, and the resulting $u(x, t)$ satisfies (111).

1.6.2 Initial-Value Problem on $-\infty < x < \infty$

Let us study how we can use the preceding result to solve the initial-value problem for Burgers' equation:

$$u_t + uu_x - \varepsilon u_{xx} = 0, \qquad -\infty < x < \infty \tag{114a}$$

$$u(x,0) = f(x) \tag{114b}$$

According to (112), the new variable $v(x,t)$ must initially satisfy

$$f(x) = -\frac{2\varepsilon v_x(x,0)}{v(x,0)} \tag{115}$$

But this is a linear first-order ordinary differential equation for $v(x,0)$ and has the general solution

$$v(x,0) = ce^{(-1/2\varepsilon)\int_0^x f(s)\,ds} \equiv cg(x), \qquad c = \text{constant}$$

Thus, for a given $f(x)$, we compute $g(x)$ by quadrature. Of course, it is understood that the integral $\int_0^x f(s)\,ds$ exists. So, we need to solve the *linear problem* for $v(x,t)$:

$$v_t - \varepsilon v_{xx} = 0, \qquad -\infty < x < \infty \tag{116a}$$

$$v(x,0) = cg(x) \tag{116b}$$

This is the second problem for equation (40) and has the solution (43) after replacing $u \to v$, $f \to cg$, $t \to \varepsilon t$:

$$v(x,t) = \frac{c}{\sqrt{4\pi\varepsilon t}} \int_{-\infty}^{\infty} g(\xi)e^{-(x-\xi)^2/4\varepsilon t}\,d\xi \tag{117}$$

It then follows that

$$v_x(x,t) = \frac{-c}{\sqrt{4\pi\varepsilon t}} \int_{-\infty}^{\infty} \frac{g(\xi)(x-\xi)}{2\varepsilon t} e^{-(x-\xi)^2/4\varepsilon t}\,d\xi$$

Therefore, using (112) to compute $u(x,t)$ gives

$$u(x,t) = \frac{\displaystyle\int_{-\infty}^{\infty} g(\xi)\frac{(x-\xi)}{t}e^{-(x-\xi)^2/4\varepsilon t}\,d\xi}{\displaystyle\int_{-\infty}^{\infty} g(\xi)e^{-(x-\xi)^2/4\varepsilon t}\,d\xi}, \tag{118}$$

in which the constant c cancels out.

For further discussion of this result, in particular the asymptotic behavior as $\varepsilon \to 0$, see the cited texts. We shall use this formula in discussing discontinuous solutions of the first-order equation

$$u_t + uu_x = 0 \tag{119}$$

in Chapter 5. In particular, we compute (118) explicitly for the case where $f(x)$ is piecewise constant in Section 5.3.6.

1.6.3 Boundary-Value Problem on $0 < x < \infty$

Encouraged by the success of solving the initial-value problem (114), we now parallel our discussion of Section 1.4.6 for the linear case and study how the Cole-Hopf transformation applies to the boundary-value problem

$$u_t + uu_x - \varepsilon u_{xx} = 0, \qquad 0 \le x < \infty \tag{120a}$$

$$u(x,0) = 0 \tag{120b}$$

$$u(0,t) = h(t), \qquad t > 0 \tag{120c}$$

Unfortunately, the Cole-Hopf transformation does not give the appropriate boundary condition for $v(0,t)$. So, we are faced with solving

$$v_t - \varepsilon v_{xx} = 0 \qquad 0 \le x < \infty \tag{121a}$$

$$v(x,0^+) = c = \text{constant} \tag{121b}$$

$$v(0,t) = k(t), \qquad t > 0 \tag{121c}$$

where $k(t)$ is unknown and c is an arbitrary constant. This is problem (55) with $u \to v - c$, $t \to \varepsilon t$, and $g \to k - c$. Therefore, using (64a), we have

$$v(x,t) = c + \int_0^t \dot{k}(\tau) \text{erfc}\left(\frac{x}{2\sqrt{\varepsilon(t-\tau)}}\right) d\tau + [k(0^+) - c]\text{erfc}\left(\frac{x}{2\sqrt{\varepsilon t}}\right) \tag{122}$$

where $k(t)$ is still unknown. We now compute

$$v_x(x,t) = -\frac{1}{\sqrt{\pi\varepsilon}} \int_0^t \frac{\dot{k}(\tau)}{\sqrt{t-\tau}} e^{-x^2/4\varepsilon(t-\tau)} d\tau - \left[\frac{k(0^+) - c}{\sqrt{\pi\varepsilon t}}\right] e^{-x^2/4\varepsilon t} \tag{123}$$

where we have used the identity

$$\frac{d}{dz}(\text{erfc}\, z) = -\frac{2}{\sqrt{\pi}} e^{-z^2} \tag{124}$$

which follows from (54) and (62).

In the first term of (123), note the occurrence of the *integrable* singularity proportional to $1/\sqrt{t-\tau}$ at $t = \tau$. Had we used the form (64b) for the solution, the corresponding singularity would have been proportional to $x^3(t-\tau)^{-5/2}$, requiring further manipulations to derive a well-behaved result for $x = 0$.

Using the expressions (122) and (123) for v and v_x, we have the solution $u(x,t)$ of (120) from

$$u(x,t) = -2\varepsilon \frac{v_x(x,t)}{v(x,t)}$$

Now, in order to satisfy the boundary condition (120c), we must set

$$u(0,t)v(0,t) = -2\varepsilon v_x(0,t)$$

or

$$h(t)\left[\int_0^t \dot{k}(\tau)\, d\tau + k(0^+)\right] = 2\sqrt{\frac{\varepsilon}{\pi}}\left[\int_0^t \frac{\dot{k}(\tau)\, d\tau}{\sqrt{t-\tau}} + \frac{k(0^+) - c}{\sqrt{t}}\right] \tag{125a}$$

Since c is arbitrary, we may take $c = k(0^+)$, and (125a) simplifies to:

$$h(t)k(t) = 2\sqrt{\frac{\varepsilon}{\pi}} \int_0^t \frac{\dot{k}(\tau)\, d\tau}{\sqrt{t-\tau}} \tag{125b}$$

Again, we do not integrate the right-hand side by parts to avoid the singularity at $\tau = t$.

In contrast to the initial-value problem, where the initial condition for v was expressed by a quadrature, we now have to solve the integral equation (125) to obtain the appropriate boundary condition $k(t)$. Once this is done, the solution for $u(x, t)$ follows from the transformation relation (112).

The perturbation solution of the boundary-value problem is discussed in Section 8.3.5.

Review Problems

1. This is a review problem to illustrate separation of variables and Fourier series. Work this problem before attempting Problem 15.

    We want to solve the one-dimensional diffusion equation in the unit interval $0 \leq x \leq 1$ with a prescribed source distribution and initial value. The boundary condition is that one end has a fixed value of u, whereas the other has $u_x = 0$. A simple model problem is

    $$u_t - u_{xx} = x \sin t, \qquad 0 \leq x \leq 1, 0 \leq t \tag{126}$$

    $$u(x, 0) = x(1 - x) \tag{127}$$

    $$u(0, t) = u_x(1, t) = 0 \qquad \text{if } t > 0 \tag{128}$$

    a. Look for a solution of the homogeneous equation (126) in the separated form, $u(x, t) = X(x)T(t)$, and show that X must be one of the eigenfunctions (compare with (86))

    $$X_n(x) = \alpha_n \sin \lambda_n x \tag{129}$$

    where the eigenvalues are $\lambda_n = (2n+1)\pi/2$ for $n = 0, 1, 2, \ldots$ and $\alpha_n =$ constant.

    b. Based on this result assume a solution of (126)–(128) in the form of a series of eigenfunctions:

    $$u(x, t) = \sum_{n=0}^{\infty} A_n(t) \sin \lambda_n x \tag{130}$$

    where the $A_n(t)$ are functions of t to be specified. Also, expand the right-hand side of (126) in a series of eigenfunctions

    $$x \sin t = \left(\sum_{n=0}^{\infty} b_n \sin \lambda_n x \right) \sin t \tag{131}$$

    Regard (131) as a Fourier sine series for x over the interval $0 \leq x \leq 2$, and use orthogonality to show that $b_n = 4/(2n + 1)\pi$.

    Now substitute (130) into (126) with (131) for its right-hand side to show that the $A_n(t)$ satisfy

    $$\frac{dA_n}{dt} + \lambda_n^2 A_n = b_n \sin t \tag{132}$$

c. Solve (132) to find

$$A_n(t) = A_n(0)e^{-\lambda_n^2 t} + \frac{b_n}{\lambda_n^4 + 1}(\lambda_n^2 \sin t - \cos t + e^{-\lambda_n^2 t}) \tag{133}$$

d. Use (130) with $A_n(t)$ given by (133) in the initial condition (127) to find

$$A_n(0) = \frac{32 - 4\pi^2(2n + 1)^2}{\pi^3(2n + 1)^3} \tag{134}$$

2. This problem gives a review of integration in the complex plane for a function with a branch-point, as encountered, for example, when Laplace transforms are used.

   The integral in the complex s-plane defining the fundamental solution $u(x, t)$ in (34) is

$$u(x, t) = \frac{1}{2\pi i} \int_\Gamma f(s; t, x)\, ds, \qquad f(s; t, x) = \frac{e^{st - s^{1/2}|x|}}{2s^{1/2}} \tag{135}$$

where Γ is any vertical path to the right of all singularities of $f(s; t, x)$. Now $s^{1/2}$ (and $f(s; t, x)$) has a branch-point at $s = 0$ and $s = \infty$. In (135) we must use the branch of $s^{1/2}$, which is positive if s is real and positive, so it is convenient to cut the s-plane along the negative real axis. This is accomplished by setting $s = re^{i\theta}$ with $-\pi < \theta < \pi, 0 < r$. See Figure 1.9.

   Moreover, as the integrand in (135) is analytic everywhere in the s-plane except along the branch cut, the path of integration consisting of the segments: $C_R^-, C_B^-, C_\varepsilon, C_B^+, C_R^+$, traversed in the indicated directions, is equivalent to Γ in the limit $R \to \infty, \varepsilon \to 0$. Thus we may write (135) as

$$u(x, t) = \frac{1}{2\pi i} \lim_{\substack{R \to \infty \\ \varepsilon \to 0}} \left(\int_{C_R^-} f\, ds + \int_{C_B^-} f\, ds + \int_{C_\varepsilon} f\, ds + \int_{C_B^+} f\, ds + \int_{C_R^+} f\, ds \right)$$

$$\tag{136}$$

a. Show that

$$\int_{C_B^+} f\, ds = i \int_\varepsilon^R \frac{e^{-rt - ir^{1/2}|x|}}{2r^{1/2}}\, dr$$

$$\int_{C_B^-} f\, ds = i \int_\varepsilon^R \frac{e^{-rt + ir^{1/2}|x|}}{2r^{1/2}}\, dr$$

$$\int_{C_\varepsilon} f\, ds \to 0 \quad \text{as} \quad \varepsilon \to 0$$

$$\int_{C_R^+ + C_R^-} f\, ds \to 0 \quad \text{as} \quad R \to \infty \text{ with } t > 0$$

and therefore

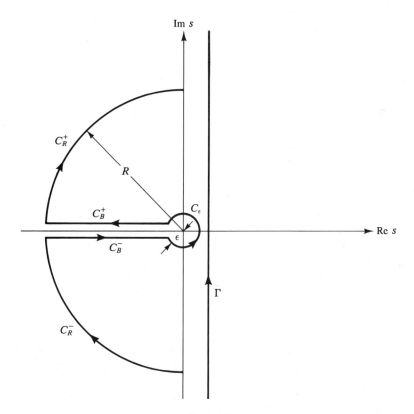

Figure 1.9

$$u(x, t) = \lim_{\substack{R \to \infty \\ \varepsilon \to 0}} \frac{1}{2\pi i} \int_{C_B^+ + C_B^-} f\, ds = \frac{1}{2\pi} \int_0^\infty e^{-rt} r^{-1/2} \cos(r^{1/2} x)\, dr$$

$$= \frac{1}{\pi} \int_0^\infty e^{-k^2 t} \cos kx\, dk \qquad (137)$$

b. Split the integral in (39) into its two contributions for $k > 0$ and $k < 0$; then change the sign of the variable of integration for $k < 0$ to obtain (137). Thus, the inversion integral for the solution using either Laplace transforms with respect to t or Fourier transforms with respect to x reduces to the real integral (137).

c. To evaluate (137), change variables again and set $k\sqrt{t} \equiv \xi$. Therefore,

$$u(x, t) = \frac{1}{\pi t^{1/2}} \int_0^\infty e^{-\xi^2} \cos\left(\frac{\xi x}{t^{1/2}}\right) d\xi$$

$$= \frac{1}{2\pi t^{1/2}} \int_{-\infty}^\infty e^{-\xi^2} \cos 2\alpha\xi\, d\xi \qquad (138)$$

where $\alpha \equiv x/2t^{1/2}$.

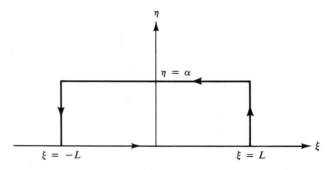

Figure 1.10

Now let ζ denote the complex variable $\zeta = \xi + i\eta$. Since $e^{-\zeta^2}$ is analytic everywhere, its integral around any closed contour vanishes. In particular,

$$\oint_C e^{-\zeta^2} \, d\zeta = 0 \tag{139}$$

where C is the rectangular contour indicated in Figure 1.10.

Decomposing (139) into its four components gives

$$\int_{-L}^{L} e^{-\xi^2} \, d\xi - \int_{-L}^{L} e^{-(\xi+i\alpha)^2} \, d\xi + \int_0^\alpha e^{-(L+i\eta)^2} i \, d\eta - \int_0^\alpha e^{-(-L+i\eta)^2} i \, d\eta = 0 \tag{140}$$

Show that as $L \to \infty$ the first term in (140) tends to $\pi^{1/2}$, that the third and fourth terms vanish, and that the second term equals $e^{\alpha^2} \int_{-\infty}^{\infty} e^{-\xi^2} \cos 2\alpha\xi \, d\xi$. Therefore, (138) gives the result derived in (29).

3. In this problem we review the idea of Green's function and superposition for an ordinary differential equation.

Consider the steady-state diffusion equation corresponding to $\partial/\partial t \equiv 0$, with a time-independent source distribution on the unit interval

$$-\frac{d^2u}{dx^2} = p(x), \qquad 0 \le x \le 1 \tag{141}$$

The boundary conditions are

$$u(0) = u(1) = 0 \tag{142}$$

This problem can also be interpreted as the static deflection u of a string constrained to have no deflection at $x = 0$ and $x = 1$ and subjected to the loading $p(x)$. See Sections 3.1 and 8.2.1.

Green's function $G(x; \xi)$ for the preceding boundary-value problem satisfies

$$-\frac{\partial^2 G}{\partial x^2} = \delta(x - \xi), \qquad 0 \le x \le 1 \tag{143}$$

$$G(0; \xi) = G(1; \xi) = 0 \tag{144}$$

where ξ is a fixed constant in the internal $0 < \xi < 1$.

a. Solve the homogeneous equation (143) in the form

$$G(x; \xi) = \begin{cases} A(\xi)x + B(\xi), & 0 \le x < \xi \\ C(\xi)x + D(\xi), & \xi < x \le 1 \end{cases} \tag{145}$$

and impose the two boundary conditions (144) to obtain

$$G(x; \xi) = \begin{cases} A(\xi)x, & 0 \le x < \xi \\ C(\xi)(x - 1), & \xi < x \le 1. \end{cases} \tag{146}$$

Integrate (143) over the interval $\xi - \varepsilon \le x \le \xi + \varepsilon$ for $\varepsilon > 0$; then take the limit $\varepsilon \downarrow 0$ to obtain the jump condition

$$-\frac{\partial G}{\partial x}(\xi^+, \xi) + \frac{\partial G}{\partial x}(\xi^-, \xi) = 1 \tag{147}$$

Use (147) together with the condition that G is continuous at $x = \xi$ to evaluate A and C and obtain

$$G(x; \xi) = \begin{cases} (1 - \xi)x, & 0 \le x < \xi \\ \xi(1 - x), & \xi < x \le 1 \end{cases} \tag{148}$$

b. Using superposition, the solution of (141)–(142) is

$$u(x) = \int_0^1 p(\xi)G(x; \xi) \, d\xi \tag{149}$$

Noting that G is symmetric, that is, $G(x; \xi) \equiv G(\xi, x)$, rewrite (149) in the explicit form

$$u(x) = (1 - x) \int_0^x \xi p(\xi) \, d\xi + x \int_x^1 (1 - \xi)p(\xi) \, d\xi \tag{150}$$

Now, solve (141)–(142) using the method of variation of parameters by assuming

$$u(x) = \alpha(x)x + \beta(x)(1 - x) \tag{151}$$

for unknown functions $\alpha(x)$ and $\beta(x)$. The boundary conditions (142) imply $\alpha(1) = 0$, $\beta(0) = 0$. Substitute (151) into (141) and solve the first-order equations that result for α and β to find (150).

Problems

1. Consider the diffusion equation with variable coefficient

$$2xu_t - u_{xx} = 0, \qquad 0 \le x < \infty, t \ge 0 \tag{152}$$

with boundary conditions

$$u(0, t) = C_1 = \text{constant} \qquad \text{if } t > 0 \tag{153}$$

$$u(\infty, t) = C_2 = \text{constant} \qquad \text{if } t > 0 \tag{154}$$

and initial condition

$$u(x,0) = C_3 = \text{constant} \tag{155}$$

a. What is the most general choice for the constants C_1, C_2, and C_3 for which the solution of the above initial- and boundary-value problem can be obtained in similarity form?

b. For the choice of constants obtained in part (a), calculate the solution and evaluate all integration constants explicitly.

2. Use similarity to reduce the following initial- and boundary-value problem for a nonlinear diffusion equation to an ordinary differential equation and corresponding boundary conditions

$$u_{xx} - uu_t = 0, \qquad 0 \le x, 0 \le t \tag{156}$$

$$u(0,t) = 0 \tag{157}$$

$$u(\infty,t) = 1 \tag{158}$$

$$u(x,0) = 1 \tag{159}$$

Discuss the behavior of the solution.

3. Verify by direct substitution that the sum of the expressions given by (41) and (43) solves the initial-value problem (40).

4. a. Verify by direct substitution that (48) solves (47) and that (51) solves (49).

b. Verify by direct substitution that the expression for $u(x,t)$ given by (64b) solves the initial- and boundary-value problem (55).

5. Consider the linear diffusion equation

$$u_t - u_{xx} = 0 \qquad 0 \le x, 0 \le t \tag{160}$$

with initial condition

$$u(x,0) = 0 \tag{161}$$

and boundary conditions at $x = 0$ of

$$u(0,t) = Ct^n, \qquad t > 0 \tag{162}$$

where n is a nonnegative constant and C is a positive constant. As usual, (161) implies the boundary condition

$$u(\infty,t) = 0, \qquad t \ge 0 \tag{163}$$

a. Use the result (64b) to express the solution in the form

$$u(x,t) = Ct^n f(\theta) \tag{164}$$

where

$$\theta \equiv \frac{x}{2t^{1/2}} \tag{165}$$

and

$$f(\theta) = \frac{2}{\sqrt{\pi}} \int_\theta^\infty \left(1 - \frac{\theta^2}{s^2}\right)^n e^{-s^2}\, ds \tag{166}$$

b. Show that the similarity form (164) satisfies the problem (160)–(163), and derive the following differential equation and boundary conditions for $f(\theta)$:

$$f'' + 2\theta f' - 4nf = 0 \tag{167}$$

$$f(0) = 1 \tag{168}$$

$$f(\infty) = 0 \tag{169}$$

Show that the solution of (167)–(169) gives (166).

c. Now consider the nonlinear diffusion equation

$$u_t - [k(u)u_x]_x = 0, \qquad 0 \le x, 0 \le t \tag{170}$$

where $k(u)$ is a prescribed function of u.

The initial condition is (161), and the boundary condition at $x = \infty$ is (163), whereas at $x = 0$ we have

$$u(0, t) = f(t), \qquad t > 0 \tag{171}$$

for some prescribed function $f(t)$. This problem is discussed in [3].

i. If $k(u) = \lambda u^v$, where λ and v are positive constants, show that the most general $f(t)$ for which a similarity solution exists is

$$f(t) = Ct^n$$

where C and n are constants as in (162). In this case, the similarity form is

$$u(x, t) = t^n \phi(\zeta), \qquad \zeta \equiv \frac{x}{t^{(vn+1)/2}} \tag{172}$$

and ϕ obeys

$$\lambda \frac{d}{d\zeta}\left(\phi^v \frac{d\phi}{d\zeta}\right) + \frac{vn + 1}{2}\zeta \frac{d\phi}{d\zeta} - n\phi = 0 \tag{173}$$

subject to the boundary conditions

$$\phi(0) = C, \qquad \phi(\infty) = 0 \tag{174}$$

ii. If $k(u)$ is prescribed arbitrarily, show that the most general $f(t)$ for which a similarity solution exists is $f(t) = C = $ constant. In this case the similarity form is

$$u(x, t) = \phi(\theta), \qquad \theta = \frac{x}{t^{1/2}} \tag{175}$$

and ϕ obeys

$$\frac{d}{d\theta}\left[k(\theta)\frac{d\phi}{d\theta}\right] + \frac{\theta}{2}\frac{d\phi}{d\theta} = 0 \tag{176}$$

with boundary conditions

$$\phi(0) = C \qquad \phi(\infty) = 0 \tag{177}$$

6. a. Use Laplace transforms and the convolution integral to show that the solution of (55) is given by (64b).
   b. Assume that the solution of (55) on the positive axis may be regarded as the response due to a source of unknown strength $q(t)$ at the origin. Therefore, $u(x, t)$ may be expressed in the form (41) with $p = \delta(x)q(t)$. In this case (41) reduces to

$$u(x, t) = \frac{1}{2\sqrt{\pi}} \int_0^t q(\tau) \frac{e^{-x^2/4(t-\tau)}}{\sqrt{t-\tau}} d\tau \qquad (178)$$

But, in order to satisfy the boundary condition (55c), we must have

$$g(t) = \frac{1}{2\sqrt{\pi}} \int_0^t \frac{q(\tau)\, d\tau}{\sqrt{t-\tau}} \qquad (179)$$

This is an integral equation (solved by Abel) for the unknown $q(t)$ in terms of the known $g(t)$.

Use Laplace transforms and the convolution integral to show that

$$q(t) = \frac{2}{\sqrt{\pi}} \frac{d}{dt} \int_0^t \frac{g(\tau)\, d\tau}{\sqrt{t-\tau}} \qquad (180)$$

Therefore, the solution of (55) may also be expressed in the form

$$u(x, t) = \frac{1}{\pi} \int_0^t \left[\frac{e^{-x^2/4(t-\tau)}}{\sqrt{t-\tau}} \frac{d}{d\tau} \int_0^\tau \frac{g(s)\, ds}{\sqrt{\tau-s}} \right] d\tau \qquad (181)$$

Show that this expression reduces to (64b).
   c. For the case $g(t) = 1$, we have shown that (181) reduces to $u(x, t) = \text{erfc}(x/2t^{1/2})$. Suppose that we wish to regard this solution in $0 \le x < \infty$, $0 \le t < \infty$ as being produced by an *unknown initial specification* of u of the form

$$u(x, 0) = \begin{cases} 0 & \text{if } x \ge 0 \\ f(x) & \text{if } x < 0 \end{cases} \qquad (182)$$

for the same diffusion equation (55a) over the infinite interval $-\infty < x < \infty$. With $\tilde{f}(x) = f(-x)$ show that $\tilde{f}$ obeys the integral equation

$$\int_0^\infty \tilde{f}(\xi) e^{-\xi^2/4t} \, d\xi = 2\sqrt{\pi t} \qquad (183)$$

Use Laplace transforms to show that $\tilde{f}(\xi) = 2$.
7. a. Modify the calculations leading to (64) so that you obtain the solution of (55) with (55b) replaced by the arbitrary initial condition

$$u(x, 0^+) = f(x) \qquad (184)$$

   b. Specialize the results in (a) to the case $f = \text{constant} = u_1$, and express the solution in a form such that $u_x(0^+, t)$ is free of singularities. (*Note*: (64b) has an apparent singularity at $x = 0$, whereas (64a) does not.)

c. Consider two semi-infinite rods with initial temperatures $u = u_1 =$ constant and $u = u_2 =$ constant, thermal diffusivities (compare with (8)) $\kappa_1^2 =$ constant and $\kappa_2^2 =$ constant, and thermal conductivities $k_1 =$ constant and $k_2 =$ constant. Suddenly, at $t = 0$, the two conductors are brought into perfect contact at $x = 0$. Let the first conductor be on $0 \le x < \infty$ and let the second conductor lie on $-\infty < x \le 0$.

It follows from the integral conservation law (5) with $A =$ constant that the interface conditions for $t > 0$ are $u(0^+, t) = u(0^-, t)$ and $k_1 u_x(0^+, t) = k_2 u_x(0^-, t)$. Show this. Use the result in (b) to show that the heat flow $k_1 u_x(0^+, t)$ (or $k_2 u_x(0^-, t)$) at the point of contact and $t > 0$ is given by

$$F(t) = \frac{1}{\sqrt{\pi t}} \frac{k_1}{\kappa_1}(u_1 - c) \tag{185}$$

where c is the constant temperature at $x = 0$:

$$c = \frac{u_2 - \alpha u_1}{1 - \alpha}, \qquad \alpha = -\frac{k_1 \kappa_2}{k_2 \kappa_1} \tag{186}$$

d. Now, consider the situation where these two rods are initially at zero temperature and in perfect thermal contact. Use the method of images to calculate the fundamental solution; that is, solve

$$u_t - \kappa^2 u_{xx} = \delta(t)\delta(x - \xi), \qquad 0 < \xi \tag{187}$$

on $-\infty < x < \infty$ with $u(x, 0^-) = 0$, where $\kappa = \kappa_1$ if $x > 0$ and $\kappa = \kappa_2$ if $x < 0$. Use the interface conditions $u(0^+, t) = u(0^-, t)$ and $k_1 u_x(0^+, t) = k_2 u_x(0^-, t)$. *Hint*: Assume that in the domain $x < 0$, the solution $u_2(x, t)$ may be regarded as the response to a source of unknown strength B and unknown location ($\xi_1 > 0$) in an inifinite medium with the uniform properties κ_2, k_2 *throughout.* Thus, $u_2(x, t)$ corresponds to a "transmitted" temperature due to the primary source at $x = \xi$ and $t = 0$. For the solution $u_1(x, t)$ in the domain $x > 0$, assume that, in addition to the response due to the primary source, there is a "reflected" contribution, which may be regarded as the response to an image source of unknown strength A located at the unknown point $x = \xi_2 < 0$ in an infinite rod with properties k_1 and κ_1 throughout. Use the interface conditions to determine A, B, ξ_1, and ξ_2. Verify that in the limits $(k_2/k_1) \to 1$ and $(\kappa_2/\kappa_1) \to 1$, $A \to 0$ and $B \to 1$.

8. Consider the diffusion equation

$$u_t - u_{xx} = 0, \qquad 0 \le t < \infty \tag{188}$$

on the *time-dependent* domain: $at \le x < \infty$, where a is a constant. We wish to solve the initial- and boundary-value problem having

$$u(x, 0^+) = 0 \tag{189}$$

$$u(at, t) = g(t) \tag{190}$$

for $t > 0$ and a prescribed $g(t)$. Thus, u is prescribed as a function of time on the left boundary that moves at a constant speed a.

a. Introduce the transformation of variables $\bar{x} = x - at, \bar{t} = t$ and solve the resulting problem by Laplace transforms.

b. Calculate the appropriate Green's function for the problem in $\bar{x}, \bar{t}$ variables and rederive the solution using this.

9. Using expression (67) for G_2, simplify the solution in (74a) to the form given by (74b), and rederive the same result using Laplace transforms.

10. To study (75), we first note that we may set $f(x) = 0$ with no loss of generality, since it is always possible to transform the dependent variable $u \to u - f$.

a. With $f = 0$, a and b constant, and $c(t)$ and $p(x, t)$ prescribed in general, calculate $U(x, s)$ (the Laplace transform of the solution). Comment on how you would handle the case where a and b are functions of t.

b. A second approach for solving (75), when a and b are constants, is to transform the dependent variable $u \to v$ such that v is a solution of the problem discussed in Section 1.4. Thus, let

$$au(x, t) + bu_x(x, t) \equiv v(x, t) \tag{191}$$

Show that v obeys

$$v_t - v_{xx} = ap(x, t) + bp_x(x, t) \equiv q(x, t)$$
$$v(x, 0) = 0 \tag{192}$$
$$v(0, t) = c(t)$$

Therefore, using the results in Section 1.4, we have

$$v(x, t) = \frac{x}{2\sqrt{\pi}} \int_0^t \tau^{-3/2} c(t - \tau) e^{-x^2/4\tau} \, d\tau$$
$$+ \int_0^t d\tau \int_0^\infty q(\xi, \tau) G_1(x, \xi, t - \tau) \, d\xi \tag{193}$$

where G_1 is defined by (46).

Knowing $v(x, t)$, we computes $u(x, t)$ by integrating (191). This gives

$$u(x, t) = \phi(t) e^{-ax/b} + \frac{e^{-ax/b}}{b} \int_0^x v(\xi, t) e^{a\xi/b} \, d\xi \tag{194}$$

where $\phi(t)$ is an arbitrary function.

Use the initial condition $u(x, 0) = 0$ to show that $\phi(0) = 0$. Show also that (194) satisfies the boundary condition (75c) identically. In order to determine ϕ, substitute (194) into (75a) and show that $\phi(t)$ obeys the first-order equation

$$\dot{\phi} - \left(\frac{a^2}{b^2}\right)\phi = \frac{ac(t)}{b^2} + p(0, t) + \frac{v_x(0, t)}{b} \tag{195}$$

Thus, the solution of (195) with $\phi(0) = 0$ defines $\phi(t)$ and (194) solves the problem when this value of $\phi(t)$ is used.

Verify your results for the special case $p(x, t) = 0$, in which case

$$v(x, t) = c \operatorname{erfc}\left(\frac{x}{2t^{1/2}}\right) \tag{196}$$

11. Use symmetry arguments to show that Green's function for the diffusion equation

$$u_t - u_{xx} = \delta(t - \tau)\delta(x - \xi) \tag{197}$$

with zero initial condition and each of the following three types of homogeneous boundary conditions, is given in the specified form.

a. $u(0, t) = u_x(1, t) = 0$ has

$$G_2(x, \xi, t - \tau) = \sum_{n=-\infty}^{\infty} (-1)^n \{F[x - (2n + \xi), t - \tau]$$
$$- F[x - (2n - \xi), t - \tau]\} \tag{198}$$

b. $u_x(0, t) = u(1, t) = 0$ has

$$G_3(x, \xi, t - \tau) = \sum_{n=-\infty}^{\infty} (-1)^n \{F[x - (2n + \xi), t - \tau]$$
$$+ F[x - (2n - \xi), t - \tau]\} \tag{199}$$

c. $u_x(0, t) = u_x(1, t) = 0$ has

$$G_4(x, \xi, t - \tau) = \sum_{n=-\infty}^{\infty} \{F[x - (2n + \xi), t - \tau]$$
$$+ F[x - (2n - \xi), t - \tau]\} \tag{200}$$

What symmetry properties, if any, can you uncover for G_2, G_3, and G_4 if $x \to \xi$, $\xi \to x$?

12. a. Evaluate (80) for the special case where $\rho = \delta(x - \xi)$, where ξ is a fixed constant on $0 < \xi < 1$. Show that as $t \to \infty$, your result reduces to (148), Green's function for the steady-state problem calculated in Review Problem 3.

b. Evaluate (82) for $f = 1$ and show that the resulting series is the same as (100).

13. Solve

$$u_t - u_{xx} = p(x, t), \qquad 0 \le x \le 1, 0 \le t \tag{201}$$

$$u(x, 0^+) = f(x) \tag{202}$$

$$u(0, t) = g(t), \qquad u_x(1, t) = h(t) \tag{203}$$

using Green's function and separation of variables after having transformed to a homogeneous boundary-value problem.

14. Consider the spherically symmetric diffusion equation with a unit source, that is,

$$u_t - u_{rr} - \frac{(n-1)}{r} u_r = \delta(t)\delta_n(r) \tag{204}$$

on $0 \leq t, 0 \leq r$, where $n = 1, 2, 3, \ldots$.

Note that $\delta_n(r)$ is the n-dimensional delta function with the property

$$\int_\varepsilon \cdots \int \delta_n(r)\,dV = \begin{cases} 1 & \text{if } \varepsilon \text{ contains } r = 0 \\ 0 & \text{otherwise} \end{cases} \tag{205}$$

where dV is the element of volume in n dimensions and ε is any n-dimensional domain. The notation $\int \cdots \int$ indicates an n-tuple integral. Thus, for example, if we take ε to be an n-dimensional "sphere" of radius ε centered at the origin, we have for $n = 1, 2, 3$:

$$n = 1: \quad \int_{-\varepsilon}^{\varepsilon} \delta_1(x)\,dx = 1 \tag{206}$$

$$n = 2: \quad \int_{r=0}^{\varepsilon} \int_{\theta=0}^{2\pi} \delta_2(r) r\,dr\,d\theta = 1 \tag{207}$$

Here r and θ are polar coordinates in the plane defined by $x = r\cos\theta$, $y = r\sin\theta$.

$$n = 3: \quad \int_{r=0}^{\varepsilon} \int_{\theta=0}^{2\pi} \int_{\phi=0}^{\pi} \delta_3(r) r^2 \sin\phi\,dr\,d\theta\,d\phi = 1 \tag{208}$$

Here r, θ and ϕ are spherical polar coordinates in a three-dimensional space defined by $x = r\sin\phi\cos\theta$, $y = r\sin\phi\sin\theta$, $z = r\cos\phi$. Show that the homogeneous solution of (204) with the appropriate singularity at $r = 0$, $t = 0$ is $Ct^{-n/2}e^{-r^2/4t}$, where C is a constant. Refer to Problem 2b of Chapter 2, where the surface area of the n-dimensional unit sphere is derived as $\omega_n = 2\pi^{n/2}/\Gamma(n/2)$, where Γ is the Gamma function defined by

$$\Gamma(x) \equiv \int_0^\infty e^{-t}t^{x-1}\,dt; \quad x > 0 \tag{209}$$

Use this result to show that $C = 1/2^n\pi^{n/2}$.

15. Consider the inhomogeneous, two-dimensional diffusion equation

$$u_t - u_{xx} - u_{yy} = e^{-|x|}y\sin\omega t \tag{210}$$

with $\omega = \text{constant}$ on the infinite strip $-\infty < x < \infty$, $0 \leq y \leq \pi$. The boundary conditions are

$$u(x, 0, t) = u_y(x, \pi, t) = 0, \quad t > 0 \tag{211}$$

$$u \to 0 \quad \text{as} \quad |x| \to \infty \tag{212}$$

and the initial condition is

$$u(x, y, 0) = \begin{cases} a = \text{constant} & \text{if } 0 \leq y \leq \pi, |x| \leq 1 \\ 0 & \text{otherwise} \end{cases} \tag{213}$$

Use Fourier transforms with respect to x and eigenfunction expansions with respect to y to derive an integral representation of the form

$$u(x, y, t) = \frac{1}{\sqrt{2\pi}} \int_{-\infty}^{\infty} \bar{u}(k, y, t) e^{-ikx} \, dk \tag{214}$$

for the solution. Do not attempt to evaluate this inversion integral explicitly. However, give the explicit series form for $\bar{u}(k, y, t)$.

References

1. B. Friedman, *Principles and Techniques of Applied Mathematics*, Wiley, New York, 1956.
2. P. A. Lagerstrom, "Laminar Flow Theory," Section B in *Theory of Laminar Flows*, ed. by F. K. Moore, Princeton University Press, Princeton, N.J., 1964, pp. 20–285.
3. G. W. Bluman, and J. D. Cole, *Similarity Methods for Differential Equations*, Springer, New York, 1974.
4. L. I. Sedov, *Similarity and Dimensional Methods*, Academic Press, New York, 1959.
5. G. F. Carrier, M. Krook, and C. E. Pearson, *Functions of a Complex Variable, Theory and Technique*, McGraw-Hill, New York, 1966.
6. R. Courant, and D. Hilbert, *Methods of Mathematical Physics*, Vol. I, Interscience Publishers, New York, 1953.
7. G. B. Whitham, *Linear and Nonlinear Waves*, Wiley-Interscience, New York, 1974.
8. C. L. Frenzen, and J. Kevorkian, "A Review of the Multiple Scale and Reductive Perturbation Methods for Deriving Uncoupled Nonlinear Evolution Equations," *Wave Motion*, 7, no. 1 (1985): 25–42.
9. J. Kevorkian, and J. D. Cole, *Perturbation Methods in Applied Mathematics*, Springer, New York, 1981.

CHAPTER 2

Laplace's Equation

2.1 Applications

There are numerous physical applications that are modeled by the inhomogeneous Laplace equation (Poisson equation)

$$\Delta u \equiv u_{xx} + u_{yy} + u_{zz} = Q(x, y, z)$$

Some of the standard examples are given in the following table.

Application	u	Q
Steady-state temperature in a solid	u = temperature	Heat source strength/unit volume
Static deflection of a thin membrane in two dimensions	u = deflection	Pressure
Electrostatics	u = electrostatic potential electric field = $\mathbf{E}$ = grad u	Charge/unit volume
Incompressible irrotational flow in two or three dimensions	u = velocity potential velocity = $\mathbf{q}$ = grad u	Mass source strength/unit volume
Two-dimensional incompressible steady flow	u = stream function = ψ velocity = $\mathbf{q} = \psi_y \mathbf{i} - \psi_x \mathbf{j}$	$-$ Vorticity
Newtonian gravitation	u = gravitational potential Force of gravity = $\mathbf{F} = -$ grad u	Mass density

Henceforth, to standardize terminology, we shall refer to u as the *potential* even though in some applications it is not a potential and one is interested in the value of u itself rather than in its gradient. Also, we shall refer to a real function that satisfies Laplace's equation in some domain D as being harmonic in D.

We have already shown that, for the steady-state problem of heat conduction in a material with constant properties, the temperature field satisfies Laplace's equation [see (14b) of Chapter 1].

A derivation of Laplace's equation for the deflection of a membrane (in the limit of small amplitudes) may be found on pp. 214–215 of [1].

In electrostatics, the potential due to a stationary distribution of charges follows directly from Maxwell's equation. For example, see p. 100 of [2]. A derivation of the gravitational potential for an arbitrary solid is given in Section 2.4.1 and a discussion of applications for incompressible flow follows next.

2.1.1 Incompressible Irrotational Flow

Consider the flow of a fluid having density $\rho(x, y, z, t)$ (g/cm^3) and defined by the vector velocity field $\mathbf{q}(x, y, z, t)$ (cm/s). As in Section 1.1, we can derive an integral law of mass conservation by equating the rate of change of mass inside a given fixed domain G to the net *inflow* of mass. If we also have an arbitrary distribution of mass sources of strength/unit volume equal to $Q(x, y, z, t)$ $(g/cm^3\ s)$, the integral law of mass conservation analogous to (11) of Chapter 1 becomes

$$\frac{d}{dt} \iiint_G \rho \, dV = - \iint_S \rho \mathbf{q} \cdot \mathbf{n} \, dA + \iiint_G Q \, dV \tag{1}$$

where again, $\mathbf{n}$ is the outward unit normal on the surface S bounding G. For smooth flows, (1) gives

$$\rho_t + \mathrm{div}(\rho \mathbf{q}) = Q \tag{2}$$

Now, if the density is a constant (incompressible flow), (2) reduces to

$$\mathrm{div}\,\mathbf{q} = \tilde{Q} \equiv Q/\rho \tag{3}$$

If, in addition, one assumes that the flow is irrotational—that is, curl $\mathbf{q} = 0$—it follows from vector calculus that $\mathbf{q}$ is the gradient of a scalar potential: $u(x, y, z, t)$ (cm^2/s); that is,

$$\mathbf{q} = \mathrm{grad}\,u \tag{4}$$

Combining (3) and (4), we obtain the Poisson equation

$$\mathrm{div}\,\mathrm{grad}\,u \equiv \Delta u = \tilde{Q} \tag{5}$$

2.1.2 Two-Dimensional Incompressible Flow

A flow is two-dimensional if the velocity field is independent of z, for instance. Consider such a flow and assume also that it is steady (independent of time), has a constant density, but is not necessarily irrotational. Mass conservation [equation (3)] gives

$$q_{1_x} + q_{2_y} = 0 \tag{6}$$

where $\mathbf{q} = q_1\mathbf{i} + q_2\mathbf{j}$ and $\mathbf{i}, \mathbf{j}$ are Cartesian unit vectors along x and y, respectively.

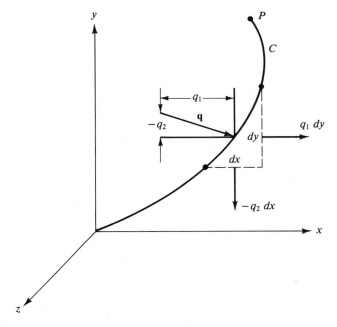

Figure 2.1

Consider an arbitrary simple curve C joining the origin to the point P as shown in Figure 2.1. The flow rate per unit depth across a given element $ds \equiv (dx^2 + dy^2)^{1/2}$ is $dM \equiv \rho(q_1\,dy - q_2\,dx)$ (gm/cm s). Therefore, the total flow/unit depth across the arc C is the line integral

$$\frac{M}{\rho} \equiv \int_C (q_1\,dy - q_2\,dx)\;(\text{cm}^2/\text{s}) \tag{7}$$

Introduce the vector $\mathbf{F} \equiv -q_2\mathbf{i} + q_1\mathbf{j}$, and note that curl $\mathbf{F} = (q_{1_x} + q_{2_y})\mathbf{k} = 0$ because of (6). Here $\mathbf{k}$ is the Cartesian unit vector in the z direction. Therefore, the line integral (7), which may also be written as

$$\frac{M}{\rho} = \int_C \mathbf{F}\cdot d\mathbf{r}; \qquad d\mathbf{r} = dx\,\mathbf{i} + dy\,\mathbf{j} \tag{8}$$

is a function only of the endpoint P and does not depend on the path C. Denote $M/\rho = \psi(x, y)$, where ψ is called the *stream function* for the flow. Thus, at any point P, ψ measures the total mass flow between P and the origin.

It follows from the fact that curl $\mathbf{F} = 0$ that

$$\mathbf{F} = \text{grad}\,\psi \equiv \psi_x\mathbf{i} + \psi_y\mathbf{j} \tag{9a}$$

Therefore,

$$q_1 = \psi_y; \qquad q_2 = -\psi_x \tag{9b}$$

that is, the velocity vector $\mathbf{q}$ at any point is tangent to the curve $\psi = $ constant passing through that point. The curves $\psi = $ constant are called *streamlines* (see Figure 2.2) and measure loci of constant mass flow relative to a reference point

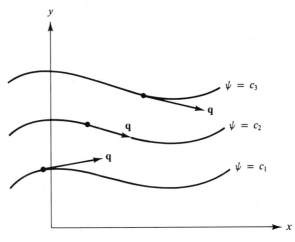

Figure 2.2

in the sense just discussed. In particular, the mass flow between any two curves $\psi = c_1$ and $\psi = c_2$ remains constant. Thus, if the distance between these two curves narrows down, the velocity must increase to conserve mass flow.

For the velocity field defined by $\mathbf{q}$, let us denote

$$\text{curl } \mathbf{q} \equiv \mathbf{\Omega} = \Omega(x, y)\mathbf{k} \tag{10}$$

The vector $\mathbf{\Omega}$ is called the vorticity and corresponds to twice the average angular velocity of a fluid element. (For example, see p. 158 of [1].)

Now, by definition

$$\text{curl } \mathbf{q} = (q_{2_x} - q_{1_y})\mathbf{k}$$

and using (9b) and (10), we obtain

$$\text{curl } \mathbf{q} = -\Delta\psi\mathbf{k} = \Omega(x, y)\mathbf{k} \tag{11}$$

Thus, for a steady, incompressible, sourceless, two dimensional, possibly rotational flow, the stream function $\psi(x, y)$ obeys

$$\Delta\psi = -\Omega(x, y) = -\text{vorticity} \tag{12a}$$

If the flow is irrotational, we have

$$\Delta\psi = 0 \tag{12b}$$

2.2   The Two-Dimensional Problem; Conformal Mapping

The solution of Laplace's equation in two-dimensional domains is intimately related to the theory of analytic functions of a complex variable. In fact, this topic occupies a significant portion of texts on complex variables and will therefore only be outlined in this section.

2.2.1 Mapping of Harmonic Functions

We restrict out discussion to simply connected domains—that is, domains D for which every simple closed curve within D encloses only points of D.

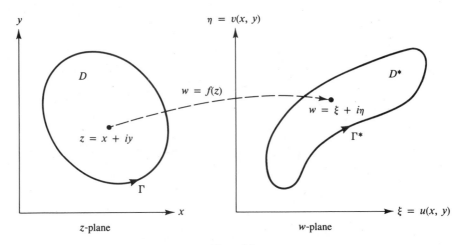

Figure 2.3

The first theorem we invoke asserts that the real and imaginary parts of an analytic function are harmonic; that is, if

$$F(z) = \phi(x, y) + i\psi(x, y); \qquad z = x + iy \tag{13}$$

is analytic, then

$$\phi_{xx} + \phi_{yy} = 0 \tag{14a}$$

$$\psi_{xx} + \psi_{yy} = 0 \tag{14b}$$

A second result asserts that if $w = f(z)$ is analytic in some domain D of the z-plane (and hence defines a conformal mapping of D to the domain D^* in the w-plane), then a harmonic function $\phi(x, y)$ defined on D maps to a harmonic function $\Phi(\xi, \eta)$ on D^*, and vice versa (see Figure 2.3). More precisely, consider the mapping $z \to w$ defined by:

$$w = u(x, y) + iv(x, y) = f(z) \tag{15}$$

where $f(z)$ is analytic in D. With $w = \xi + i\eta$, regard the pair of equations $\xi = u(x, y)$, $\eta = v(x, y)$, as a coordinate transformation $(x, y) \leftrightarrow (\xi, \eta)$. If $\Phi(\xi, \eta)$ satisfies, $\Phi_{\xi\xi} + \Phi_{\eta\eta} = 0$ in D^*, then

$$\phi(x, y) \equiv \Phi(u(x, y), v(x, y)) \tag{16}$$

satisfies $\phi_{xx} + \phi_{yy} = 0$ in D, and vice versa. (See Review Problem 1.)

2.2.2 Transformation of Boundary Conditions

An important question associated with this mapping concerns the transformation of a boundary condition on Γ, the boundary of D, to Γ^*, the boundary of D^*. To fix ideas, let us parameterize Γ in the form

$$x = \alpha(s); \qquad y = \beta(s) \tag{17}$$

where $s: 0 \leq s \leq s_0$ is a parameter that varies montonically along Γ and $\alpha(s_0) =$

$\alpha(0)$; $\beta(s_0) = \beta(0)$ if D is bounded. Moreover, let D lie to the left as Γ is traversed in the direction of increasing s. Now, Γ^*, the boundary of D^*, is defined by:

$$\xi = u(\alpha(s), \beta(s)) \equiv \lambda(s)$$
$$\eta = v(\alpha(s), \beta(s)) \equiv \mu(s) \tag{18}$$

and is also traversed in the same sense.

Consider now a general linear boundary condition on Γ of the form

$$A(s)\phi(\alpha, \beta) + B(s)\phi_n(\alpha, \beta) = C(s) \tag{19}$$

where $\alpha(s)$ and $\beta(s)$ are given in (17), the functions A, B, C are prescribed, and ϕ_n denotes the outward normal derivative of ϕ on Γ; that is,

$$\phi_n(\alpha, \beta) \equiv \phi_x(\alpha, \beta)n_1(\dot\alpha, \dot\beta) + \phi_y(\alpha, \beta)n_2(\dot\alpha, \dot\beta) \tag{20a}$$

with n_1 equal to the x-component and n_2 equal to the y-component of the unit outward normal; that is,

$$n_1 \equiv \frac{\dot\beta}{(\dot\alpha^2 + \dot\beta^2)^{1/2}}; \qquad n_2 \equiv -\frac{\dot\alpha}{(\dot\alpha^2 + \dot\beta^2)^{1/2}} \tag{20b}$$

It is easy to show that the boundary condition (19) transforms to

$$A(s)\Phi(\lambda, \mu) + B(s)\left(\frac{\dot\lambda^2 + \dot\mu^2}{\dot\alpha^2 + \dot\beta^2}\right)^{1/2}\Phi_N(\lambda, \mu) = C(s) \tag{21}$$

where, again, Φ_N is the outward normal derivative of Φ on Γ^*; that is,

$$\Phi_N(\lambda, \mu) \equiv \Phi_\xi(\lambda, \mu)N_1(\dot\lambda, \dot\mu) + \Phi_\eta(\lambda, \mu)N_2(\dot\lambda, \dot\mu) \tag{22a}$$

with

$$N_1 \equiv \frac{\dot\mu}{(\dot\lambda^2 + \dot\mu^2)^{1/2}}; \qquad N_2 \equiv -\frac{\dot\lambda}{(\dot\lambda^2 + \dot\mu^2)^{1/2}} \tag{22b}$$

In particular, if $B \equiv 0$ (Dirichlet problem), the boundary values *of ϕ map unchanged to boundary values of Φ at corresponding points.* If $A \equiv 0$ (Neumann problem), boundary values of the normal derivative at corresponding points are scaled by the factor $(\dot\lambda^2 + \dot\mu^2)^{1/2}/(\dot\alpha^2 + \dot\beta^2)^{1/2}$. A boundary condition $\phi_n \equiv 0$ is mapped unchanged to a boundary condition $\Phi_N \equiv 0$.

2.2.3 Example, Solution in a "Simpler" Transformed Domain

To illustrate an application of the preceding ideas, consider the Dirichlet problem for Laplace's equation in the corner domain sketched in Figure 2.4.

The boundary values for ϕ are specified in terms of the two functions $a(y)$ and $b(y)$.

For the case $a(y) \equiv a =$ constant, $b(y) \equiv b =$ constant $\neq a$, it is convenient to map D onto the strip domain D^* using $w = \log z$. Here, we use the principal branch of $\log z$—that is, $-\pi < \arg z < \pi$—and cut the z-plane along the negative

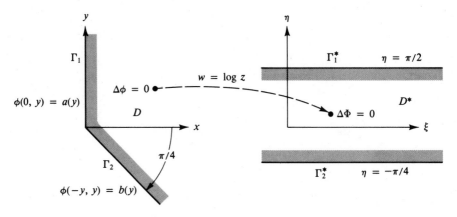

Figure 2.4

real axis. It then follows that

$$\xi = \log(x^2 + y^2)^{1/2}; \qquad \eta = \tan^{-1}\left(\frac{y}{x}\right) \tag{23}$$

Now Γ_1 maps to Γ_1^*, the horizontal line $\eta = \pi/2$, and Γ_2 maps to Γ_2^*, the horizontal line $\eta = -\pi/4$. Moreover, the boundary condition on Φ becomes

$$\Phi\left(\xi, \frac{\pi}{2}\right) = a, \qquad \Phi\left(\xi, -\frac{\pi}{4}\right) = b \tag{24}$$

Thus, we need to solve $\Delta\Phi = 0$ in the strip domain D^* subject to boundary conditions *that do not depend on ξ*. It then follows that Φ does not depend on ξ, and we need to solve only $\partial^2\phi/\partial\eta^2 = 0$—that is,

$$\Phi = C_1 + C_2\eta \tag{25}$$

where C_1 and C_2 are constants. Imposing the boundary condition (24) determines C_1 and C_2, and we find:

$$\Phi = \frac{2b + a}{3} + \frac{4(a - b)}{3\pi}\eta \tag{26a}$$

Therefore, the solution in D is

$$\phi(x, y) = \frac{2b + a}{3} + \frac{4(a - b)}{3\pi}\tan^{-1}\left(\frac{y}{x}\right) \tag{26b}$$

The success of this approach is directly due to the fact that we were able to write the solution for Φ in D^* by inspection. This, in turn, is a direct consequence of the simple boundary conditions. In fact, if a and b are given functions of y, the transformation (23) to D^* certainly still holds, but (25) cannot, in general, satisfy the boundary conditions because Φ will depend on ξ along the two boundaries Γ_1^* and Γ_2^*. If we attempt to accommodate the variable boundary conditions by assuming $C_1(\xi)$ and $C_2(\xi)$, we immediately discover that this is solution only if

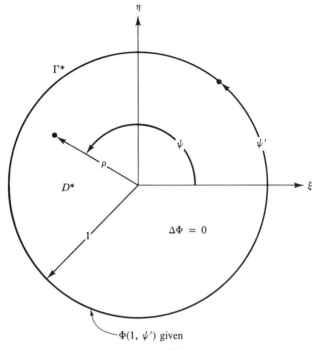

Figure 2.5

$C_1'' \equiv 0, C_2'' \equiv 0$. Thus, we can handle only the case for which a and b transform to linear functions on Γ_1^* and Γ_2^*.

To make any progress for the case where $a(y)$, $b(y)$ are arbitrarily prescribed, we need to transform D into a domain D^* *for which we can solve the Dirichlet problem for arbitrary boundary conditions.* Recall from your study of complex variables that we can solve Laplace's equation in the interior of the unit circle with arbitrary boundary values for Φ on the circumference of the unit circle. This is Poisson's formula, given next for the geometry sketched in Figure 2.5. (See for example, p. 47 of [3] for a derivation of this result using Cauchy's integral formula or (165) for a derivation using Green's function).

$$\Phi(\rho, \psi) = \frac{1 - \rho^2}{2\pi} \int_0^{2\pi} \frac{\Phi(1, \psi')\,d\psi'}{1 + \rho^2 - 2\rho\cos(\psi - \psi')} \tag{27}$$

Therefore, if we can transform D to the interior of the unit circle, we will have solved the problem. There is a famous theorem due to Riemann that asserts that any simply connected domain whose boundary consists of more than one point can be mapped conformally onto the interior of the unit circle. It is also possible to make an arbitrary point in D and a direction through this point correspond, respectively, to the origin and the positive real axis. If this is done, the mapping is unique. For a proof of this theorem, see Chapter 5 of [4].

Although this result is reassuring, it does not provide us with the mapping itself. We shall see in Section 2.10 that *finding the mapping is equivalent to finding Green's function* for the domain.

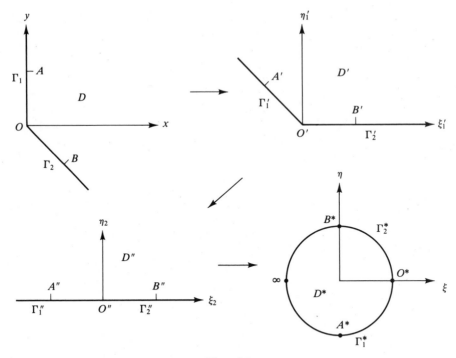

Figure 2.6

Let us use our knowledge of the mapping properties of simple functions to map D into the unit circle via the sequence of simple mappings sketched in Figure 2.6.

The first mapping, $z \to w_1$, is a rotation by $\pi/4$ and therefore obeys

$$w_1 = ze^{i\pi/4} \tag{28a}$$

The second mapping, $w_1 \to w_2$, bends Γ_1', the map of Γ_1, (which is now given by arg $w_1 = 3\pi/4$) to the negative real axis of w_2 but leaves Γ_2', the map of Γ_2, unchanged. Therefore,

$$w_2 = w_1^{4/3} \tag{28b}$$

The third mapping must be a linear fractional transformation (sometimes also called a bilinear transformation) and we derive it in the form:

$$w = (i - w_2)/(i + w_2) \tag{28c}$$

See Review Problem 2.

Combining these and simplifying gives

$$w = \frac{1 - e^{i\pi/6}z^{4/3}}{1 + e^{i\pi/6}z^{4/3}} \tag{29a}$$

or using the polar form $z = re^{i\theta}$, we find:

$$w = \frac{1 - r^{8/3} - 2ir^{4/3}\sin\left(\dfrac{4\theta}{3} - \dfrac{\pi}{6}\right)}{1 + r^{8/3} + 2r^{4/3}\cos\left(\dfrac{4\theta}{3} - \dfrac{\pi}{6}\right)} \tag{29b}$$

Therefore, the ξ and η components are defined by

$$\xi = \frac{(1 - r^{8/3})}{D(r,\theta)} \equiv u(r,\theta) \tag{30a}$$

$$\eta = -\frac{2r^{4/3}\sin\left(\dfrac{4\theta}{3} - \dfrac{\pi}{6}\right)}{D(r,\theta)} \equiv v(r,\theta) \tag{30b}$$

where $D(r,\theta)$ is the denominator of (29b):

$$D(r,\theta) \equiv 1 + r^{8/3} + 2r^{4/3}\cos\left(\frac{4\theta}{3} - \frac{\pi}{6}\right) \tag{30c}$$

In order to specify the boundary conditions on Γ^*, we must transform the given $a(y)$ and $b(y)$ to obtain $\Phi(1, \psi)$. The details are straightforward but laborious and are therefore omitted. Knowing $\Phi(1, \psi)$ defines $\Phi(\rho, \psi)$ according to (27), and $\phi(r, \theta)$ is then given by

$$\phi(r,\theta) = \Phi\left(\sqrt{u^2 + v^2}, \tan^{-1}\frac{v}{u}\right) \tag{31}$$

where $u(r, \theta)$ and $v(r, \theta)$ are defined in (30).

Actually the transformation (29a) is not needed for this problem and was worked out for purposes of illustration only; the intermediate transformation to the upper half-plane defined by (28a) and (28b) suffices because the solution of the Dirichlet problem there is also available for arbitrary boundary values on the real axis (see Problems 1 and 13). Usually, the transformation to the interior of the unit circle is appropriate for bounded domains D.

The reader will find a number of interesting and challenging examples in Chapter 4 of [3]. The material in the present chapter does not dwell on techniques that are appropriate only for the two-dimensional Laplacian; rather, whenever the opportunity arises, we point out how general results specialize to two dimensions and relate to ideas from complex variables.

2.3   Fundamental Solution; Dipole Potential

As in Chapter 1, we begin our discussion of solution techniques by studying the influence of a unit source in the infinite domain. This is the fundamental solution that is the basic building block for constructing the solutions of more general boundary-value problems.

2.3.1 Point Source in Three Dimensions

Consider the potential at a point $P = (x, y, z)$ due to a unit positive source located at the origin; that is, we seek the solution of

$$\Delta u = \delta(x)\delta(y)\delta(z) \tag{32}$$

Because of symmetry, we need consider only the spherically symmetric Laplacian and solve:

$$\frac{d^2u}{dr^2} + \frac{2}{r}\frac{du}{dr} = \delta_3(r) \tag{33}$$

It is important to bear in mind that $\delta_3(r) \equiv \delta(x)\delta(y)\delta(z)$. Thus,

$$\iiint_G \delta_3(r)\,dV = \begin{cases} 1, & \text{if } G \text{ contains origin} \\ 0, & \text{otherwise} \end{cases} \tag{34}$$

where G is a given three-dimensional domain, and dV is the element of volume in G.

Solving the homogeneous version of (33) gives

$$u = \frac{C_1}{r} + C_2 \tag{35}$$

and we discard C_2 by normalizing $u(\infty) = 0$.

To evaluate C_1, we integrate (32) over G_ε, the interior of a sphere of radius ε centered at the origin. Hence,

$$\iiint_{G_\varepsilon} \Delta u\,dV = 1 \tag{36}$$

The left-hand side of (36) can be computed using Gauss' theorem; that is,

$$\iiint_G \operatorname{div} \mathbf{F}\,dV = \iint_\Gamma \mathbf{F}\cdot\mathbf{n}\,dA \tag{37a}$$

where $\mathbf{F}$ is a prescribed vector field in the domain G with boundary Γ, outward unit normal $\mathbf{n}$, element of volume dV, and element of area dA. Let G in (37a) be the interior of the ε-sphere and set $\mathbf{F} = \operatorname{grad} u$, where $u = C_1/r$. Therefore, $\mathbf{F}\cdot\mathbf{n} = du/dr$ and (37a) gives

$$\iiint_{G_\varepsilon} \Delta u\,dV = \iint_{\Gamma_\varepsilon} \frac{du}{dr}\,dA = \int_{\theta=0}^{\pi}\int_{\psi=0}^{2\pi}\left(\frac{-C_1}{\varepsilon^2}\right)\varepsilon^2\sin\theta\,d\theta\,d\psi = -4\pi C_1 \tag{37b}$$

Thus, $C_1 = -1/4\pi$ and the solution of (32) is $u = -1/4\pi r$. More generally, if the source is located at the point $Q = (\xi, \eta, \zeta)$ and we denote

$$r_{PQ}^2 \equiv (x - \xi)^2 + (y - \eta)^2 + (z - \zeta)^2 \tag{38a}$$

$$\delta_3(P, Q) \equiv \delta(x - \xi)\delta(y - \eta)\delta(z - \zeta) \tag{38b}$$

the solution of

$$\Delta u = \delta_3(P, Q) \tag{39}$$

is

$$u = -\frac{1}{4\pi r_{PQ}} \tag{40}$$

The fundamental solution for the n-dimensional Laplacian is discussed in Problem 2.

2.3.2 Fundamental Solution in Two Dimensions; Descent

The case $n = 2$ is special [see (241)] and is discussed next. The axially symmetric equivalent of (32) with $r^2 \equiv x^2 + y^2$ is

$$\frac{d^2 u}{dr^2} + \frac{1}{r}\frac{du}{dr} = \delta_2(r) \tag{41}$$

where

$$\delta_2(r) \equiv \delta(x)\delta(y)$$

and

$$\iint_A \delta_2(r)\, dS = \begin{cases} 1, & \text{if } A \text{ contains the origin} \\ 0, & \text{otherwise} \end{cases} \tag{42}$$

Here, A is a given planar domain and dS is the area element.

The homogeneous solution of (41) is

$$u = C_1 \log r + C_2 \tag{43}$$

and it is no longer possible to normalize $u(\infty) = 0$. However, set $C_2 = 0$ by requiring $u(1) = 0$. Now, we can evaluate C_1 by applying the two-dimensional Gauss theorem:

$$\iint_A \operatorname{div} \mathbf{F}\, dS = \int_C \mathbf{F} \cdot \mathbf{n}\, ds \tag{44}$$

where $\mathbf{F}$ is a two-dimensional vector field in the planar domain A with element of area dS, boundary C with element of arc ds, and outward unit normal $\mathbf{n}$. Again, integrating (41) over A_ε, the interior of a circle of radius ε centered at the origin, gives

$$\iint_{A_\varepsilon} \Delta u\, dS = 1 \tag{45}$$

Now use $u = C_1 \log r$ and set $\mathbf{F} = \operatorname{grad} u$ in (44) to obtain $C_1 = 1/2\pi$. Thus

$$u = \frac{1}{2\pi} \log r \tag{46}$$

More generally, the fundamental solution of

$$u_{xx} + u_{yy} = \delta(x - \xi)\delta(y - \eta) \tag{47}$$

is

$$u = \frac{1}{2\pi}\log\sqrt{(x - \xi)^2 + (y - \eta)^2} \tag{48}$$

It is interesting to derive (46) by "descent" from the three-dimensional result. Since a two-dimensional unit source is just a distribution of sources of unit strength along an infinite straight line, it must be possible to obtain the result (46) as a solution of the following three-dimensional problem:

$$u_{xx} + u_{yy} + u_{zz} = \delta(x)\delta(y) \tag{49}$$

By superposition of the fundamental solution (40), the potential at the point $r = \sqrt{x^2 + y^2}$, $z = 0$ is simply

$$u(r) = -\frac{1}{4\pi}\int\!\!\!\int\!\!\!\int_{-\infty}^{\infty}\frac{\delta(\xi)\delta(\eta)\,d\xi\,d\eta\,d\zeta}{\sqrt{(x - \xi)^2 + (y - \eta)^2 + \zeta^2}}$$

$$= -\frac{1}{4\pi}\int_{-\infty}^{\infty}\frac{d\zeta}{\sqrt{r^2 + \zeta^2}} \tag{50}$$

(*Note*: There is no loss of generality in setting $z = 0$.

The integral (50) diverges, which is not surprising because the value of u itself need not be bounded. Thus, if u in (49) is interpreted as the steady temperature due to the continuous distribution of point sources of heat along an infinite straight line, the temperature at any finite radius r is indeed infinite.

However, if u is interpreted as a potential, it is only its gradient which is of interest, and in computing this gradient, the contribution due to a large additive constant may be ignored. Therefore, we interpret (50) in the sense of Hadamard's definition of the "finite part" of a divergent integral, denoted by the integral sign $\unicode{x2A0D}$. For more details concerning divergent integrals, see Appendix D, Equation 14.1, of [5]. Thus, we regard u to be

$$u = -\frac{1}{4\pi}\unicode{x2A0D}_{-\infty}^{\infty}\frac{d\zeta}{\sqrt{r^2 + \zeta^2}} \equiv -\frac{1}{4\pi}\int_{1}^{r}\left[\int_{-\infty}^{\infty}\frac{\partial}{\partial\rho}\left(\frac{1}{\sqrt{\rho^2 + \zeta^2}}\right)d\zeta\right]d\rho \tag{51}$$

Since the integral in the square brackets converges—in fact,

$$\int_{-\infty}^{\infty}\frac{\partial}{\partial\rho}\left[\frac{1}{\sqrt{\rho^2 + \zeta^2}}\right]d\zeta = -\int_{-\infty}^{\infty}\frac{\rho\,d\zeta}{(\rho^2 + \zeta^2)^{3/2}} = -\frac{2}{\rho}$$

we find $u = (1/2\pi)\log r$, as before. The definition (51) for the finite part of u involves two steps. First, we filter out the infinite constant in (50) by evaluating the derivative of u given by the *convergent* integral

$$u_r = \frac{1}{4\pi}\int_{-\infty}^{\infty}\frac{r\,d\zeta}{(r^2 + \zeta^2)^{3/2}} = \frac{1}{2\pi r} \tag{52}$$

We then integrate this expression for u_r using the normalization $u(1) = 0$.

2.3.3 Effect of Lower-Derivative Terms

If we consider the n-dimensional Laplacian, to which are added arbitrary linear terms in the $\partial u/\partial x_i$ and u, we obtain the following general equation:

$$\sum_{i=1}^{n}\left(\frac{\partial^2 u}{\partial x_i^2} + a_i\frac{\partial u}{\partial x_i}\right) + au = 0 \tag{53}$$

Here, the a_i and a are constants.

The first-derivative terms in (53) can be removed by the transformation of dependent variable $u \to w$ defined by

$$u(x_1, \ldots, x_n, t) \equiv w(x_1, \ldots, x_n, t)\exp\left\{-\frac{1}{2}\sum_{i=1}^{n} a_i x_i\right\} \tag{54}$$

An easy calculation shows that if u satisfies (53), then w obeys

$$\sum_{i=1}^{n}\frac{\partial^2 w}{\partial x_i^2} + \lambda w = 0 \tag{55a}$$

where λ is the constant

$$\lambda = a - \frac{1}{4}\sum_{i=1}^{n} a_i^2 \tag{55b}$$

With $\lambda > 0$, (55a) is the Helmholtz equation (and if $\lambda < 0$, it is denoted as the "modified" Helmholtz equation), which also arises when the time derivative is eliminated from a diffusion, or wave, equation [compare with (110) of Chapter 1].

To calculate the fundamental solution of (55a) with $\lambda > 0$, we need to solve [compare with (240)]:

$$\frac{d^2 w}{dr^2} + \frac{n-1}{r}\frac{dw}{dr} + \lambda w = \delta_n(r) \tag{56}$$

where $r \equiv (\sum_{i=1}^{n} x_i^2)^{1/2}$ and δ_n is the n-dimensional delta function (see Problem 14 of Chapter 1).

A second transformation of the dependent variable $w \to v$ is indicated. Set

$$w(r) = r^{-(n-2)/2}v(\theta); \qquad \theta \equiv \lambda^{1/2}r \tag{57}$$

to obtain

$$\frac{d^2 v}{d\theta^2} + \frac{1}{\theta}\frac{dv}{d\theta} + \left[1 - \left(\frac{n-2}{2\theta}\right)^2\right]v = 0 \tag{58}$$

corresponding to the homogeneous version of (56).

This is Bessel's equation of order $(n-2)/2$. We note that if n is even, $(n-2)/2$ is an integer, and the two linearly independent solutions of (58) are $J_{(n-2)/2}(\theta)$ and $Y_{(n-2)/2}(\theta)$, where J_p and Y_p are the Bessel functions (of order p) of the first and second kind, respectively. Conversely, if n is odd, $(n-2)/2$ is not an integer, and the two solutions $J_{(n-2)/2}(\theta)$ and $J_{-(n-2)/2}(\theta)$ are also independent and convenient to use.

We recall that J_p, J_{-p} and Y_p have the following behavior as $\theta \to 0$ (for example, see [6]):

$$J_p(\theta) \sim \frac{\theta^p}{2^p \Gamma(1 + p)} \tag{59a}$$

$$J_{-p}(\theta) \sim \frac{2^p}{\theta^p \Gamma(1 - p)} \tag{59b}$$

$$Y_p(\theta) \sim -\frac{2^p (p - 1)!}{\pi \theta^p}, \qquad p = \text{integer} \neq 0 \tag{59c}$$

$$Y_0(\theta) \sim \frac{2}{\pi} \log \theta \tag{59d}$$

We discard the solution of (58), which, upon multiplication by $r^{-(n-2)/2}$, is finite at $r = 0$ and obtain

$$w(r) = B_n r^{-(n-2)/2} Y_{(n-2)/2}(\lambda^{1/2} r), \qquad n = \text{even} \tag{60a}$$

$$w(r) = C_n r^{-(n-2)/2} J_{-(n-2)/2}(\lambda^{1/2} r), \qquad n = \text{odd} \tag{60b}$$

where B_n and C_n are constants to be determined from evaluating (56) over the interior of an n-dimensional sphere of radius ε centered at the origin. For ε small, the term λw in (56) does not contribute, and using the n-dimension Gauss theorem we obtain

$$\int \cdots \int_\varepsilon \frac{dw}{dr} dA = 1 \tag{61}$$

where the integral is evaluated on the surface of the ε-sphere.

For example, with n equal to an odd integer and using the expression in (59b) for $J_{-(n-2)/2}$, (61) gives:

$$C_n \left(\frac{2}{\lambda^{1/2}}\right)^{(n-2)/2} \frac{(2 - n)}{\Gamma\left(\dfrac{4 - n}{2}\right)} \omega_n = 1 \tag{62}$$

where ω_n is the surface area of the unit n-sphere, as defined in (247), and Γ is the Gamma function defined in (246). Using this expression for ω_n and solving for C_n gives

$$C_n = \frac{\Gamma\left(\dfrac{4 - n}{2}\right) \Gamma\left(\dfrac{n}{2}\right)}{2(2 - n)\pi^{n/2}} \left(\frac{\lambda^{1/2}}{2}\right)^{(n-2)/2} \tag{63}$$

It is easily verified that for $\lambda \to 0$, (60b) reduces to

$$w = \frac{1}{(2 - n)r^{n-2}\omega_n} \tag{64}$$

in agreement with the result derived in Problem 2 for Laplace's equation.

2.3.4 Potential Due to a Dipole

In a number of applications to be discussed later, we need to solve (32) with a higher-order singularity on the right-hand side. This is discussed next.

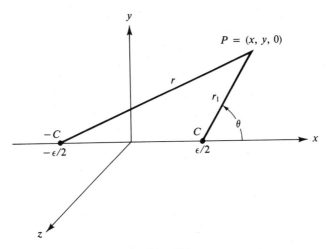

Figure 2.7

Consider the potential at a point $P = (x, y, 0)$ due to a source of strength C located at $x = \varepsilon/2$, $y = 0$, $z = 0$ superposed on the potential of a source of strength $-C$ located at $x = -\varepsilon/2$, $y = 0$, $z = 0$. If we denote the sum of the two potentials by w, we have

$$w = \frac{C}{4\pi}\left[\frac{1}{r} - \frac{1}{r_1}\right] \tag{65}$$

where, as is shown in Figure 2.7,

$$r^2 \equiv \left(x + \frac{\varepsilon}{2}\right)^2 + y^2$$

$$r_1^2 \equiv \left(x - \frac{\varepsilon}{2}\right)^2 + y^2$$

Now, if $0 < \varepsilon \ll 1$, we find*

$$r_1 = r - \varepsilon\cos\theta + O(\varepsilon^2) \tag{66}$$

Hence,

$$w = -\frac{C\varepsilon}{4\pi r^2}\cos\theta + O(\varepsilon^2) \tag{67}$$

If we consider the limit $\varepsilon \to O$ with $C\varepsilon \equiv D$ fixed, the potential reduces to

$$w = -\frac{D}{4\pi r^2}\cos\theta = -\frac{Dx}{4\pi r^3} \tag{68}$$

This limiting configuration is called a *dipole* (sometimes also called a *doublet*)

---

*The notations $\ll$ and O are defined carefully in (3) and (1), respectively, of Chapter 8. For the present purposes, we shall interpret $\ll$ to mean "very much less than" and O to mean "proportional to."

of *strength D* located at $x = 0$ and *oriented* along the positive x-axis. Note that the orientation of the dipole is determined by the relative location of the positive and negative sources.

For a dipole of strength D located at $x = \xi_0$, $y = z = 0$, and oriented in the positive x-direction, we have the potential

$$w(x, y, z) = -\frac{D}{4\pi} \frac{(x - \xi_0)}{[(x - \xi_0)^2 + y^2 + z^2]^{3/2}} \tag{69a}$$

at the point $P = (x, y, z)$. We note that this potential is just

$$w = \frac{\partial}{\partial \xi} \left\{ -\frac{D}{4\pi} \frac{1}{[(x - \xi)^2 + (y - \eta)^2 + (z - \zeta)^2]^{1/2}} \right\} \bigg|_{\xi = \xi_0, \eta = 0, \zeta = 0} \tag{69b}$$

In general, the potential at $P = (x, y, z)$ due to a unit dipole located at $Q_0 = (\xi_0, \eta_0, \zeta_0)$ and oriented along the unit vector $\mathbf{a} = a_1\mathbf{i} + a_2\mathbf{j} + a_3\mathbf{k}$ is

$$w = -\frac{1}{4\pi} \mathrm{grad}_Q \left(\frac{1}{r_{PQ}}\right) \cdot \mathbf{a} \bigg|_{Q = Q_0} \tag{70}$$

where, as usual, $r_{PQ}^2 \equiv (x - \xi)^2 + (y - \eta)^2 + (z - \zeta)^2$, and grad_Q means that in evaluating the gradient, partial derivatives are taken with respect to $Q = (\xi, \eta, \zeta)$.

It is also interesting to observe that the result (68) is the solution of

$$\Delta w = -D\delta'(x)\delta(y)\delta(z) \tag{71}$$

To see this, note that the solution of $\Delta w = C[\delta(x - \varepsilon/2) - \delta(x + \varepsilon/2)]\delta(y)\delta(z)$ is just (65), and in the limit as $\varepsilon \to 0$ with $\varepsilon C \equiv D = $ fixed, this solution tends to (68). But,

$$\lim_{\substack{\varepsilon \to 0 \\ \varepsilon C \equiv D = \text{fixed}}} C\left\{\delta\left(x - \frac{\varepsilon}{2}\right) - \delta\left(x + \frac{\varepsilon}{2}\right)\right\} = -D\delta'(x) \tag{72}$$

as can be verified using any representation of the delta function—for example,

$$\delta(x) \approx \frac{1}{2\sqrt{\pi\alpha}} e^{-x^2/4\alpha} \tag{73}$$

with α small [see (32) of Chapter 1]. Therefore, (68) solves (71).

Although we can construct more complicated limiting singularities from 4, 6, ... sources of zero total strength in various configurations, these do not play a role in solving (1). As we shall see later on, source and dipole distributions are crucial in describing solutions of the two main boundary-value problems for Laplace's equation.

2.4   Potential Due to Volume, Surface, and Line Distributions of Sources and Dipoles

In this section, we study the effect of distributing sources and dipoles in various configurations. These distributions may directly represent an actual physical state. For example, as discussed in Section 2.4.1, a continuous distribution of mass sources of variable strength defines the gravitational field of a given body or a distribution of stationary charges in space defines an electrostatic field. In other applications, a distribution of singularities may be used to simulate a given physical situation. For example, the flow past a nonlifting body of revolution may be represented by an appropriate distribution of positive and negative sources of mass along the axis. This simple problem is worked out in Section 2.4.3, and the more general case is discussed in Section 4.12.

2.4.1 Volume Distribution of Sources

To fix ideas, consider the force of gravity acting on a point of mass m located at $P = (x, y, z)$ due to a point of mass μ located at $Q = (\xi, \eta, \zeta)$. According to Newton's law of gravitation, this force $\mathbf{f}$ is given by

$$\mathbf{f} = -\frac{\gamma m \mu}{r_{PQ}^2} \frac{\mathbf{r}_{PQ}}{r_{PQ}} \tag{74}$$

where γ (6.67×10^{-8} dyne cm^2 g^{-2}) is the universal gravitational constant and $\mathbf{r}_{PQ}$ is the displacement vector from Q to P. Thus, $\mathbf{f}$ is in the direction opposite $\mathbf{r}_{PQ}$ in Figure 2.8.

Clearly, the specific force $\mathbf{f}/m$ can be derived from the potential V according to

$$\frac{\mathbf{f}}{m} = -\text{grad}_P V(P, Q) \tag{75}$$

where the subscript P indicates that partial derivatives are taken with respect to the x, y, z coordinates and

$$V(P, Q) \equiv -\frac{\gamma \mu}{r_{PQ}} \tag{76}$$

Thus, V obeys [compare with (39) and (40)]

$$\Delta_P V = 4\pi \mu \gamma \delta(P, Q) \tag{77}$$

If we now have an arbitrary distribution of mass (density $= \rho(x, y, z)$) in

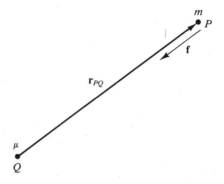

Figure 2.8

some domain G, the gravitational potential at some point P obeys

$$\Delta V = 4\pi\gamma\rho(x, y, z) \tag{78}$$

Hence, the solution for V is given in the form of a volume distribution of (mass) sources of strength/unit volume (density) ρ:

$$V(x, y, z) = -\gamma \iiint_G \frac{\rho(\xi, \eta, \zeta)\, d\xi\, d\eta\, d\zeta}{\sqrt{(x - \xi)^2 + (y - \eta)^2 + (z - \zeta)^2}} \tag{79}$$

Thus, having the fundamental solution, we can construct the solution due to an arbitrary right-hand side in (77) by superposition. Some examples are outlined in Problems 4 and 5.

2.4.2 Surface and Line Distribution of Sources or Dipoles

Consider now a distribution of sources on a prescribed surface S defined parametrically by

$$x = f(s_1, s_2); \qquad y = g(s_1, s_2); \qquad z = h(s_1, s_2) \tag{80}$$

See Review Problem 3. If the source strength/unit area is $q(s_1, s_2)$, the potential at a point $P = (x, y, z)$ is just

$$u(x, y, z) = -\frac{1}{4\pi} \iint_S \frac{q(s_1, s_2)\, dA}{\sqrt{(x - f)^2 + (y - g)^2 + (z - h)^2}} \tag{81}$$

where dA is the element of area on S; that is,

$$dA = |\mathbf{b}_1 \times \mathbf{b}_2|\, ds_1\, ds_2 \tag{82}$$

with $\mathbf{b}_1$ and $\mathbf{b}_2$ the tangent vectors

$$\mathbf{b}_1 = \frac{\partial f}{\partial s_1}\mathbf{i} + \frac{\partial g}{\partial s_1}\mathbf{j} + \frac{\partial h}{\partial s_1}\mathbf{k} \tag{83a}$$

$$\mathbf{b}_2 = \frac{\partial f}{\partial s_2}\mathbf{i} + \frac{\partial g}{\partial s_2}\mathbf{j} + \frac{\partial h}{\partial s_2}\mathbf{k} \tag{83b}$$

If we have a surface distribution of dipoles of strength $p(s_1, s_2)$/unit area, oriented along the unit vector $\mathbf{a}(s_1, s_2) = a_1\mathbf{i} + a_2\mathbf{j} + a_3\mathbf{k}$, the potential w is

$$\begin{aligned} w &= -\frac{1}{4\pi} \iint_S p(s_1, s_2)\, \mathrm{grad}_Q\left(\frac{1}{r_{PQ}}\right) \cdot \mathbf{a}\, dA \\ &= -\frac{1}{4\pi} \iint_S \frac{p(s_1, s_2)[(x - f)a_1 + (y - g)a_2 + (z - h)a_3]\, dA}{[(x - f)^2 + (y - g)^2 + (z - h)^2]^{3/2}} \end{aligned} \tag{84}$$

Similarly, if C is the curve defined by

$$x = f(s), \qquad y = g(s), \qquad z = h(s) \tag{85}$$

we can compute the potential due to a distribution of sources or dipoles along C in the following forms:

1. Sources of strength $q(s)$/unit length:

$$u = -\frac{1}{4\pi} \int_C \frac{q(s)\, ds}{[(x - f)^2 + (y - g)^2 + (z - h)^2]^{1/2}} \tag{86}$$

2. Dipoles of strength $p(s)$/unit length oriented along $\mathbf{a}(s)$:

$$w = -\frac{1}{4\pi} \int_C \frac{p(s)[(x - f)a_1 + (y - g)a_2 + (z - h)a_3]\, ds}{[(x - f)^2 + (y - g)^2 + (z - h)^2]^{3/2}} \tag{87}$$

where ds is the element of length along C.

2.4.3 An Example: Flow Over a Nonlifting Body of Revolution

As an illustration of the use of (86), consider the problem of computing the flow of an incompressible irrotational fluid outside a body of revolution defined by $r \equiv \sqrt{z^2 + y^2} = F(x)$. (See Figure 2.9.) We assume the flow at $x = -\infty$ is uniform, $U = \mathbf{i}$ in dimensionless units, and we represent the velocity potential for the flow outside the body by its uniform part: $u = x$, plus a disturbance potential in the form

$$u(x, r) = x - \frac{1}{4\pi} \int_0^1 \frac{S(\xi)\, d\xi}{[(x - \xi)^2 + r^2]^{1/2}} \tag{88}$$

Thus, the disturbance potential, which need not be small for finite x and r, is assumed to be due to an *axial* distribution of sources of strength $S(x)$ per unit length along the unit interval. In this example, the sources, which are located inside the body, are outside the domain where we wish to solve for u. Hence, the potential defined by (88) satisfies $\Delta u = 0$ outside the body, and it is also easily seen that $u \to x$ as $r \to \infty$ or $|x| \to \infty$.

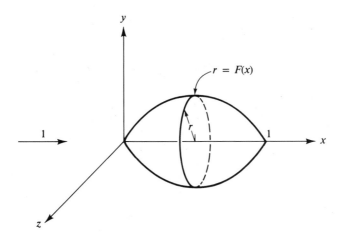

Figure 2.9

We shall see in Section 2.12 that the problem we wish to solve is a special case of a general boundary-value problem (Neumann problem) for which the assumed form of solution is appropriate. The question now is how to choose $S(x)$ so that the flow defined by the potential (88) is tangent to the given surface $r = F(x)$—that is,

$$\frac{u_r(x, F(x))}{u_x(x, F(x))} = F'(x) \tag{89}$$

This boundary condition is clearly necessary for a frictionless flow at a solid boundary. Using (88) gives

$$\int_0^1 \frac{[F(x) - F'(x)(x - \xi)]S(\xi)}{[(x - \xi)^2 + F^2(x)]^{3/2}} d\xi = 4\pi F'(x) \tag{90}$$

Equation (90) is an integral equation for the unknown $S(\xi)$ in the form

$$\int_0^1 k(x, \xi)S(\xi) d\xi = G(x) \tag{91}$$

where the kernel k is the following given function of x and ξ:

$$k(x, \xi) \equiv \frac{F(x) - F'(x)(x - \xi)}{[(x - \xi)^2 + F^2(x)]^{3/2}} \tag{92}$$

and $G(x) \equiv 4\pi F'(x)$ is also given.

One can solve (91) numerically by discretizing the integral. Subdivide the unit interval $0 \le x \le 1$ into N equal parts and denote

$$x_i \equiv \frac{i}{N}; \qquad i = 1, \dots, N - 1$$

$$S_i \equiv S(\xi_i); \qquad G_i \equiv G(x_i) \tag{93}$$

$$k_{ij} \equiv k(x_i, \xi_j)$$

Using the trapezoidal rule gives the following system of $N - 1$ linear algebraic equations:

$$\sum_{j=1}^{N-1} k_{ij}S_j = NG_i; \qquad i = 1, \dots, N - 1 \tag{94}$$

Solving these determines the $N - 1$ unknowns S_j; as a check, we verify that $\sum_{j=1}^{N-1} S_j = 0$, since we have a closed body. *Note*: $S(0) = S(1) = 0$.

This result is studied for the case $N = 4$ in Problem 6. The perturbation solution for the case of a slender body is given in Section 8.3.4.

2.4.4 Limiting Surface Values for Source and Dipole Distributions

In the applications discussed in Section 2.12, we shall represent the potential at a point $P = (x, y, z)$ by a surface distribution of sources or dipoles as given by the integral expressions (81) or (84), respectively. In these problems we shall need to satisfy a specified boundary condition on the surface S itself. For example, we require either the potential or its normal derivative to equal a specified function on S. We note, however, that if P is on S, the integrals (81) and (84) (as well as the expressions that result from these for the normal derivative of the potential) become improper because the denominators vanish at P. Corresponding singularities also occur in the integrals (86) and (87) for source or dipole distributions on a curve C. This difficulty did not arise in the example worked out in Section 2.4.3 because sources were distributed on the x-axis, whereas the boundary condition was evaluated on the surface $r = F(x) \geq 0$, which is off the axis in the interval $0 < x < 1$. In general, we cannot avoid evaluating a boundary condition on the surface over which sources or dipoles are distributed.

Even though the expression that results when P is taken on S is singular, the actual potential (or its normal derivative) is well behaved in this limit. In fact, if we first evaluate the integral for P, not on S, and then let P approach S, the result is perfectly well defined; it is only when the limit is imposed on the integral representation that we encounter a difficulty.

To illustrate this situation, consider the special case of a dipole distribution having a strength per unit area equal to a constant p_0 over the entire $z = 0$ plane. The axes of the dipoles are taken in the $+z$ direction, so (84) (with $p = p_0 =$ constant, $f = s_1 = \xi, g = s_2 = \eta, h = 0, a_1 = a_2 = 0, a_3 = 1$) specializes to

$$w(x, y, z) = -\frac{p_0 z}{4\pi} \int_{\xi=-\infty}^{\infty} \int_{\eta=-\infty}^{\infty} \frac{d\xi \, d\eta}{[(x - \xi)^2 + (y - \eta)^2 + z^2]^{3/2}} \tag{95a}$$

Now, if we set $z = 0$ in (95a), we encounter an improper integral because the expression for the double integral has a nonintegrable singularity at $\xi = x, \eta = y$ (if $z = 0$). This expression is then multipled by z, and it is not helpful to set $z = 0$ directly in (95a). However, w can be evaluated exactly if $z \neq 0$ for this simple example. First we introduce $\xi - x$ and $\eta - y$ as integration variables and observe, as expected, that w does not depend on x or y; that is,

$$w(x, y, z) = -\frac{p_0 z}{4\pi} \int_{-\infty}^{\infty} \int_{-\infty}^{\infty} \frac{d\xi \, d\eta}{(\xi^2 + \eta^2 + z^2)^{3/2}}; \qquad z \neq 0$$

or using polar coordinates (ρ, ϕ) defined by $\xi = \rho \cos \phi, \eta = \rho \sin \phi$, we find

$$w(x, y, z) = -\frac{p_0 z}{4\pi} \int_{\rho=0}^{\infty} \int_{\phi=0}^{2\pi} \frac{\rho \, d\rho \, d\phi}{(\rho^2 + z^2)^{3/2}}; \qquad z \neq 0$$

$$= -\frac{p_0 z}{2} \int_{\rho=0}^{\infty} \frac{\rho \, d\rho}{(\rho^2 + z^2)^{3/2}} = \frac{p_0 z}{2} \frac{1}{(\rho^2 + z^2)^{1/2}} \Bigg|_{\rho=0}^{\rho=\infty}$$

$$= -\frac{p_0 z}{2} \frac{1}{|z|} = \begin{cases} -p_0/2 & \text{if } z > 0 \\ p_0/2 & \text{if } z < 0 \end{cases} \tag{95b}$$

Thus, w is a constant in each half-space, and the limiting value of w as $z \to 0^+$ or $z \to 0^-$ is well defined; it is just $-p_0/2$ or $p_0/2$, respectively.

In this simple example it was possible to evaluate w explicitly and then take the limit as P approaches S. In general, an explicit result will not be feasible and we shall need to calculate the limiting expression for the potential or its normal

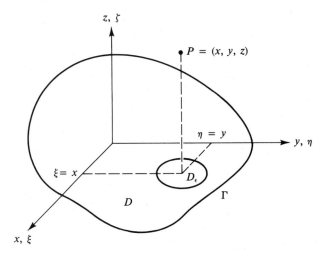

Figure 2.10

derivative on S by a construction that involves the integral expression itself. To illustrate ideas, let us consider the more general problem of a dipole distribution in the simply connected domain D, which lies in the $z = 0$ plane and has the boundary Γ. Again, we assume that the dipole axes are all normal to the plane D and let their strength per unit area be a specified function $p(x, y)$. The potential w at $P = (x, y, z)$ is then given by

$$w(x, y, z) = -\frac{z}{4\pi} \iint\limits_{D} \frac{p(\xi, \eta)\, d\xi\, d\eta}{[(x - \xi)^2 + (y - \eta)^2 + z^2]^{3/2}} \tag{96}$$

which generalizes (95a).

As shown in Figure 2.10, we subdivide the integration domain D into two parts: (1) the interior D_ε of a circle of radius ε centered at $\xi = x, \eta = y, \zeta = 0$, with ε sufficiently small so that D_ε is entirely contained in D, and (2) the remainder $D_a = D - D_\varepsilon$. Thus, (96) will now involve two contributions, one from D_ε, denoted by w_ε, and the other from D_a, denoted by w_a; that is,

$$w(x, y, z) = w_\varepsilon(x, y, z; \varepsilon) + w_a(x, y, z; \varepsilon) \tag{97}$$

We introduce a local polar coordinate system

$$\xi = x + \rho \cos \phi; \qquad \eta = y + \rho \sin \phi$$

and express w_ε and w_a in the form

$$w_\varepsilon(x, y, z; \varepsilon) = -\frac{z}{4\pi} \int_0^{2\pi} \int_0^\varepsilon \frac{p(x + \rho \cos \phi, y + \rho \sin \phi)}{(\rho^2 + z^2)^{3/2}}\, \rho\, d\rho\, d\phi \tag{98a}$$

$$w_a(x, y, z; \varepsilon) = -\frac{z}{4\pi} \int_0^{2\pi} \int_\varepsilon^{R(\phi, x, y)} \frac{p(x + \rho \cos \phi, y + \rho \sin \phi)}{(\rho^2 + z^2)^{3/2}}\, \rho\, d\rho\, d\phi \tag{98b}$$

Here $\rho = R(\phi, x, y)$ is the expression defining the distance between the fixed point

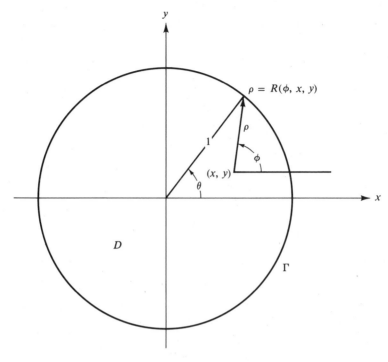

Figure 2.11

(x, y) and a point on Γ in polar coordinates. For example, if Γ is the unit circle centered at the origin, we see from Figure 2.11 that

$$\rho \sin \phi + y = \sin \theta$$

$$\rho \cos \phi + x = \cos \theta$$

Therefore, eliminating θ and solving the resulting quadratic for ρ defines R in the form $R(\phi, x, y) = -x \cos \phi - y \sin \phi + [1 - (x \sin \phi - y \cos \phi)^2]^{1/2}$, where $R > 0$ if (x, y) is inside D.

The individual contributions w_ε and w_a depend on ε, but their sum does not. The basic idea for our calculation of the limiting value of w as z approaches zero is to regard z as some as yet unspecified function α of ε, to be chosen so as to simplify the calculation of the integrals (98) when z is small. We then obtain the limiting value of w from

$$w(x, y, 0) = \lim_{\varepsilon \to 0} \{w_\varepsilon(x, y, \alpha(\varepsilon); \varepsilon) + w_a(x, y, \alpha(\varepsilon); \varepsilon)\} \qquad (99)$$

where $\alpha \to 0$ as $\varepsilon \to 0$.

Consider first (98a) for w_ε. Since the maximum value of ρ is ε, we rescale this variable by setting $\rho = \varepsilon \tilde{\rho}$, so that the $\tilde{\rho}$ now varies over the interval $(0, 1)$. This gives

$$w_\varepsilon(x, y, \alpha; \varepsilon) = -\frac{(\alpha/\varepsilon)}{4\pi} \int_0^{2\pi} \int_0^1 \frac{p(x + \varepsilon \tilde{\rho} \cos \phi, y + \varepsilon \tilde{\rho} \sin \phi)}{[\tilde{\rho}^2 + (\alpha/\varepsilon)^2]^{3/2}} \tilde{\rho} \, d\tilde{\rho} \, d\phi \qquad (100)$$

Assuming that p is analytic, we can expand

$$p(x + \varepsilon\tilde{\rho}\cos\phi, y + \varepsilon\tilde{\rho}\sin\phi) = p(x, y) + \varepsilon p_x(x, y)\tilde{\rho}\cos\phi$$
$$+ \varepsilon p_y(x, y)\tilde{\rho}\sin\phi + O(\varepsilon^2\tilde{\rho}^2)$$

Now interchanging the order of integration in (100) shows that the terms proportional to $\sin\phi$ and $\cos\phi$ do not contribute. (In fact, if the higher-order terms in the series for p are included, only the averages of the various products of trigometric functions will contribute.) Therefore, the integral (100) has the approximation

$$w_\varepsilon(x, y, \alpha; \varepsilon) = -\frac{(\alpha/\varepsilon)}{2} \int_0^1 \frac{\tilde{\rho}\, d\tilde{\rho}}{[\tilde{\rho}^2 + (\alpha/\varepsilon)^2]^{3/2}} [p(x, y) + O(\varepsilon^2\tilde{\rho}^2)] \tag{101a}$$

We can evaluate this integral [see (95b)] and find

$$w_\varepsilon(x, y, \alpha; \varepsilon) = \frac{p(x, y)}{2}\left\{ \frac{(\alpha/\varepsilon)}{[1 + (\alpha/\varepsilon)^2]^{1/2}} - \frac{(\alpha/\varepsilon)}{|\alpha/\varepsilon|} \right\} + O(\alpha\varepsilon) \tag{101b}$$

So far, we have made no assumptions regarding $\alpha(\varepsilon)$. It is clear from (101b) that if $(\alpha/\varepsilon) \to 0$ as $\varepsilon \to 0$, the limiting value of w_ε will be independent of (α/ε) and is simply given by

$$w_\varepsilon(x, y, 0^+; \varepsilon) = -\frac{p(x, y)}{2} \tag{102a}$$

$$w_\varepsilon(x, y, 0^-; \varepsilon) = +\frac{p(x, y)}{2} \tag{102b}$$

We now show that the choice $(\alpha/\varepsilon) \to 0$ also simplifies the calculation for the limiting value of w_a. Changing variables from ρ to $\tilde{\rho}$ in (98b) gives

$$w_a(x, y, \alpha; \varepsilon) = -\frac{(\alpha/\varepsilon)}{4\pi} \int_0^{2\pi} \int_1^{R(\phi, x, y)/\varepsilon} \frac{p(x + \varepsilon\tilde{\rho}\cos\phi, y + \varepsilon\tilde{\rho}\sin\phi)}{[\tilde{\rho}^2 + (\alpha/\varepsilon)^2]^{3/2}} \tilde{\rho}\, d\tilde{\rho}\, d\phi$$

For $\varepsilon \neq 0$, the denominator in the integrand does not vanish over the interval $1 \le \rho \le R/\varepsilon$, so that the double integral exists and the factor (α/ε) in front implies that $w_a = O(\alpha/\varepsilon)$ as $\varepsilon \to 0$. Therefore, the choice $(\alpha/\varepsilon) \to 0$ gives

$$w_a(x, y, 0; 0) = 0$$

and we conclude that

$$w(x, y, 0^+) = -\frac{p(x, y)}{2} \tag{103a}$$

$$w(x, y, 0^-) = +\frac{p(x, y)}{2} \tag{103b}$$

We reiterate that the final result (103) does not depend on the limiting behavior of α; the choice $(\alpha/\varepsilon) \to 0$ is made to simplify the calculations. We illustrate this point by reconsidering the simple example $p = p_0 = $ constant, $R = \infty$ discussed earlier. Expressing (95b) in terms of the decomposition (97) gives the exact result

$$w_\varepsilon(x, y, \alpha; \varepsilon) = \frac{p_0}{2} \left\{ \frac{(\alpha/\varepsilon)}{[1 + (\alpha/\varepsilon)^2]^{3/2}} - \frac{(\alpha/\varepsilon)}{|\alpha/\varepsilon|} \right\} \tag{104a}$$

$$w_a(x, y, \alpha; \varepsilon) = -\frac{p_0}{2} \frac{(\alpha/\varepsilon)}{[1 + (\alpha/\varepsilon)^2]^{3/2}} \tag{104b}$$

We see that regardless of the choice of the limiting value of (α/ε), the first term in each of the expressions on the right-hand sides of (104) have opposite signs and cancel in the sum $w_\varepsilon + w_a$. For the choice $(\alpha/\varepsilon) \to 0$, $w_a \to 0$, and the limiting value of w is the same as the limiting value of w_ε.

We have shown in (103) that for a planar dipole distribution, the potential at any point on the surface depends on the *local* dipole strength $p(x, y)$ only. This feature does not persist in general for non-planar dipole distributions; the integrated contribution corresponding to w_a for this case will not vanish in general (see Problem 7b).

Consider now the potential due to a planar distribution of sources of strength per unit area equal to $q(\xi, \eta)$. The expression (81) specializes to the following integral analogous to (96):

$$u(x, y, z) = -\frac{1}{4\pi} \iint_D \frac{q(\xi, \eta)\, d\xi\, d\eta}{[(x - \xi)^2 + (y - \eta)^2 + z^2]^{1/2}} \tag{105}$$

and if we decompose D and introduce local polar coordinates, we find

$$u_\varepsilon(x, y, \alpha; \varepsilon) = -\frac{1}{4\pi} \int_0^{2\pi} \int_0^\varepsilon \frac{q(x + \rho \cos \phi, y + \rho \sin \phi)}{(\rho^2 + \alpha^2)^{1/2}} \rho\, d\rho\, d\phi \tag{106a}$$

$$u_a(x, y, \alpha; \varepsilon) = -\frac{1}{4\pi} \int_0^{2\pi} \int_\varepsilon^{R(\phi, x, y)} \frac{q(x + \rho \cos \phi, y + \rho \sin \phi)}{(\rho^2 + \alpha^2)^{1/2}} \rho\, d\rho\, d\phi \tag{106b}$$

Again, we change the ρ variable to $\varepsilon\tilde{\rho}$ and find that u_ε has the approximation

$$u_\varepsilon(x, y, \alpha; \varepsilon) = -\frac{\varepsilon}{4\pi} \int_0^{2\pi} \int_0^1 \frac{1}{[\tilde{\rho}^2 + (\alpha/\varepsilon)^2]^{1/2}} \{q(x, y)$$
$$+ \varepsilon q_x(x, y)\tilde{\rho} \cos \phi + \varepsilon q_y(x, y)\tilde{\rho} \sin \phi + O(\varepsilon^2 \tilde{\rho}^2)\} \tilde{\rho}\, d\tilde{\rho}\, d\theta \tag{107}$$

if q is analytic. Integrating with respect to ϕ first shows that the $O(\varepsilon)$ terms in the integrand give no contributions, and integrating with respect to $\tilde{\rho}$ gives

$$u_\varepsilon(x, y, \alpha; \varepsilon) = O(\varepsilon) + O(\alpha) \tag{108}$$

Thus, u_ε gives no contribution as long as α and ε both vanish in the limit. The limit of the ratio (α/ε) does not affect this result and can therefore not affect the limiting value of u_a. In fact, this value is uniquely given by

$$u_a(x, y, 0; 0) = -\frac{1}{4\pi} \int_0^{2\pi} \int_0^{R(\phi, x, y)} q(x + \rho \cos \phi, y + \rho \sin \phi)\, d\rho\, d\phi \equiv u(x, y, 0) \tag{109}$$

In contrast with the result (103) for dipoles, we see that the potential at a point on the surface depends on the entire distribution.

For the special case $q = q_0 = $ constant over the unit disc centered at the origin, (109) becomes:

$$u(x, y, 0) = -\frac{q_0}{4\pi} \int_0^{2\pi} \{-x\cos\phi - y\sin\phi + [1 - (x\sin\phi - y\cos\phi)^2]^{1/2}\} \, d\phi$$

which simplifies to a function of $r \equiv (x^2 + y^2)^{1/2}$ only:

$$u(r, 0) = -\frac{q_0}{\pi} \int_0^{\pi/2} (1 - r^2 \sin^2\phi)^{1/2} \, d\phi$$

$$= -\frac{q_0}{\pi} E(r^2) \tag{110a}$$

Here E is the complete elliptic integral of the second kind (see pp. 590–591 of [6]). Using the expansion for $E(r^2)$, valid if $r^2 < 1$, gives

$$u(r, 0) = -\frac{q_0}{2}\left[1 - \left(\frac{1}{2}\right)^2 \frac{r^2}{1} - \left(\frac{1}{2}\cdot\frac{3}{4}\right)^2 \frac{r^4}{3} - \left(\frac{1}{2}\cdot\frac{3}{4}\cdot\frac{5}{6}\right)^2 \frac{r^6}{5} - \cdots\right] \tag{110b}$$

and, in particular, $u(0, 0) = -q_0/2$.

The normal derivate of the potential (105) on the $z = 0$ plane is just the z-derivative, which for $z \neq 0$ is

$$u_z(x, y, z) = \frac{z}{4\pi} \iint_D \frac{q(\xi, \eta) \, d\xi \, d\eta}{[(x - \xi)^2 + (y - \eta)^2 + z^2]^{3/2}} \tag{111}$$

This is formally the same expression as the potential due to a dipole distribution of strength $-q$ [see (96)]. Therefore, it follows from (103) that

$$u_z(x, y, 0^+) = \frac{q(x, y)}{2} \tag{112a}$$

$$u_z(x, y, 0^-) = -\frac{q(x, y)}{2} \tag{112b}$$

Finally, let us study the limiting value (as $z = 0^+$) of the normal derivative of the potential for the dipole distribution (96). It follows from (101b) that

$$\frac{\partial w_\varepsilon}{\partial z} = \frac{p(x, y)}{2} \frac{\varepsilon^2}{(\varepsilon^2 + z^2)^{3/2}} + O(\varepsilon), \qquad z > 0$$

which implies that for $\varepsilon \to 0$, we have the singular behavior

$$\frac{\partial w_\varepsilon}{\partial z} = \frac{p(x, y)}{2\varepsilon} + O(\varepsilon) \qquad \text{as } \varepsilon \to 0 \tag{113}$$

if $(z/\varepsilon) \to 0^+$. We therefore anticipate finding a corresponding singularity of opposite sign in $\partial w_a/\partial z$.

Now, as $z = \alpha \to 0$, with $(\alpha/\varepsilon) \to 0$, $\partial w_a/\partial z$ is of the form

$$\frac{\partial w_a(x, y, \alpha; \varepsilon)}{\partial z} = -\frac{1}{4\pi} \int_0^{2\pi} \int_\varepsilon^{R(\phi, x, y)} \frac{p(x + \rho\cos\phi, y + \rho\sin\phi)}{\rho^2} \, d\rho \, d\phi + O(\alpha). \tag{114a}$$

To exhibit the singular behavior near the lower limit $\rho = \varepsilon$, we subtract and add the first term in the expansion of p to obtain the following expression, which is identical with (114a):

$$
\frac{\partial w_a(x, y, \alpha; \varepsilon)}{\partial z} = -\frac{1}{4\pi} \int_0^{2\pi} \int_\varepsilon^{R(\phi, x, y)} \frac{1}{\rho^2} [p(x + \rho \cos \phi, y + \rho \sin \phi)
$$
$$
- p(x, y)] \, d\rho \, d\phi - \frac{p(x, y)}{4\pi} \int_0^{2\pi} \int_\varepsilon^{R(\phi, x, y)} \frac{1}{\rho^2} d\rho \, d\phi + O(\alpha)
$$

$$(114b)$$

Now, as $\varepsilon \to 0$, the first integral is well behaved. In fact, it is $O(1)$ as $\varepsilon \to 0$ because the second term in the development of p gives a zero contribution when integrated with respect to ϕ. Evaluating the second integral gives

$$
\frac{\partial w_a(x, y, \alpha; \varepsilon)}{\partial z} = -\frac{1}{4\pi} \int_0^{2\pi} \int_\varepsilon^{R(\phi, x, y)} \frac{1}{\rho^2} [p(x + \rho \cos \phi, y + \rho \sin \phi) - p(x, y)] \, d\rho \, d\phi
$$
$$
+ \frac{p(x, y)}{4\pi} \int_0^{2\pi} \frac{d\phi}{R(\phi, x, y)} - \frac{p(x, y)}{2\varepsilon}
$$

$$(114c)$$

and we exhibit the needed $O(\varepsilon^{-1})$ singularity.

Thus, adding (113) to (114c) and taking the limit as $\varepsilon \to 0$ gives

$$
w_z(x, y, 0^+) = -\frac{1}{4\pi} \int_0^{2\pi} \int_0^{R(\phi, x, y)} l(\rho, \phi, x, y) \, d\rho \, d\phi + \frac{p(x, y)}{4\pi} \int_0^{2\pi} \frac{d\phi}{R(\phi, x, y)}
$$

$$(115)$$

where

$$
l(\rho, \phi, x, y) \equiv \frac{1}{\rho^2} [p(x + \rho \cos \phi, y + \rho \sin \phi) - p(x, y)]
$$

$$(116)$$

For the special case $p = p_0 = \text{constant}$, we have $l = 0$, and (115) simplifies to

$$
w_z(x, y, 0^+) = \frac{p_0}{4\pi} \int_0^{2\pi} \frac{d\phi}{R(\phi, x, y)}
$$

$$(117)$$

2.5   Green's Formula and Applications

In this section, we derive a number of general results concerning properties of solutions of Laplace's equation. The starting point is the familiar Gauss theorem of vector calculus.

Consider a one-valued vector field $\mathbf{F}$ defined in a domain G with boundary Γ on which $\mathbf{n}$ is an outward unit normal. Gauss' theorem states that if $\mathbf{F}$ has continuous first partial derivatives in G, then

$$
\iiint_G \operatorname{div} \mathbf{F} \, dV = \iint_\Gamma \mathbf{F} \cdot \mathbf{n} \, dA
$$

$$(118)$$

This result is valid for multiplied connected domains as long as Γ includes all the boundaries of G. If G is infinite, we assume that $|\mathbf{F}| \ll r^{-2}$, as $r \to \infty$, where r is the scalar distance, in order to ensure the existence of the integrals.

To derive *Green's formula*, we choose $\mathbf{F} = v \operatorname{grad} u$ for prescribed scalar functions $u(x, y, z)$ and $v(x, y, z)$ having continuous second partial derivatives. Then $\operatorname{div} \mathbf{F} = \operatorname{div}(v \operatorname{grad} u) = (\operatorname{grad} u \cdot \operatorname{grad} v) + v \Delta u$, and (118) reduces to:

$$\iiint\limits_{G} [\operatorname{grad} u \cdot \operatorname{grad} v + v \Delta u]\, dV = \iint\limits_{\Gamma} v \frac{\partial u}{\partial n}\, dA \tag{119}$$

where $\partial u / \partial n$ is the directional derivative of u in the outward normal direction

$$\frac{\partial u}{\partial n} \equiv \operatorname{grad} u \cdot \mathbf{n} \tag{120}$$

Interchanging u and v in (119) and subtracting the result gives the *symmetric form of Green's formula*:

$$\iiint\limits_{G} [u \Delta v - v \Delta u]\, dV = \iint\limits_{\Gamma} \left(u \frac{\partial v}{\partial n} - v \frac{\partial u}{\partial n} \right) dA \tag{121}$$

2.5.1 Gauss' Integral Theorem

If $\Delta u = 0$ and $v = 1$, (119) gives

$$\iint\limits_{\Gamma} \frac{\partial u}{\partial n}\, dA = 0 \tag{122}$$

Thus, the integral of the normal derivative of a harmonic function vanishes, as is intuitively obvious if we interpret u as a velocity potential. Equation (122) states that for an incompressible, irrotational flow with no sources, the *net* mass flow through a prescribed boundary Γ is zero.

2.5.2 Energy Theorem and Corollaries

Setting $u = v$ with $\Delta u = 0$ in (119) gives the energy theorem

$$\frac{1}{2} \iiint\limits_{G} (\operatorname{grad} u)^2\, dV = \frac{1}{2} \iint\limits_{\Gamma} u \frac{\partial u}{\partial n}\, dA \tag{123}$$

relating the total kinetic energy in the interior to the integral of $u(\partial u / \partial n)$ on the boundary.

It follows from (123) that if u vanishes on Γ, then $\operatorname{grad} u$ must vanish everywhere in the interior, and this combined with the fact that $u = 0$ in Γ implies that u is identically equal to zero in the interior of G.

Similarly, if $\partial u / \partial n$ vanishes on Γ, then u must be a constant throughout the interior. It is important to keep in mind that this result applies only to *one-valued* functions u. For example, consider the two-dimensional Laplacian in the annular region $D: 0 < r_1 \leq r \leq r_2$ contained between the two concentric circles with radii $r = r_1$ and $r = r_2$ centered at the origin. With r and θ denoting polar coordinates, we note that the multivalued function $w = c\theta \equiv c \tan^{-1}(y/x)$ (where c is an arbitrary constant) is harmonic in D. The normal derivative of w on the boundary circles is the radial derivative, and this vanishes. Because w is multivalued, our assertion ($\Delta w = 0$ in D and $\partial w/\partial n = 0$ on Γ implies that w is a constant) does not apply. Of course, we may render w one-valued by introducing the radial cut along $\theta = \theta_1$ and restricting allowable values of θ according to $\theta_1 \le \theta < \theta_1 + 2\pi$. In this case, the barrier at $\theta = \theta_1$ *becomes part of the boundary* on which w satisfies

$$\frac{\partial w}{\partial n} = \frac{1}{r}\frac{\partial w}{\partial \theta} = \frac{c}{r}$$

Hence, in order to have the normal derivative of w vanish on *all* the boundaries, we must set $c = 0$, and this result is indeed consistent with our claim.

2.5.3 Uniqueness Theorems

As a direct consequence of the foregoing results concerning harmonic functions with zero boundary conditions, we now prove the following uniqueness theorems for certain boundary-value problems satisfying $\Delta u = 0$ in G.

> *Dirichlet problem.* If two functions u_1 and u_2 are harmonic in G and coincide on Γ, they are identical in G. To see this, we note that $u = u_1 - u_2$ is also harmonic in G and vanishes on Γ; therefore, $u = 0$ and hence $u_1 = u_2$ in G.
>
> *Neumann Problem.* If two one-valued functions u_1 and u_2 are harmonic in G and their normal derivatives coincide on Γ, then u_1 and u_2 differ by at most a constant in G. This is also an immediate consequence of the second corollary to (123).
>
> *Mixed boundary-value problem.* If two functions u_1 and u_2 are harmonic in G and satisfy the same mixed boundary condition (where u is prescribed as part of Γ and $\partial u/\partial n$ is prescribed on the remainder), then u_1 and u_2 coincide in G. The specification of u_1 on part of the boundary eliminates the arbitrary constant that arises if only $\partial u/\partial n$ is prescribed.

2.5.4 Mean-Value Theorem

Let $P(x, y, z)$ be a fixed point inside G and let $\Delta u = 0$ in G. Consider (121) with respect to the integration variables $Q = (\xi, \eta, \zeta)$ and regard u as a function of Q, that is, $\Delta u = \Delta_Q u \equiv u_{\xi\xi} + u_{\eta\eta} + u_{\zeta\zeta} = 0$. Also, in (121), let $v(P, Q) = -1/4\pi r_{PQ}$, where $r_{PQ}^2 \equiv (x - \xi)^2 + (y - \eta)^2 + (z - \zeta)^2$. Let G_1 be a sphere of radius R centered at P with boundary Γ_1 lying entirely inside G. Clearly, $\Delta_Q v = \Delta_P v = \delta_3(P, Q)$ [see (39)–(40)]. Therefore, the left-hand side of (121) evaluated inside G_1 reduces to

$$\iiint\limits_{G_1} [u(Q)\Delta_Q v - v(P, Q)\Delta_Q u]\, dV_Q = \iiint\limits_{G_1} u(Q)\delta_3(P, Q)\, dV_Q = u(P) \qquad (124)$$

The right-hand side is

$$\iint\limits_{\Gamma_1} \left(u\frac{\partial v}{\partial n_Q} - v\frac{\partial u}{\partial n_Q}\right) dA_Q = \iint\limits_{\Gamma_1} u\frac{\partial}{\partial \rho}\left(-\frac{1}{4\pi\rho}\right)\Bigg|_{\rho = R} dA_Q \qquad (125)$$

since the second term reduces to

$$\frac{1}{4\pi R} \iint\limits_{\Gamma_1} \frac{\partial u}{\partial n_Q} dA_Q = 0 \tag{126}$$

according to Gauss' integral theorem (122). Setting

$$\frac{\partial}{\partial \rho}\left(-\frac{1}{4\pi\rho}\right)\bigg|_{\rho=R} = \frac{1}{4\pi R^2}$$

in (125) gives

$$u(P) = \frac{1}{4\pi R^2} \iint\limits_{\Gamma_1} u \, dA \tag{127}$$

Thus, the value of a harmonic function at any point P is the average of the values it takes on any sphere surrounding that point.

As a corollary of (127), called the *maximum-minimum theorem*, it follows that the maximum and minimum values of a harmonic function must occur on the boundary; in particular, if u is constant on the boundary, then it is constant everywhere in G.

2.5.5 Surface Distribution of Sources and Dipoles

A harmonic function in G can be represented by a distribution of sources and dipoles on the boundary Γ.

In the derivation of the mean-value theorem, we restricted our attention to a spherical domain G_1 inside G, and this implied (126). If we repeat the calculations for the entire domain G and boundary Γ, (124) still holds, but the right-hand side is more complicated. In fact, we find

$$u(P) = \iint\limits_{\Gamma} \left(-\frac{\partial u}{\partial n_Q}\right)\left(-\frac{1}{4\pi r_{PQ}}\right) dA_Q + \iint\limits_{\Gamma} u \frac{\partial}{\partial n_Q}\left(-\frac{1}{4\pi r_{PQ}}\right) dA_Q \tag{128}$$

which means that $u(P)$ can be regarded as the potential due to a surface distribution of sources of strength $-\partial u/\partial n$, as given by the first term on the right-hand side of (128), plus a distribution of dipoles oriented along the outward normal to Γ having strength u. Since arbitrarily prescribing both u and $\partial u/\partial n$ on the boundary leads to an ill-posed problem for Laplace's equation (see the two-dimensional example discussed in Section 4.4.5), (128) does not provide the solution of a realistic boundary-value problem. Rather, it should be interpreted as an integral equation for $\partial u/\partial n$ on the boundary if u is prescribed there (or vice versa). This point of view is discussed in more detail in Section 2.12. We shall also use the general result (128) in interpreting solutions of Dirichlet's and Neumann's problems for simple geometries in Section 2.8.

Strictly speaking, the derivation leading to (127) or (128) is suspect because we have used Gauss' theorem for functions that are singular at $P = Q$. Let us verify that (128) is indeed correct using a more careful derivation that avoids delta functions. In subsequent derivations, we shall again rely on formal derivations with the aid of delta functions without justification, as these calculations are significantly simpler.

Let G_ε be a sphere of radius ε centered at P and lying entirely in G. We apply Green's formula now in the domain *outside* G_ε and inside G. Taking u and v in (121) again to be $u(\xi, \eta, \zeta)$: $\Delta_Q u = 0$ and $v = -1/4\pi r_{PQ}$, the left-hand side of (121) vanishes because the Laplacians of both u and v are equal to zero in $G - G_\varepsilon$. To compute the right-hand side, we include the boundary contributions for both Γ and Γ_ε and obtain

$$0 = \iint_\Gamma \left[u \frac{\partial}{\partial n_Q} \left(-\frac{1}{4\pi r_{PQ}} \right) + \frac{1}{4\pi r_{PQ}} \frac{\partial u}{\partial n_Q} \right] dA_Q$$

$$- \iint_{\Gamma_\varepsilon} \left[u \frac{\partial}{\partial n_Q} \left(-\frac{1}{4\pi r_{PQ}} \right) + \frac{1}{4\pi r_{PQ}} \frac{\partial u}{\partial n_Q} \right] dA_Q \qquad (129)$$

In the second integral of (129), we must use the direction of *increasing* radius as the normal n_Q since we have introduced the minus sign in front.

To evaluate the integral over Γ_ε, we change variables from $Q = (\xi, \eta, \zeta)$ to spherical polar coordinates ρ, ϕ, θ centered at $P = (x, y, z)$—that is,

$$\xi = x + \rho \sin \phi \cos \theta$$

$$\eta = y + \rho \sin \phi \sin \theta$$

$$\zeta = z + \rho \cos \phi$$

Therefore,

$$\iint_{\Gamma_\varepsilon} \left[u \frac{\partial}{\partial n_Q} \left(-\frac{1}{4\pi r_{PQ}} \right) + \frac{1}{4\pi r_{PQ}} \frac{\partial u}{\partial n_Q} \right] dA_Q$$

$$= \int_{\phi=0}^{\pi} \int_{\theta=0}^{2\pi} \left[\frac{u(x + \varepsilon \sin \phi \cos \theta, \, y + \varepsilon \sin \phi \sin \theta, \, z + \varepsilon \cos \phi)}{4\pi \varepsilon^2} \right.$$

$$\left. + \frac{1}{4\pi \varepsilon} \left(\frac{\partial u}{\partial \rho} \right)_{\rho=\varepsilon} \right] \varepsilon^2 \sin \phi \, d\phi \, d\theta \qquad (130)$$

As $\varepsilon \to 0$, the second term on the right-hand side does not contribute, and the first term tends to $u(x, y, z)$. Using this result in (129) gives (128).

2.5.6 Potential Due to a Dipole Distribution of Unit Strength

As an application of result (128), consider a given surface Γ, not necessarily closed, and let P be a point not on Γ. If Γ is closed, let C be an arbitrary simple closed curve on Γ, and if Γ is open, let C be its boundary. We generate a cone by running a straight line from P along C (Figure 2.12). Exclude from this cone the spherical cap K_ε generated by the sphere of radius ε centered at P, and denote what remains by Ω. In the region G bounded by Γ, Ω, K_ε, the function $u = -1/4\pi r_{PQ}$ is harmonic because $\Delta_Q u = \delta_3(Q, P)$ and P is outside G. Thus, using

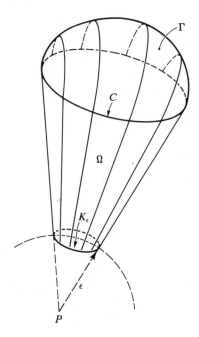

Figure 2.12

Gauss' integral theorem (122), we have

$$\iint\limits_{\Gamma+\Omega+K_\varepsilon} \frac{\partial u}{\partial n}\,dA = 0 \tag{131}$$

Now, on Γ,

$$\iint\limits_{\Gamma} \frac{\partial u}{\partial n}\,dA = \iint\limits_{\Gamma} \frac{\partial}{\partial n}\left(-\frac{1}{4\pi r_{PQ}}\right) dA \equiv v(P) \tag{132}$$

where it is clear from the definition of v that it is the potential at P due to a distribution of *dipoles of unit strength* oriented along the outward normal on Γ. The contribution to (131) from Ω vanishes,

$$\iint\limits_{\Omega} \frac{\partial u}{\partial n}\,dA = 0$$

because on Ω, $\operatorname{grad}(1/r_{PQ})$ is perpendicular to the normal to Ω. To evaluate the contribution from K_ε, we note that the outward normal on K_ε is in the direction of decreasing radius. Therefore, on K_ε

$$\frac{\partial u}{\partial n} = \frac{\partial}{\partial(-\rho)}\left(-\frac{1}{4\pi\rho}\right)\bigg|_{\rho=\varepsilon} = -\frac{1}{4\pi\varepsilon^2}$$

Using this together with the definition of v in (132) gives

$$v(P) = \frac{1}{4\pi\varepsilon^2} \iint\limits_{K_\varepsilon} \varepsilon^2 \, d\omega, \tag{133}$$

where $d\omega$ is the area element on the unit sphere centered at P. In other words, $v(P)$ equals the ratio of the area of K_ε to the area $4\pi\varepsilon^2$ of the entire ε sphere—that is, the solid angle subtended by C.

This result also implies that the actual shape of Γ is irrelevant; the potential at P is the same for all possible surface distributions having a given constant strength and boundary C. If C lies in a plane, the solid angle tends to $1/2$ as P approaches this plane and is independent of C. This is just a special case, with $p = \pm 1$, of our earlier result (103). If Γ is a closed surface, we can generate C by intersecting Γ with an arbitrary plane. In this case, if P is an interior point, the contributions to v from the two portions of Γ add, and we have $v = 1$. Conversely, if P is outside Γ, these contributions cancel, and we have $v = 0$. Thus, for a unit dipole distribution on a closed surface, v approaches 1 or 0 as P approaches the surface from inside or outside respectively.

2.6   Green's and Neumann's Functions

2.6.1   Green's Function

Given a domain G with boundary Γ and two points P, Q in G, *Green's* function $K(P,Q)$ satisfies

$$\Delta_P K(P,Q) = \delta_3(P,Q) \tag{134}$$

with boundary condition

$$K(P_\Gamma, Q) = 0 \tag{135}$$

where P_Γ denotes a point P on the boundary Γ. Thus, K consists of the fundamental solution plus a harmonic function chosen so that it cancels out the value of the fundamental solution on Γ. In particular, finding K is very much dependent on how complicated the domain G is.

It is easy to show that K is *symmetric*; that is,

$$K(P,Q) = K(Q,P). \tag{136}$$

To prove (136), we use the symmetric form of Green's formula (121) with respect to the integration variables $R = (\alpha, \beta, \gamma)$ and regard P and Q as *fixed* points in G. Let $u = K(R,P)$ and $v = K(R,Q)$ in (121). Since δ_3 is the product of three one-dimensional delta functions, each of which is an even function of its argument, we have $\delta_3(R,Q) = \delta_3(Q,R)$ and $\delta_3(R,P) = \delta_3(P,R)$. Therefore $\Delta_R K(R,Q) = \delta_3(R,Q) = \delta_3(Q,R)$, $\Delta_R K(R,P) = \delta_3(R,P) = \delta_3(P,R)$, and the left-hand side of (121) becomes

$$\iiint\limits_{G} [K(R,P)\delta_3(Q,R) - K(R,Q)\delta_3(P,R)] \, dV_R = K(Q,P) - K(P,Q)$$

and the right-hand side of (121) vanishes because $K = 0$ on the boundary, Q.E.D.

Green's function for a given domain is *unique*, as can be seen by assuming the contrary. If there exist two Green's functions K_1 and K_2 for a given domain G, the difference between K_1 and K_2 is harmonic *everywhere* inside G even though K_1 and K_2 individually fail to be harmonic at $P = Q$. Therefore, according to the first corollary to the energy theorem, $K_1 - K_2 = 0$ in G. Q.E.D.

2.6.2 Neumann's Function

Neumann's function is denoted by $N(P, Q)$ and satisfies

$$\Delta_P N(P, Q) = \delta_3(P, Q) \tag{137}$$

in G with boundary condition

$$\frac{\partial N}{\partial n_P}(P_\Gamma, Q) = C = \text{constant} \tag{138}$$

on Γ.

Actually, the constant C is not arbitrary because integrating the left-hand side of (137) over the interior of G gives, according to Gauss' theorem,

$$\iiint_G \text{div grad}_P N(P, Q) \, dV_P = \iint_\Gamma \frac{\partial N}{\partial n_P} \, dA_P = C \iint_\Gamma dA_P$$

But the volume integral of the right-hand side is just $\iiint_G \delta_3(P, Q) \, dV_P = 1$. Therefore,

$$C = \frac{1}{\text{area of } \Gamma} \tag{139}$$

We also note that the solution of the boundary-value problem (137)–(138) is not unique in the sense that given one solution N_1, we can define a second solution N_2, which differs from N_1 by an arbitrary constant. This result, analogous to the uniqueness theorem of Section 2.5.3, also follows immediately from the second corollary to the energy theorem applied to the difference $N_1 - N_2$. It will be convenient to make Neumann's function unique by appending the normalizing condition:

$$\iint_\Gamma N(P_\Gamma, Q) \, dA_P = 0 \tag{140}$$

whenever the integral (140) exists

As in (136), it is also easy to show that Neumann's function is symmetric; that is,

$$N(P, Q) = N(Q, P) \tag{141}$$

2.7   Dirichlet's and Neumann's Problems

Dirichlet's problem, also called the boundary-value problem of the first kind, consists of solving $\Delta u = 0$ subject to prescribed values of u on the boundary; that is,

$$\Delta u = 0 \quad \text{in } G$$
$$u = f = \text{prescribed on } \Gamma \tag{142}$$

If G is infinite, we assume that $u \to 0$ at infinity.

Once Green's function for G is known, we can write down the solution of (142) immediately. To see this, recall (121), the symmetric form of Green's formula, and let the coordinates Q be the integration variables, while P is regarded as a fixed point in G. In (121), let $u(Q)$ be the solution of (142) and let $v = K(P,Q) = K(Q,P)$ be Green's function for G. We then have

$$\iiint_G [u(Q)\delta_3(P,Q) - K(Q,P)\Delta_Q u]\,dV_Q$$
$$= \iint_\Gamma \left[u(Q_\Gamma)\frac{\partial K}{\partial n_Q}(P,Q) - K(P_\Gamma,Q)\frac{\partial u}{\partial n_Q} \right] dA_Q$$

Since $\Delta_Q u = 0$ and $K(P_\Gamma,Q) = 0$, this reduces to

$$u(P) = \iint_\Gamma f(Q)\frac{\partial K}{\partial n_Q}(P,Q)\,dA_Q \tag{143}$$

which is called the *generalized Poisson formula*. It gives u at any interior point P by quadrature once the boundary values f are prescribed *as long as K is known for G*.

Neumann's problem consists of

$$\Delta u = 0 \quad \text{in } G$$
$$\frac{\partial u}{\partial n} = g = \text{prescribed on } \Gamma \tag{144}$$

Here, the function g cannot be prescribed arbitrarily, as (122) requires that $\iint_\Gamma g\,dA = 0$. Also, as in (140), we introduce the normalizing condition (if possible)

$$\iint_\Gamma u\,dA = 0 \tag{145}$$

to make the solution unique.

Assuming that we have found Neumann's function $N(P,Q)$ for the given domain, we again use (121) with $u = u(Q)$, $v = N(P,Q)$ to obtain

$$\iiint_G [u(Q)\delta_3(P,Q) - N(Q,P)\Delta_Q u]\,dV_Q = \iint_\Gamma \left[u(Q)\frac{\partial N}{\partial n_Q} - N(P,Q)\frac{\partial u}{\partial n_Q} \right] dA_Q$$

The left-hand side reduces to $u(P)$, and the first integral in the right-hand side is the constant $C \iint_\Gamma u(Q)\,dA$, which vanishes for the choice of normalization (145). If the domain is infinite and $\iint_\Gamma u(Q)\,dA$ does not exist, $C \to 0$ as it is the reciprocal area of Γ, and the product vanishes.

Thus, we find

$$u(P) = -\iint_\Gamma g(Q)N(P,Q)\,dA_Q \tag{146}$$

Again, we emphasize that the solutions (143) and (146) are quadratures, once K or N has been derived for the given domain. Moreover, faced with the problem of computing u for a given domain and different boundary data, these formulas are most convenient as one need only compute K or N once.

2.8   Examples of Green's and Neumann's Functions

2.8.1 Upper Half-Plane, $y \geq 0$ (Two Dimensions)

Green's function for the upper half-plane may be interpreted as the deflection of a membrane (which is clamped all along the x-axis) as measured at a point $P = (x, y)$ due to a unit concentrated force at $Q = (\xi, \eta)$. It is clear from symmetry that the zero boundary condition on $y = 0$ can be achieved by adding an image force of unit negative strength at the point $\bar{Q} = (\xi, -\eta)$ (see Figure 2.13). Thus, we have

$$K(P,Q) = \frac{1}{2\pi}\log r_{PQ} - \frac{1}{2\pi}\log r_{P\bar{Q}} \tag{147}$$

where

$$r_{PQ}^2 \equiv (x - \xi)^2 + (y - \eta)^2; \qquad r_{P\bar{Q}}^2 \equiv (x - \xi)^2 + (y + \eta)^2 \tag{148}$$

Note that in (147), K consists of the fundamental solution $(1/2\pi)\log r_{PQ}$, plus a

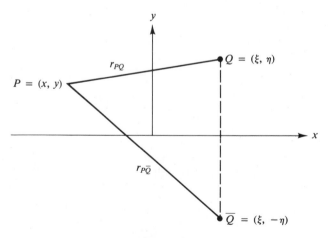

Figure 2.13

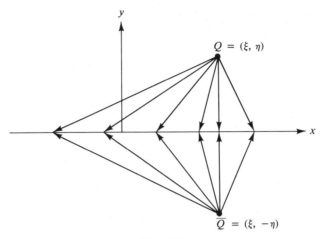

Figure 2.14

harmonic function in the upper half-plane (because the singularity of $\log r_{P\bar{Q}}$ is in the lower half-plane), which cancels out the values of the fundamental solution on the boundary and thus satisfies the requirement $K(P_\Gamma, Q) = 0$.

Neumann's function in the upper half-plane is simply

$$N(P, Q) = \frac{1}{2\pi} \log r_{PQ} + \frac{1}{2\pi} \log r_{P\bar{Q}} \tag{149}$$

and may be interpreted as the velocity potential due to a two-dimensional unit source at Q and an image source of equal strength at $\bar{Q}$. Now, $\partial N/\partial n = -\partial N/\partial y$, and it is clear that on $y = 0$ the normal components of velocity due to the two sources are exactly opposite and cancel out (see Figure 2.14). Note that since the domain is infinite, integral (140) does not exist and we have normalized N by requiring its minimum value on Γ to occur at $P = (\xi, 0)$.

2.8.2   Upper Half-Space, $z \geq 0$ (Three Dimensions)

Here we replace (147)–(149) by the appropriate expressions using three-dimensional sources—that is,

$$K(P, Q) = -\frac{1}{4\pi} \left(\frac{1}{r_{PQ}} - \frac{1}{r_{P\bar{Q}}} \right) \tag{150a}$$

$$N(P, Q) = -\frac{1}{4\pi} \left(\frac{1}{r_{PQ}} + \frac{1}{r_{P\bar{Q}}} \right) \tag{150b}$$

where

$$r_{PQ}^2 = (x - \xi)^2 + (y - \eta)^2 + (z - \zeta)^2; \quad r_{P\bar{Q}} = (x - \xi)^2 + (y - \eta)^2 + (z + \zeta)^2$$

Suppose we wish to solve Dirichlet's problem in the upper half-space $z \geq 0$—that is,

$$\Delta u = 0 \quad \text{in} \quad z \geq 0 \tag{151a}$$

with boundary condition

$$u(x, y, 0^+) = f(x, y) = \text{prescribed} \tag{151b}$$

using (143). We need to compute $\partial K / \partial n_Q$—that is, $-\partial K / \partial \zeta$—on $\zeta = 0$. Using (150a), we find

$$\frac{\partial K}{\partial \zeta} = -\frac{1}{4\pi} \left(\frac{z - \zeta}{r_{PQ}^3} + \frac{z + \zeta}{r_{P\bar{Q}}^3} \right)$$

Therefore, on the boundary

$$-\frac{\partial K}{\partial \zeta} = \frac{z}{2\pi} \frac{1}{[(x - \xi)^2 + (y - \eta)^2 + z^2]^{3/2}}$$

and (143) becomes

$$u(x, y, z) = \frac{z}{2\pi} \int_{-\infty}^{\infty} \int_{-\infty}^{\infty} \frac{f(\xi, \eta) \, d\xi \, d\eta}{[(x - \xi)^2 + (y - \eta)^2 + z^2]^{3/2}} \tag{152}$$

This result may be interpreted as the potential at $P = (x, y, z)$ due to a distribution of dipoles on the $z = 0$ plane. The strength of this distribution is $-2f(x, y)$ and the axes of the dipoles are all along the $+z$ direction; see (96). Moreover, (103a) confirms that the boundary condition (151b) on $z = 0$ is indeed satisfied. In Section 2.12, we shall extend this idea to general Dirichlet problems and seek a representation of the solution in terms of a dipole distribution of unknown strength on the boundary surface.

It is instructive to rederive (152) and the corresponding formula for the Neumann problem in the upper half-space directly using (128). Assume $\Delta u = 0$ in $z \geq 0$, and let $P = (x, y, z)$ and $Q = (\xi, \eta, \zeta)$ be two interior points in $z > 0$, whereas $Q_0 = (\xi, \eta, 0)$ is a point on the boundary. Specializing the integral formula (128) for this case, we have

$$r_{PQ} \equiv [(x - \xi)^2 + (y - \eta)^2 + (z - \zeta)^2]^{1/2}$$

$$r_{PQ_0} \equiv [(x - \xi)^2 + (y - \eta)^2 + z^2]^{1/2}$$

We compute

$$\frac{\partial}{\partial n_Q} \left(\frac{1}{r_{PQ}} \right) = \frac{-1}{r_{PQ}^2} \frac{\partial r_{PQ}}{\partial (-\zeta)} = -\frac{(z - \zeta)}{r_{PQ}^3}$$

$$\left[\frac{\partial}{\partial n_Q} \left(\frac{1}{r_{PQ}} \right) \right]_{Q = Q_0} = -\frac{z}{r_{PQ_0}^3}$$

Therefore, (128) becomes

$$u(x, y, z) = \frac{1}{4\pi} \int\!\!\int_{-\infty}^{\infty} \left[\frac{u(\xi, \eta, 0^+) z}{r_{PQ_0}^3} - \frac{\frac{\partial u}{\partial \zeta}(\xi, \eta, 0^+)}{r_{PQ_0}} \right] d\xi \, d\eta \tag{153}$$

Now consider (121) with $u(Q)$ denoting the solution of (151a) and $v = 1/4\pi r_{P\bar{Q}}$. Here $\bar{Q} = (\xi, \eta, -\zeta)$ is the mirror image of Q with respect to the $z = 0$ plane, and

$$r_{P\bar{Q}} \equiv [(x - \xi)^2 + (y - \eta)^2 + (z + \zeta)^2]^{1/2}$$

Since $\bar{Q}$ is in the lower half-space, $r_{P\bar{Q}}$ does not vanish for $z \geq 0$ and $\Delta_Q v = 0$. Thus, the left-hand side of (121) equals zero. To compute the right-hand side, we note that

$$\left(\frac{\partial u}{\partial n_Q}\right)_{Q=Q_0} = \frac{\partial u(\xi, \eta, 0^+)}{\partial(-\zeta)} = -\frac{\partial u(\xi, \eta, 0^+)}{\partial \zeta}$$

$$\left(\frac{\partial v}{\partial n_Q}\right)_{Q=Q_0} = \frac{1}{4\pi}\left[\frac{\partial}{\partial(-\zeta)}\left(\frac{1}{r_{P\bar{Q}}}\right)\right]_{Q=Q_0} = \frac{1}{4\pi}\frac{z}{r_{PQ_0}^3}$$

Therefore, (121) reduces to

$$0 = \frac{1}{4\pi}\int\int\limits_{-\infty}^{\infty}\left[\frac{u(\xi, \eta, 0^+)z}{r_{PQ_0}^3} + \frac{\frac{\partial u}{\partial \zeta}(\xi, \eta, 0^+)}{r_{PQ_0}}\right]d\xi\,d\eta \tag{154}$$

Adding (153) and (154) gives Poisson's formula (152), and substracting gives

$$u(x, y, z) = \frac{-1}{2\pi}\int\int\limits_{-\infty}^{\infty}\frac{\frac{\partial u}{\partial \zeta}(\xi, \eta, 0^+)}{r_{PQ_0}}d\xi\,d\eta \tag{155}$$

which is the solution of the following Neumann problem in the upper half-space (see Problem 10):

$$\Delta u = 0 \quad \text{in} \quad z \geq 0 \tag{156a}$$

$$\frac{\partial u}{\partial z}(x, y, 0^+) = g(x, y) = \text{prescribed} \tag{156b}$$

Result (155) may be interpreted as the potential at $P = (x, y, z)$ due to a distribution of sources of strength $q(x, y) = 2g(x, y)$ on the $z = 0$ plane [compare (155) with (105) and (156b) with (112a) to confirm that the boundary condition is satisfied]. In Section 2.12, we shall also extend this idea to solve Neumann's problem in a more general domain with a prescribed value of the normal derivative on the surface by using an unknown distribution of sources on the boundary.

2.8.3 Interior (Exterior) of Unit Sphere or Circle

Consider a unit sphere, two points $P = (r, \theta, \phi)$ and $Q = (\rho, \theta', \phi')$ in the interior, and the point $\bar{Q} = (1/\rho, \theta', \phi')$ outside the sphere, as shown in Figure 2.15. If we denote the angle POQ by γ, the distance PQ by r_{PQ}, and the distance $P\bar{Q}$ by $r_{P\bar{Q}}$, we have

$$\cos\gamma \equiv \cos\theta\cos\theta' + \sin\theta\sin\theta'\cos(\phi - \phi') \tag{157a}$$

$$r_{PQ}^2 \equiv r^2 + \rho^2 - 2r\rho\cos\gamma \tag{157b}$$

$$r_{P\bar{Q}}^2 \equiv r^2 + \frac{1}{\rho^2} - \frac{2r}{\rho}\cos\gamma \tag{157c}$$

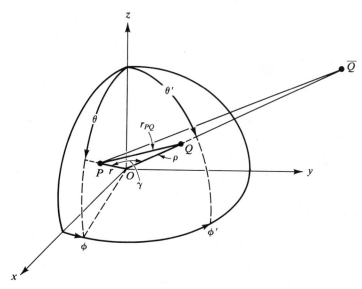

Figure 2.15

It is easily seen from (157) that if P is on the surface of the sphere—that is, if $r = 1$—then

$$\frac{r_{PQ}}{r_{P\bar{Q}}}\bigg|_{r=1} = \rho \tag{158}$$

Using this result, we can immediately derive the following Green's functions:

Interior of Unit Circle

$$K = \frac{1}{2\pi}\log\frac{r_{PQ}}{\rho r_{P\bar{Q}}}; \qquad \phi = \phi' = 0 \tag{159}$$

Exterior of Unit Circle

$$K = \frac{1}{2\pi}\log\frac{\rho r_{P\bar{Q}}}{r_{PQ}}; \qquad \phi = \phi' = 0 \tag{160}$$

Interior of Unit Sphere

$$K = -\frac{1}{4\pi}\left[\frac{1}{r_{PQ}} - \frac{1}{\rho r_{P\bar{Q}}}\right] \tag{161}$$

Exterior of Unit Sphere

$$K = -\frac{1}{4\pi}\left[\frac{1}{r_{P\bar{Q}}} - \frac{\rho}{r_{PQ}}\right] \tag{162}$$

Poisson's Formula for the Sphere   To solve

$$\Delta u \equiv u_{rr} + \frac{2}{r}u_r + \frac{1}{r^2}u_{\theta\theta} + \frac{1}{r^2\sin^2\theta}u_{\phi\phi} + \frac{\cot\theta}{r^2}u_\theta = 0$$

$$u(1,\theta,\phi) = f(\theta,\phi) = \text{prescribed}$$

in the interior domain $r \leq 1$, we use (161) and calculate

$$\frac{\partial K}{\partial \eta_Q} = \frac{\partial K}{\partial \rho} = \frac{1}{4\pi}\left(\frac{\rho - r\cos\gamma}{r_{PQ}^3} - \frac{\rho r^2 - r\cos\gamma}{\rho^3 r_{P\bar{Q}}^3}\right)$$

Therefore, on the boundary $\rho = 1$, we have

$$\frac{\partial K}{\partial \rho}\bigg|_{\rho=1} = \frac{1 - r^2}{4\pi r_B^3}; \qquad r_B^2 \equiv 1 + r^2 - 2r\cos\gamma$$

and (143) becomes

$$u(r, \theta, \phi) = \frac{1 - r^2}{4\pi}\int_{\theta'=0}^{\pi}\int_{\phi'=0}^{2\pi}\frac{f(\theta', \phi')\sin\theta'\,d\theta'\,d\phi'}{[1 + r^2 - 2r\cos\gamma]^{3/2}} \tag{163}$$

This is called *Poisson's formula*. Now, unlike the result (152) for the planar problem, (163) is not the potential only of a surface distribution of dipoles; it also includes a surface distribution of sources and is a special case of the general result (128). In fact, it is easy to show that (163) consists of a surface dipole distribution of strength $2f(\theta', \phi')$ plus a surface source distribution of strength $f(\theta', \phi')$ (see Problem 11).

Kelvin Inversion

It is easy to show by direct substitution that if $u = F(r, \theta, \phi)$ is harmonic inside the unit sphere, then $v = (1/r)F(1/r, \theta, \phi)$ is harmonic outside, and $v \to 0$ as $r \to \infty$ (see Problem 15).

Using this result, or explicit calculation with K given by (162), we find the following Poisson formula for the exterior problem:

$$u(r, \theta, \phi) = \frac{r^2 - 1}{4\pi}\int_{\theta'=0}^{\pi}\int_{\phi'=0}^{2\pi}\frac{f(\theta', \phi')\sin\theta'\,d\theta'\,d\phi'}{[1 + r^2 - 2r\cos\gamma]^{3/2}} \tag{164}$$

which only differs from (163) by a minus sign.

The corresponding result for the interior of the circle is [see (27)]

$$u(r, \phi) = \frac{1 - r^2}{2\pi}\int_0^{2\pi}\frac{u(1, \phi')\,d\phi'}{1 + r^2 - 2r\cos(\phi - \phi')} \tag{165}$$

and a change in sign gives the formula for the exterior problem.

The formulas for the Neumann problems inside or outside the sphere and circle are left as an exercise (Problem 12).

2.9   Estimates; Harnack's Inequality

Consider an arbitrary *bounded* domain G with $P = (x, y, z)$ and $Q = (\xi, \eta, \zeta)$ as interior points. Surround G with the smallest sphere that contains it entirely, and let $K(P, Q)$ be Green's function for G while $K_S(P, Q)$ is Green's function for the surrounding sphere (for the same points P, Q). Normalize the radius of this surrounding sphere to be unity so that K_S is given by (161). Since $\rho r_{P\bar{Q}} > r_{PQ}$ for all $r < 1$, $K_S < 0$ inside the surrounding sphere. Now, $K_S - K$ is harmonic in G

because the singular terms in K and K_S contributed by the fundamental solutions cancel out exactly. Since $K = 0$ on the boundary Γ of G and K_S is negative there, we have $K_S - K < 0$ on Γ. It then follows from the maximum-minimum theorem for harmonic functions that $K_S - K < 0$ in G. Similarly, if we surround Q by a sphere of radius ε contained entirely in G, we have $K < 0$ on the surface of the ε-sphere if ε is sufficiently small. Since $K = 0$ on Γ, the maximum-minimum theorem for harmonic functions implies that $K < 0$ in the domain outside the ε-sphere and inside Γ. Therefore,

$$K_S < K < 0 \qquad \text{in } G \tag{166}$$

and this gives an estimate for the Green's function in an arbitrary bounded domain.

Next, consider the kernel for Poisson's formula (163), denoted by J:

$$J \equiv \frac{1 - r^2}{[1 + r^2 - 2r \cos \gamma]^{3/2}}$$

The minimum and maximum values for J occur when $\cos \gamma = -1$ and $\cos \gamma = +1$, respectively, in which case

$$J_{\min} = \frac{1 - r^2}{(1 + r)^3} = \frac{1 - r}{(1 + r)^2}$$

$$J_{\max} = \frac{1 - r^2}{(1 - r)^3} = \frac{1 + r}{(1 - r)^2}$$

It then follows that the potential $u(P)$ for the interior of the unit sphere is bounded above and below in the form:

$$\frac{1 - r}{4\pi(1 + r)^2} \iint_{r=1} u\, dA \leq u(P) \leq \frac{1 + r}{4\pi(1 - r)^2} \iint_{r=1} u\, dA$$

But according to the mean-value theorem, $1/4\pi \iint_{r=1} u\, dA = u(0)$, and we obtain *Harnack's inequality*:

$$\frac{1 - r}{(1 + r)^2} u(0) \leq u(P) \leq \frac{1 + r}{(1 - r)^2} u(0) \tag{167}$$

bounding the value of u at any interior point in terms of $u(0)$ and the distance from the origin.

2.10   Connection between Green's Function and Conformal Mapping (Two Dimensions); Dipole-Green's Functions

Let D be a given domain in the xy plane with boundary Γ consisting of at least two points. Let (x_0, y_0) be a fixed point inside D. According to Riemann's mapping theorem (for example, see p. 175 of [4]), there exists a conformal map of D onto the interior of the unit circle in the ζ-plane that sends the point

$z_0 \equiv x_0 + iy_0$ to the origin $\zeta = 0$. Denote this conformal mapping by

$$\zeta = f(z, z_0); \qquad \zeta \equiv \xi + i\eta; \qquad z \equiv x + iy$$

Clearly, Green's function for the interior of the unit circle with source at the origin is $(1/2\pi) \log |\zeta|$. Therefore, Green's function for D is

$$K(x, y, x_0, y_0) = \frac{1}{2\pi} \log |f(z, z_0)| \qquad (168)$$

Thus, knowing Green's function for a two-dimensional domain is equivalent to knowing the mapping of the domain to the interior of a unit circle and vice versa (see Problem 16).

For an arbitrary source point $\zeta = \xi + i\eta$, Green's function for the interior of the unit circle is the real part of the complex function:

$$L(z, \zeta, \bar{\zeta}) = \frac{1}{2\pi} \log \frac{(z - \zeta)}{z\bar{\zeta} - 1} \equiv L_1 + iL_2; \qquad \bar{\zeta} \equiv \xi - i\eta$$

If we now fix z and ζ and regard L as an analytic function of $\bar{\zeta}$, the Cauchy-Riemann equations associated with the existence of the derivative $dL/d\bar{\zeta} = \partial L/\partial \xi$ read

$$\frac{\partial L_1}{\partial \xi} = -\frac{\partial L_2}{\partial \eta}; \qquad \frac{\partial L_1}{\partial \eta} = \frac{\partial L_2}{\partial \xi}$$

Denote

$$\mathrm{Re}\left(\frac{dL}{d\bar{\zeta}}\right) \equiv H_1 = \mathrm{Re}\left(\frac{\partial L}{\partial \xi}\right)$$

We find

$$H_1 = \frac{1}{2\pi} \mathrm{Re}\left\{ \frac{z\bar{\zeta} - 1}{z - \zeta} \cdot \frac{\partial}{\partial \xi}\left(\frac{z - \zeta}{z\bar{\zeta} - 1}\right)\right\} = \frac{1}{2\pi} \mathrm{Re}\left\{ \frac{z\bar{\zeta} - 1}{z - \zeta} \cdot \frac{[-(z\bar{\zeta} - 1) - (z - \zeta)z]}{(z\bar{\zeta} - 1)^2}\right\}$$

$$= \frac{1}{2\pi} \mathrm{Re}\left\{ \frac{z^2 + z(\bar{\zeta} - \zeta) - 1}{(\zeta - z)(z\bar{\zeta} - 1)}\right\} \qquad (169)$$

It is easy to see that when z is on the unit circle—that is, $|z| = 1$—$H_1 = 0$.

It was shown earlier that we can compute the potential due to a dipole by differentiating the potential due to a source. Here, we have computed H_1 by differentiating L_1 with respect to ξ. But L_1 is Green's function and consists of the fundamental solution plus an appropriate harmonic function to satisfy the boundary condition. Therefore, $\partial L_1/\partial \xi$ is the analog of Green's function but with a *unit dipole* located at ζ and oriented along the x-axis; that is, H_1 satisfies [see (71)]

$$\Delta H_1 = -\delta'(x - \xi)\delta(y - \eta)$$

$$H_1 = 0 \qquad \text{on } x^2 + y^2 = 1$$

Similarly, for a unit dipole located at ζ and oriented along the y-axis, we find

$$H_2 \equiv \frac{\partial L_1}{\partial \eta} = \frac{1}{2\pi} \mathrm{Im}\left\{ \frac{1 + z^2 - z(\zeta + \bar{\zeta})}{(\zeta - z)(\bar{\zeta}z - 1)}\right\} \qquad (170)$$

While formulas (169) and (170) are interesting in their own merit, they are not as fundamental as Green's or Neumann's functions, which are the essential ingredients for solving the two basic boundary-value problems for Laplace's equation.

2.11   Series Representations; Connection with Separation of Variables

Consider Poisson's formula (163) for the interior problem. We wish to develop this result in series form valid for $r < 1$. A useful identity for the kernel in Poisson's formula is

$$J \equiv \frac{1 - r^2}{[1 + r^2 - 2r \cos \gamma]^{3/2}} = \left[-\frac{1}{r_{PQ}} - 2 \frac{\partial}{\partial \rho} \left(\frac{1}{r_{PQ}} \right) \right]_{\rho=1} \tag{171}$$

which can be verified immediately using the definition for r_{PQ} in (157b).

Recall the generating function for Legendre polynomials (for example, see pp. 102–103 of [3]):

$$\frac{1}{r_{PQ}} = \frac{1}{\rho \left[1 - \frac{2r}{\rho} \cos \gamma + \left(\frac{r}{\rho} \right)^2 \right]^{1/2}} = \frac{1}{\rho} \sum_{n=0}^{\infty} P_n(\cos \gamma) \left(\frac{r}{\rho} \right)^n; \quad \left(\frac{r}{\rho} \right) < 1 \tag{172}$$

where P_n denotes the Legendre polynomials

$$P_0 \equiv 1, \qquad P_1 \equiv \cos \gamma, \qquad P_2 \equiv \tfrac{3}{2}(\cos^2 \gamma - \tfrac{1}{3}), \dots$$

We compute

$$\frac{\partial}{\partial \rho} \left(\frac{1}{r_{PQ}} \right) = -\frac{1}{\rho^2} \sum_{n=0}^{\infty} P_n(\cos \gamma) \left(\frac{r}{\rho} \right)^n (n + 1)$$

Therefore, using identity (171) gives

$$J = \sum_{n=0}^{\infty} (2n + 1) P_n(\cos \gamma) r^n \tag{173}$$

This series converges uniformly if $r < 1$, and we can interchange the order of summation and integration in (163) to obtain

$$u(r, \theta, \phi) = \sum_{n=0}^{\infty} \frac{r^n (2n + 1)}{4\pi} \left\{ \int_{\theta'=0}^{\pi} \int_{\phi'=0}^{2\pi} f(\theta', \phi') P_n(\cos \gamma) \sin \theta' \, d\theta' \, d\phi' \right\} \tag{174}$$

This is a power series in r^n with coefficients C_n given by the double integral times $(2n + 1)/4\pi$.

If the prescribed data are axially symmetric—that is,

$$f(\theta', \phi') = g(\theta')$$

—the expression in (174) involves ϕ' only through the P_n term. We can use the identity (for example, see pp. 326–328 of [7])

$$\frac{1}{2\pi} \int_0^{2\pi} P_n(\cos \gamma) \, d\phi' = P_n(\cos \theta) P_n(\cos \theta')$$

to carry out the ϕ' integration, and (179) reduces to

$$u(r, \theta) = \sum_{n=0}^{\infty} \frac{r^n(2n+1)}{2} \left\{ \int_0^{\pi} g(\theta') P_n(\cos \theta') \sin \theta' \, d\theta' \right\} P_n(\cos \theta) \tag{175}$$

With the change of variables $s = \cos \theta'$, this becomes

$$u(r, \theta) = \sum_{n=0}^{\infty} C_n P_n(\cos \theta) r^n \tag{176a}$$

where

$$C_n = \frac{2n+1}{2} \int_{-1}^{1} g(\cos^{-1} s) P_n(s) \, ds \tag{176b}$$

This result also follows directly by solving Laplace's equation with axial symmetry ($u_\phi = 0$); that is,

$$u_{rr} + \frac{2}{r} u_r + \frac{1}{r^2} u_{\theta\theta} + \frac{\cot \theta}{r^2} u_\theta = 0 \tag{177}$$

by separation of variables.

Assuming $u(r, \theta) = A(r)B(\theta)$ gives

$$\frac{r^2 A''}{A} + 2r \frac{A'}{A} = -\frac{B''}{B} - \cot \theta \frac{B'}{B} = \lambda = \text{constant} \tag{178}$$

Therefore,

$$B'' + (\cot \theta)B' + \lambda B = 0 \tag{179}$$

which is Legendre's equation written in terms of θ. The conventional form of Legendre's equation is in terms of the independent variable $x = \cos \theta$ and has the form (with $B(\theta) = y(x)$)

$$(1 - x^2)y'' - 2xy' + \lambda y = 0 \tag{180}$$

Bounded solutions exist only for $\lambda = n(n+1)$ with $n = $ integer. Then

$$B(\theta) = P_n(\cos \theta)$$

The equation for A is equidimensional, with solutions r^n and r^{-n-1}. We discard the r^{-n-1} solutions for the interior problem and have the series:

$$u(r, \theta) = \sum_{n=0}^{\infty} C_n r^n P_n(\cos \theta)$$

which corresponds to (176a). Using the boundary condition gives

$$u(1, \theta) = g(\theta) = \sum_{n=0}^{\infty} C_n P_n(\cos \theta)$$

and the orthogonality of the P_n gives

$$C_n = \frac{2n+1}{2} \int_0^{\pi} g(\theta') \sin \theta' P_n(\cos \theta') \, d\theta'$$

which reduces to (176b) if we set $s = \cos \theta'$.

2.12   Solutions in Terms of Integral Equations

In Sections 2.6–2.8, we saw that the solution of Dirichlet's and Neumann's problems could be expressed as a quadrature in terms of Green's function K, or Neumann's function N, respectively. The principal task in solving either of these boundary-value problems is the actual computation of K or N for the given domain. For the simple geometries considered in Section 2.8 (see also Problem 16), symmetry arguments easily defined K or N. We also observed that for planar boundaries, our final result for u in Dirichlet's problem turned out to be the potential due to a distribution of dipoles only and in Neumann's problem a distribution of sources only. For the case of a spherical boundary, our result for Dirichlet's problem involved a surface distribution of dipoles as well as sources. In summary, for the simple geometries that we considered, it was easy to obtain, a posteriori, expressions linking the strengths of these distributions to the boundary data.

In this section, we propose to bypass the calculation of K and N (as these may be impractical for a nontrivial geometry) and to proceed *directly* to representations in terms of dipole or source distributions of *unknown* strength. We shall show that one can then derive *integral equations* linking these unknown distribution strengths to the given boundary data.

2.12.1 Dirichlet's Problem

Consider Dirichlet's problem in an interior domain G bounded by the smooth surface Γ. We wish to solve

$$\Delta u = 0 \quad \text{in} \quad G \tag{181a}$$

subject to

$$u = f \tag{181b}$$

a prescribed function on Γ. Assume that the solution can be represented by a surface distribution of dipoles of strength μ oriented in the outward normal direction on Γ—that is,

$$u(P) = \iint_{\Gamma} \mu(Q_{\Gamma}) \frac{\partial}{\partial n_Q} \left(\frac{-1}{4\pi r_{PQ}} \right) dA_Q \tag{182}$$

In order to apply the boundary condition $u = f$ on Γ, we must evaluate (182) for $P \rightarrow p$, a point on the boundary. Because $r_{PQ} \rightarrow r_{pQ}$ and $r_{pQ} \rightarrow 0$ for $Q = p$, (182) leads to an improper integral, as discussed in Section 2.4.4 for a planar boundary. We show next how one may evaluate $u(p)$ by a process similar to the one used for the planar boundary.

Consider a small sphere of radius ε centered at p, as shown in Figure 2.16. This sphere intersects Γ along a curve C and excludes the small "disc" Γ_ε from Γ. Now with $\varepsilon \neq 0$, subdivide the integration over Γ in (182) into a contribution due to Γ_ε and one due to $\Gamma - \Gamma_\varepsilon$; then let $\varepsilon \rightarrow 0$ and $P \rightarrow p$. If $\alpha(\varepsilon)$, the distance between P and p, tends to zero faster than ε, the potential at p may be expressed

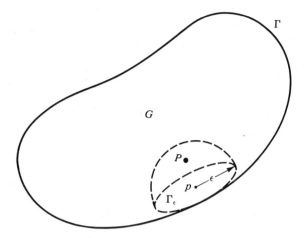

Figure 2.16

in the form

$$u(p) = \lim_{\varepsilon \to 0} \left\{ \iint_{\Gamma_\varepsilon} \mu(Q) \frac{\partial}{\partial n_Q} \left(-\frac{1}{4\pi r_{PQ}} \right) dA_Q + \iint_{\Gamma - \Gamma_\varepsilon} \mu(Q) \frac{\partial}{\partial n_Q} \left(-\frac{1}{4\pi r_{PQ}} \right) dA_Q \right\}$$

(183)

Note carefully the distinction between r_{PQ} and r_{pQ} in (183). In particular, it is important to note that we evaluate both integrals with ε fixed and not equal to 0. In the integral over Γ_ε, the observer is at $P(\varepsilon) \neq p$ with $\varepsilon \neq 0$, whereas in the integral over $\Gamma - \Gamma_\varepsilon$, we have set $P = p$ in the integrand. We then take the limit $\varepsilon \to 0$ for the two expressions. It is easy to verify that this approach gives exactly the same result for $u(p)$ as would be derived from the more elaborate limit process discussed in Section 2.4.4 for the case of a planar boundary. In effect, setting $P = p$ in the second integrand in (183) corresponds to choosing the function $P(\varepsilon)$, such that $(P/\varepsilon) \to 0$.

We recall the result (133) for the potential of a dipole distribution of unit strength and note that as $\varepsilon \to 0$ (with $\alpha/\varepsilon \to 0$ also), the solid angle subtended from P tends to $\frac{1}{2}$ if Γ is smooth at p. Therefore, the contribution to (183) from Γ_ε is

$$\lim_{\varepsilon \to 0} \iint_{\Gamma_\varepsilon} \mu(Q) \frac{\partial}{\partial n_Q} \left(-\frac{1}{4\pi r_{PQ}} \right) dA_Q = \frac{\mu(p)}{2}$$

This result also follows directly from (103b), since for ε sufficiently small and Γ smooth, Γ_ε is nearly planar, and the axis of the dipole at p points outside G for $\mu(p) > 0$.

Consider now the contribution to $u(p)$ from $\Gamma - \Gamma_\varepsilon$. Since the disc Γ_ε is excluded from the second integral, $r_{pQ} > 0$ and the integrand is not singular as long as $\varepsilon > 0$. The question then arises as to the behavior of $(\partial/\partial n_Q)(-1/4\pi r_{pQ})$

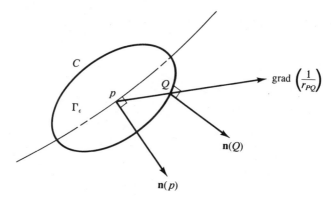

Figure 2.17

along the critical curve C as $\varepsilon \to 0$ (see Figure 2.17). Now,

$$\frac{\partial}{\partial n_Q}\left(-\frac{1}{4\pi r_{pQ}}\right) = -\frac{1}{4\pi}\,\mathrm{grad}\left(\frac{1}{r_{pQ}}\right)\cdot \mathbf{n} \tag{184}$$

where n_Q is the unit outward normal to Γ along C. Clearly, as $\varepsilon \to 0$, $\mathrm{grad}(1/r_{pQ})$ and $\mathbf{n}$ become orthogonal and (184) vanishes on C, so there is no difficulty associated with the contribution to (184) near C, as C shrinks to the point p.

We have shown that the integral equation to be solved is

$$f = \frac{\mu(p)}{2} + \iint\limits_{\Gamma} \mu(Q)\frac{\partial}{\partial n_Q}\left(-\frac{1}{4\pi r_{pQ}}\right)dA_Q \tag{185}$$

where the integral is to be interpreted in the limiting sense just discussed of having the point p "removed" from the surface of integration Γ. For the exterior problem and the same representation (182), we find the analogous result

$$f = -\frac{\mu(p)}{2} + \iint\limits_{\Gamma} \mu(Q)\frac{\partial}{\partial n_Q}\left(-\frac{1}{4\pi r_{pQ}}\right)dA_Q \tag{186}$$

with the opposite sign for the contribution due to Γ_ε. Once f is known, the solution can be obtained from (182) by quadrature. Note that if Γ is not smooth at p, the contribution of the first term on the right-hand side of (185) or (186) is not $\pm\mu(p)/2$ and must be worked out in detail. Also, if the entire surface Γ is nearly planar, we expect the contribution due to the surface integral in (185) or (186), to be small in absolute value relative to $|\mu(p)/2|$.

Let us next verify that the result (185) gives the correct solution (152) for Dirichlet's problem in the upper half-space. In this case, Γ is the plane $z = 0$, and $\partial/\partial n_Q = -\partial/\partial\zeta$. The assumed solution (182) using a surface ($z = 0$) distribution of dipoles with axes oriented along the $-z$ direction and with strength μ becomes:

$$u(x, y, z) = \frac{1}{4\pi} \int\limits_{-\infty}^{\infty}\!\!\int \mu(\xi, \eta) \left[\frac{\partial}{\partial \zeta} \left(\frac{1}{\sqrt{(x - \xi)^2 + (y - \eta)^2 + (z - \zeta)^2}} \right) \right]_{\zeta=0} d\xi \, d\eta$$

$$= \frac{z}{4\pi} \int\limits_{-\infty}^{\infty}\!\!\int \frac{\mu(\xi, \eta) \, d\xi \, d\eta}{[(x - \xi)^2 + (y - \eta)^2 + z^2]^{3/2}} \tag{187}$$

This is the same expression as (96) with $p = -\mu$. The sign change is needed because we have the dipole axes in (187) pointing along the negative z direction, whereas those in (96) are along $+z$.

To evaluate (187) on the boundary, we need to let $z \to 0^+$; we obtain

$$u(x, y, 0^+) = \frac{\mu(x, y)}{2} \tag{188}$$

according to (103a). Thus, for the planar problem, the double integral in (185) or (186) equals zero, and we need not solve an integral equation; the dipole strength μ is given explicitly by

$$\mu(x, y) = 2f(x, y) \tag{189}$$

For nearly planar surfaces we expect the contributions of the integrals in (185) or (186) to be small in magnitude compared with $|\mu/2|$. This is, in fact, why we chose to solve Dirichlet's problem using a surface dipole distribution. We could equally well have used a surface source distribution, but the expression corresponding to (189) for the source strength is more complicated.

To see this, consider the general result (128) as it applies to the solution of the planar boundary-value problem (151). The value of $u(x, y, 0^+)$ is specified but $(\partial u/\partial z)(x, y, 0^+)$ is not known. Therefore, the second term on the right-hand side of (128) is known, but the first is not. In fact, (128) reduces to

$$u(x, y, z) = -\frac{1}{4\pi} \int\limits_{-\infty}^{\infty}\!\!\int \frac{\lambda(\xi, \eta) \, d\xi \, d\eta}{[(x - \xi)^2 + (y - \eta)^2 + z^2]^{1/2}}$$

$$+ \frac{z}{4\pi} \int\limits_{-\infty}^{\infty}\!\!\int \frac{f(\xi, \eta) \, d\xi \, d\eta}{[(x - \xi)^2 + (y - \eta)^2 + z^2]^{3/2}} \tag{190}$$

where f is the prescribed boundary condition (151b) and $\lambda(x, y) = (\partial u/\partial z)(x, y, 0^+)$ is unknown.

To determine $\lambda(x, y)$, we evaluate (190) on the $z = 0^+$ plane. The left-hand side must equal $f(x, y)$ if u is to solve (151), and the second term on the right-hand side tends to $f(x, y)/2$ according to (103a) [see (96) with $p = -f$]. Therefore, $\lambda(x, y)$ satisfies the integral equation

$$f(x, y) = -\frac{1}{2\pi} \int\limits_{-\infty}^{\infty}\!\!\int \frac{\lambda(\xi, \eta) \, d\xi \, d\eta}{[(x - \xi)^2 + (y - \eta^2)]^{1/2}} \tag{191}$$

Recognizing that the second term on the right-hand side of (190) is just $u(x, y, z)/2$, we note that (190) gives the following representation of the solution of (151) in terms of a surface distribution of sources

$$u(x, y, z) = -\frac{1}{2\pi} \int\limits_{-\infty}^{\infty}\int \frac{\lambda(\xi, \eta)\, d\xi\, d\eta}{[(x - \xi)^2 + (y - \eta)^2 + z^2]^{1/2}} \tag{192}$$

Clearly, for this example, the dipole representation (187) is significantly simpler to implement than the source representation (192). This motivates the use of dipoles for the general Dirichlet problem (181). We could, in principle, choose an *arbitrary mix* of surface distributions of dipoles and sources to represent the solution. In fact, we saw in interpreting (163) that the solution of Dirichlet's problem for the sphere is most easily represented by a particular combination of dipoles and sources on the surface.

2.12.2 Neumann's Problem

To solve Neumann's problem—that is,

$$\Delta u = 0 \qquad \text{in } G \tag{193}$$

$$\frac{\partial u}{\partial n} = g \qquad \text{on } \Gamma; \qquad \iint\limits_{\Gamma} g\, dA = 0$$

—we assume an unknown source distribution of strength $v(Q)$ on Γ. Hence, let

$$u(P) = \iint\limits_{\Gamma} v(Q)\left(-\frac{1}{4\pi r_{PQ}}\right) dA_Q \tag{194}$$

Now, applying the boundary condition gives [see (112b)]

$$g(P) = -\frac{v(P)}{2} + \iint\limits_{\Gamma} v(Q)\frac{\partial}{\partial n_P}\left(-\frac{1}{4\pi r_{PQ}}\right) dA_Q \tag{195}$$

for the interior problem, where $\partial/\partial n_P$ is now with respect to the coordinates P. For the exterior problem, we have the same formulas except the first term on the right-hand side of (195) now has a positive sign.

For the simple example with axial symmetry, discussed earlier in Section 2.4.3, the axial distribution of sources suffices. In fact, it is intuitively obvious that for any *axially symmetric surface* distribution of sources, there corresponds an equivalent unique *axial* distribution such that the potential at every point in the domain G is identical in *both cases and satisfies* the same boundary condition (see Problem 20).

The techniques used in aerodynamic applications for the numerical approximation of the integral equations (186) and (195) are called *panel methods* because the surface Γ is subdivided into a number of planar panels, over which the surface integral is approximated by the area of the panel times the value of the integrand at its centroid.

The idea of using an unknown distribution of appropriate singularities can also be used for problems where u is prescribed on part of the boundary and $\partial u/\partial n$ on the remainder. Some simple examples are discussed in Problem 20.

Review Problems

1. This problem illustrates the fact that under an arbitrary coordinate transformation the Cartesian Laplacian form $\Delta \equiv \partial^2/\partial x^2 + \partial^2/\partial y^2$ is not necessarily preserved. However, if the mapping is defined by an analytic function of the complex variable $z = x + iy$, then a harmonic function ϕ of x, y maps to a function $\Phi(\xi, \eta)$ which satisfies $\Phi_{\xi\xi} + \Phi_{\eta\eta} = 0$.

   Consider Laplace's equation

   $$\phi_{xx} + \phi_{yy} = 0 \tag{196}$$

   in some domain D of the xy-plane. Let ξ and η be general curvilinear coordinates in D defined by

   $$\xi = u(x, y) \tag{197a}$$

   $$\eta = v(x, y) \tag{197b}$$

   for prescribed functions $u(x, y)$, $v(x, y)$. Assume that u, v, u_x, v_x, u_y, v_y, u_{xx}, u_{xy}, u_{yy}, v_{xx}, v_{xy}, v_{yy} are all continuous in D and that the mapping is one-to-one in D.
   a. Define $\Phi(\xi, \eta)$, the image of $\phi(x, y)$ under mapping (197), by (16). Show that

   $$\phi_{xx} + \phi_{yy} = (u_x^2 + u_y^2)\Phi_{\xi\xi} + 2(u_x v_x + u_y v_y)\Phi_{\xi\eta} + (v_x^2 + v_y^2)\Phi_{\eta\eta}$$
   $$+ (u_{xx} + u_{yy})\Phi_\xi + (v_{xx} + v_{yy})\Phi_\eta \tag{198}$$

   Thus, if (196) holds, $\Phi(\xi, \eta)$ does not necessarily satisfy the Cartesian form $\Phi_{\xi\xi} + \Phi_{\eta\eta} = 0$.
   b. Take the special case where (197) defines polar coordinates in the plane—that is,

   $$\xi = r \equiv (x^2 + y^2)^{1/2} \tag{199a}$$

   $$\eta = \theta \equiv \tan^{-1}\frac{y}{x} \tag{199b}$$

   Verify that the mapping is one to one except at $r = 0$. Show that (196) and (198) imply

   $$\phi_{xx} + \phi_{yy} = \Phi_{rr} + \frac{1}{r}\Phi_r + \frac{1}{r^2}\Phi_{\theta\theta} = 0 \tag{200}$$

   Thus, the Cartesian form of the Laplacian is not preserved.
   c. Now assume that u and v in (197) are the real and imaginary parts, respectively, of an analytic function, as in (15). Use the Cauchy-Riemann conditions

   $$u_x = v_y \tag{201a}$$

   $$v_x = -u_y \tag{201b}$$

which must hold in this case, to conclude that (196) and (198) imply

$$\Phi_{\xi\xi} + \Phi_{\eta\eta} = 0 \tag{202}$$

Specialize your results for the case where $f(z)$ in (15) is

$$f(z) = z^2 \tag{203}$$

Verify that the Cauchy-Riemann conditions (201) do not hold for the pair of functions $u(x, y)$, $v(x, y)$ defined by (199).

2. In this problem we review some properties of linear fractional transformations. A linear fractional transformation is defined by

$$w = \frac{az + b}{cz + d} \tag{204}$$

for complex constants a, b, c, d. If $c = 0$, (204) reduces to a linear transformation. Factoring a and c from the numerator and denominator, respectively, shows that the right-hand side of (204) reduces to the constant a/c if $ad - bc = 0$. We exclude the cases $c = 0$ and $ad = bc$.

a. Show that (204) is the composition of the following five sequential mappings

$$w^{(1)} = z + \frac{d}{c} \qquad \text{translation} \tag{205a}$$

$$w^{(2)} = cw^{(1)} \qquad \text{rotation and dilation} \tag{205b}$$

$$w^{(3)} = \frac{1}{w^{(2)}} \qquad \text{inversion} \tag{205c}$$

$$w^{(4)} = \frac{(bc - ad)}{c} w^{(3)} \qquad \text{rotation and dilation} \tag{205d}$$

$$w = \frac{a}{c} + w^{(4)} \qquad \text{translation.} \tag{205e}$$

b. Show that each of these mapings takes circles and straight lines into circles and straight lines. Therefore, (204) has this property.

c. Show that if z, z_1, z_2, z_3 are four points in the finite z-plane with images under (204) given by w, w_1, w_2, w_3, respectively, then the following identity holds:

$$\frac{(w_1 - w_2)(w_3 - w)}{(w_1 - w)(w_3 - w_2)} = \frac{(z_1 - z_2)(z_3 - z)}{(z_1 - z)(z_3 - z_2)} \tag{206}$$

Also, (206) is a linear fractional transformation. If any point w_i or z_i is the point at infinity, we suppress the factor containing that point from both numerator and denominator; for example, if $w_3 = \infty$, the left-hand side of (206) becomes $(w_1 - w_2)/(w_1 - w)$. Argue why the linear fractional transformation that takes the circle or straight line passing through three given points in the z-plane to the circle or straight line passing through three given points in the w-plane is unique. Thus, (206) may be regarded as a generating formula for linear fractional transformations.

d. Show that the linear fractional transformation taking
   i. $z_1 = -1, z_2 = 0, z_3 = 1$ to $w_1 = 0, w_2 = i, w_3 = 3i$ is

$$w = -3i\frac{z+1}{z-3} \tag{207a}$$

   ii. $z_1 = -1, z_2 = 0, z_3 = 1$ to $w_1 = -i, w_2 = 1, w_3 = i$ is [see (28c)]

$$w = \frac{i-z}{i+z} \tag{207b}$$

   iii. $z_1 = 0, z_2 = 1, z_3 = \infty$ to $w_1 = -1, w_2 = -i, w_3 = 1$ is

$$w = \frac{(1-i)z - (1+i)}{(1-i)z + (1+i)} \tag{207c}$$

3. In this problem we review some ideas from vector calculus for curvilinear coordinates. Let the functions

$$x = f(\xi, \eta, \zeta) \tag{208a}$$

$$y = g(\xi, \eta, \zeta) \tag{208b}$$

$$z = h(\xi, \eta, \zeta) \tag{208c}$$

be prescribed in some domain in which the Jacobian determinant does not vanish—that is,

$$J \equiv \det\begin{pmatrix} f_\xi & f_\eta & f_\zeta \\ g_\xi & g_\eta & g_\zeta \\ h_\xi & h_\eta & h_\zeta \end{pmatrix} \neq 0 \tag{209}$$

We may regard ξ, η, ζ as curvilinear coordinates in the Cartesian xyz-space. Holding any of the coordinates ξ, η or ζ constant in (208) defines a surface in xyz-space. Holding any pair of curvilinear coordinates constant in (208) defines a curve in xyz-space.

a. Describe geometrically the following curvilinear coordinate systems:

$$x = \xi \cos \eta; \qquad y = \xi \sin \eta; \qquad z = \zeta \tag{210a}$$

$$x = \xi \sin \eta \cos \zeta; \qquad y = \xi \sin \eta \sin \zeta; \qquad z = \xi \cos \eta \tag{210b}$$

$$x = a \cosh \xi \cos \eta; \qquad y = a \sinh \xi \sin \eta \sin \zeta;$$

$$z = a \sinh \xi \sin \eta \cos \zeta; \qquad a = \text{const} > 0 \tag{210c}$$

$$x = \eta - \xi^3; \qquad y = \xi + \eta; \qquad z = \zeta \tag{210d}$$

b. Show that the infinitesimal displacement vector from the point $P \equiv (\xi_0, \eta_0, \zeta_0)$ to the neighboring point $Q \equiv (\xi_0 + \Delta\xi, \eta_0, \zeta_0)$ along the direction of increasing ξ is given by

$$\mathbf{PQ} \equiv \{f_{\xi}(\xi_0, \eta_0, \zeta_0)\mathbf{i} + g_{\xi}(\xi_0, \eta_0, \zeta_0)\mathbf{j} + h_{\xi}(\xi_0, \eta_0, \zeta_0)\mathbf{k}\}\, \Delta\xi + O(\Delta\xi^2)$$
$$(211)$$

where $\mathbf{i}, \mathbf{j}, \mathbf{k}$ are Cartesian unit vectors in the x, y, and z directions respectively.

Therefore,

$$\mathbf{b}_1(\xi_0, \eta_0, \zeta_0) \equiv \lim_{\Delta\xi \to 0} \frac{\mathbf{PQ}}{\Delta\xi} = f_{\xi}(\xi_0, \eta_0, \zeta_0)\mathbf{i} + g_{\xi}(\xi_0, \eta_0, \zeta_0)\mathbf{j}$$
$$+ h_{\xi}(\xi_0, \eta_0, \zeta_0)\mathbf{k} \qquad (212a)$$

is a tangent vector to the curve $\xi = \xi_0 = $ constant, $\eta = \eta_0 = $ constant at P; that is, $\mathbf{b}_1$ is a tangent vector in the direction of increasing ξ. Similarly

$$\mathbf{b}_2(\xi_0, \eta_0, \zeta_0) = f_{\eta}(\xi_0, \eta_0, \zeta_0)\mathbf{i} + g_{\eta}(\xi_0, \eta_0, \zeta_0)\mathbf{j} + h_{\eta}(\xi_0, \eta_0, \zeta_0)\mathbf{k} \quad (212b)$$

and

$$\mathbf{b}_3(\xi_0, \eta_0, \zeta_0) = f_{\zeta}(\xi_0, \eta_0, \zeta_0)\mathbf{i} + g_{\zeta}(\xi_0, \eta_0, \zeta_0)\mathbf{j} + h_{\zeta}(\xi_0, \eta_0, \zeta_0)\mathbf{k} \quad (212c)$$

are tangent vectors in the η and ζ directions, respectively.

Denote the scalar product

$$\mathbf{b}_i \cdot \mathbf{b}_j \equiv g_{ij}; \qquad i, j = 1, 2, 3 \tag{213}$$

Thus, $g_{ij} = g_{ji}$ in general.

Show that the tangent vectors associated with a curvilinear coordinate system (208) satisfying (209) are linearly independent and may therefore be regarded as a local basis at each P. A curvilinear system is said to be orthogonal if $g_{ij} \equiv 0$ for $i \neq j$. One can then define an orthonormal basis $\mathbf{e}_1, \mathbf{e}_2, \mathbf{e}_3$, characterized by

$$\mathbf{e}_i \cdot \mathbf{e}_j = \delta_{ij}; \qquad \delta_{ij} = \text{Kronecker delta} \tag{214a}$$

by setting

$$\mathbf{e}_i = \frac{\mathbf{b}_i}{g_{ii}^{1/2}}; \qquad i = 1, 2, 3 \tag{214b}$$

Calculate the $\{\mathbf{b}_i\}$ basis for each of the curvilinear coordinate systems (210), evaluate the g_{ij} in each case, and indicate which of these coordinate systems is orthogonal. Calculate the orthonormal basis associated with each orthogonal basis.

c. The infinitesimal displacement vector $d\mathbf{r}$ at a point P in an arbitrary direction has Cartesian components dx, dy, dz; that is,

$$d\mathbf{r} = dx\mathbf{i} + dy\mathbf{j} + dz\mathbf{k} \tag{215a}$$

Show that the components of $d\mathbf{r}$ with respect to the basis $\mathbf{b}_1, \mathbf{b}_2, \mathbf{b}_3$ defined in (212) are $d\xi$, $d\eta$, $d\zeta$; that is,

$$d\mathbf{r} = d\xi\mathbf{b}_1 + d\eta\mathbf{b}_2 + d\zeta\mathbf{b}_3 \tag{215b}$$

In view of the formal similarity between (215a) and (215b), the vectors $\mathbf{b}_1$,

$\mathbf{b}_2$, $\mathbf{b}_3$ are sometimes denoted as the *natural basis* for the curvilinear coordinate system (208). Express $d\mathbf{r}$ in terms of the $\mathbf{e}_1$, $\mathbf{e}_2$, $\mathbf{e}_3$ basis for each of the orthogonal systems in (210).

d. Consider the surface

$$x = f(\xi_1, \xi_2) \tag{216a}$$

$$y = g(\xi_1, \xi_2) \tag{216b}$$

$$z = h(\xi_1, \xi_2), \tag{216c}$$

defined in parametric form (with the parameters ξ_1, ξ_2) in xyz-space. Such a surface could be defined, for example, by setting one of the curvilinear coordinates ξ, η, or ζ equal to zero in (208), in which case ξ_1 and ξ_2 describe the remaining two coordinates that vary on the surface. We define the tangent vectors $\mathbf{b}_1$ and $\mathbf{b}_2$ on this surface in the usual way [see (212)]. Show that the infinitesimal area element dA on this surface is given by

$$dA \equiv |\mathbf{b}_1 \times \mathbf{b}_2| \, d\xi_1 \, d\xi_2 \tag{217}$$

and verify that this result gives

$$dA = \xi^2 \sin \eta \, d\eta \, d\zeta$$

for the surface defined by holding $\xi = $ constant in the coordinate system (210b).

Therefore, given a function $F(x, y, z)$, the surface integral of this function on the portion S of the surface (216) is

$$I \equiv \iint_S F(f(\xi_1, \xi_2), g(\xi_1, \xi_2), h(\xi_1, \xi_2)) \, dA \tag{218}$$

where dA is given in (217). Use this result and (210b) to calculate I on the surface of the unit sphere for the case where $F = xy$.

e. Show that the infinitesimal volume element dV for the curvilinear system (208) is given by

$$dV \equiv |\mathbf{b}_1 \cdot \mathbf{b}_2 \times \mathbf{b}_3| \, d\xi \, d\eta \, d\zeta \tag{219}$$

Therefore, the volume integral of some given function $F(x, y, z)$ over some domain G is given by

$$K \equiv \iiint_G F(f(\xi, \eta, \zeta), g(\xi, \eta, \zeta), h(\xi, \eta, \zeta)) \, dV \tag{220}$$

where dV is defined in (219). Specialize (220) to calculate the integral of $F = xy$ over the interior of the unit sphere.

f. Let F be a scalar function of position. The gradient of F (denoted by grad F) is a vector that is defined independently of the choice of coordinates by

$$dF = \text{grad } F \cdot d\mathbf{r} \tag{221}$$

Here dF is the differential of F, $d\mathbf{r}$ is the infinitesimal displacement vector (215), and a dot denotes the scalar product.

Show that using Cartesian coordinates, setting $F = \phi(x, y, z)$, and expressing grad F in terms of its components with respect to the $\mathbf{i}, \mathbf{j}, \mathbf{k}$ basis in (221) gives

$$\text{grad } F = \phi_x \mathbf{i} + \phi_y \mathbf{j} + \phi_z \mathbf{k} \tag{222}$$

Now express F in terms of the curvilinear coordinates (208)

$$F = \psi(\xi, \eta, \zeta) \tag{223}$$

and denote

$$\text{grad } F = \Gamma_1 \mathbf{b}_1 + \Gamma_2 \mathbf{b}_2 + \Gamma_3 \mathbf{b}_3 \tag{224}$$

Substituting (223), (224) and (215b) into (221) shows that we must have

$$\begin{aligned}
\psi_\xi \, d\xi + \psi_\eta \, d\eta + \psi_\zeta \, d\zeta &= (\Gamma_1 \mathbf{b}_1 + \Gamma_2 \mathbf{b}_2 + \Gamma_3 \mathbf{b}_3) \cdot (d\xi \mathbf{b}_1 + d\eta \mathbf{b}_2 + d\zeta \mathbf{b}_3) \\
&= (\Gamma_1 g_{11} + \Gamma_2 g_{12} + \Gamma_3 g_{13}) \, d\xi \\
&\quad + (\Gamma_1 g_{12} + \Gamma_2 g_{22} + \Gamma_3 g_{33}) \, d\eta \\
&\quad + (\Gamma_1 g_{13} + \Gamma_2 g_{23} + \Gamma_3 g_{33}) \, d\zeta
\end{aligned} \tag{225}$$

Identifying the multipliers of $d\xi$, $d\eta$, and $d\zeta$ on both sides of (225) gives three linear algebraic equations for the three unknowns Γ_1, Γ_2, Γ_3. Solve these to obtain

$$\begin{aligned}
\Gamma_1 = \frac{1}{M} \big[(g_{22} g_{33} - g_{23}^2) \psi_\xi &+ (g_{13} g_{23} - g_{12} g_{33}) \psi_\eta \\
&+ (g_{12} g_{23} - g_{13} g_{22}) \psi_\zeta \big]
\end{aligned} \tag{226a}$$

where M is the determinant of the matrix $\{g_{ij}\}$; that is,

$$M \equiv g_{11} g_{22} g_{33} - (g_{11} g_{23}^2 + g_{22} g_{13}^2 + g_{33} g_{12}^2) + 2 g_{12} g_{13} g_{23}$$

Permute indices $1 \to 2$, $2 \to 3$, $3 \to 1$ and variables $\xi \to \eta$, $\eta \to \zeta$, $\zeta \to \xi$ to obtain

$$\begin{aligned}
\Gamma_2 = \frac{1}{M} \big[(g_{23} g_{13} - g_{12} g_{33}) \psi_\xi &+ (g_{11} g_{33} - g_{13}^2) \psi_\eta \\
&+ (g_{13} g_{12} - g_{11} g_{23}) \psi_\zeta \big]
\end{aligned} \tag{226b}$$

A second permutation gives

$$\begin{aligned}
\Gamma_3 = \frac{1}{M} \big[(g_{12} g_{23} - g_{22} g_{13}) \psi_\xi &+ (g_{13} g_{12} - g_{23} g_{11}) \psi_\eta \\
&+ (g_{22} g_{11} - g_{12}^2) \psi_\zeta \big]
\end{aligned} \tag{226c}$$

Verify that for an orthogonal curvilinear system (226) simplifies to give

$$\text{grad } F = \frac{1}{g_{11}} \psi_\xi \mathbf{b}_1 + \frac{1}{g_{22}} \psi_\eta \mathbf{b}_2 + \frac{1}{g_{33}} \psi_\zeta \mathbf{b}_3 \tag{227a}$$

or

$$\text{grad } F = \frac{1}{\sqrt{g_{11}}}\psi_\xi\mathbf{e}_1 + \frac{1}{\sqrt{g_{22}}}\psi_\eta\mathbf{e}_2 + \frac{1}{\sqrt{g_{33}}}\psi_\zeta\mathbf{e}_3 \tag{227b}$$

in terms of the $\mathbf{e}_1, \mathbf{e}_2, \mathbf{e}_3$, basis.

Calculate the gradient of $x^2 + 2xy^2 + z^3$ in the curvilinear coordinates (210b) and (210d).

g. Suppose the vector $\mathbf{a} = a_1\mathbf{i} + a_2\mathbf{j} + a_3\mathbf{k}$ is an arbitrarily specified unit vector at the point $P = (x, y, z)$. Let $F = \phi(x, y, z)$ be a given scalar function. Consider a point Q infinitesimally close to P *along* the vector $\mathbf{a}$ from P; that is, $\mathbf{PQ} = \mathbf{a}\,\Delta s$, where Δs is small. Now, the derivative of ϕ in the $\mathbf{a}$ direction is by definition

$$\lim_{\Delta s\to 0}\frac{1}{\Delta s}[\phi(x + a_1\,\Delta s, y + a_2\,\Delta s, z + a_3\,\Delta s) - \phi(x, y, z)]$$

$$= \phi_x a_1 + \phi_y a_2 + \phi_z a_3 \equiv \frac{d\phi}{da} \tag{228a}$$

We call $d\phi/da$ the directional derivative of ϕ in the $\mathbf{a}$ direction. It follows from (228a) and (222) that $d\phi/da$ is

$$\frac{d\phi}{da} \equiv \text{grad }\phi\cdot\mathbf{a}, \tag{228b}$$

independently of the choice of coordinates.

Use (228) to argue that $\text{grad }\phi$ is a vector normal to the surface $\phi = $ constant at each point on this surface.

h. The divergence of a vector field $\mathbf{W}$ is a scalar field denoted by div $\mathbf{W}$ and defined for each point P in the following form, which is independent of the choice of coordinates:

$$\text{div }\mathbf{W} \equiv \lim_{\tau\to 0}\frac{1}{\tau}\iint_S \mathbf{W}\cdot\mathbf{n}\,dA \tag{229}$$

Here τ is the volume of an arbitrary domain containing P, S is the surface of τ, $\mathbf{n}$ is the unit outward normal to S, and dA is the infinitesimal area element on S.

Let $\mathbf{W}$ have Cartesian components W_1, W_2, W_3—that is,

$$\mathbf{W} = W_1(x, y, z)\mathbf{i} + W_2(x, y, z)\mathbf{j} + W_3(x, y, z)\mathbf{k}. \tag{230}$$

Choose a prism with sides $\Delta x, \Delta y, \Delta z$ surrounding P to show that

$$\text{div }\mathbf{W} = \frac{\partial W_1}{\partial x} + \frac{\partial W_2}{\partial y} + \frac{\partial W_3}{\partial z} \tag{231}$$

The definition (229) may be used to derive the expression for div $\mathbf{W}$ when $\mathbf{W}$ is expressed in terms of its components with respect to an arbitrary curvilinear coordinate system. Consider here the special case where the curvilinear system (208) is orthogonal and denote

$$\mathbf{W} = W_1^*(\xi, \eta, \zeta)\mathbf{e}_1 + W_2^*(\xi, \eta, \zeta)\mathbf{e}_2 + W_3^*(\xi, \eta, \zeta)\mathbf{e}_3 \tag{232}$$

where the $\mathbf{e}_1, \mathbf{e}_2, \mathbf{e}_3$ are the orthonormal unit vectors in the ξ, η, ζ directions defined by (214b). Denote

$$h_i \equiv (g_{ii})^{1/2}; \qquad i = 1, 2, 3$$

and show that

$$\operatorname{div} \mathbf{W} = \frac{1}{h_1 h_2 h_3}\left[\frac{\partial}{\partial \xi}(h_2 h_3 W_1^*) + \frac{\partial}{\partial \eta}(h_1 h_3 W_2^*) + \frac{\partial}{\partial \zeta}(h_1 h_2 W_3^*)\right] \tag{233}$$

Specialize (233) for the cases (210b) and (210c).

i. The Laplacian of a scalar field F is denoted by ΔF and is defined by

$$\Delta F \equiv \operatorname{div}(\operatorname{grad} F) \tag{234}$$

Again assume that (208) is orthogonal and show that [see (227)]

$$\Delta F = \frac{1}{h_1 h_2 h_3}\left[\frac{\partial}{\partial \xi}\left(\frac{h_2 h_3}{h_1}\frac{\partial \psi}{\partial \xi}\right) + \frac{\partial}{\partial \eta}\left(\frac{h_1 h_3}{h_2}\frac{\partial \psi}{\partial \eta}\right) + \frac{\partial}{\partial \zeta}\left(\frac{h_1 h_2}{h_3}\frac{\partial \psi}{\partial \zeta}\right)\right] \tag{235}$$

Specialize this result to the cases (210b) and (210c).

Problems

1. We wish to solve

$$u_{xx} + u_{yy} = 0 \tag{236}$$

in the domain $y \geq 0$, subject to

$$u(x, 0) = f(x); \qquad f \to 0 \quad \text{as} \quad |x| \to \infty \tag{237}$$

$$u(x, \infty) = 0 \tag{238}$$

a. Use the mapping (28c) and Poisson's formula (27) to show that $u(x, y)$ is given in the form

$$u(x, y) = \frac{1}{\pi}\int_{-\infty}^{\infty} \frac{y f(s)\, ds}{(s - x)^2 + y^2} \tag{239}$$

This result may also be observed directly (see Problem 13).

b. Use (239) and the mapping defined by combining (28a) with (28b) to solve (236) in the corner domain of Figure 2.4 with $a = e^{-y}$ on $y \geq 0$ and $b = (\sin y)/y$ on $y \leq 0$.

2. The fundamental solution of the n-dimensional Laplace equation solves

$$\frac{d^2 u}{dr^2} + \frac{n-1}{r}\frac{du}{dr} = \delta_n(r) \tag{240}$$

where δ_n is the n-dimensional delta function (see Problem 14 of Chapter 1).

a. Show that if $u(\infty) = 0$, the homogeneous equation gives

$$u = \frac{C_n}{r^{n-2}}; \qquad n \neq 2 \tag{241}$$

where C_n is a constant. Regarding $\Delta \equiv \text{div grad}$ and using the n-dimensional Gauss theorem to evaluate the left-hand side of (240) gives

$$C_n \int \cdots \int_\varepsilon \frac{d}{dr}(r^{2-n}) \, dA = 1 \tag{242a}$$

where the integral is evaluated on the surface of the n-dimensional sphere of radius $r = \varepsilon$. Therefore,

$$C_n = \frac{1}{(2-n)\omega_n}; \qquad n \geq 3 \tag{242b}$$

where ω_n is the surface area of the n-dimensional unit sphere.

b. To evaluate ω_n, consider the following identity for the n-tuple integral of a function $f(r)$:

$$\int \cdots \int_{R_n} f(r) \, dx_1 \, dx_2 \cdots dx_n = \int_0^\infty f(r) r^{n-1} \omega_n \, dr \tag{243}$$

where R_n is the entire n-dimensional space and

$$r \equiv \left(\sum_{i=1}^n x_i^2 \right)^{1/2} \tag{244}$$

Of course, it is assumed that $f(r)$ decays appropriately fast as $r \to \infty$ so that the integrals in (243) exist. For example, if we take

$$f(r) = e^{-r^2} = e^{-(x_1^2 + x_2^2 + \cdots x_n^2)} \tag{245}$$

we can evaluate both sides of (243). Recalling the definition of the Gamma function,

$$\Gamma(x) \equiv \int_0^\infty e^{-t} t^{x-1} \, dt; \qquad x > 0 \tag{246}$$

show that

$$\omega_n = 2\pi^{n/2}/\Gamma(n/2) \tag{247}$$

3. Calculate the constant B_n in (60a) including the special case $n = 2$, and verify that as $\lambda \to 0$, your results agree with those derived for Laplace's equation. Repeat the calculations for the fundamental solution of (56) for the case $\lambda < 0$.

4. a. Consider the spherically symmetric density distribution

$$\rho = \begin{cases} f(r), & 0 \leq r \leq R \\ 0, & r > R \end{cases} \tag{248}$$

Calculate the force of gravity on a point P of mass m when P is either outside or inside the sphere $0 \leq r \leq R$. Show that this force is exactly the same as if the entire mass of the portion of the sphere below P were concentrated at the origin. Specialize your result to the case of a hollow sphere where $f(r) = 0$ for all $r: 0 \leq r \leq a < R$.

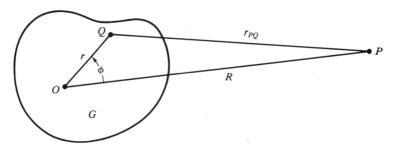

Figure 2.18

b. If the density distribution is axially symmetric—that is,

$$\rho = \begin{cases} f(r,\phi) = \text{prescribed}, & 0 \le r \le R; 0 \le \phi \le \pi \\ 0, & r > R \end{cases} \tag{249}$$

derive a series expansion for the potential in powers of r^{-1} for the case $(r/R) > 1$. Use the identity corresponding to (172) to expand $1/r_{PQ}$.

5. Consider a body occupying the finite domain G and having a prescribed density distribution. If $P(x,y,z)$ is a point at a distance R from O, the center of mass of G, and R is large compared to the dimensions of G, show that the gravitational potential at P is given approximately by

$$V = -\frac{\gamma M}{R} - \frac{\gamma}{2R^3}(A + B + C - 3I), \tag{250}$$

where M is the mass of G, A, B, and C are the principal moments of inertia, and I is the moment of inertia about the axis OP.

Hint: Introduce spherical polar coordinates about the OP axis: $\xi = r\sin\phi\cos\theta$; $\eta = r\sin\phi\sin\theta$; $\zeta = r\cos\phi$; then expand (79) for small values of r/R using $r_{PQ} \equiv \sqrt{R^2 + r^2 - 2rR\cos\phi}$ for the distance between P and the variable point of integration Q, as shown in Figure 2.18.

6. As an application of result (94), consider the case of an ellipsoid of revolution with $F(x) = 2b[x(1-x)]^{1/2}$; $0 < b \le 1$. Introduce three point sources at $x = \frac{1}{4}, \frac{1}{2}, \frac{3}{4}$ with strengths S_1, S_2, S_3, respectively. Solve (94) for S_1, S_2, S_3 and calculate $u(x,r)$ by discretizing the integral in (88). How well does the result satisfy the boundary condition (89) over the entire surface $r = F(x)$?

7. a. Let x, r, θ be the cylindrical polar coordinates of the point $P = (x, y, z)$ and let ξ, ρ, ϕ be the corresponding cylindrical polar coordinates of the point $Q = (\xi, \eta, \zeta)$. See Figure 2.19.

$$\begin{aligned} y &= r\cos\theta \\ z &= r\sin\theta \end{aligned} \tag{251}$$

$$\begin{aligned} \eta &= \rho\cos\phi \\ \zeta &= \rho\sin\phi \end{aligned} \tag{252}$$

Let D be the domain *outside* the cylindrically symmetric surface Γ of unit length defined by $\rho = F(\xi)$, with $F(0) = F(1) = 0$, for a given function F.

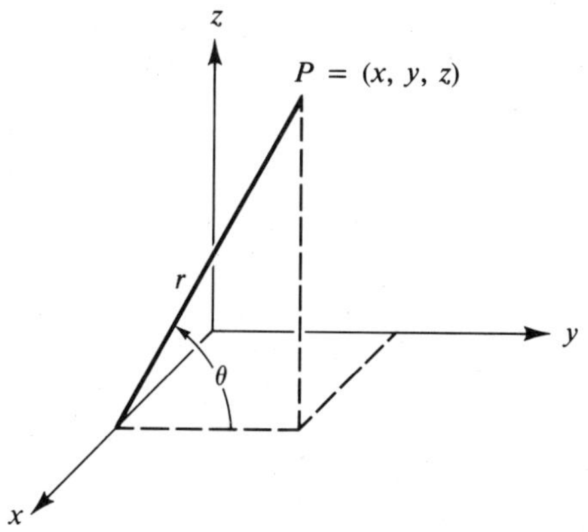

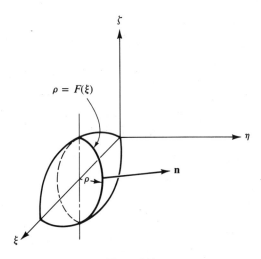

Figure 2.19

Figure 2.20

Let **n** be the unit normal in the direction indicated (Figure 2.20). What is the potential at P due to a dipole distribution of strength per unit area equal to $\mu(\xi, \phi)$ on Γ? The dipole axes are oriented along **n**. Derive an explicit formula for the kernel L in the expression

$$u(x, r, \theta) = \int_{\xi=0}^{1} \int_{\phi=0}^{2\pi} \mu(\xi, \phi) L(x, r, \theta, \xi, \phi) \, d\xi \, d\phi \qquad (253)$$

7. b. Consider a uniform distribution of dipoles on the surface of the unit hemisphere $z = (1 - x^2 - y^2)^{1/2}$. Assume that the dipole strength per unit area equals unity and that the axes are oriented along the outward normal to the surface. Show that the potential $w(0, 0, z)$ along the z-axis, as given by (84), reduces to

$$w(0,0,z) = -\frac{1}{2}\int_0^{\pi/2} \frac{(z\cos\phi - 1)\sin\phi}{[1 + z^2 - 2z\cos\phi]^{3/2}} d\phi$$

$$= \frac{1}{2}\left[\frac{1-z}{|1-z|} + \frac{z}{(1+z^2)^{1/2}}\right]$$

Verify that this result also follows from (133).

The preceding implies that as $z \to 1^{\pm}$, we have $w(0,0,1^+) = -\frac{1}{2}(1 - 2^{-1/2})$ and $w(0,0,1^-) = \frac{1}{2}(1 + 2^{-1/2})$. Denote

$$w_\varepsilon = -\frac{1}{2}\int_0^\varepsilon \frac{(z\cos\phi - 1)\sin\phi}{[1 + z^2 - 2z\cos\phi]^{3/2}} d\phi$$

$$w_a = -\frac{1}{2}\int_\varepsilon^{\pi/2} \frac{(z\cos\phi - 1)\sin\phi}{[1 + z^2 - 2z\cos\phi]^{3/2}} d\phi$$

and show that as $\varepsilon \to 0^+$ and $z - 1 \equiv \alpha \to 0^{\pm}$, we have the following behavior

$$w_\varepsilon = \mp\frac{1}{2} + \frac{1}{2}\frac{\varepsilon/2 + (\alpha/\varepsilon)}{\sqrt{1 + (\alpha/\varepsilon)^2}} + \cdots$$

$$w_a = \frac{1}{2^{3/2}} - \frac{1}{2}\frac{\varepsilon/2 + (\alpha/\varepsilon)}{\sqrt{1 + (\alpha/\varepsilon)^2}} + \cdots$$

Thus, if $(\alpha/\varepsilon) \to 0$, $w_\varepsilon \to \mp\frac{1}{2}$, and this is the same limiting value that we calculated for a planar distribution. But now we have $w_a = 2^{-3/2} \neq 0$ unlike the planar case. In fact, the only choice for which $w_a = 0$ is to require $(\alpha/\varepsilon) \to 1$ as $\varepsilon \to 0$.

8. We wish to solve the two-dimensional Laplace equation in the rectangular domain $0 \leq x \leq \ell; 0 \leq y \leq h$, subject to prescribed values of u on the perimeter of the rectangle. Use a finite-difference scheme based on the mean-value theorem and let $h = M\xi$ and $\ell = N\xi$, where ξ is a small constant and M and N are positive integers. Subdivide the rectangle into $M \times N$ squares and denote

$$u(i\xi, j\xi) \equiv u_{ij} \tag{254}$$

Derive a linear system of $(M - 1) \times (N - 1)$ algebraic equations for the interior values of u_{ij} in terms of the known boundary values.

9. This problem concerns solutions of the Helmholtz equation

$$\Delta u + \lambda u = 0 \tag{255}$$

a. Derive the fundamental solution of (255) with $\lambda > 0$ directly [that is, without using the identity for $J_{-1/2}$ in (60b)] in the form

$$F(P,Q) = -\frac{1}{4\pi}\frac{\cos\lambda^{1/2}r_{PQ}}{r_{PQ}} \tag{256}$$

where $r_{PQ}^2 \equiv (x - \xi)^2 + (y - \eta)^2 + (z - \zeta)^2$.

b. Use the fundamental solution to show that every solution of (255) with $\lambda \geq 0$ satisfies the mean-value theorem generalizing (127):

$$\frac{\sin R\sqrt{\lambda}}{R\sqrt{\lambda}} u(P) = \frac{1}{4\pi R^2} \iint_{\Gamma_1} u\, dA \tag{257}$$

where Γ_1 is the surface of a sphere of radius R.

c. Use the method of descent to show that the fundamental solution of the two-dimensional Helmholtz equation for a source located at the origin is given by

$$u(r) = -\frac{1}{2\pi} \int_0^\infty \frac{\cos[\lambda(r^2 + \zeta^2)]^{1/2}\, d\zeta}{[r^2 + \zeta^2]^{1/2}} \tag{258}$$

Use the integral representation for $Y_0(\lambda^{1/2} r)$:

$$Y_0(\lambda^{1/2} r) \equiv -\frac{2}{\pi} \int_0^\infty \cos(\lambda^{1/2} r \cosh s)\, ds; \qquad r > 0 \tag{259}$$

to show that (258) reduces to (60a) for $n = 2$.

d. Consider the boundary-value problem of Dirichlet type for (255) inside a domain G with boundary Γ; that is,

$$\Delta u + \lambda u = 0 \qquad \text{in } G \tag{260}$$

$$u = f \qquad \text{on } \Gamma \tag{261}$$

Assume that you have computed Green's function for G; that is, you have solved

$$\Delta_P K(P, Q) + \lambda K(P, Q) = \delta_3(P, Q) \qquad \text{in } G \tag{262}$$

$$K(P, Q) = 0 \qquad \text{if } P \text{ is on } \Gamma \tag{263}$$

for points Q in G and P in $G + \Gamma$.

Multiply (260) by $-K$ and (262) by u, add, and integrate the result over G; then use (121) to show that the solution of (260)–(261) is given by the same generalized Poisson formula (143) as for Laplace's equation.

e. Calculate Green's function for the upper half-space $z \geq 0$ and use this result to obtain an integral representation of the solution of (260)–(261) for this case.

10. Specialize (146) to the case where G is the upper half-space, and use (150b) to show that (146) reduces to (155).

11. Specialize (84) to spherical polar coordinates r, θ, ϕ (see Figure 2.15) and a distribution of dipoles of strength $p(\theta', \phi')$ oriented along the outward normal to the surface of a unit sphere to show that it reduces to

$$w(r, \theta, \phi) = \frac{1}{4\pi} \int_{\theta'=0}^\pi \int_{\phi'=0}^{2\pi} \frac{(1 - r\cos\gamma)p(\theta', \phi')\sin\theta'\, d\theta'\, d\phi'}{[1 + r^2 - 2r\cos\gamma]^{3/2}} \tag{264}$$

Specialize (81) to the same geometry and show that it reduces to

$$u(r, \theta, \phi) = -\frac{1}{4\pi} \int_{\theta'=0}^\pi \int_{\phi'=0}^{2\pi} \frac{q(\theta', \phi')\sin\theta'\, d\theta'\, d\phi'}{[1 + r^2 - 2r\cos\gamma]^{1/2}} \tag{265}$$

Use this to show that Poisson's formula (163) is the potential due to a surface distribution of dipoles of strength $p(\theta', \phi') = 2f(\theta', \phi')$ plus the potential due to surface distribution of sources of strength $q(\theta', \phi') = f(\theta', \phi')$.

12. a. Show that Neumann's function for the interior of the unit circle is

$$N(P, Q) = \frac{1}{2\pi} \log r_1 r_2 \rho \tag{266}$$

where, as usual, P has the polar coordinates r, θ and Q has the polar coordinates ρ, θ'. Also,

$$r_1^2 \equiv r^2 + \rho^2 - 2r\rho \cos(\theta - \theta') \tag{267}$$

$$r_2^2 \equiv r^2 + \frac{1}{\rho^2} - \frac{2r}{\rho} \cos(\theta - \theta') \tag{268}$$

Thus,

$$\Delta_P N(P, Q) = \delta_2(P, Q) \tag{269}$$

and

$$\frac{\partial N}{\partial r} = \frac{1}{2\pi} \qquad \text{for } P \text{ on } r = 1 \tag{270}$$

What is N for the exterior problem?

b. Show that Neumann's function for the interior of the unit sphere is

$$N(P,Q) = -\frac{1}{4\pi} \left[\frac{1}{r_{PQ}} + \frac{1}{\rho r_{P\bar{Q}}} + \log \frac{C}{1 - \rho r \cos \gamma + (r^2\rho^2 + 1 - 2r\rho \cos \gamma)^{1/2}} \right] \tag{271}$$

where the notation is defined in (157) and where C is an arbitrary constant that may be chosen to satisfy (140).

13. Consider the Dirichlet problem for the two-dimensional Laplacian in the upper half-plane—that is,

$$u_{xx} + u_{yy} = 0; \qquad -\infty < x < \infty; 0 \le y < \infty \tag{272a}$$

with boundary conditions

$$u(x, 0) = f(x); \qquad f \to 0 \quad \text{as} \quad |x| \to \infty \tag{272b}$$

$$u(x, \infty) = 0 \tag{272c}$$

Derive the solution in the form (see Problem 1):

$$u(x, y) = \frac{y}{\pi} \int_{-\infty}^{\infty} \frac{f(\xi) \, d\xi}{(x - \xi)^2 + y^2} \tag{273}$$

using

a. Green's function (147) in the two-dimensional version of (143).

b. Green's function (147), and superposition for Poisson's equation with homogeneous boundary condition $w(x, 0) = 0$ which results for $w(x, y) \equiv u(x, y) - f(x)$.

c. Fourier transforms with respect to x

14. a. Use Green's and Neumann's functions to solve the Dirichlet and Neumann problems, respectively, for the interior of the unit circle; that is,
show that when (159) is used in the two-dimensional version of (143), you
obtain Poisson's formula:

$$u(r, \theta) = \frac{1 - r^2}{2\pi} \int_0^{2\pi} \frac{u(1, \theta') \, d\theta'}{1 + r^2 - 2r \cos(\theta - \theta')} \tag{274}$$

which is the solution of

$$\Delta u \equiv u_{rr} + \frac{1}{r} u_r + \frac{1}{r^2} u_{\theta\theta} = 0, \qquad \text{in } r \le 1 \tag{275}$$

with the boundary condition

$$u(1, \theta) = f(\theta) = \text{prescribed} \tag{276}$$

b. Solve (275)–(276) using separation of variables to obtain the Fourier
series

$$u(r, \theta) = \frac{a_0}{2} + \sum_{n=1}^{\infty} r^n (a_n \cos n\theta + b_n \sin n\theta) \tag{277}$$

where

$$a_n = \frac{1}{\pi} \int_0^{2\pi} f(\theta) \cos n\theta \, d\theta \tag{278a}$$

$$b_n = \frac{1}{\pi} \int_0^{2\pi} f(\theta) \sin n\theta \, d\theta \tag{278b}$$

Show that this result follows from (274) when the integrand is expanded
in a power series in r and integrated term by term.

c. Calculate $u_r(r, \theta)$ using (274), and simplify the improper integral which
results as $r \to 1^-$ to derive a well behaved integral representation for
$u_r(1^-, \theta)$ in terms of $f(\theta)$. Verify that your result is correct for the special
cases $u(r, \theta) = r^n \cos n\theta$ and $r^n \sin n\theta$, where n is an integer.

d. Show that when $N(P, Q)$, as defined in (266), is used in (146), you obtain

$$u(r, \theta) = -\frac{1}{2\pi} \int_0^{2\pi} g(\theta') \log[1 + r^2 - 2r \cos(\theta - \theta')] \, d\theta' \tag{279}$$

Verify that (279) solves

$$\Delta u = 0, \qquad \text{in } r \le 1 \tag{280}$$

with boundary condition

$$\frac{\partial u}{\partial r}(1, \theta) = g(\theta) = \text{prescribed:} \int_0^{2\pi} g(\theta) \, d\theta = 0 \tag{281}$$

and the solution is made unique by requiring

$$\int_0^{2\pi} u(1, \theta') \, d\theta' = 0 \tag{282}$$

In verifying the boundary condition on $r = 1$, evaluate the improper integral which arises using the approach discussed in Section 2.4.4. In particular, show that the only contribution to u_r as $r \to 1$ occurs from the integral

$$I(r, \theta; \varepsilon) \equiv \int_{\theta-\varepsilon}^{\theta+\varepsilon} g(\theta') \frac{r - \cos(\theta - \theta')}{1 + r^2 - 2r\cos(\theta - \theta')} d\theta'$$

if $(1 - r)/\varepsilon \to 0$ as $\varepsilon \to 0$.

15. Carry out the details of the proof for Kelvin's inversion; that is, show that if

$$u = F(r, \theta, \phi) \tag{283}$$

satisfies

$$\Delta u \equiv \frac{1}{r^2} \left[\frac{\partial}{\partial r} \left(r^2 \frac{\partial u}{\partial r} \right) + \frac{1}{\sin\theta} \frac{\partial}{\partial\theta} \left(\sin\theta \frac{\partial u}{\partial\theta} \right) + \frac{1}{\sin^2\theta} \frac{\partial^2 u}{\partial\phi^2} \right] = 0 \tag{284}$$

then $v(r, \theta, \phi) \equiv (1/r) F(1/r, \theta, \phi)$ satisfies $\Delta v = 0$.

16. This problem concerns the connection between conformal mapping and Green's function of the first kind for various two-dimensional domains D—that is, solutions of

$$u_{xx} + u_{yy} = \delta(x - \xi)\delta(y - \eta) \qquad \text{in } D \tag{285}$$

$$u = 0 \qquad \text{on the boundary of } D \tag{286}$$

where ξ and η are the coordinates of a point in D.

a. Consider the corner domain D_1: $x \geq 0$, $y \geq 0$, and use image sources to show that Green's function is given by

$$K(x, y, \xi, \eta) = \frac{1}{2\pi} \log \frac{[(x-\xi)^2 + (y-\eta)^2]^{1/2} [(x-\xi)^2 + (y+\eta)^2]^{1/2}}{[(x-\xi)^2 + (y+\eta)^2]^{1/2} [(x-\xi)^2 + (y-\eta)^2]^{1/2}} \tag{287}$$

where $\xi > 0$, $\eta > 0$. Use the mapping

$$w = z^2 \tag{288}$$

of D_1 onto the upper half-plane $y \geq 0$, and Green's function (147) for $y \geq 0$ to derive (287).

b. Now consider the strip domain D_2: $-\infty < x < \infty$, $0 \leq y \leq \pi$. Use (147) and the mapping

$$w = \log z \tag{289}$$

of D_2 onto the upper half-plane to derive Green's function for D_2. Next calculate Green's function in D_2 using image sources and compare your two results to deduce the identity

$$\frac{\cosh x - \cos y}{\cosh x - \cos z} = \prod_{n=-\infty}^{\infty} \frac{[x^2 + (y - 2n\pi)^2]}{[x^2 + (z - 2n\pi)^2]} \tag{290}$$

for real variables x, y, z.

   c. Replace the right-hand side of (285) with a dipole of unit strength located at $x = \xi$, $y = \eta$ and oriented in the positive x-direction. What is the solution of (285), as just modified, subject to (286)?

   d. Use the results in part (b) and the mapping defined by

$$w = \log \frac{z^2 + 1}{2z} \tag{291}$$

   to solve (285)–(286) in the interior of the half disc D_3: $x^2 + y^2 \le 1$, $y \ge 0$.

17. Solve the two-dimensional Laplacian in $y \ge 0$,

$$u_{xx} + u_{yy} = 0; \qquad y \ge 0 \tag{292}$$

with boundary condition

$$u(x, 0) = \begin{cases} 1, & -a \le x \le b \\ 0, & \text{otherwise} \end{cases} \tag{293}$$

using Poisson's formula for the upper half-plane (Problem 13), and relate your result to Cauchy's integral formula.

18. Consider the three-dimensional Laplacian in $z \ge 0$ with the following boundary condition:

$$u_z(x, y, 0^+) = \begin{cases} 1, & \text{on } r \\ 0, & \text{on } r \end{cases}; r^2 \equiv x^2 + y^2 \tag{294}$$

Solve for $u(x, y, z)$ on $z > 0$, and derive an expansion for u valid near ∞.

19. Use the Poisson formula (163) for the interior of the unit sphere to calculate the potential on the z-axis when $u = \text{sign } z$ on the surface.

20. Use an axially symmetric *surface* distribution of sources to solve the problem of Section 2.4.3. What is the integral equation governing the strength of this distribution? Show that the flow field outside the body is identical to the one defined by (88) and (90).

20. a. Show that the mixed boundary-value problem for the cylindrically symmetric Laplacian

$$\Delta u = 0; \qquad \text{on } z \ge 0 \tag{295a}$$

$$u(r, 0^+) = 0; \qquad \text{on } r \ge 1; r^2 = x^2 + y^2 \tag{295b}$$

$$u_z(r, 0^+) = f(r); \qquad \text{on } r < 1 \tag{295c}$$

may be solved using a distribution of dipoles of strength $h(r)$ on the unit disc: $r \le 1$, $z = 0$. Derive the integral equation governing h.

   b. Solve the dual boundary-value problem with (295b) and (295c) replaced by

$$u_z(r, 0^+) = 0; \qquad r \ge 1 \tag{296a}$$

$$u(r, 0^+) = f(r); \qquad \text{on } r < 1 \tag{296b}$$

using a source distribution on the unit disc.

c. The two-dimensional problem corresponding to (295) is

$$u_{xx} + u_{yy} = 0; \qquad \text{on } y \geq 0 \tag{297a}$$

$$u(x, 0^+) = 0; \qquad \text{on } |x| \geq 1 \tag{297b}$$

$$u_y(x, 0^+) = f(x); \qquad \text{on } |x| < 1 \tag{297c}$$

Solve this using a two-dimensional dipole distribution on the interval $-1 \leq x \leq 1$, and derive the integral equation governing the dipole strength with particular attention to the evaluation of improper integrals.

References

1. M. D. Greenberg, *Foundations of Applied Mathematics*, Prentice Hall, Englewood Cliffs, N.J., 1978.
2. L. D. Landau, and E. M. Lifshitz, *The Classical Theory of Fields*, Addison-Wesley, Reading, Mass., 1962.
3. G. F. Carrier, M. Krook, and C. E. Pearson, *Functions of a Complex Variable, Theory and Technique*, McGraw-Hill, New York, 1966.
4. Z. Nehari, *Conformal Mapping*, McGraw-Hill, New York, 1952.
5. M. A. Heaslet and H. Lomax, "Supersonic and Transonic Small Perturbation Theory," Section D of *General Theory of High Speed Aerodynamics*, ed. by W. R. Sears, Princeton University Press, Princeton, N.J., 1954, pp. 122–344.
6. M. Abramowitz and I. A. Stegun, *Handbook of Mathematical Functions*, National Bureau of Standards, 1964.
7. E. T. Whittaker, and G. N. Watson, *Modern Analysis*, Cambridge University Press, New York, 1952.

C H A P T E R   3

The Wave Equation

In Chapters 1 and 2, the applications that we studied were directly modeled by a linear equation (the diffusion equation or Laplace's equation). Here, however, most of the meaningful physical applications are governed by essentially non-linear partial differential equations. We begin our discussion by considering in some detail the mathematical modeling for three such problems. We shall study the solution of these nonlinear equations in Chapters 7 and 8, and we devote the remainder of this chapter to solving the linear wave equations that result when we assume a small disturbance approximation.

3.1 The Vibrating String

Consider a string of infinitesimal thickness stretched initially under constant tension τ_0 (g cm/s^2) in the interval $0 \leq X \leq L_0$. Let the density of the string be $\rho \equiv \rho_0 r(X/L_0)$ (g/cm). Assume that the string deflects under the influence of a prescribed loading $P \equiv P_0 p(X/L_0, T/T_0)$ (g/s^2), which always remains vertical. Here, T_0 is a characteristic time scale, to be defined later, and capital letters indicate dimensional quantities. Thus, r and p are prescribed dimensionless functions that define the variation of the density and loading, respectively. Focus on a particular increment of length ΔX in the equilibrium state, as shown on the top of Figure 3.1, and examine the state of this increment when in motion. In general, the increment will have translated so that its left end has moved a horizontal distance U and a vertical distance V, as shown on the bottom of Figure 3.1. Also, the element will have rotated by an angle θ with respect to the horizontal. The deflected length of the element is denoted by ΔL, and the tension τ in the deflected state acts at the angle θ (that is, along the direction of the local tangent).

The equations of motion (force = mass $\times$ acceleration) for the horizontal and vertical deflections are therefore given by [neglecting terms of order $(\Delta X)^2$]:

$$\rho_0 r \frac{\partial^2 U}{\partial T^2} = \frac{\partial}{\partial X}(\tau \cos \theta) \tag{1a}$$

$$\rho_0 r \frac{\partial^2 V}{\partial T^2} = P_0 p\left(\frac{x}{L_0} + U, \frac{T}{T_0}\right)\left(1 + \frac{\partial U}{\partial X}\right) + \frac{\partial}{\partial X}(\tau \sin \theta) \tag{1b}$$

Now, the tension is a function of the strain σ, the change in length divided by the original length of a given element. Thus, we denote $\tau \equiv \tau_0[1 + f(\sigma)]$, where $f(\sigma_0) = 0$ (see Figure 3.2).

We have $\sigma - \sigma_0 = (\Delta L - \Delta X)/\Delta X$, and referring to Figure 3.1, we see that

$$\sigma - \sigma_0 = \left[\left(1 + \frac{\partial U}{\partial X}\right)^2 + \left(\frac{\partial V}{\partial X}\right)^2\right]^{1/2} - 1 \tag{2}$$

$$\sin \theta = \frac{(\partial V/\partial X)}{[(1 + \partial U/\partial X)^2 + (\partial V/\partial X)^2]^{1/2}} \tag{3}$$

$$\cos \theta = \frac{1 + (\partial U/\partial X)}{[(1 + \partial U/\partial X)^2 + (\partial V/\partial X)^2]^{1/2}} \tag{4}$$

Let us introduce dimensionless variables:

$$u(x, t) \equiv \frac{U}{L_0}; \qquad v(x, t) \equiv \frac{V}{L_0} \tag{5}$$

$$x \equiv \frac{X}{L_0}; \qquad t \equiv \frac{T}{T_0} \tag{6}$$

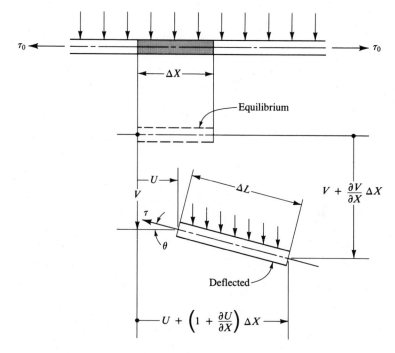

Figure 3.1

where $T_0 \equiv L_0\sqrt{\rho_0/\tau_0}$; that is, we have used the characteristic speed $\sqrt{\tau_0/\rho_0}$ associated with the equilibrium state. The dimensionless equations of motion become:

$$r(x)u_{tt} = \{[1 + f(\sigma)]\cos\theta\}_x \tag{7a}$$

$$r(x)v_{tt} = \varepsilon p(x + u, t)(1 + u_x) + \{[1 + f(\sigma)]\sin\theta\}_x \tag{7b}$$

where

$$\sigma = \sigma_0 + [(1 + u_x)^2 + v_x^2]^{1/2} - 1 \tag{8}$$

$$\sin\theta = \frac{v_x}{[(1 + u_x)^2 + v_x^2]^{1/2}} \tag{9a}$$

$$\cos\theta = \frac{1 + u_x}{[(1 + u_x)^2 + v_x^2]^{1/2}} \tag{9b}$$

and the dimensionless parameter $\varepsilon \equiv P_0 L_0/\tau_0$ gives a measure of the magnitude of the load in comparison with the equilibrium tension. Thus, for weak loading, $\varepsilon \ll 1$.

If the string is initially at rest and $\varepsilon = 0$, the solution is the trivial equilibrium state $u = v = 0$. Thus, if ε and the initial motion are small, we look for a perturbation expansion, valid for ε small, in the form

$$u(x, t; \varepsilon) = \varepsilon u_1(x, t) + \varepsilon^2 u_2(x, t) + O(\varepsilon^3) \tag{10a}$$

$$v(x, t; \varepsilon) = \varepsilon v_1(x, t) + \varepsilon^2 v_2(x, t) + O(\varepsilon^3) \tag{10b}$$

A more precise interpretation of these perturbation expansions is given in Chapter 8. Here, we proceed formally and substitute these series into (7) to find that u_1 and v_1 satisfy

$$r(x)u_{1_{tt}} - f'(\sigma_0)u_{1_{xx}} = 0 \tag{11a}$$

$$r(x)v_{1_{tt}} - v_{1_{xx}} = p(x, t) \tag{11b}$$

The equations for u_2 and v_2 may also be derived (see Problem 1) by keeping track of the $O(\varepsilon^2)$ terms. We see that the horizontal (u_1) and vertical (v_1) motions are decoupled to $O(\varepsilon)$ and obey one-dimensional wave equations.

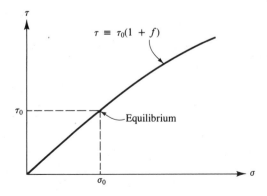

Figure 3.2

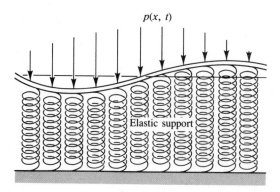

$p(x, t)$

Elastic support

Figure 3.3

For future reference, let us also consider the effect of introducing an elastic restoring force for the vertical component of the motion. For example, we can imagine the stretched string in equilibrium to be resting on a continuous elastic support, as shown in Figure 3.3. The support exerts a force per unit length given by $F \equiv F_0 \phi(v)$ (g/s^2), in the direction opposite v. This force depends only on v with $\phi(0) = 0$. In this case, we must add $(-F)$ to the right-hand side of (1b), and this results in the added term $-\lambda\phi(v)$ to the right-hand side of (7b), where λ is the dimensionless parameter $\lambda \equiv F_0 L_0/\tau_0$ analogous to ε. The perturbation term v_1 then obeys the linear wave-equation

$$r(x)v_{1_{tt}} - v_{1_{xx}} + \lambda\phi'(0)v_1 = p(x, t) \tag{12}$$

which we shall study in Section 3.7 for the case $r = 1$.

3.2   Shallow-Water Waves

The problem of wave propagation in shallow water is perhaps the simplest problem in fluid mechanics that exhibits all the features that we intend to discuss in this and subsequent chapters.

3.2.1 Assumptions

The basic equations follow from elementary physical concepts under the following assumptions.

The *shallow-water* or *long-wave* approximation is characterized by disturbances that have a long wavelength L_0 compared to the undisturbed depth H. In particular, this implies that vertical motions are ignored, and we have hydrostatic balance in the vertical, Y, direction. In addition, we neglect surface tension and viscosity and we take the density ρ to be a constant. We shall also assume that flow quantities do not vary in the lateral direction (one-dimensional flow) and we simplify the geometry by taking a flat horizontal bottom. The last two assumptions are not necessary for deriving a shallow-water theory. Thus, in terms of an average (over depth) velocity vector, we may study a two-dimensional problem and also account for a variable bottom (see Problem 2). The geometry for our simple one-dimensional problem is sketched in Figure 3.4.

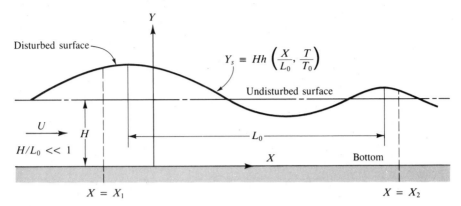

Figure 3.4

The characteristic length and time scales L_0 and T_0 will be defined for some specific examples later on.

3.2.2 Hydrostatic Balance

Hydrostatic balance means that any element of water with volume $\Delta X \Delta Y \cdot 1$ is in static equilibrium in the vertical direction under the influence of gravity and pressure forces (see Figure 3.5). Thus, the net upward force due to the pressure difference on the upper and lower surfaces must be balanced by the weight of the element; that is,

$$P(X, Y, T)\,\Delta X - P(X, Y + \Delta Y, T)\,\Delta X - \rho g\,\Delta X\,\Delta Y = 0 \qquad (13a)$$

where P is the pressure and g is the acceleration of gravity. In the limit as $\Delta Y \to 0$, we obtain

$$\frac{\partial P}{\partial Y} = -\rho g = \text{constant} \qquad (13b)$$

and integrating this expression from the free surface $Y_s = Hh$, where $P = 0$, to Y

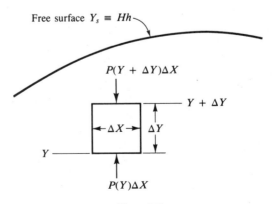

Figure 3.5

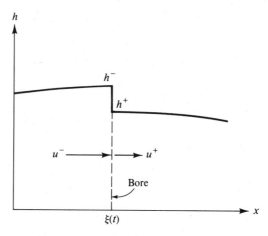

Figure 3.6

gives

$$\int_0^{P(X,Y,T)} dP' = -\rho g \int_{Hh}^{Y} dY'$$

or

$$P(X, Y, T) = \rho g(Hh - Y) \tag{13c}$$

3.2.3 Conservation of Mass

Let X_1 and X_2 be two fixed positions $(X_1 < X_2)$ through which the water flows. Denote the domain bounded by the bottom, the free surface, and the fixed vertical planes $X = X_1$ and $X = X_2$ by G. In order to satisfy mass conservation in G, we must require that the time rate of change of mass be equal to the net inflow of mass through the boundaries of G. Since mass flows only in (or out) from X_1 and X_2, we have:

$$\frac{d}{dT} \int_{X_1}^{X_2} \rho Hh \, dX = -\rho U Hh \Big|_{X=X_1}^{X=X_2} \tag{14}$$

where U is the vertically averaged component of velocity in the X direction.

It is convenient to introduce dimensionless variables:

$$x \equiv \frac{X}{L_0}; \quad y \equiv \frac{Y}{H}; \quad t \equiv \frac{T}{T_0}; \quad u \equiv \frac{U}{(gH)^{1/2}}; \quad p \equiv \frac{P}{\rho g H} \tag{15}$$

and to choose $T_0 \equiv L_0/(gH)^{1/2}$. Then (14) becomes

$$\frac{d}{dt} \int_{x_1}^{x_2} h(x, t) \, dx + u(x_2, t)h(x_2, t) - u(x_1, t)h(x_1, t) = 0 \tag{16}$$

Note that (16) is valid even if u and h are not continuous inside (x_1, x_2). For example, consider a typical discontinuity in u and h at some value of ξ: $x_1 < \xi < x_2$, as sketched in Figure 3.6. Such a discontinuity is called a *bore* and is physically quite relevant, as is discussed in Chapters 5, 7, and 8. The values of u and h on either side of the bore, indicated by $\pm$ superscripts, are not equal. Moreover, the bore will be moving—that is, $\xi(t)$. In Chapter 5, we shall use (16) to derive relations linking u^+, u^-, h^+, h^- with $d\xi/dt$.

3.2.4 Conservation of Momentum in the X Direction

According to Newton's law of momentum conservation, the time rate of change of momentum in G is balanced by the net inflow of momentum through the boundaries plus the forces exerted by the boundaries on the water contained in G.

Since the pressure is zero on the free surface, we have no force there. Also, the stress is normal to the bottom (because we have an inviscid fluid and there is no shear stress on the bottom). Hence the pressure on a flat horizontal bottom does not contribute a horizontal force; this is no longer true if the bottom varies (see Problem 2). In the present case, the only horizontal forces acting on the boundaries of G are the pressure forces on the vertical planes $X = X_1$ and $X = X_2$. The integral law of horizontal momentum conservation is therefore given by:

$$\frac{d}{dT} \int_{X_1}^{X_2} \rho U H h \, dX + \rho U^2 H h \bigg|_{X=X_1}^{X=X_2} + \int_0^{Hh_2} P_2 \, dY - \int_0^{Hh_1} P_1 \, dY = 0 \qquad (17a)$$

where $h_i \equiv h(x_i, t)$, $P_i \equiv P(X_i, Y, T)$, $i = 1, 2$, and so on. In dimensionless form, this becomes

$$\frac{d}{dt} \int_{x_1}^{x_2} uh \, dx + u^2 h \bigg|_{x=x_1}^{x=x_2} + \int_0^{h_2} p_2 \, dy - \int_0^{h_1} p_1 \, dy = 0 \qquad (17b)$$

Using the dimensionless form of (13c)—that is,

$$p = h - y \qquad (18)$$

—we calculate:

$$\int_0^{h_i} p_i \, dy = \int_0^{h_i} (h_i - y) \, dy = \left(\frac{h_i y - y^2}{2} \right) \bigg|_{y=0}^{y=h_i} = \frac{h_i^2}{2}; \qquad i = 1, 2 \qquad (19)$$

Therefore, the integral law of momentum conservation reduces to

$$\frac{d}{dt} \int_{x_1}^{x_2} uh \, dx + \left(u^2 h + \frac{h^2}{2} \right) \bigg|_{x=x_1}^{x=x_2} = 0 \qquad (20)$$

Here again (20) remains valid for the discontinuous solutions in G, as sketched in Figure 3.6.

3.2.5 Smooth Solutions

If u and h are smooth (that is, continuous and have continuous first partial derivatives with respect to x and t), (16) and (20) reduce to the differential conservation relations:

$$h_t + (uh)_x = 0 \qquad (21a)$$

$$(uh)_t + \left(u^2h + \frac{h^2}{2}\right)_x = 0 \tag{21b}$$

Equation (21b) simplifies if h_t is eliminated using (21a) and we find

$$h_t + (uh)_x = 0 \tag{22a}$$

$$u_t + h_x + uu_x = 0 \tag{22b}$$

This pair of quasilinear (derivatives occur in a linear way) equations will be discussed in detail in Chapter 7 when we study hyperbolic systems of equations. It is important to bear in mind that since we have ignored vertical motions and have averaged the horizontal velocity, (20) provides only an approximate description of the flow. A more systematic derivation of these equations as the leading approximation for a long-wave (that is, $HL_0^{-1} \ll 1$) theory can be found in Section 5.2 of [1], where it is shown that terms of order $(H/L_0)^2$ have been ignored in (20). See also Section 8.4.4.

In the present chapter, we are primarily concerned with the linearized version of the equations that result when we assume small disturbances (see Section 3.2.9). We shall also see in Section 3.3 that analogous equations describe the one-dimensional flow of a perfect compressible gas.

3.2.6 Energy Conservation

It is natural to ask at this point whether the flow described by the solution of the system (22) is compatible with the law of energy conservation. Note that in the ensuing derivation, we are restricted to a narrow definition of energy, since our model does not admit dissipation.

Consider a column of water of width ΔX and unit breadth extending from the bottom $Y = 0$ to the free surface $Y = Hh$. The kinetic energy of this column is

$$\text{K.E.} \equiv \frac{\rho \Delta X H h U^2}{2} \tag{23}$$

The potential energy is

$$\text{P.E.} \equiv \frac{\rho g \Delta X H^2 h^2}{2} \tag{24}$$

In dimensionless form, the total mechanical energy is

$$\text{K.E.} + \text{P.E.} = \int_{x_1}^{x_2} \left(\frac{u^2h}{2} + \frac{h^2}{2}\right) dx$$

Thus, equating the rate of change of energy to the flow of energy plus the work done on the water in G by the pressure forces acting at x_1 and x_2, we obtain the integral law of conservation of energy:

$$\frac{d}{dt}\int_{x_1}^{x_2} \left(\frac{u^2h}{2} + \frac{h^2}{2}\right) dx + \frac{u^3h}{2}\bigg|_{x=x_1}^{x=x_2} + \frac{uh^2}{2}\bigg|_{x=x_1}^{x=x_2} + \int_0^{h_2} p_2 u_2 \, dy - \int_0^{h_1} p_1 u_1 \, dy = 0 \tag{25}$$

Using (19), this reduces to

$$\frac{d}{dt} \int_{x_1}^{x_2} \left(\frac{u^2 h}{2} + \frac{h^2}{2} \right) dx + \frac{u^3 h}{2} \bigg|_{x=x_1}^{x=x_2} + uh^2 \bigg|_{x=x_1}^{x=x_2} = 0 \tag{26}$$

The corresponding differential conservation relation for smooth solutions is

$$\left\{ \frac{u^2 h}{2} + \frac{h^2}{2} \right\}_t + \left\{ \frac{u^3 h}{2} + uh^2 \right\}_x = 0 \tag{27}$$

It is easy to verify that if (22) is used in (27), this reduces to an identity. Thus, under the assumption of continuous partial derivatives, conservation of mass and momentum imply *conservation of mechanical energy*. We shall explore this result further when we discuss discontinuous solutions in Section 5.3.4.

3.2.7 Initial-Value Problem

We now consider an initial state of water of infinite extent, $-\infty < x < \infty$, which will define the solution for all later times. It is intuitively obvious that if we specify the entire field u and h at $t = 0$, we then ought to be able to use (22) to determine u and h for all subsequent times and all x. A very simple initial state is to have the water at rest ($u = 0$) in some nonequilibrium state. For instance, imagine pressing down on the free surface with a solid sheet of width L_0 and shape defined by $Y = H + Af(x/L_0)$ with $Af < 0$. Here A is a constant characterizing the amplitude of the surface distortion. Thus the water, while still at rest, is forced below the equilibrium level H by the amount Af over an interval L_0 wide. Now at $t = 0$, we suddenly remove the surface constraint.

We use the width L_0 of the initial surface disturbance in the normalizations (15) and conclude that the initial conditions are

$$u(x, 0; \varepsilon) = 0; \qquad h(x, 0; \varepsilon) = 1 + \varepsilon f(x) \tag{28}$$

where $\varepsilon \equiv A/H$, and we exhibit the dependence of the solution on ε explicitly by the notation $u(x, t; \varepsilon)$, $h(x, t; \varepsilon)$. A number of other initial disturbance mechanisms are possible.

3.2.8 Signaling Problem

Another means of generating a disturbance is to install a vertical wavemaker located initially at some point, say $X = 0$, and to consider the motion generated in the water to one side of the wavemaker as this moves in some prescribed fashion (see Figure 3.7).

The wavemaker displacement, X_w, as a function of time may be defined in the form $X_w \equiv L_1 s(T/T_0)$; a possible curve for X_w is sketched in Figure 3.8. Using $L_0 \equiv \sqrt{gHT_0}$ and T_0 as the characteristic length and time scales in (15), the dimensionless form of the wavemaker displacement becomes:

$$x_w \equiv \frac{L_1}{\sqrt{gHT_0}} s(t) \tag{29}$$

Now, the appropriate boundary condition at the wavemaker is that the velocity

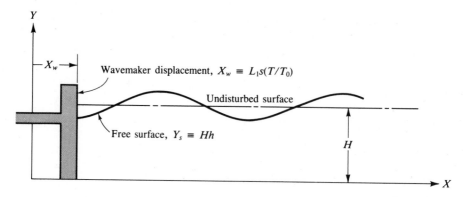

Figure 3.7

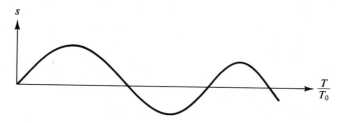

Figure 3.8

of the water u must equal the velocity of the wavemaker dx_w/dt. We compute

$$\frac{dx_w}{dt} = v\dot{s}(t); \qquad v \equiv \frac{L_1}{\sqrt{gHT_0}}$$

where the dimensionless parameter v is the ratio of the wavemaker characteristic speed (L_1/T_0) to the characteristic flow speed $\sqrt{gH}$.

Thus, we have initial conditions

$$u(x, 0; v) = 0; \qquad h(x, 0; v) = 1 \tag{30}$$

and the boundary condition

$$u(vs(t), t; v) = v\dot{s}(t), \qquad t > 0 \tag{31}$$

3.2.9 Small-Amplitude Theory

In each of the preceding examples, the solution involves one dimensionless parameter (ε or v). Henceforth, we denote v by ε. The case $\varepsilon \ll 1$ corresponds to small-amplitude initial (boundary) disturbances. Since, for $\varepsilon = 0$, the solution is $u(x, t; \varepsilon) = 0$, $h(x, t; \varepsilon) = 1$, we again seek to represent the solution for small ε in a perturbation series:

$$u(x, t; \varepsilon) = \varepsilon u_1(x, t) + \varepsilon^2 u_2(x, t) + O(\varepsilon^3) \tag{32a}$$

$$h(x, t; \varepsilon) = 1 + \varepsilon h_1(x, t) + \varepsilon^2 h_2(x, t) + O(\varepsilon^3) \tag{32b}$$

Actually, it only makes sense to look at u_1 and h_1 as predicted by this model because if we want accuracies of order ε^2, we must retain a term in (22b) that got thrown out by our crude physical model. This point is discussed on pages 518–519 of [I] and in Section 8.4.4. However, for the sake of simplicity, and in order to illustrate the procedure for deriving the equations for the higher-order terms, let us regard (22) as an "exact" model of the physical situation.

To derive governing equations for u_1, u_2, h_1, h_2, ..., we substitute the expansions (32) into (22) and collect terms of order ε, ε^2, ... The result is

$$\varepsilon(h_{1_t} + u_{1_x}) + \varepsilon^2[h_{2_t} + u_{2_x} + (u_1 h_1)_x] = O(\varepsilon^3) \tag{33a}$$

$$\varepsilon(u_{1_t} + h_{1_x}) + \varepsilon^2[u_{2_t} + h_{2_x} + \tfrac{1}{2}(u_1^2)_x] = O(\varepsilon^3) \tag{33b}$$

This perturbation series must be valid for ε arbitrary (but small). Therefore, the coefficient of each power of ε must vanish, and we obtain the following linear first-order system governing u_1, h_1:

$$h_{1_t} + u_{1_x} = 0 \tag{34a}$$

$$u_{1_t} + h_{1_x} = 0 \tag{34b}$$

The perturbation terms u_2, h_2 obey

$$h_{2_t} + u_{2_x} = -(u_1 h_1)_x \tag{35a}$$

$$u_{2_t} + h_{2_x} = -\tfrac{1}{2}(u_1^2)_x \tag{35b}$$

and so on. The homogeneous operator is the same linear one to all orders, and the right-hand sides of the equations for u_n, h_n depend only on u_{n-1}, u_{n-2}, ..., u_1, h_{n-1}, h_{n-2}, ..., h_1.

Initial-value Problem

Consider first the initial-value problem (28). Using the initial conditions (28) in (32), evaluated at $t = 0$, implies that

$$u_1(x,0) = 0 \tag{36a}$$

$$h_1(x,0) = f(x) \tag{36b}$$

$$u_2(x,0) = 0 \tag{37a}$$

$$h_2(x,0) = 0 \tag{37b}$$

Thus, we first need to solve (34) subject to the initial conditions (36). One approach is to eliminate either u_1 or h_1 to derive a second-order equation for the other variable. For example, if we differentiate (34b) with respect to x and subtract the result from the partial derivative of (34a) with respect to t, we obtain the *wave equation* for h_1

$$h_{1_{tt}} - h_{1_{xx}} = 0; \qquad -\infty < x < \infty; t \geq 0 \tag{38}$$

Based on our experience so far in choosing appropriate initial conditions, we expect (38) to require *two* conditions at $t = 0$ as the operator is second-order in t. In Chapter 4, this will be fully justified. Now, one initial condition is given in (36b), and we obtain the other from evaluating (34a) at $t = 0$ and using the condition (36a). Since $u_1(x,0) = 0$ implies $u_{1_x}(x,0) = 0$, we have

$$h_{1_t}(x,0) = 0 \tag{39}$$

The solution of (38), subject to (36b) and (39), is discussed in Section 3.4.2. Once $h_1(x, t)$ is known, we can compute $u_1(x, t)$ from (34b) by quadrature in the form

$$u_1(x, t) = -\int_0^t h_{1_x}(x, \tau)\, d\tau \tag{40}$$

since $u_1(x, 0) = 0$.

A second alternative for solving (34) is to eliminate h_1 from (34) to derive the following initial-value problem for u_1:

$$u_{1_{tt}} - u_{1_{xx}} = 0; \qquad -\infty < x < \infty; t \geq 0 \tag{41a}$$

$$u_1(x, 0) = 0 \tag{41b}$$

$$u_{1_t}(x, 0) = -f'(x) \tag{41c}$$

Again, the solution of this problem is discussed in Section 3.4.2.

It turns out that when the governing equations have the special canonical form (34) or if we can transform them to this form, then it is more efficient to compute the solution directly proceeding from the system of first-order equations. This so-called method of characteristics is illustrated in Section 3.4.3 for the wave equation (34) and a particular choice of initial data. It is discussed in more generality in Section 4.5.

We can now also derive equations for the perturbation terms u_2 and h_2 in a similar manner. Thus, eliminating u_2 from (35) gives

$$h_{2_{tt}} - h_{2_{xx}} = \{\tfrac{1}{2}(u_1^2)_x - (u_1 h_1)_t\}_x; \qquad -\infty < x < \infty; t \geq 0 \tag{42a}$$

and using (34), the right-hand side may also be written in the alternate form

$$h_{2_{tt}} - h_{2_{xx}} = \{\tfrac{1}{2}(h_1^2)_x - 2u_1 h_{1_t}\}_x \tag{42b}$$

The initial conditions are now given by (37b), and

$$h_{2_t}(x, 0) = -u_{2_x}(x, 0) - \{u_1(x, 0)h_1(x, 0)\}_x = 0 \tag{43}$$

Having calculated u_1 and h_1, the right-hand side of (42) is a known function of x, t, and we have to solve an inhomogeneous wave equation for h_2 (or u_2) with zero initial conditions. This is also discussed in Section 3.4.2. Once h_2 is known, u_2 can be calculated by quadrature with respect to t using (35b). Of course, the alternative approaches of deriving a wave equation for u_2 or proceeding from the system (35) directly, also apply.

Signaling Problem

For the signaling problem formulated in Section 3.2.8, the boundary condition at the left end involves u; hence it is more convenient to eliminate h_1 from (34). Thus, we need to solve

$$u_{1_{tt}} - u_{1_{xx}} = 0; \qquad 0 \leq x < \infty; 0 \leq t \tag{44}$$

subject to the zero initial conditions

$$u_1(x, 0) = u_{1_t}(x, 0) = 0 \tag{45}$$

which result from (30) and (34b). The boundary condition for u_1 [with $v = \varepsilon$ in (31)] is just

$$u_1(0, t) = \dot{s}(t); \qquad t \geq 0 \tag{46}$$

when the leading term of the expansion (32a) is used. The solution of (44)–(46) is worked out in Section 3.5.4 using Green's functions and the method of characteristics. See also Problem 3(a), where the solution by Laplace transforms is outlined. Knowing u_1, we obtain h_1 from (34a) by quadrature in the form

$$h_1(x, t) = -\int_0^t u_{1_x}(x, \tau)\, d\tau \tag{47}$$

since $h_1(x, 0) = 0$.

To derive the equation governing u_2, we eliminate h_2 from the system (35) and find

$$u_{2_{tt}} - u_{2_{xx}} = \{(u_1 h_1)_x - \tfrac{1}{2}(u_1^2)_t\}_x; \qquad 0 \leq x; 0 \leq t \tag{48}$$

The right-hand side of (48) is a known function of x, t once u_1 and h_1 have been calculated. The initial conditions for u_2 are also zero:

$$u_2(x, 0) = u_{2_t}(x, 0) = 0 \tag{49}$$

but the boundary condition at the left end must be derived with some care. The exact boundary condition (31) involves ε, first as a multiplier of s and $\dot{s}$, and also because u itself depends on ε. Thus, expanding u according to (32a) implies that

$$\varepsilon u_1(\varepsilon s, t) + \varepsilon^2 u_2(\varepsilon s, t) + \cdots = \varepsilon \dot{s} \tag{50a}$$

Now, expanding u_1 and u_2 around $x = 0$ gives

$$\varepsilon[u_1(0, t) + u_{1_x}(0, t)\varepsilon s + \cdots] + \varepsilon^2 u_2(0, t) + \cdots = \varepsilon \dot{s} \tag{50b}$$

Therefore, the $O(\varepsilon^2)$ boundary condition is

$$u_2(0, t) = -u_{1_x}(0, t)s(t); \qquad t > 0 \tag{51}$$

and the right-hand side is known once $u_1(x, t)$ has been determined. The solution of (48) subject to (49) and (51) is also discussed in Section 3.5.4. We observe that each u_n obeys an inhomogeneous wave equation with zero initial conditions and a prescribed boundary condition similar to (51) involving lower-order terms $u_{n-1}, u_{n-2}, \ldots, u_1$.

Although a perturbation solution of the form (32) may be derived to any order, in principle, we must be aware of various stipulations regarding the validity of such a solution. As pointed out earlier, the mathematical model (22) is not accurate to $O(\varepsilon^2)$. This shortcoming is easily remedied by inclusion of the appropriate correction term on the right-hand side of (22b). [See (235b) of Chapter 8.] However, it is also pointed out in [1] and Chapter 8 that a perturbation expansion in the form (22) breaks down when x (or t) become as large as $O(\varepsilon^{-1})$. The evidence for this breakdown is the occurrence of terms proportional to x (or t) in u_2 and h_2. (See the details of the solution in Section 3.5.4.) A perturbation series that remains valid for x and t large will be derived for the initial-value problem (28) in Section 8.4.4.

3.3   Compressible Flow

In this section, we study a third application area, which leads to the wave equation in the limit of small disturbances. Again, we proceed from the general laws of mass, momentum, and energy conservation.

3.3.1 Conservation Laws

Consider a domain G that is fixed in space and time, has a bounding surface S, and is occupied by a gas with density $\rho(\mathbf{x}, t)$ and velocity $\mathbf{u}(\mathbf{x}, t)$. Let $\mathbf{n}$ denote the outward unit normal vector on S and let $\mathbf{F}(\mathbf{x}, t)$ denote the body force (usually gravity) per unit mass. Denote the stress by $\boldsymbol{\tau}(\mathbf{x}, t)$, and let the internal energy per unit mass for the molecular motion be $e(\mathbf{x}, t)$. Let the heat flow per unit area across the boundary be denoted by $\mathbf{q}(\mathbf{x}, t)$.

The laws of conservation of mass, momentum and energy are (see Section 6.1 of [2]):

$$\frac{d}{dt} \iiint_G \rho \, dV + \iint_S \rho u_j n_j \, dS = 0 \qquad \text{(mass)} \qquad (52a)$$

$$\frac{d}{dt} \iiint_V \rho u_i \, dV + \iint_S (\rho u_i n_j u_j - \tau_i) \, dS = \iiint_V \rho F_i \, dV,$$

$$i = 1, 2, 3 \qquad \text{(momentum)} \quad (52b)$$

$$\frac{d}{dt} \iiint_V \left(\tfrac{1}{2}\rho u_i^2 + \rho e\right) dV + \iint_S \left\{\tfrac{1}{2}\rho u_i^2 + \rho e)n_j u_j - \tau_i u_i + n_j q_j\right\} dS$$

$$= \iiint_V \rho F_i u_i \, dV \qquad \text{(energy)} \qquad (52c)$$

In the preceding, we are using Cartesian tensor notation and summing over repeated indices. Thus, $\mathbf{u} = (u_1, u_2, u_3)$, $u_i n_i \equiv \sum_{i=1}^{3} u_i n_i$, and so on.

Equations (52) provide five equations for the eleven quantities ρ, u_i, τ_i, q_i, e, $(i = 1, 2, 3)$. To complete the system, additional relations between the flow variables must be specified.

3.3.2 One-Dimensional Ideal Gas

For one-dimensional flow, the stress τ_1 and the heat flux q_1 are given by

$$\tau_1 \equiv -p + \tfrac{4}{3}\mu u_x \qquad (53a)$$

$$q_1 \equiv -\lambda \theta_x \qquad (53b)$$

where

p = pressure

θ = temperature

μ = coefficient of viscosity

λ = thermal conductivity

An *ideal gas* is one that obeys the *equation of state*

$$p = \rho R \theta \tag{54a}$$

where

R = gas constant $\equiv C_p - C_v$

C_p = specific heat at constant pressure

C_v = specific heat at constant volume

If one assumes that C_p and C_v are both constant for an ideal gas, one can show that the internal energy e is just

$$e = C_v \theta \tag{54b}$$

In all our work here and later on, we shall assume that (53)–(54) hold with constant values for μ, λ, C_p, and C_v. In particular, if body forces are negligible, the integral conservation laws of mass, momentum, and energy (52) reduce to the following form for one-dimensional flow:

$$\frac{d}{dt} \int_{x_1}^{x_2} \rho \, dx + \rho u \Big|_{x=x_1}^{x=x_2} = 0 \tag{55a}$$

$$\frac{d}{dt} \int_{x_1}^{x_2} \rho u \, dx + \left(\rho u^2 + p - \frac{4}{3} \mu u_x \right) \Big|_{x=x_1}^{x=x_2} = 0 \tag{55b}$$

$$\frac{d}{dt} \int_{x_1}^{x_2} \rho \left(\frac{u^2}{2} + C_v \theta \right) dx + \left[\left(\frac{u^2}{2} + C_v \theta \right) \rho u + pu - \frac{4}{3} \mu u u_x - \lambda \theta_x \right] \Big|_{x=x_1}^{x=x_2} = 0 \tag{55c}$$

Equations (55) combined with the equation of state (54a) are four relations governing the four dependent variables ρ, u, p, and θ.

For smooth solutions (that is, with ρ, u, p, θ having continuous first partial derivatives with respect to x and t), the preceding conservation laws reduce to the differential conservation relations

$$\rho_t + (\rho u)_x = 0 \tag{56a}$$

$$(\rho u)_t + (\rho u^2 + p - \tfrac{4}{3} \mu u_x)_x = 0 \tag{56b}$$

$$\left(\frac{1}{2} \rho u^2 + \rho C_v \theta \right)_t + \left[\left(\frac{u^2}{2} + C_v \theta \right) \rho u + pu - \frac{4}{3} \mu u u_x - \lambda \theta_x \right]_x = 0 \tag{56c}$$

Using (56a), (56b) simplifes to

$$\rho u_t + \rho u u_x + p_x - \tfrac{4}{3} \mu u_{xx} = 0 \tag{57a}$$

Using (56a) and (57a), (56c) simplifies to

$$\rho C_v (\theta_t + u \theta_x) + (p - \tfrac{4}{3} \mu u_x) u_x - \lambda \theta_{xx} = 0 \tag{57b}$$

Thus, (54a), (56a), (57a), and (57b) provide three partial differential equations plus one algebraic equation governing the four unknowns ρ, u, p, θ.

3.3.3 Signaling Problem for One-Dimensional Flow

We wish to study the signaling problem of generating a disturbance in a semi-infinite region that is initially at rest. For the sake of simplicity, assume that the disturbance is modeled by a piston with a prescribed temperature θ_p moving to the right or left according to

$$x_p = L_0 f\left(\frac{t}{T_0}\right) \tag{58}$$

where L_0 and T_0 are characteristic length and time scales, respectively, associated with the piston displacement, as sketched in Figure 3.9.

Suppose that the ambient properties of the gas are p_0, ρ_0, θ_0, λ_0, μ, C_p, C_v, and $u_0 = 0$. We introduce dimensionless variables denoted by an asterisk as follows:

$$u^* \equiv \frac{u}{a_0}; \qquad p^* \equiv \frac{p}{p_0}; \qquad \theta^* \equiv \frac{\theta}{\theta_0}; \qquad \rho^* \equiv \frac{\rho}{\rho_0}; \qquad \theta_p^* \equiv \frac{\theta_p}{\theta_0} \tag{59}$$

$$x^* \equiv \frac{x}{a_0 T_0}; \qquad t^* \equiv \frac{t}{T_0}; \qquad x_p^* \equiv \frac{x_p}{a_0 T_0} \tag{60}$$

where a_0 is the ambient speed of sound defined by

$$a_0 \equiv \left(\frac{\gamma p_0}{\rho_0}\right)^{1/2} \tag{61}$$

and γ is the ratio of specific heats

$$\gamma \equiv \frac{C_p}{C_v} \tag{62}$$

The dimensionless version of (56a), (57a), and (57b) becomes (dropping

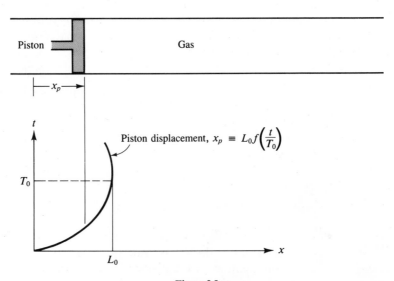

Figure 3.9

asterisks for simplicity of notation)

$$\rho_t + (\rho u)_x = 0 \tag{63a}$$

$$\rho u_t + \rho u u_x + \frac{p_x}{\gamma} - \frac{4}{3Re} u_{xx} = 0 \tag{63b}$$

$$\rho \theta_t + \rho u \theta_x + \left[(\gamma - 1)p - \frac{4\gamma(\gamma - 1)}{3Re} u_x \right] u_x - \frac{\gamma}{RePr} \theta_{xx} = 0 \tag{63c}$$

and involves the three dimensionless parameters γ, Re, and Pr, where γ is defined in (62) and

$$Re \equiv \frac{\rho_0 a_0^2 T_0}{\mu} = \text{Reynolds number} \tag{64}$$

$$Pr \equiv \mu C_p / \lambda = \text{Prandtl number} \tag{65}$$

In addition, the equation of state, (54a), becomes:

$$p = \rho \theta \tag{66}$$

The boundary conditions at the piston are

$$u(\varepsilon x_p, t) = \varepsilon \frac{dx_p}{dt}; \qquad t > 0 \tag{67}$$

$$\theta(\varepsilon x_p, t) = \theta_p(t); \qquad t > 0 \tag{68}$$

where

$$\varepsilon \equiv \frac{L_0}{a_0 T_0}$$

Thus, ε is the ratio of the piston characteristic speed (L_0/T_0) to the ambient speed of sound a_0.

3.3.4 Inviscid, Non-Heat-Conducting Gas; Analogy with Shallow-Water Waves

For an inviscid ($\mu = 0$), non-heat-conducting ($\lambda = 0$) gas, we have $Re \to \infty$, $Pr \to \infty$, and our system reduces to:

$$\rho_t + (\rho u)_x = 0 \tag{69}$$

$$\rho u_t + \rho u u_x + \frac{p_x}{\gamma} = 0 \tag{70}$$

$$\rho \theta_t + \rho u \theta_x + (\gamma - 1)p u_x = 0 \tag{71}$$

As discussed on p. 156 of [2], (71) is equivalent to

$$\left(\frac{p}{\rho^\gamma} \right)_t + u \left(\frac{p}{\rho^\gamma} \right)_x = 0 \tag{72}$$

if all flow variables have continuous derivatives (Problem 4). Thus, for smooth solutions, conservation of energy implies that p/ρ^γ remains constant along particle paths. These are paths defined by $(dx/dt) = u(x, t)$. Since the entropy S is a function of p/ρ^{γ}, we conclude that S is constant along particle paths; such a flow is called *adiabatic*. See also Section 5.3.4 and Section 7.5.1.

If, in addition, p/ρ^{γ} is initially constant when the gas is at rest (as is the case for our signaling problem where $p/\rho^{\gamma} = 1$ at $t = 0$), it will remain constant as long as no discontinuities occur in the flow. This is called *isentropic* flow.

Using $p = \rho^{\gamma}$ to eliminate p from (70) gives

$$u_t + uu_x + \rho^{\gamma-2}p_x = 0 \tag{73}$$

The two equations (69) and (73) are exact analogs of equations (22) for shallow-water waves if we identify u in both cases, set $h = \rho$, and take $\gamma = 2$. Or, equivalently, using the dimensionless local speed of sound a defined by

$$a^2 \equiv \frac{\partial p}{\partial \rho} = \gamma\rho^{\gamma-1} \tag{74}$$

instead of ρ gives the alternate form

$$a_t + ua_x + \frac{\gamma - 1}{2}au_x = 0 \tag{75a}$$

for mass conservation and

$$u_t + uu_x + \frac{2}{\gamma - 1}aa_x = 0 \tag{75b}$$

for momentum conservation. Now, comparing these with (22), we identify u in both cases, set $a = \sqrt{h}$, and take $\gamma = 2$ to obtain the formal analogy.

Unfortunately, $\gamma = 1.4$ for air, so that it is not possible to have an accurate quantitative analog for compressible flow using a hydraulic model. Nevertheless, all the qualitative features of compressible flow in one dimension may be easily (and inexpensively) duplicated using the hydraulic analogy in a laboratory. This question is disucssed further in Section 5.3.4.

3.3.5 Small-Disturbance Theory in One-Dimensional Flow (Signaling Problem)

Assume that $\varepsilon \ll 1$—that is, that the characteristic speed (L_0/T_0) associated with the piston displacement is very small compared with the ambient speed of sound a_0. Assume further that the piston temperature that is prescribed does not differ much from the ambient temperature. This means that θ_p is of the form

$$\theta_p = 1 + \nu\theta_p^*(t)$$

where ν is a small dimensionless parameter of the order of ε, say $\nu = c\varepsilon$ ($c =$ constant).

Clearly, if $\varepsilon = 0$, the solution must be the ambient state: $u = 0, p = \rho = \theta = 1$. Thus, we assume a perturbation expansion in the form

$$u(x, t; \varepsilon) = \varepsilon u_1(x, t) + O(\varepsilon^2) \tag{76a}$$

$$p(x, t; \varepsilon) = 1 + \varepsilon p_1(x, t) + O(\varepsilon^2) \tag{76b}$$

$$\rho(x, t; \varepsilon) = 1 + \varepsilon\rho_1(x, t) + O(\varepsilon^2) \tag{76c}$$

$$\theta(x, t; \varepsilon) = 1 + \varepsilon\theta_1(x, t) + O(\varepsilon^2) \tag{76d}$$

and calculate the following linear system of equations for u_1, ρ_1, p_1, and θ_1:

$$\rho_{1_t} + u_{1_x} = 0 \tag{77a}$$

$$u_{1_t} + \frac{p_{1_x}}{\gamma} - \frac{4}{3Re}u_{1_{xx}} = 0 \tag{77b}$$

$$\theta_{1_t} + (\gamma - 1)u_{1_x} - \frac{\gamma}{RePr}\theta_{1_{xx}} = 0 \tag{77c}$$

where the equation of state (66) implies that

$$p_1 = \rho_1 + \theta_1 \tag{77d}$$

The boundary conditions (67)–(68) at the piston for $t > 0$ reduce to

$$u_1(0, t) = \frac{dx_p}{dt} \equiv h(t) = \text{prescribed;} \quad t > 0 \tag{78a}$$

$$\theta_1(0, t) = c\theta_p^*(t) = \text{prescribed;} \quad t > 0 \tag{78b}$$

whereas the initial conditions are

$$u_1(x, 0) = p_1(x, 0) = \rho_1(x, 0) = \theta_1(x, 0) = 0 \tag{79}$$

We do not study system (77) here. We could use Laplace transforms with respect to t and then solve the resulting linear system of ordinary differential equations. For a discussion, see [3].

In the inviscid, non-heat-conducting case ($Re = Pr = \infty$), we find after using (77d) to eliminate p_1 that

$$\rho_{1_t} + u_{1_x} = 0 \tag{80a}$$

$$u_{1_t} + \frac{1}{\gamma}(\rho_{1_x} + \theta_{1_x}) = 0 \tag{80b}$$

$$\theta_{1_t} + (\gamma - 1)u_{1_x} = 0 \tag{80c}$$

Using (80a) in (80c) gives

$$\theta_{1_t} - (\gamma - 1)\rho_{1_t} = 0 \tag{81a}$$

or upon integration with respect to t, we obtain

$$\theta_1 - (\gamma - 1)\rho_1 = f(x) \tag{81b}$$

But since $\theta_1(x, 0) = \rho_1(x, 0) = 0$, we conclude that $f(x) = 0$. Hence,

$$\theta_1(x, t) = (\gamma - 1)\rho_1(x, t) \tag{82}$$

and this defines θ_1 once ρ_1 is known. Moreover, (77d) gives

$$p_1(x, t) = \gamma\rho_1(x, t) \tag{83}$$

which also defines p_1 once ρ_1 is known.

Using (82) in (80b) reduces the problem to the pair of equations

$$\rho_{1_t} + u_{1_x} = 0 \tag{84a}$$

$$u_{1_t} + \rho_{1_x} = 0 \tag{84b}$$

If we now eliminate ρ_1, we obtain the wave equation

$$u_{1_{tt}} - u_{1_{xx}} = 0 \tag{85}$$

subject to the boundary condition (78a).

The two initial conditions needed at $t = 0$ are

$$u_1(x, 0) = 0 \tag{86a}$$

from (79), and

$$u_{1_t}(x, 0) = 0 \tag{86b}$$

which follows from (84b), since $\rho_{1_x}(x, 0) = 0$.

3.3.6 Small-Disturbance Theory in Three-Dimensional, Inviscid Non-Heat-Conducting Flow

This limit for three-dimensional flows is called *acoustics*. To begin with, we neglect viscosity, heat conduction, and body forces; for smooth solutions, equations (52a) and (52b) imply that

$$\rho_t + \text{div}(\rho \mathbf{u}) = 0 \tag{87}$$

$$(\rho \mathbf{u})_t + \text{div}(\rho \mathbf{u} \circ \mathbf{u}) + \text{grad}\, p = 0 \tag{88}$$

where $\rho \mathbf{u} \circ \mathbf{u}$ is the flow of momentum tensor, expressed as the diadic product of $\rho \mathbf{u}$ with $\mathbf{u}$ (the ith component of div $(\mathbf{a} \circ \mathbf{b})$ is $(\partial / \partial x_k)(a_i b_k)$ for Cartesian tensors).

In this limit, the energy equation is equivalent to

$$\left(\frac{p}{\rho^\gamma}\right)_t + \mathbf{u} \cdot \text{grad}\left(\frac{p}{\rho^\gamma}\right) = 0 \tag{89}$$

and for a problem where p/ρ^γ is initially constant everywhere, we conclude from (89) that

$$\frac{p}{\rho^\gamma} = \text{constant} \tag{90}$$

in space and time. Equation (90) allows us to eliminate p in favor of ρ or vice versa.

Consider now a signaling or initial-value problem in the small-disturbance limit, where we perturb an ambient state $\mathbf{u} = 0$. We can proceed as in the one-dimensional case to define dimensionless variables and a small parameter appropriate to either problem. However, in the interest of brevity, we do not specify the small parameter ε. Let

$$\frac{\mathbf{u}}{a_0} = \varepsilon \tilde{\mathbf{u}}; \qquad \frac{p}{p_0} = 1 + \varepsilon \tilde{p}; \qquad \frac{\rho}{\rho_0} = 1 + \varepsilon \tilde{\rho} \tag{91}$$

where the tildes denote dimensionless perturbation quantities that are $O(1)$. Zero subscripts denote ambient values, and according to (74), the ambient speed of sound is

$$a_0 \equiv \sqrt{\frac{\gamma p_0}{\rho_0}} = \sqrt{\gamma \rho_0^{\gamma - 1}} \tag{92}$$

The linearized version of (87) and (88) is

$$\tilde{\rho}_t + \text{div } \tilde{\mathbf{u}} = 0 \tag{93}$$

$$\gamma \tilde{\mathbf{u}}_t + \text{grad } \tilde{p} = 0 \tag{94}$$

Here, we have normalized lengths by dividing by a characteristic length L_0, and we have normalized the time by L_0/a_0.

Taking the curl of (94) and noting that curl grad $\tilde{p} = 0$, we have curl $\gamma \tilde{\mathbf{u}}_t = \gamma (\partial/\partial t)(\text{curl } \tilde{\mathbf{u}}) = 0$. Thus, curl $\tilde{\mathbf{u}}$ is independent of time. But since $\tilde{\mathbf{u}} = 0$ initially, we conclude that curl $\tilde{\mathbf{u}} = 0$ for all space and time. Therefore, $\tilde{\mathbf{u}}$ is an irrotational vector field and we must have

$$\tilde{\mathbf{u}} = \text{grad } \phi \tag{95}$$

for a scalar potential $\phi(x, y, z, t)$.

Equation (94) can now be written as

$$\text{grad}\,(\gamma \phi_t + \tilde{p}) = 0 \tag{96}$$

Thus, $\gamma \phi_t + \tilde{p}$ can only be a function of time and, evaluating it at infinity, we conclude that

$$\gamma \phi_t + \tilde{p} = 0 \tag{97}$$

Using the isentropy condition (90) gives

$$1 + \varepsilon \tilde{p} = (1 + \varepsilon \tilde{\rho})^\gamma \tag{98}$$

Expanding the right-hand side for small ε gives

$$\tilde{p} = \gamma \tilde{\rho} + \cdots \tag{99}$$

Therefore, using (99), we can also write (97) as

$$\phi_t + \tilde{\rho} = 0 \tag{100}$$

For future reference, note that the pressure or density perturbations are proportional to $-\phi_t$.

Finally, if we differentiate (100) with respect to t and use (93) to eliminate $\tilde{\rho}_t$, we obtain the three-dimensional wave equation

$$\phi_{tt} - \Delta \phi = 0 \tag{101}$$

where, as usual, $\Delta \equiv \text{div grad}$ and

$$\Delta \equiv \frac{\partial^2}{\partial x^2} + \frac{\partial^2}{\partial y^2} + \frac{\partial^2}{\partial z^2} \tag{102}$$

for Cartesian coordinates.

3.4   The One-Dimensional Problem in the Infinite Domain

In this section we parallel the discussion of Sections 1.2–1.3 to study

$$u_{tt} - u_{xx} = p(x, t); \qquad -\infty < x < \infty; t \geq 0 \tag{103a}$$

$$u(x, 0^+) = f(x) \tag{103b}$$

$$u_t(x, 0^+) = h(x) \tag{103c}$$

As discussed in 3.1–3.3, equation (103a) with $p = 0$ may represent the perturbation velocity (or surface height) for small-amplitude water waves or the perturbation velocity or density for small-amplitude disturbances in a one-dimensional compressible gas. A third interpretation, discussed in Section 3.1, has u representing either the lateral or axial displacement of a vibrating string in tension in the limit of small-amplitude oscillations. We shall appeal to these physical models to interpret better the results that we derive in this and subsequent sections.

3.4.1 Fundamental Solution

As in Chapter 1, we begin our study with the derivation of the fundamental solution—that is,

$$u_{tt} - u_{xx} = \delta(x)\delta(t); \qquad -\infty < x < \infty; t \geq 0 \tag{104a}$$

$$u(x, 0^-) = u_t(x, 0^-) = 0 \tag{104b,c}$$

A crude justification for imposing two initial conditions (104b)–(104c) (rather than one, as in the diffusion equation) is to argue that (104a) is second order in t and therefore requires two conditions in order to define a solution uniquely. Using physical reasoning, for example, for the vibrating string, we would argue that in order to define the state of a dynamical system, we must initially specify both the *displacement* and the *velocity*. When we study the solution of (103) or more general hyperbolic equations in Chapter 4, we shall develop more systematic criteria for determining what constitutes a "well-posed" problem.

In solving (104), it is convenient to consider the equivalent homogeneous equation

$$u_{tt} - u_{xx} = 0 \tag{105a}$$

on $-\infty < x < \infty$ valid for $t > 0$ with the following initial conditions imposed at $= 0^+$:

$$u(x, 0^+) = 0 \tag{105b}$$

$$u_t(x, 0^+) = \delta(x) \tag{105c}$$

To see that (104) and (105) are equivalent, we integrate (104a) with respect to t from $t = 0^-$ to $t = 0^+$ and use the initial conditions (104b)–(104c) to derive the corresponding initial conditions (105b)–(105c).

The homogeneous equation (105a) is exceptional (among linear second-order partial differential equations) in the sense that every solution can be derived in the *D'Alembert form*:

$$u(x,t) = \phi(x+t) + \psi(x-t) \tag{106}$$

for appropriate functions ϕ and ψ. To see this, we first transform variables $x, t \to \zeta, \sigma$, where $\zeta \equiv x+t$ and $\sigma \equiv x-t$ and regard $u(x,t) = u((\zeta+\sigma)/2, (\zeta-\sigma)/2) \equiv U(\zeta,\sigma)$. Equation (105a) transforms to $U_{\zeta\sigma} = 0$. Therefore, integrating once with respect to σ implies that U_ζ is a function of ζ alone, say $U_\zeta = \phi'(\zeta)$, and integrating this with respect to ζ gives (106).

To determine the functions ϕ and ψ, we impose the initial conditions. Equation (105b) implies that $\phi(x) + \psi(x) = 0$ for all x. Therefore, $\psi = -\phi$, and (106) now reads

$$u(x,t) = \phi(x+t) - \phi(x-t) \tag{107}$$

The second initial condition (105c) applied to (107) gives $\delta(x) = 2\phi'(x)$. Therefore, $\phi(x) = \frac{1}{2}H(x)$, where H is the Heaviside function, and the solution of (105) or (104) is

$$u(x,t) = \frac{1}{2}[H(x+t) - H(x-t)] \tag{108}$$

This represents a uniform front of height $u = \frac{1}{2}$ propagating with constant unit speed in the $+x$ and $-x$ directions, as sketched in Figure 3.10. If we

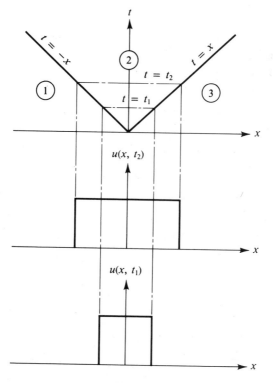

Figure 3.10

subdivide the xt-plane into the three regions as indicated, we note that in region ①, $x + t < 0$, $x - t < 0$. Therefore, $H(x + t) = H(x - t) = 0$, and hence $u = 0$. In region ②, $x + t > 0$, $x - t < 0$. Therefore, $H(x + t) = 1$, $H(x - t) = 0$, and hence, $u = \frac{1}{2}$. Finally, in region ③, $x + t > 0$, $x - t > 0$. Therefore, $H(x + t) = H(x - t) = 1$, and $u = 0$. The triangular domain ② is called the *zone of influence* of the source at $(0, 0)$.

In contrast with the diffusion equation of Chapter 1, (104a) has a distinct "disturbance wave front," which propagates with unit speed (for our dimensionless formulation) and separates regions of disturbed and undisturbed "motion."

The result (108) also follows less directly using Laplace transforms with respect to t or Fourier transforms with respect to x (Problem 6). For a physical discussion of (108) in terms of water waves, see Problem 8.

The fundamental solution for an arbitrary source location $x = \xi$ and switch-on time $t = \tau$ is obtained from (108) by translation. Thus, the solution of

$$u_{tt} - u_{xx} = \delta(x - \xi)\delta(t - \tau); \qquad -\infty < x < \infty; t \geq \tau \tag{109a}$$

with $\xi = $ constant, $\tau = $ constant, and initial conditions

$$u(x, \tau^-) = u_t(x, \tau^-) = 0 \tag{109b, c}$$

is

$$F(x - \xi, t - \tau) = \frac{1}{2}[H(x - \xi + t - \tau) - H(x - \xi - t + \tau)] \tag{110}$$

3.4.2 General Initial-Value Problem on $-\infty < x < \infty$

The general initial-value problem is called a Cauchy problem and obeys

$$u_{tt} - u_{xx} = p(x, t); \qquad -\infty < x < \infty; 0 \leq t \tag{111a}$$

$$u(x, 0^+) = f(x) \tag{111b}$$

$$u_t(x, 0^+) = h(x) \tag{111c}$$

Linearly allows us to split this up into the following three problems:

1. $p(x, t) = $ prescribed; $\quad u(x, 0^-) = u_t(x, 0^-) = 0$
2. $p(x, t) = 0$; $\quad u(x, 0^+) = 0$; $\quad u_t(x, 0^+) = h(x)$
3. $p(x, t) = 0$; $\quad u(x, 0^+) = f(x)$; $\quad u_t(x, 0^+) = 0$.

The sum of the three solutions solves (111).

The solution for problem 1 follows from (110) superposition, and we find:

$$u(x, t) = \int_0^t d\tau \int_{-\infty}^{\infty} F(x - \xi, t - \tau)p(\xi, \tau)\, d\xi \tag{112}$$

Now, for any *fixed* point $P = (x, t)$, consider the triangular domain D bounded by the ξ-axis and the two straight lines

$$\xi = x + (t - \tau) \tag{113a}$$

$$\xi = x - (t - \tau) \tag{113b}$$

as sketched in Figure 3.11. In (112) the integration with respect to ξ occurs

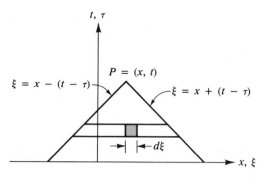

Figure 3.11

for a fixed τ, as shown in Figure 3.11. But, the expression for F vanishes whenever $\xi > x + t - \tau$ and $\xi < x - (t - \tau)$, and in the interval $x - (t - \tau) < \xi < x + (t - \tau)$, the value of F equals $\frac{1}{2}$, according to (110). Therefore, (112) reduces to

$$u(x,t) = \frac{1}{2} \int_0^t d\tau \int_{x-(t-\tau)}^{x+(t-\tau)} p(\xi, \tau) \, d\xi \qquad (114)$$

Thus, only the values of $p(x, t)$ defined in the triangular domain D can influence the value of u at P. The domain D is called the *domain of dependence* of the point P.

To solve problem 2, we note that it is equivalent to

$$u_{tt} - u_{xx} = \delta(t)h(x) \qquad (115a)$$

$$u(x, 0^-) = 0 \qquad (115b)$$

$$u_t(x, 0^-) = 0 \qquad (115c)$$

which is a special case of problem 1 with $p(x, t) = \delta(t)h(x)$. After changing the order of integration, the solution given by (114) for this value of p is:

$$u(x, t) = \frac{1}{2} \int_{\xi=x-t}^x h(\xi) \left[\int_{\tau=0^-}^{\xi-x+t} \delta(\tau) \, d\tau \right] d\xi + \frac{1}{2} \int_{\xi=x}^{x+t} h(\xi) \left[\int_{\tau=0^-}^{x+t-\xi} \delta(\tau) \, d\tau \right] d\xi$$

or

$$u(x, t) = \frac{1}{2} \int_{x-t}^{x+t} h(\xi) \, d\xi \qquad (116)$$

Thus, only the portion of initial data that lies in domain of dependence of the point x, t influences the solution there. Information outside the interval $x - t \le \xi \le x + t$ does not affect the solution at x, t (see Figure 3.12). In particular, two initial-value problems for which h coincides on some interval $x_1 \le x \le x_2, t = 0$ but not outside, will be identical for all x, t in the triangular domain $t + x_1 \le x \le x_2 - t$.

Finally, problem 3 may be reduced to the type 2 by introducing the following

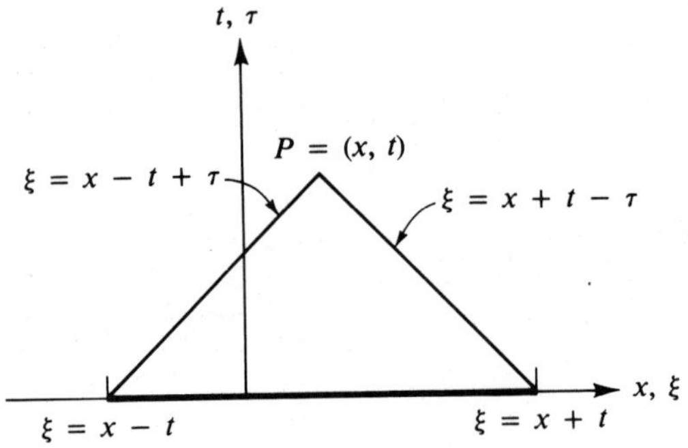

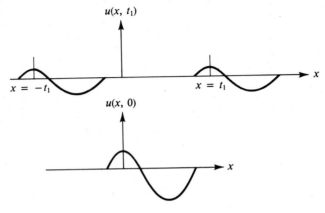

Figure 3.12

Figure 3.13

transformation of dependent variable $u \to v$:

$$v(x, t) \equiv \int_0^t u(x, \tau) \, d\tau \tag{117}$$

If $u(x, t)$ is a solution of problem 3, then

$$v_t = u(x, t), \qquad v_{tt} = u_t(x, t)$$

$$v_{xx} = \int_0^t u_{xx}(x, \tau) \, d\tau = \int_0^t u_{tt}(x, \tau) \, d\tau = u_t(x, t) - u_t(x, 0) = u_t(x, t) = v_{tt} \tag{118}$$

Therefore, $v(x, t)$ satisfies

$$v_{tt} - v_{xx} = 0 \tag{119a}$$

$$v(x, 0^+) = 0 \tag{119b}$$

$$v_t(x, 0^+) = u(x, 0^+) = f(x) \tag{119c}$$

It then follows from (116) that $v(x, t)$ is given by

$$v(x, t) = \tfrac{1}{2} \int_{x-t}^{x+t} f(\xi) \, d\xi$$

and $u(x, t)$ is

$$u(x, t) = v_t(x, t) = \tfrac{1}{2}[f(x + t) + f(x - t)] \tag{120}$$

Thus, the initial disturbance splits up into two identical *half-scale* shapes, which propagate to the left and right undistorted with constant speed unity, as shown in Figure 3.13.

To describe the solution of the general initial-value problem, we combine results (114), (116), and (120). This is called D'Alembert's solution.

3.4.3 An Example

To illustrate a particular application of the preceding results, let us study the propagation of two initial discontinuities in the surface height and speed of shallow water, as sketched in Figure 3.14. In this chapter, we shall consider only the small-amplitude case, as exhibited by the occurrence of the small parameter ε in the initial conditions. The nonlinear problem $[\varepsilon = O(1)]$ is discussed in Chapters 5, 7 and 8.

We expand u and h as in (32):

$$u(x, t; \varepsilon) = \varepsilon u_1(x, t) + \cdots \tag{121a}$$

$$h(x, t; \varepsilon) = 1 + \varepsilon h_1(x, t) + \cdots \tag{121b}$$

and obtain the linear equations (34) for u_1 and h_1—that is,

$$h_{1_t} + u_{1_x} = 0 \tag{122a}$$

$$u_{1_t} + h_{1_x} = 0 \tag{122b}$$

The initial conditions indicated in the sketch translate to:

$$u_1(x, 0) = \tilde{u}(x) \equiv \begin{cases} \tilde{u}_1 = \text{constant} > 0, & \text{if } -\infty < x < 0 \\ 0, & \text{if } 0 < x < 1 \\ \tilde{u}_2 = \text{constant} < 0, & \text{if } 1 < x < \infty \end{cases} \tag{123a}$$

$$h_1(x, 0) = \tilde{h}(x) \equiv \begin{cases} \tilde{h}_1 = \text{constant} > 0, & \text{if } -\infty < x < 0 \\ 0, & \text{if } 0 < x < 1 \\ \tilde{h}_2 = \text{constant} > \tilde{h}_1, & \text{if } 1 < x < \infty \end{cases} \tag{123b}$$

We have a number of options for calculating the solution. One is to exploit the fact that the solution must depend on the so-called characteristic coordinates [see (107)]

$$\zeta = x + t \tag{124a}$$

$$\sigma = x - t \tag{124b}$$

Thus, transforming (122) to the ζ, σ variables should simplify the calculations. This is a special case of the *method of characteristics* that we discuss in general in Chapters 4–7; this method of solution is particularly well suited to solving coupled systems of first order, as in (122). A second approach is to eliminate u_1 or h_1 from (122) and solve the resulting wave equation using the results of the

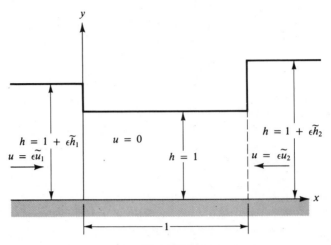

Figure 3.14

previous section. These two approaches are illustrated next. One could also use Laplace transforms with respect to t or Fourier transforms with respect to x on (122), but the details are not worked out here.

Method of Characteristics

Using (124), we obtain the transformation rules

$$\frac{\partial}{\partial t} = \frac{\partial}{\partial \zeta} - \frac{\partial}{\partial \sigma}; \qquad \frac{\partial}{\partial x} = \frac{\partial}{\partial \zeta} + \frac{\partial}{\partial \sigma}$$

Therefore, letting $h_1(x, t) = h_1((\zeta + \sigma)/2, (\zeta - \sigma)/2) \equiv H_1(\zeta, \sigma)$ and $u_1(x, t) = u_1((\zeta + \sigma)/2, (\zeta - \sigma)/2) \equiv U_1(\zeta, \sigma)$, we find that (122) transform to

$$H_{1_\zeta} - H_{1_\sigma} + U_{1_\zeta} + U_{1_\sigma} = 0 \qquad (125a)$$

$$H_{1_\zeta} + H_{1_\sigma} + U_{1_\zeta} - U_{1_\sigma} = 0 \qquad (125b)$$

Adding and subtracting gives

$$(H_1 + U_1)_\zeta = 0 \qquad (126a)$$

$$(H_1 - U_1)_\sigma = 0 \qquad (126b)$$

and we conclude that

$$H_1 + U_1 = F(\sigma) \qquad (127a)$$

$$H_1 - U_1 = G(\zeta) \qquad (127b)$$

for functions F and G to be specified. Solving for H_1 and U_1 gives

$$H_1(\zeta, \sigma) = \tfrac{1}{2}[F(\sigma) + G(\zeta)] \qquad (128a)$$

$$U_1(\zeta, \sigma) = \tfrac{1}{2}[F(\sigma) - G(\zeta)] \qquad (128b)$$

Using the initial conditions, (123) defines F and G in the form:

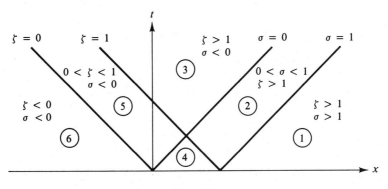

Figure 3.15

$$F(\sigma) = \tilde{h}(\sigma) + \tilde{u}(\sigma) \tag{129a}$$

$$G(\zeta) = \tilde{h}(\zeta) - \tilde{u}(\zeta) \tag{129b}$$

Therefore, the solution for H_1 and U_1 is

$$H_1(\zeta, \sigma) = \tfrac{1}{2}\{\tilde{h}(\sigma) + \tilde{u}(\sigma) + \tilde{h}(\zeta) - \tilde{u}(\zeta)\} \tag{130a}$$

$$U_1(\zeta, \sigma) = \tfrac{1}{2}\{\tilde{h}(\sigma) + \tilde{u}(\sigma) - \tilde{h}(\zeta) + \tilde{u}(\zeta)\} \tag{130b}$$

Since the functions $\tilde{h}(x)$ and $\tilde{u}(x)$ change value at $x = 0$ and $x = 1$, we must focus on the four characteristic lines $\sigma = 0$, $\sigma = 1$, $\zeta = 0$ and $\zeta = 1$ at which the functions $\tilde{h}(\sigma)$, $\tilde{h}(\zeta)$, $\tilde{u}(\sigma)$, and $\tilde{u}(\zeta)$ will also switch values. The four characteristic lines subdivide the upper half-xt-plane into the six regions sketched in Figure 3.15.

The solution (130) takes on the following values in the six regions:

①: $\zeta > 1, \sigma > 1$
$$H_1 = \tfrac{1}{2}\{\tilde{h}_2 + \tilde{u}_2 + \tilde{h}_2 - \tilde{u}_2\} = \tilde{h}_2$$
$$U_1 = \tfrac{1}{2}\{\tilde{h}_2 + \tilde{u}_2 - \tilde{h}_2 + \tilde{u}_2\} = \tilde{u}_2$$

②: $\zeta > 1, 0 < \sigma < 1$
$$H_1 = \tfrac{1}{2}\{0 + 0 + \tilde{h}_2 - \tilde{u}_2\} = \tfrac{1}{2}(\tilde{h}_2 - \tilde{u}_2)$$
$$U_1 = \tfrac{1}{2}\{0 + 0 - \tilde{h}_2 + \tilde{u}_2\} = \tfrac{1}{2}(\tilde{u}_2 - \tilde{h}_2)$$

③: $\zeta > 1, \sigma < 0$
$$H_1 = \tfrac{1}{2}\{\tilde{h}_1 + \tilde{u}_1 + \tilde{h}_2 - \tilde{u}_2\}$$
$$U_1 = \tfrac{1}{2}\{\tilde{h}_1 + \tilde{u}_1 - \tilde{h}_2 + \tilde{u}_2\}$$

④: $0 < \zeta < 1, 0 < \sigma < 1$
$$H_1 = \tfrac{1}{2}\{0 + 0 + 0 - 0\} = 0$$
$$U_1 = \tfrac{1}{2}\{0 + 0 - 0 + 0\} = 0$$

⑤: $0 < \zeta < 1,\, \sigma < 0$

$H_1 = \frac{1}{2}\{\tilde{h}_1 + \tilde{u}_1 + 0 - 0\} = \frac{1}{2}(\tilde{h}_1 + \tilde{u}_1)$

$U_1 = \frac{1}{2}\{\tilde{h}_1 + \tilde{u}_1 - 0 + 0\} = \frac{1}{2}(\tilde{h}_1 + \tilde{u}_1)$

⑥: $\zeta < 0,\, \sigma < 0$

$H_1 = \frac{1}{2}\{\tilde{h}_1 + \tilde{u}_1 + \tilde{h}_1 - \tilde{u}_1\} = \tilde{h}_1$

$U_1 = \frac{1}{2}\{\tilde{h}_1 + \tilde{u}_1 - \tilde{h}_1 + \tilde{u}_1\} = \tilde{u}_1$

D'Alembert Solution

We eliminate u_1 from (122) and obtain

$$h_{1_{tt}} - h_{1_{xx}} = 0 \tag{131}$$

subject to (123b). In order to derive the initial condition on h_{1_t}, we evaluate (122a) at $t = 0$ and use the derivative of (123a) to calculate $u_{1_x}(x, 0)$. Since we are dealing with a discontinuous function, we interpret the derivative in the symbolic sense. Thus, if $f(x)$ is a function that is defined and has a continuous derivative everywhere except at $x = x_0$, then the symbolic derivative $f_s'(x)$ is defined as (see pp. 141–143 of [4])

$$f_s'(x) \equiv f'(x) + [f(x_0^+) - f(x_0^-)]\delta(x - x_0) \tag{132}$$

Therefore,

$$h_{1_t}(x, 0) = -u_{1_x}(x, 0) = -\tilde{u}_s'(x) \tag{133}$$

and since $\tilde{u}(x) = 0$ if $x \neq 0$ and $x \neq 1$, we have, according to (123a):

$$h_{1_t}(x, 0) = \delta(x)\tilde{u}_1 - \delta(x - 1)\tilde{u}_2 \tag{134}$$

The solution of (131) subject to (123b) and (134) is then given by combining

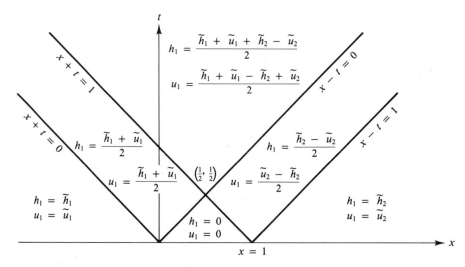

Figure 3.16

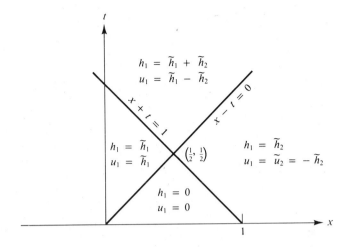

Figure 3.17

(116) and (120)—that is,

$$h_1(x,t) = \tfrac{1}{2}\{\tilde{h}(x+t) + \tilde{h}(x-t)\} + \tfrac{1}{2} \int_{x-t}^{x+t} \{\delta(\xi)\tilde{u}_1 - \delta(\xi - 1)\tilde{u}_2\}\, d\xi \qquad (135)$$

It is left as an exercise to show that

$$u_1(x,t) = \tfrac{1}{2}\{\tilde{u}(x+t) + \tilde{u}(x-t)\} + \tfrac{1}{2} \int_{x-t}^{x+t} [\delta(\xi)\tilde{h}_1 - \delta(\xi - 1)\tilde{h}_2]\, d\xi \qquad (136)$$

and that the results (135) and (136) reduce to the expressions obtained from (130) in the six regions (Problem 7). The solution is summarized in Figure 3.16.

Let us view our results for arbitrary choices of the four initial constants $\tilde{u}_1$, $\tilde{u}_2$, $\tilde{h}_1$, and $\tilde{h}_2$. We see that at time $t = 0^+$, outward propagating disturbances arise at $x = 0$ and $x = 1$ in addition to the initial inward propagating disturbances there. In particular, the initial discontinuity at $x = 1$ splits up into two waves, one propagating to the right along $x - t = 1$ and a second propagating to the left along $x + t = 1$. A similar situation occurs for the initial discontinuity at $x = 0$. Thus, arbitrary values of these four constants do not constitute a "steady configuration" in the sense of producing only one wave (which is the translation of the initial discontinuity at $x = 1$) moving to the left and a second wave (the translation of the initial discontinuity at $x = 0$) moving to the right. To achieve this steady configuration, we must require the solutions in regions ① and ② to coincide and the solutions in ⑤ and ⑥ to coincide. This will be true only if we choose

$$\tilde{u}_2 \equiv -\tilde{h}_2 \qquad (137a)$$

$$\tilde{u}_1 \equiv \tilde{h}_1 \qquad (137b)$$

In this case, the solution in region ③ is given by

$$h_1 = \tilde{h}_1 + \tilde{h}_2 \qquad (138a)$$

$$u_1 = \tilde{h}_1 - \tilde{h}_2 \qquad (138b)$$

and consists of the initial discontinuities alone, as sketched in Figure 3.17.

We see that in this linear problem, discontinuities propagate along the characteristics $\sigma = $ constant and $\zeta = $ constant. This feature of linear hyperbolic problems is explored more fully in Chapter 4. The situation is significantly more complicated for nonlinear problems where discontinuities (bores or shocks) still occur but are propagated along certain curves that are not characteristics. Moreover, these shocks do interact, unlike the situation described in Figures 3.16 and 3.17. See Problem 6 of Chapter 7.

3.5   Initial- and Boundary-Value Problems on the Semi-infinite Interval; Green's Functions

We can appeal to symmetry arguments, as in Chapters 1 and 2, to construct Green's functions for various homogeneous boundary-value problems. These, in turn, may be used to solve both homogeneous and inhomogeneous problems on the semi-infinite interval $0 \leq x < \infty$.

3.5.1 Green's Function of the First Kind

Green's function of the first kind solves the problem

$$u_{tt} - u_{xx} = \delta(x - \xi)\delta(t - \tau); \qquad 0 \leq x; \tau \leq t \tag{139a}$$

with zero initial conditions at $t = \tau$ of

$$u(x, \tau^-) = u_t(x, \tau^-) = 0 \tag{139b,c}$$

and zero boundary value for u at $x = 0$ of

$$u(0, t) = 0; \qquad t > \tau \tag{139d}$$

Here, ξ and τ are arbitrary positive constants.

It is clear by symmetry that the solution consists of the sum of the two fundamental solutions given next (see (45) of Chapter 1):

$$G_1(x, \xi, t - \tau) \equiv F(x - \xi, t - \tau) - F(x + \xi, t - \tau) \tag{140}$$

where F is defined by (110).

Since the two sources are switched on at $t = \tau$, the value of G_1 in region ①—that is, $t < \tau$—is equal to zero for all x (see Figure 3.18). Now, for a *fixed source location* (ξ, τ) and image source location $(-\xi, \tau)$, consider the value that G_1 takes on at different points in the xt-plane. This value depends on whether the point (x, t) is in the zone of influence of the primary source alone; the image source alone; both; or neither. Thus, if (x, t) is in ②, ⑦ and ⑥, it is outside the zone of influence of either the primary or image source, and $G_1 = 0$. In region ⑤, $G_1 = \frac{1}{2}$ because (x, t) is influenced only by the primary, and in region ③, $G_1 = -\frac{1}{2}$ because (x, t) is influenced only by the image source. Finally, in region ④, (x, t) is influenced by both sources and its value is zero, since the two contributions cancel.

It is also useful to look at (140) with (x, t) *fixed* for different values of ξ and τ to determine the domain of dependence of (x, t). That is, the set of points (ξ, τ) in the plane at which a primary source [with a mirror source located at $(-\xi, \tau)$] will result in a nonvanishing value of G_1 at (x, t). And, since (x, t) is in the first quadrant, the value of G_1 will equal $\frac{1}{2}$.

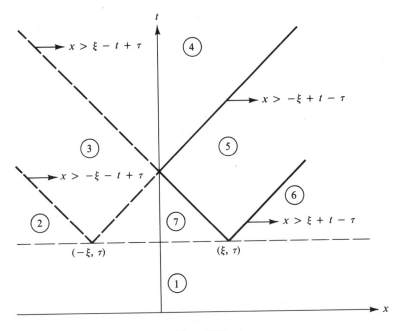

Figure 3.18

We distinguish two cases, $x < t$ and $x > t$, as sketched in Figure 3.19. Keeping in mind that (x, t) is a fixed point and that we are studying varying values of ξ and τ, we locate the four critical lines

$$\xi = x + t - \tau$$
$$\xi = -x + t - \tau$$
$$\xi = x - t + \tau$$
$$\xi = -x - t + \tau$$

at which the arguments of the two Heaviside functions in (140) equal zero. It then follows that the domain of dependence of (x, t) is the region bounded by the straight lines through $PQRSP$ for $x < t$. For $x > t$, this region is the triangle $PRSP$.

In summary, for a fixed P, the value of G_1 equals $\frac{1}{2}$ only if the primary source is located somewhere inside the shaded domain of dependence of P. Otherwise, $G_1 = 0$.

3.5.2 Homogeneous Boundary Condition, Nonzero Initial Conditions

As in (47) of Chapter 1, we can use (140) to solve:

$$u_{tt} - u_{xx} = p(x, t); \qquad 0 \le x; 0 \le t \tag{141a}$$

$$u(x, 0^-) = u_t(x, 0^-) = 0 \tag{141b,c}$$

$$u(0, t) = 0; \qquad t > 0 \tag{141d}$$

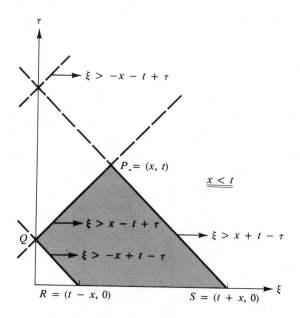

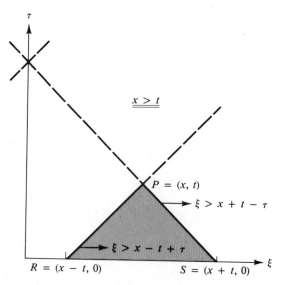

Figure 3.19

by superposition in the form

$$u(x,t) = \int_0^t d\tau \int_0^\infty p(\xi, \tau) G_1(x, \xi, t - \tau) \, d\xi \tag{142}$$

Now, for a fixed (x, t), G_1, regarded as a function of ξ and τ, vanishes outside the domain of dependence of (x, t) and equals $\frac{1}{2}$ inside. Therefore, (142) has the more explicit form:

$$u(x,t) = \tfrac{1}{2} \int_{\tau=0}^{t} \left[\int_{\xi=x-t+\tau}^{x+t-\tau} p(\xi,\tau)\,d\xi \right] d\tau; \qquad t < x \tag{143a}$$

$$u(x,t) = \tfrac{1}{2} \int_{\tau=0}^{t-x} \left[\int_{\xi=-x+t-\tau}^{x+t-\tau} p(\xi,\tau)\,d\xi \right] d\tau + \tfrac{1}{2} \int_{\tau=t-x}^{t} \left[\int_{\xi=x-t+\tau}^{x+t-\tau} p(\xi,\tau)\,d\xi \right] d\tau; t > x \tag{143b}$$

Now, suppose we wish to solve the homogeneous wave equation with homogeneous boundary condition on the left but with a prescribed value of u_t at $t = 0$—that is,

$$u_{tt} - u_{xx} = 0; \qquad 0 \le x; 0 \le t \tag{144a}$$

$$u(x,0^+) = 0 \tag{144b}$$

$$u_t(x,0^+) = h(x) \tag{144c}$$

$$u(0,t) = 0; \qquad t > 0 \tag{144d}$$

We convert this to the equivalent inhomogeneous equation

$$u_{tt} - u_{xx} = \delta(t)h(x); \qquad 0 \le x; 0 \le t \tag{145a}$$

with zero initial conditions,

$$u(x,0^-) = u_t(x,0^-) = 0 \tag{145b,c}$$

and the same zero-boundary condition (144d).

Therefore, we can use the results in (143) with $p = \delta(t)h(x)$ to obtain:

$$u(x,t) = \tfrac{1}{2} \int_{x-t}^{x+t} h(\xi)\,d\xi; \qquad t < x \tag{146a}$$

$$u(x,t) = \tfrac{1}{2} \int_{-x+t}^{x+t} h(\xi)\,d\xi; \qquad t > x \tag{146b}$$

The result (146a) for $t < x$ is the same as that in (116) for the problem on the infinite interval. This is because the boundary $x = 0$ is outside the domain of dependence of the point (x, t) for $t < x$. In other words, the observer is unaware of the boundary for times $t < x$ because reflected disturbances from $x = 0$ have not arrived yet. The result (146b), which was obtained formally from (143b) and in which the second integral of (143b) gives no contribution, can also be deduced directly using the following symmetry arguments.

To solve (144), consider the equivalent initial-value problem on *the entire axis*, $-\infty < x < \infty$, where the initial velocity $u_t(x,0^+)$ is now defined by:

$$u_t(x,0^+) = h^*(x) \equiv \begin{cases} h(x), & \text{if } x > 0 \\ -h(-x), & \text{if } x < 0 \end{cases}$$

Thus, we extend the definition of the given initial velocity $h(x)$, available on $0 \le x < \infty$, to the negative axis in such a way as to ensure a zero value of u at the origin.

The solution of the "extended initial-value problem" is, according to (116),

$$u(x,t) = \frac{1}{2}\int_{x-t}^{x+t} h^*(\xi)\,d\xi = \frac{1}{2}\left[\int_{x-t}^{0} h^*(\xi)\,d\xi + \int_{0}^{x+t} h^*(\xi)\,d\xi\right], \qquad \text{if } t > x$$

$$= \frac{1}{2}\int_{x-t}^{0} -h(-\xi)\,d\xi + \frac{1}{2}\int_{0}^{x+t} h(\xi)\,d\xi$$

$$= \frac{1}{2}\int_{-x+t}^{0} h(\xi)\,d\xi + \frac{1}{2}\int_{0}^{x+t} h(\xi)\,d\xi = \frac{1}{2}\int_{-x+t}^{x+t} h(\xi)\,d\xi$$

in agreement with (146b).

To complete our listing of solutions of the homogeneous boundary-value problem, we need to consider:

$$u_{tt} - u_{xx} = 0; \qquad 0 \le x; 0 \le t \tag{147a}$$

$$u(x,0^+) = f(x) \tag{147b}$$

$$u_t(x,0^+) = 0 \tag{147c}$$

$$u(0,t) = 0; \qquad t > 0 \tag{147d}$$

As in our discussion of the initial-value problem on the infinite x-axis, we introduce the transformation (117):

$$v(x,t) \equiv \int_{0}^{t} u(x,\tau)\,d\tau \tag{148}$$

to obtain

$$v_{tt} - v_{xx} = 0 \tag{149a}$$

$$v(x,0^+) = 0 \tag{149b}$$

$$v_t(x,0^+) = f(x) \tag{149c}$$

and the boundary condition at $x = 0$ is still zero, because

$$v(0,t) = \int_{0}^{t} u(0,\tau)\,d\tau = 0 \tag{149d}$$

Therefore, the solution for v is given by (146) with h replaced by f, and taking the partial derivative of this result with respect to t gives u; that is,

$$u(x,t) = \tfrac{1}{2}[f(x+t) + f(x-t)]; \qquad t < x \tag{150a}$$

$$u(x,t) = \tfrac{1}{2}[f(x+t) - f(-x+t)]; \qquad t > x \tag{150b}$$

Here again, the idea of an "image" initial condition,

$$u(x,0) = f^*(x) = \begin{cases} f(x), & \text{if } x > 0 \\ -f(-x), & x < 0 \end{cases}$$

used in (120) gives the desired result. We have, according to (120):

$$u(x,t) = \tfrac{1}{2}[f^*(x+t) + f^*(x-t)] = \tfrac{1}{2}[f(x+t) + f(x-t)] \qquad \text{if } t < x$$

and

$$u(x,t) = \tfrac{1}{2}[f(x+t) - f(-x+t)] \qquad \text{if } t > x$$

in agreement with (150).

3.5.3 Inhomogeneous Boundary Condition $u(0, t) = g(t)$

We wish to solve the problem [see the linearized wavemaker problem (44)–(46) or the linearized piston problem defined by (85), (86), and (78a)]:

$$u_{tt} - u_{xx} = 0; \qquad 0 \le x; 0 \le t \tag{151a}$$

with zero initial conditions

$$u(x, 0^+) = 0; \qquad u_t(x, 0^+) = 0 \tag{151b,c}$$

and a prescribed boundary condition

$$u(0, t) = g(t); \qquad t > 0 \tag{151d}$$

We introduce the homogenizing transformation $w(x, t) \equiv u(x, t) - g(t)$, as in (56) of Chapter 1, then solve the problem for w using the results in the previous section.

We find that $w(x, t)$ obeys

$$w_{tt} - w_{xx} = -\ddot{g}(t) \tag{152a}$$

$$w(x, 0^+) = -g(0^+) \tag{152b}$$

$$w_t(x, 0^+) = -\dot{g}(0^+) \tag{152c}$$

$$w(0, t) = 0; \qquad t > 0 \tag{152d}$$

For $t < x$, using the results in (143a) with $p = -\ddot{g}$, (146a) with $h = -\dot{g}(0^+)$, and (150a) with $f = -g(0^+)$ gives

$$w(x, t) = \frac{1}{2} \int_{0^+}^t \left[\int_{x-t+\tau}^{x+t-\tau} \ddot{g}(\tau)\, d\xi \right] d\tau - \frac{1}{2} \int_{x-t}^{x+t} \dot{g}(0^+)\, d\xi - g(0^+); \qquad t < x \tag{153}$$

Evaluating the various integrals gives $w(x, t) = -g(t)$, or $u(x, t) = 0$ if $t < x$. For $t > x$, we have

$$w(x, t) = -\frac{1}{2} \int_{0^+}^{t-x} \left[\int_{-x+t-\tau}^{x+t-\tau} \ddot{g}(\tau)\, d\xi \right] d\tau - \frac{1}{2} \int_{t-x}^{t} \left[\int_{x-t+\tau}^{x+t-\tau} \ddot{g}(\tau)\, d\xi \right] d\tau$$

$$- \frac{1}{2} \int_{-x+t}^{x+t} \dot{g}(0^+)\, d\xi; \qquad t > x. \tag{154}$$

Evaluating the integrals in (154) results in

$$w(x, t) = -g(t) + g(t - x); \qquad t > x$$

Therefore,

$$u(x, t) = \begin{cases} 0; & \text{if } t < x \\ g(t - x); & \text{if } t > x \end{cases} \tag{155}$$

as derived using Laplace transforms in Problem 3.

As discussed in Problem 10 of Chapter 1, we can use Green's functions to solve certain mixed boundary-value problems after an appropriate transformation of the dependent variable. This idea of "transform and conquer" is further illustrated in Problem 10 of this chapter.

3.5.4 An Example

We illustrate the results of Sections 3.5.1–3.5.3 by studying in detail the solution of the signaling problem for water waves as formulated in Section 3.2.9. We have shown that the $O(\varepsilon)$ problem satisfies [see (34), (45), and (46)]

$$h_{1_t} + u_{1_x} = 0 \tag{156a}$$

$$u_{1_t} + h_{1_x} = 0 \tag{156b}$$

with zero initial conditions

$$u_1(x, 0) = 0 \tag{156c}$$

$$h_1(x, 0) = 0 \tag{156d}$$

and the wavemaker boundary condition

$$u_1(0, t) = \dot{s}(t), \qquad t > 0 \tag{156e}$$

The $O(\varepsilon^2)$ problem is governed by [see (35), (49) and (51)]

$$h_{2_t} + u_{2_x} = -(u_1 h_1)_x \tag{157a}$$

$$u_{2_t} + h_{2_x} = -\tfrac{1}{2}(u_1^2)_x \tag{157b}$$

with zero initial conditions

$$u_2(x, 0) = 0 \tag{157c}$$

$$h_2(x, 0) = 0 \tag{157d}$$

and the wavemaker boundary condition

$$u_2(0, t) = -u_{1_x}(0, t)s(t) \tag{157e}$$

As in Section 3.4.3, we shall solve these signaling problems first directly using the method of characteristics and then in terms of Green's function for the wave equations for u_1 and u_2 that result when h_1 and h_2 are eliminated from the governing equations.

Method of Characteristics

For the signaling problem, it is convenient to change the sign of (124b), and we introduce the two characteristic variables

$$\zeta = t + x \tag{158a}$$

$$\mu = t - x \tag{158b}$$

which imply the following transformations for the x and t derivatives.

$$\frac{\partial}{\partial x} = \frac{\partial}{\partial \zeta} - \frac{\partial}{\partial \mu} \tag{158c}$$

$$\frac{\partial}{\partial t} = \frac{\partial}{\partial \zeta} + \frac{\partial}{\partial \mu} \tag{158d}$$

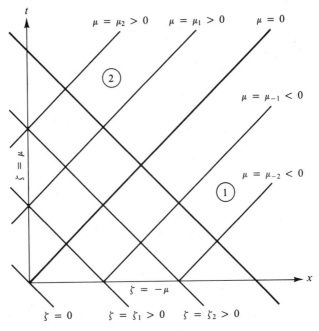

Figure 3.20

Denoting $u_i(x, t) = u_i((\zeta - \mu)/2, (\zeta + \mu)/2) \equiv U_i(\zeta, \mu)$ and $h_i(x, t) = h_i((\zeta - \mu)/2,$ $(\zeta + \mu)/2) \equiv H_i(\zeta, \mu)$ for $i = 1, 2$, we find that (156a)–(156b) imply that U_1 and H_1 satisfy

$$(H_1 + U_1)_\zeta = 0 \tag{159a}$$

$$(H_1 - U_1)_\mu = 0 \tag{159b}$$

The characteristic lines $\zeta = \text{constant}$ and $\mu = \text{constant}$ are sketched in Figure 3.20, and we are interested in the domain $\zeta - \mu > 0$, $\zeta + \mu > 0$. The initial conditions (156c)–(156d) imply that

$$U_1(\zeta, -\zeta) = H_1(\zeta, -\zeta) = 0; \qquad \zeta \geq 0 \tag{160}$$

and the boundary condition (156e) becomes

$$U_1(\mu, \mu) = \dot{s}(\mu); \qquad \mu > 0 \tag{161}$$

The solution of (159) is simply

$$H_1 + U_1 = F_1(\mu) \tag{162a}$$

$$H_1 - U_1 = G_1(\zeta) \tag{162b}$$

where the functions F_1 and G_1 are to be determined by the initial and boundary conditions. Applying the initial conditions (160) first, we find that

$$F_1(\mu) = 0 \qquad \text{if } \mu \leq 0 \tag{163a}$$

$$G_1(\zeta) = 0 \qquad \text{if } \zeta \geq 0 \tag{163b}$$

Therefore, (162) reduces to $H_1 + U_1 = 0$ in ① and $H_1 - U_1 = 0$ in ① and ②. In particular, the preceding implies that

$$U_1(\zeta, \mu) = H_1(\zeta, \mu) = 0 \qquad \text{in ①} \tag{164a}$$

and that

$$H_1(\zeta, \mu) = U_1(\zeta, \mu) \qquad \text{in ②} \tag{164b}$$

To complete the solution in ②, we need to calculate $F_1(\mu)$ there for $\mu > 0$. First we set $H_1 = U_1$ in (162a) to obtain

$$2U_1(\zeta, \mu) = F_1(\mu) \qquad \text{in ②} \tag{165}$$

Applying boundary condition (161) gives

$$2\dot{s}(\mu) = F_1(\mu) \qquad \text{for } \mu > 0 \tag{166}$$

and this defines F_1 for $\mu > 0$. Therefore, (165) gives

$$U_1(\zeta, \mu) = \dot{s}(\mu) \qquad \text{in ②} \tag{167a}$$

and (164b) gives

$$H_1(\zeta, \mu) = \dot{s}(\mu) \qquad \text{in ②} \tag{167b}$$

The solution in terms of the x, t, u_1, h_1 variables is, therefore,

$$u_1(x, t) = h_1(x, t) = \begin{cases} 0, & \text{if } t \leq x \\ \dot{s}(t - x), & \text{if } t > x \end{cases} \tag{168}$$

We shall use the same procedure for deriving U_2 and H_2, but now the inhomogeneous terms in (157a)–(157b) complicate the calculations slightly. These equations imply that

$$2(H_2 + U_2)_\zeta = -\left(U_1 H_1 + \frac{U_1^2}{2}\right)_\zeta + \left(U_1 H_1 + \frac{U_1^2}{2}\right)_\mu \tag{169a}$$

$$2(H_2 - U_2)_\mu = -\left(U_1 H_1 - \frac{U_1^2}{2}\right)_\zeta + \left(U_1 H_1 - \frac{U_1^2}{2}\right)_\mu \tag{169b}$$

Substituting the known solution for U_1 and H_1 gives

$$2(H_2 + U_2)_\zeta = 3\dot{s}(\mu)\ddot{s}(\mu) \tag{170a}$$

$$2(H_2 - U_2)_\mu = \dot{s}(\mu)\ddot{s}(\mu) \tag{170b}$$

and (170) are valid in both ① and ② as long as we regard

$$\dot{s}(\mu) \equiv 0 \qquad \text{if } \mu \leq 0 \tag{170c}$$

Henceforth (170c) is tacitly assumed in our calculations and results. The initial conditions (157c)–(157d) transform to

$$U_2(\zeta, -\zeta) = H_2(\zeta, -\zeta) = 0; \qquad \zeta \geq 0 \tag{171}$$

and when the solution (167a) is used, boundary condition (157e) becomes

$$U_2(\mu, \mu) = \ddot{s}(\mu)s(\mu) \qquad \text{for } \mu > 0 \tag{172}$$

We integrate (170b) with respect to μ to obtain

$$H_2 - U_2 = \tfrac{1}{4}\dot{s}^2(\mu) + G_2(\zeta) \tag{173a}$$

where G_2 is as yet unspecified. In view of the initial conditions (171), we must have $G_2(\zeta) \equiv 0$. Therefore, one relation between U_2 and H_2 that is valid in the entire domain of interest is

$$H_2(\zeta,\mu) - U_2(\zeta,\mu) = \tfrac{1}{4}\dot{s}^2(\mu) \tag{173b}$$

We next integrate (170a) to find

$$H_2(\zeta,\mu) + U_2(\zeta,\mu) = \tfrac{3}{2}\dot{s}(\mu)\ddot{s}(\mu)\zeta + F_2(\mu) \tag{174}$$

In preparation for evaluating the unknown function $F_2(\mu)$, we use (173b) to eliminate H_2 from (174). This gives

$$2U_2(\zeta,\mu) = -\tfrac{1}{4}\dot{s}^2(\mu) + \tfrac{3}{2}\dot{s}(\mu)\ddot{s}(\mu)\zeta + F_2(\mu) \tag{175}$$

Imposing the boundary condition (172) on (175) defines $F_2(\mu)$ to be

$$F_2(\mu) = \tfrac{1}{4}\dot{s}^2(\mu) - \tfrac{3}{2}\dot{s}(\mu)\ddot{s}(\mu)\mu + 2\ddot{s}(\mu)s(\mu) \tag{176}$$

Now (175) gives

$$U_2(\zeta,\mu) = \tfrac{3}{4}\dot{s}(\mu)\ddot{s}(\mu)(\zeta - \mu) + \ddot{s}(\mu)s(\mu) \tag{177a}$$

and (173b) gives

$$H_2(\zeta,\mu) = \tfrac{1}{4}\dot{s}^2(\mu) + \tfrac{3}{4}\dot{s}(\mu)\ddot{s}(\mu)(\zeta - \mu) + \ddot{s}(\mu)s(\mu) \tag{177b}$$

This completes the solution of the $O(\varepsilon^2)$ problem, which can also be written in terms of the x, t, u_2, h_2 variables in the form

$$u_2(x,t) = \begin{cases} 0, & \text{if } t \le x \\ \tfrac{3}{2}\dot{s}(t-x)\ddot{s}(t-x)x + \ddot{s}(t-x)s(t-x), & \text{if } t > x \end{cases} \tag{178a}$$

$$h_2(x,t) = \begin{cases} 0, & \text{if } t \le x \\ \tfrac{1}{4}\dot{s}^2(t-x) + \tfrac{3}{2}\dot{s}(t-x)\ddot{s}(t-x)x + \ddot{s}(t-x)s(t-x), & \text{if } t > x \end{cases} \tag{178b}$$

Suppose the wavemaker has a periodic motion—say $s(t) \equiv \sin kt$ with $k = $ constant. We see from (178) that each of our results for u or h to $O(\varepsilon^2)$ contains the term $\tfrac{3}{2}\varepsilon^2[x\dot{s}(t-x)\ddot{s}(t-x)] = -\tfrac{3}{2}\varepsilon^2[xk^3\cos k(t-x)\sin k(t-x)]$, which has an amplitude equal to $\tfrac{3}{2}\varepsilon^2 xk^3$. Thus, no matter how small ε is, this term, which is nominally $O(\varepsilon^2)$, will grow to be of order ε if x is as large as $O(\varepsilon^{-1})$. This result is not only physically inconsistent with a periodic boundary disturbance, but it also violates the implicit ordering of terms assumed in the expansion (32). Thus, the solution we have calculated is not valid in the "far-field"; it is reasonable only as long as $x = O(1)$. See Section 8.4.4, where we discuss an expansion procedure that remains valid in the far-field.

Green's Function

Let us now rederive the preceding results using the Green's function approach discussed in Sections 3.5.1–3.5.3. As pointed out in Section 3.2.9, eliminating h_1 gives the wave equation (44) for u_1, the initial conditions (45), and the boundary condition (46). This is just the problem discussed in Section 3.5.3 with $g = \dot{s}$, and in fact the result (155) agrees with (168) for $u_1(x, t)$. To compute h_1, we substitute the preceding value of u_1 in (47) and ensure that the possible discontinuity in u_1 and u_{1_x} at $x = t$ is taken into account by writing [compare with the discussion following (131)]

$$u_1(x, t) = \dot{s}(t - x)H(t - x) \tag{179a}$$

where H is the Heaviside function. The x-derivative must then be written as

$$u_{1_x}(x, t) = -\ddot{s}(t - x)H(t - x) - \dot{s}(t - x)\delta(t - x) \tag{179b}$$

Equation (47) becomes

$$
\begin{aligned}
h_1(x, t) &= \int_{\tau=0}^{t} \ddot{s}(\tau - x)H(\tau - x)\,d\tau + \int_{\tau=0}^{t} \dot{s}(\tau - x)\delta(\tau - x)\,d\tau \\
&= \begin{cases} 0, & \text{if } t \le x \\ \int_{x}^{t} \ddot{s}(\tau - x)\,d\tau + \dot{s}(0^+) = \dot{s}(t - x), & \text{if } t > x \end{cases} \\
&= \dot{s}(t - x)H(t - x) \tag{180}
\end{aligned}
$$

which is also in agreement with our earlier result.

Consider now the $O(\varepsilon^2)$ terms governed by (157). If we substitute the expressions (179b) and (180) for u_1 and h_1 into the right-hand sides of (157a,b), we have

$$h_{2_t} = u_{2_x} = 2\dot{s}(t - x)\ddot{s}(t - x)H(t - x) + \dot{s}^2(t - x)\delta(t - x) \tag{181a}$$

$$u_{2_t} = h_{2_x} = \dot{s}(t - x)\ddot{s}(t - x)H(t - x) + \tfrac{1}{2}\dot{s}^2(t - x)\delta(t - x) \tag{181b}$$

To eliminate h_2, we differentiate (181b) with respect to t and subtract from this the derivative of (181a) with respect to x to obtain

$$u_{2_{tt}} - u_{2_{xx}} = p(t - x) \tag{182}$$

where

$$
\begin{aligned}
p(t - x) &\equiv 3[\dot{s}(t - x)\ddot{s}(t - x)]\dot{\ }H(t - x) + 6\dot{s}(t - x)\ddot{s}(t - x)\delta(t - x) \\
&\quad + \tfrac{3}{2}\dot{s}^2(t - x)\dot{\delta}(t - x) \tag{183}
\end{aligned}
$$

The initial conditions are

$$u_2(x, 0) = u_{2_t}(x, 0) = 0 \tag{184}$$

and the boundary condition (157e) becomes

$$u_2(0, t) = \ddot{s}(t)s(t), \qquad t > 0 \tag{185}$$

The solution of the homogeneous equation (182a) subject to the given boundary and initial conditions is again a special case of (155) with $g = \ddot{s}s$. Therefore, this part of the problem contributes the last term in (178a) for the solution of u_2. To compute the contribution of the inhomogeneous term p, it is convenient to change the variables of integration from ξ, τ to $\alpha = \tau + \xi, \beta = \tau - \xi$ in the general formula (143b) in order to exploit the fact that $p \equiv 0$ if $t < x$. The domain of integration is the rectangle bounded by the straight lines $\beta = 0$, $\beta = t - x$,

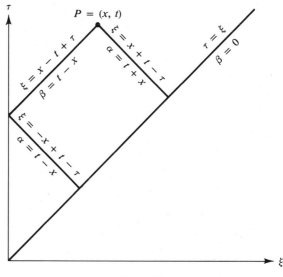

Figure 3.21

$\alpha = t - x$, $\alpha = t + x$, as shown in Figure 3.21; this is derived from the upper Figure 3.19 by deleting the shaded region below the line $\tau = \xi$. Noting that the absolute value of the Jacobian of the transformation $(\tau, \xi) \to (\alpha, \beta)$ is $\frac{1}{2}$, (143b) takes the form

$$
\begin{aligned}
u(x,t) &= \tfrac{1}{4} \int_{\alpha=t-x}^{t+x} d\alpha \int_{\beta=0}^{t-x} p(\beta)\, d\beta \\
&= \tfrac{1}{4} \int_{\alpha=t-x}^{t+x} 3\dot{s}(t-x)\ddot{s}(t-x)\, d\alpha \\
&= \tfrac{3}{2}\dot{s}(t-x)\ddot{s}(t-x)x
\end{aligned}
\tag{186}
$$

In the integration of $p(\beta)$, we have made use of the identity

$$
\int_{-\varepsilon}^{\varepsilon} f(\beta)\dot{\delta}(\beta)\, d\beta = -\dot{f}(0); \qquad \varepsilon > 0
\tag{187}
$$

which follows easily after integration by parts.

We note again that result (186) is in agreement with the first term in the right-hand side of (178a). The solution for h_2 follows by quadrature from (157a); this calculation is left as an exercise.

The reader has now a choice of which approach to use for a signaling problem. We could use Laplace transforms with respect to t, as illustrated for the $O(\varepsilon)$ solution in Problem 3. The advantage of this approach is that the solution for the transform is very easy to compute, but in general some ingenuity may be needed to invert the result and derive the explicit dependence of the solution on x, t. The method of characteristics has the advantage of a systematic and direct procedure for computing the solution. Moreover, in those cases where the equations in characteristic form cannot be solved explicitly, we have a convenient formulation for a numerical solution. This is discussed in Sections

4.4.2 and 4.5.4 for the general linear second-order problem. Finally, the result in terms of Green's function provides a compact integral expression for the solution. As we have seen, one must proceed with care to calculate an explicit result from such an expression. Also, Green's function may not be available for a more complicated problem, e.g., for a wave equation with variable coefficients, in which case a numerical solution may be the only option available.

3.5.5 A Second Example: Solution with a Fixed Interface; Reflected and Transmitted Waves

To fix ideas, consider the problem of transverse vibrations of an infinite string, half of which ($x > 0$) has a density ρ_1 and the other half ($x < 0$) has density ρ_2. The string is initially at rest with tension $\tau_0 = $ constant and is set in motion by applying a concentrated force α at $t = 0$, $x = \xi > 0$. Therefore, redimensionalizing (11b) with $p = \alpha \delta(x - \xi)\delta(t)$ and letting $v_1 = u$, we have:

$$\rho u_{tt} - (\tau_0 u_x)_x = \alpha \delta(x - \xi)\delta(t) \tag{188a}$$

or

$$u_{tt} - c^2 u_{xx} = \delta(x - \xi)\delta(t) \tag{188b}$$

where

$$c^2 = \begin{cases} c_1^2 = \dfrac{\tau_0}{\rho_1}, & \text{if } x > 0 \\[2mm] c_2^2 = \dfrac{\tau_0}{\rho_2}, & \text{if } x < 0 \end{cases}$$

For simplicity, we have chosen $\alpha = \rho_1$ and $\rho_2 = 1$. The initial conditions are $u(x, 0^-) = u_t(x, 0^-) = 0$.

The vertical component of the tension in the deflected string at any point (x, t) is $\tau_0 u_x(x, t)$, and it must be continuous everywhere, including at the interface $x = 0$. Obviously, the deflection at $x = 0$ must also be continuous. So, for this physical model, we have the following interface conditions:

$$u(0^+, t) = u(0^-, t) \tag{189a}$$

$$u_x(0^+, t) = u_x(0^-, t) \tag{189b}$$

To solve this problem, we shall use the idea of images in the right and left extended domains, as discussed in Problem 7(c) of Chapter 1.

First, note that until the disturbance initiated at $t = 0$, $x = \xi$ reaches the interface $x = 0$, we must have exactly the same response as in an infinite string with $c = c_1$ throughout.

This is the solution indicated by $u_1(x, t)$ in Figure 3.22. If we redimensionalize our result (110) for the fundamental solution, we obtain the following expression for u_1:

$$u_1(x, t) = \frac{1}{2c_1}[H(x - \xi + c_1 t) - H(x - \xi - c_1 t)] \tag{190}$$

Now, this disturbance (a uniform wave of amplitude $1/2c_1$ spreading at

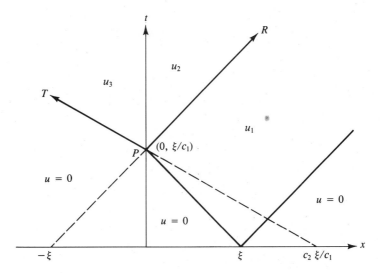

Figure 3.22

speed c_1 to the left and right) propagates unchanged until the leftward-moving front reaches $x \doteq 0$ at time $t = \xi/c_1$. At this point, we expect the interface to introduce a "reflected" disturbance propagating to the right and to allow a "transmitted" disturbance to move to the left. Since the density is constant in each of the half-spaces, the reflected and transmitted disturbances must be confined to the zone of influence of the point $P = (0, \xi/c_1)$. In the $x > 0$ portion to the zone of influence of P, we denote the solution as u_2, and in the $x < 0$ portion, we denote it as u_3.

Next, we postulate that u_2 consists of the primary disturbance u_1 defined by (190) plus a "reflected wave." In view of the region in which this reflected wave travels, we may regard it to be due to an image source of unknown strength A turned on at $t = 0$ and $x = \xi_2 = -\xi < 0$, in an infinite medium with $c = c_1$ throughout. The location of this image source is a priori obvious here (in contrast to the situation in Problem 7(c) of Chapter 1) because we know that its disturbance must produce the front PR in $x > 0$, and this front propagates with speed c_1. Thus,

$$u_2(x, t) = u_1(x, t) + \frac{A}{2c_1}[H(x + \xi + c_1 t) - H(x + \xi - c_1 t)] \tag{191}$$

Finally, we assume that the transmitted wave u_3 in $x < 0$ may be regarded as the disturbance due to an image source of unknown strength B switched on at $t = 0$ and $x = \xi_1 = c_2\xi/c_1$, in a medium with constant speed c_2 throughout. Again, the location of the image source is obvious a priori by extrapolating the front PT to the right with speed c_2. The situation depicted in Figure 3.22 corresponds to the case $(c_2/c_1) > 1$. The expression for u_3 is therefore given by

$$u_3(x, t) = \frac{B}{2c_2}[H(x - \xi_1 + c_2 t) - H(x - \xi_1 - c_2 t)] \tag{192}$$

Now, we determine A and B using the two interface conditions. It follows from (191) and (192) that as $x \to 0^+$ and 0^-, respectively, we have to equate

$$\frac{1}{2c_1} + \frac{A}{2c_1} = \frac{B}{2c_2} \tag{193}$$

The calculation of (189b), for the derivatives, requires more care. Differentiating (191) with respect to x and letting $x \to 0^+$ gives

$$u_{2_x}(0^+, t) = \frac{1}{2c_1}[\delta(-\xi + c_1 t) - \delta(-\xi - c_1 t)] + \frac{A}{2c_1}[\delta(\xi + c_1 t) - \delta(\xi - c_1 t)] \tag{194}$$

Now, for $t > \xi/c_1$, the arguments of each of the four delta functions in (194) are nonzero, so we can obtain useful information from (194) only in the limit as $t \downarrow \xi/c_1$. In view of the fact that $\delta(-\xi + c_1 t) = \delta(\xi - c_1 t)$, we find

$$u_{2_x}(0^+, t) \to \frac{1 - A}{2c_1}\delta(\xi - c_1 t) \qquad \text{as } t \to \left(\frac{\xi}{c_1}\right)^+ \tag{195}$$

A similar calculation for u_3 shows that

$$u_{3_x}(0^+, t) \to \frac{Bc_1}{2c_2^2}\delta(\xi - c_1 t) \qquad \text{as } t \to \left(\frac{\xi}{c_1}\right)^+ \tag{196}$$

It then follows from (189b) that we must set

$$\frac{1 - A}{2c_1} = \frac{Bc_1}{2c_2^2} \tag{197}$$

Solving (193) and (197) for A and B gives

$$A = \frac{c_2 - c_1}{c_1 + c_2}; \qquad B = \frac{2c_2^2}{c_1(c_1 + c_2)} \tag{198}$$

We verify that the limiting cases $c_1 = c_2$ (no interface) and $c_2 = 0$ (no deflection at $x = 0$) are correctly contained in our results. In particular, for $c_1 = c_2$, we find $A = 0$, $B = 1$; that is, $u_3 = u_2 = u_1 = 1/2c_1$; there is no reflected wave, and the transmitted wave is just the primary wave. If $c_2 = 0$, we have $A = -1$, $B = 0$, and we see that $u_3 = u_2 = 0$, whereas $u_1 = 1/2c_1$. This is just Green's function $G_1(x, \xi, c_1 t)/c_1$ derived in (140). In Problem 11, these ideas are developed for a signaling problem with a periodic boundary condition.

3.5.6 Green's Function of the Second Kind

Green's function of the second kind solves the problem

$$u_{tt} - u_{xx} = \delta(x - \xi)\delta(t - \tau); \qquad 0 \le x; \tau \le t \tag{199a}$$

with zero initial conditions at $t = \tau$ of

$$u(x, \tau^-) = u_t(x, \tau^-) = 0 \tag{199b,c}$$

and a zero boundary value for u_x at $x = 0$ of

$$u_x(0, t) = 0; \qquad t > \tau \tag{199d}$$

Here again, ξ and τ are arbitrary positive constants.

By symmetry, we need to add the solutions due to a unit source at $x = \xi$, $t = \tau$ and an image unit source that is also positive, at $x = -\xi, t = \tau$. The result is

$$G_2(x, \xi, t - \tau) \equiv F(x - \xi, t - \tau) + F(x + \xi, t - \tau) \tag{200}$$

where F is defined in (110). The zone of influence of the source point (ξ, τ) is the same as the one for G_1 sketched in Figure 3.18. Now, however, $G_2 = 1$ in region ④ of $x > 0$. To solve boundary-value problems with $u_x = 0$ or u_x prescribed at $x = 0$, we follow the procedures discussed in Sections 3.5.2 and 3.5.3 (see Problem 12).

3.6   Initial- and Boundary-Value Problems on the Finite Interval; Green's Functions

As in Section 1.5, we use the idea of an infinite array of sources located along the x-axis in such a way as to satisfy homogeneous boundary conditions at the endpoints of a finite interval.

3.6.1   Green's Function of the First Kind on $0 \le x \le 1$

As in Chapter 1, we have four Green's functions on the unit interval $0 \le x \le 1$, depending on whether $u = 0$ or $u_x = 0$ at each end. The formulas are identical to those in (77) and Problem 11 of Chapter 1 with regards to the dependence on the fundamental solution and are, therefore, not repeated (see Problem 13). Of course, in the present case F is defined by (110).

Consider, for example, G_1, which solves

$$u_{tt} - u_{xx} = \delta(x - \xi)\delta(t - \tau); \qquad 0 \le x \le 1; \qquad \tau \le t \tag{201}$$

with $0 < \xi = \text{constant} < 1$ and $0 < \tau = \text{constant}$. The initial conditions are $u(x, \tau^-) = u_t(x, \tau^-) = 0$, and the boundary conditions are $u(0, t) = u(1, t) = 0$ for $t > \tau$. Henceforth, a problem for which u is specified at both ends will be called a boundary-value problem of the *first kind*.

As in (77) of Chapter 1, we have

$$G_1(x, \xi, t - \tau) = \sum_{n=-\infty}^{\infty} \{F(x - 2n - \xi, t - \tau) - F(x - 2n + \xi, t - \tau)\} \tag{202}$$

where F is defined by (110). It is also easy to show that $G_1(x, \xi, t - \tau) = G_1(\xi, x, t - \tau)$ here. If we now deploy a triangular zone of influence for each source in (202), as discussed in Section 3.4.1, we obtain the pattern sketched in Figure 3.23.

Notice that unlike the situation in Chapter 1, where the effect of the sources in the far field were negligible initially, here *any source can be exactly ignored* until the first signal from it arrives, but then this signal is unattenuated and cannot be ignored, no matter how far the source is. Fortunately, we also notice

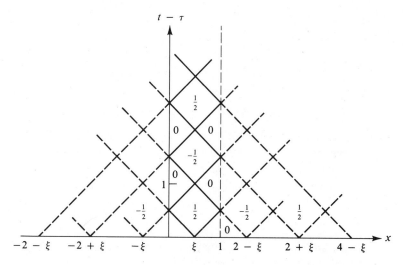

Figure 3.23

that the pattern of disturbances in $0 \le x \le t$ *is an odd periodic function of* $t - \tau$ with period 2; thus, we need only define the value of G_1 in the time interval $0 \le t - \tau \le 1$ to have it for any other time.

Next, we illustrate the application of these results to boundary-value problems of this first kind. Other applications involving G_2, G_3, G_4, and various associated boundary-value problems are not catalogued. An example is outlined in Problem 14.

3.6.2 The Inhomogeneous Problem, Nonzero Initial Conditions

First, we consider the inhomogeneous wave equation

$$u_{tt} - u_{xx} = p(x, t); \qquad 0 \le x \le 1; 0 \le t \tag{203a}$$

with zero initial conditions of

$$u(x, 0^-) = u_t(x, 0^-) = 0 \tag{203b,c}$$

and homogeneous boundary conditions of the first kind:

$$u(0, t) = u(1, t) = 0; \qquad t > 0 \tag{203d,e}$$

The solution follows immediately by superposition in terms of the Green's function G_1 of (202) in the form

$$u(x, t) = \int_0^t d\tau \left[\int_0^1 G_1(x, \xi, t - \tau) p(\xi, \tau) d\xi \right] \tag{204}$$

As in Section 3.5.2, we can use this result to also solve the homogeneous wave equation with homogeneous boundary conditions of the first kind but a nonzero initial condition. For example, note that

$$u_{tt} - u_{xx} = 0; \qquad 0 \le x \le 1, 0 \le t \tag{205a}$$

$$u(x, 0^+) = 0 \tag{205b}$$

$$u_t(x, 0^+) = h(x) \tag{205c}$$

$$u(0, t) = u(1, t) = 0; \qquad t > 0 \tag{205d,e}$$

is equivalent to

$$u_{tt} - u_{xx} = \delta(t)h(x); \qquad 0 \leq x \leq 1, 0 \leq t \tag{206a}$$

$$u(x, 0^-) = u_t(x, 0^-) = 0 \tag{206b,c}$$

$$u(0, t) = u(1, t) = 0; \qquad t > 0 \tag{206d,e}$$

Therefore, the solution can be written, using (204) with $p = \delta(t)h(x)$, in the form

$$u(x, t) = \int_0^1 h(\xi) G_1(x, \xi, t) \, d\xi \tag{207}$$

As in the case of the diffusion equation, we can derive an equivalent result (which only looks different) using separation of variables. In fact (see Problem 15a), this is found in the form

$$u(x, t) = \int_0^1 h(\xi) K(x, \xi, t) \, d\xi \tag{208}$$

where

$$K(x, \xi, t) \equiv \sum_{n=1}^{\infty} \frac{\sin n\pi t}{n\pi} [\cos n\pi(x - \xi) - \cos n\pi(x + \xi)] \tag{209}$$

It is easy to show [Problem 15(a)] that for any fixed x and ξ, K is just the Fourier series in $\sin n\pi t$ of the odd periodic function of t defined by the rectangular pattern for G_1 in the xt-plane.

A third representation of the solution of (205) may be derived using Laplace transforms [Problem 15(b)]. Just as in the case of the diffusion equation (Problem 12 of Chapter 1), when the inversion integral is approximated for s *large*, we obtain result (207).

In summary, we see that if we take account of all the sources that contribute to the value of u at a given time t, (207) is *exact*. The infinite series (209) converges to G_1 everywhere except at the discontinuities on the edges of the rectangles where G_1 changes value. Here, the infinite series converges to one-half the value of the jump. Truncating the series (209) will result in an approximation that is valid for t *large*, and the truncated series exactly satisfies the boundary conditions at $x = 0$ and $x = 1$.

The other possibility for a nonzero initial condition has

$$u_{tt} - u_{xx} = 0; \qquad 0 \leq x \leq 1, 0 \leq t \tag{210a}$$

$$u(x, 0^+) = f(x) \tag{210b}$$

$$u_t(x, 0^+) = 0 \tag{210c}$$

$$u(0, t) = u(1, t) = 0; \qquad t > 0 \tag{210d,e}$$

As in Section 3.5.2, we change variables to $v(x, t)$, defined by (148), and note that $v(x, t)$ obeys

$$v_{tt} - v_{xx} = 0 \tag{211a}$$

$$v(x, 0^+) = 0 \tag{211b}$$

$$v_t(x, 0^+) = f(x) \tag{211c}$$

$$v(0, t) = v(1, t) = 0; \quad t > 0 \tag{211d}$$

Therefore, $v(x, t)$ is defined by (207) with $h = f$, and u is given by v_t; that is,

$$u(x, t) = \int_0^1 f(\xi) \frac{\partial G_1}{\partial t} (x, \xi, t) \, d\xi \tag{212}$$

Since G_1 is a series of Heaviside functions, the expressions in (207) and (212) may be developed further. One may also derive these results by appropriate extensions of the initial data on the entire x-axis (Problem 16).

3.6.3 Inhomogeneous Boundary Conditions

Consider now the case where $u(0, t)$ and $u(1, t)$ are prescribed functions of time. We have

$$u_{tt} - u_{xx} = 0 \tag{213a}$$

$$u(x, 0^+) = u_t(x, 0^+) = 0 \tag{213b,c}$$

$$u(0, t) = g(t); \quad t > 0 \tag{213d}$$

$$u(1, t) = l(t); \quad t > 0 \tag{213e}$$

We pick the simple linear (in x) transformation to a new dependent variable $w(x, t)$ that satisfies zero boundary conditions at the endpoints:

$$w(x, t) \equiv u(x, t) + [g(t) - l(t)]x - g(t) \tag{214}$$

If $u(x, t)$ solves (213), then $w(x, t)$ must satisfy

$$w_{tt} - w_{xx} = (\ddot{g} - \ddot{l})x - \ddot{g} \equiv p(x, t) \tag{215a}$$

$$w(x, 0) = [g(0) - l(0)]x - g(0) \equiv f(x) \tag{215b}$$

$$w_t(x, 0) = [\dot{g}(0) - \dot{l}(0)]x - \dot{g}(0) \equiv h(x) \tag{215c}$$

$$w(0, t) = w(1, t) = 0 \tag{215d}$$

Therefore, the solution for $w(x, t)$ is obtained by combining (204), (207), and (212) for the expressions p, f, and h defined in (215a)–(215c). Having defined w, u follows from (214).

3.6.4 Uniqueness of the General Initial- and Boundary-Value Problem of the First Kind

The most general initial- and boundary-value problem of the first kind on the unit interval combines (203), (205), (210) and (213)—that is,

$$u_{tt} - u_{xx} = p(x,t); \qquad 0 \le x \le 1; 0 \le t \tag{216a}$$

$$u(x,0) = f(x); \qquad u_t(x,0) = h(x) \tag{216b,c}$$

$$u(0,t) = g(t); \qquad u(1,t) = l(t); t > 0 \tag{216d,e}$$

We shall now prove that (216) has a unique solution. As a preliminary step in the proof, we show that the "energy" associated with *any* solution of the homogeneous wave equation with homogeneous boundary conditions is constant. More precisely, let $\phi(x,t)$ denote any solution of

$$\phi_{tt} - \phi_{xx} = 0 \tag{217a}$$

satisfying the homogeneous boundary conditions

$$\phi(0,t) = \phi(1,t) = 0; \qquad t > 0 \tag{217b}$$

To calculate the total energy, we proceed as in dynamics and multiply (217a) by the velocity ϕ_t and integrate with respect to x over $0 \le x \le 1$. This gives

$$\int_0^1 [\phi_t \phi_{tt} - \phi_{xx} \phi_t] \, dx = 0 \tag{218}$$

We recognize the first term as the time rate of change of the kinetic energy because

$$\int_0^1 \phi_t \phi_{xx} \, dx = \frac{d}{dt} \int_0^1 \frac{\phi_t^2}{2} \, dx = \frac{d}{dt} (\text{K.E.}) \tag{219a}$$

Integrating the second term in (218) by parts gives:

$$-\int_0^1 \phi_t \phi_{tt} \, dx = -\phi_x \phi_t \Big|_0^1 + \int_0^1 \phi_x \phi_{xt} \, dx$$

But, $\phi(0,t) = \phi(1,t) = 0$; therefore, $\phi_t(0,t) = \phi_t(1,t) = 0$, and we find

$$-\int_0^1 \phi_t \phi_{xx} \, dx = \int_0^1 \phi_x \phi_{xt} \, dx = \frac{d}{dt} \int_0^1 \frac{\phi_x^2}{2} \, dx \tag{219b}$$

which is just the time rate of change of the potential energy.

Therefore, we have shown that along any solution $\phi(x,t)$ of (217), the total energy, defined as

$$E(t) \equiv \frac{1}{2} \int_0^1 (\phi_t^2 + \phi_x^2) \, dx = E(0) \tag{220}$$

is actually a constant equal to its initial value.

Now, return to problem (216), and assume that there are *two solutions*: $u_1(x,t)$ and $u_2(x,t)$, which satisfy all the conditions (216). Denote $\phi(x,t) \equiv u_1(x,t) - u_2(x,t)$. Clearly, $\phi(x,t)$ satisfies (217); therefore, $E = \text{constant} = E(0)$. But, since $\phi(x,0) = \phi_t(x,0) = 0$, we also have $E(0) = 0$. Therefore, the integral in (220) vanishes. Now, the integrand in (200) is nonnegative, so we conclude that $\phi_t(x,t) \equiv 0$ and $\phi_x(x,t) \equiv 0$. This means that $\phi(x,t)$ is a constant, and we may evaluate this constant at any (x,t), say $(x,0)$, where $\phi = 0$. Therefore, $\phi(x,t) \equiv 0$ and $u_1 = u_2$.

3.7   Effect of Lower-Derivative Terms

In many applications, the linearized wave equation that we derive in the limit of small disturbances involves terms proportional to u, u_x, and u_t. A general form would be:

$$u_{tt} - u_{xx} + au_x + bu_t + cu = 0 \qquad (221)$$

where a, b, and c are constants. The case where a, b, c, and the coefficients of u_{xx} and u_{tt} are functions of x and t is discussed in Chapter 4.

For example, with $a = b = 0$, $c > 0$, we could interpret (221) as the equation of a vibrating string on an elastic support (see Section 3.1). With $a = c = 0$, and $b > 0$, (221) is the "telegraph equation," and u gives the voltage along an electric transmission line in appropriate dimensionless variables. More generally, consider the oscillations of a chain or cable suspended from one end in the vertical direction. Here, the equilibrium tension in the cable is due to its weight and increases linearly with height from the value zero at the free end. Thus, the basic wave equation for the lateral motion is

$$u_{tt} - (xu_x)_x = 0$$

3.7.1 Transformation to D'Alembert Form; Removal of Lower-Derivative Terms

Proceeding as in Section 3.5.4, let us introduce the characteristic variables*

$$\zeta \equiv t + x \qquad (222a)$$

$$\mu = -\sigma \equiv t - x \qquad (222b)$$

With

$$u(x, t) = u\left(\frac{\zeta - \mu}{2}, \frac{\zeta + \mu}{2}\right) \equiv U(\zeta, \mu) \qquad (222c)$$

we calculate:

$$u_x = U_\zeta - U_\mu \qquad (223a)$$

$$u_t = U_\zeta + U_\mu \qquad (223b)$$

$$u_{xx} = U_{\zeta\zeta} - 2U_{\zeta\mu} + U_{\mu\mu} \qquad (224a)$$

$$u_{tt} = U_{\zeta\zeta} + 2U_{\zeta\mu} + U_{\mu\mu} \qquad (224b)$$

Therefore, after dividing by 4, (221) becomes:

$$U_{\zeta\mu} + \frac{(b + a)}{4} U_\zeta + \frac{(b - a)}{4} U_\mu + \frac{c}{4} U = 0 \qquad (225)$$

---

* As in Section 3.5.4, the choice of μ rather than σ is convenient for discussing the stability of solutions in Section 3.7.3.

We can remove the terms depending on U_ζ and U_μ by transforming the dependent variable $U \to W$ according to:

$$U(\zeta, \mu) \equiv W(\zeta, \mu) \exp\left\{ -\frac{(b-a)\zeta}{4} - \frac{(b+a)\mu}{4} \right\} \tag{226}$$

If U obeys (225), it is easily seen that W obeys

$$W_{\zeta\mu} + \lambda W = 0 \tag{227a}$$

where λ is the constant:

$$\lambda = \frac{a^2 - b^2 + 4c}{16} \tag{227b}$$

Note that if $\lambda \neq 0$, we lose the simple D'Alembert solution (106). However, the canonical form (227) is still easier to solve than the original equation (221).

3.7.2 Fundamental Solution; Stability

The fundamental solution of (227a) with the proper source strength to correspond to a unit source $\delta(x)\delta(t)$ applied to the right-hand side of (221) can be derived by similarity (Problem 20). The result for $\lambda > 0$ is

$$W(\zeta, \mu) = \begin{cases} \frac{1}{2}J_0(2\sqrt{\lambda\zeta\mu}), & \text{if } \zeta > 0; \mu > 0 \\ 0, & \text{otherwise} \end{cases} \tag{228a}$$

where J_0 is the Bessel function of the first kind of order zero. For $\lambda < 0$, we have

$$W(\zeta, \mu) = \begin{cases} \frac{1}{2}I_0(2\sqrt{-\lambda\zeta\mu}), & \text{if } \zeta > 0; \mu > 0 \\ 0, & \text{otherwise} \end{cases} \tag{228b}$$

where I_0 is the modified Bessel function of the first kind of order zero. Having defined W, we obtain u from (222) and (226). Thus, if $t < |x|$, $u = 0$, and if $t > |x|$, we have

$$u = \begin{cases} \frac{1}{2}\exp[-(b-a)(t+x)/4 - (b+a)(t-x)/4]J_0(2\sqrt{\lambda(t^2-x^2)}), & \text{if } \lambda > 0 \\ \frac{1}{2}\exp[-(b-a)(t+x)/4 - (b+a)(t-x)/4]I_0(2\sqrt{-\lambda(t^2-x^2)}), & \text{if } \lambda < 0 \end{cases} \tag{229}$$

Since $\lambda = 0$ if $a = b = c = 0$, we recover the result in (108) for this special case. For arbitrary values of a, b, and c, the behavior of the solution in the far-field ($|x|$ and t large) depends on the behavior of J_0 and I_0 and the relative importance of these compared with the exponential factors in (229).

Using standard tables (for example, (9.2.1) and (9.7.1) of [5]), we find that for z real,

$$J_0(z) = \left(\frac{2}{\pi z}\right)^{1/2}\left[\cos\left(z - \frac{\pi}{4}\right) + O(z^{-1})\right] \qquad \text{as } z \to \infty \tag{230a}$$

$$I_0(z) = \frac{e^z}{(2\pi z)^{1/2}} [1 + O(z^{-1})] \qquad \text{as } z \to \infty \tag{230b}$$

Thus, in (228), $|J_0| = O[(\zeta\mu)^{-1/4}]$ and $I_0 = O[(\zeta\mu)^{-1/4} e^{2|\lambda|^{1/2}(\zeta\mu)^{1/2}}]$ as $|\zeta\mu| \to \infty$. If we wish to consider the behavior of the solution for $|x|$ *fixed* and $t \to \infty$, we have $\zeta \to \infty$, and we see that in this case, it is sufficient to require $b > 0$ because a does not contribute to the time dependence of the exponentials in (229). Moreover, the growth of I_0 is suppressed by the stronger exponential decay $e^{-bt/2}$. Thus, *for bounded x, the solution is stable if $b > 0$* and unstable otherwise. This is physically obvious and corresponds to a positive damping.

It is more meaningful to consider the behavior of the solution as both x (or $-x$) and t become large. Thus, we need to take one limit in which $x \to \infty$ and $t = t_0 + x$ with $t_0 = \text{constant} > 0$; this corresponds to $\mu_0 = t_0 = \text{fixed}$, $\zeta \to \infty$. We also need to consider the case where $x \to -\infty$ and $t = t_0 - x$. This limit has $\zeta = t_0 = \text{fixed}$ and $\mu \to \infty$. For this stricter stability requirement, we must have $b - a > 0$ and $b + a > 0$. These two conditions are equivalent to the requirement

$$\frac{b}{|a|} > 1 \tag{231}$$

We shall rederive this result in Section 4.4.7 using the rules for the propagation of discontinuities along characteristics.

We must be on guard against using the linear problem for cases where the stability condition is violated. In such cases, either the physical model leading to (221) is inaccurate, or it is inconsistent to rely on a linear theory in the far-field because small disturbances grow there, and we must retain the nonlinear terms in order to describe the solution correctly.

3.7.3 Green's Functions; Initial- and Boundary-Value Problems

Starting with the fundamental solution (229), we can proceed as in Sections 3.4–3.6 to derive solutions on the infinite, semi-infinite, and bounded intervals. We shall not catalog the results here but merely point out that all statements regarding domains of validity, zones of influence, and so on are still true in the general case, since the disturbance due to a source at ξ, τ propagates along the straight characteristics $x = \xi - (t - \tau)$ and $x = \xi + (t - \tau)$, as before. The only difference in our formulas will be the more complicated expression (229) for the fundamental solution. Two specific examples are outlined in Problems 21 and 22. Of course, a solution using the characteristic variables is also possible, and this is discussed for the general case in Chapter 4.

3.8   Dispersive Waves on the Infinite Interval

As implied by the results of the previous section, it is sufficient to study

$$u_{tt} - u_{xx} + u = 0, \qquad -\infty < x < \infty, t \geq 0 \tag{232}$$

rather than (221); we examine various solutions of this equation, starting with the elementary special case of *uniform waves*.

3.8.1 Uniform Waves

Uniform waves are solutions of the form

$$u \equiv U(\theta) \tag{233a}$$

$$\theta \equiv kx - \omega(k)t \tag{233b}$$

for a given constant k. At this stage, the two functions U of θ and ω of k have not been specified. Geometrically, such a solution represents the uniform translation of the initial shape $u = U(kx)$ to the right (if $\omega > 0$). Since θ is a linear function of x and t, the translation speed $c(k)$, which is called the *phase speed*, remains constant and represents the speed with which each phase of the "wave" U moves. As all points on U move with the same constant speed, we have a "uniform wave" in the sense that the initial waveform does not distort as it travels (see Figure 3.24). To calculate the value of $c(k)$, consider a fixed phase A on the wave. Suppose that this phase corresponds to $u = u_0$ at $x = x_0$ and $t = 0$; that is, $u_0 \equiv U(kx_0)$. A short time Δt later, the wave shape is defined by $u = U(kx - \omega(k)\Delta t)$. Assume that the phase A has now moved to the point $x_0 + \Delta x$. Therefore, $u_0 = U(kx_0 + k\Delta x - \omega(k)\Delta t) = U(kx_0)$. This means that we must have $k\Delta x - \omega(k)\Delta t = 0$; that is, the phase speed is

$$c(k) \equiv \lim_{\Delta t \to 0} \frac{\Delta x}{\Delta t} = \frac{\omega(k)}{k} \tag{234}$$

If a wave equation—or, for that matter, any partial differential equation in the two variables (x, t)—admits a uniform wave solution, substitution of the form of solution (233) into (232) defines U and ω (see Problem 23 for another example). In our case, we find

$$(\omega^2 - k^2)\frac{d^2U}{d\theta^2} + U = 0 \tag{235}$$

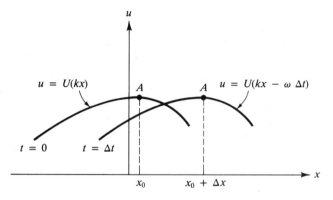

Figure 3.24

Thus, bounded solutions of the form (233) are possible on the *infinite interval* only if $\omega^2 - k^2 > 0$, say $\omega^2 - k^2 \equiv \lambda^2$, in which case

$$U = \alpha \sin\left(\frac{kx \mp \sqrt{k^2 + \lambda^2}\,t}{\lambda} + \beta\right) \tag{236a}$$

Here, α and β are arbitrary constants defining the amplitude and phase shift. Since k and λ can be chosen arbitrarily and U depends only on the ratio k/λ, it is convenient to normalize the preceding result by setting $\lambda = 1$. Thus,

$$U = \alpha \sin(kx \mp \sqrt{k^2 + 1}\,t + \beta) \tag{236b}$$

The resulting relations for $\omega(k)$ and $c(k)$ are

$$\omega(k) = \pm(k^2 + 1)^{1/2} \tag{237a}$$

$$c(k) = \frac{\omega(k)}{k} = \frac{\pm(1 + k^2)^{1/2}}{k} \tag{237b}$$

The relation (237a) defining $\omega(k)$ is called the *dispersion* relation, and solutions of the form (235) are called *dispersive waves* for reasons we shall discuss later on. Observe that if we apply the preceding ideas to the wave equation $u_{tt} - u_{xx} = 0$, the result corresponding to (235) would be $(\omega^2 - k^2)U'' = 0$. Setting $U'' = 0$ gives only unbounded solutions (or trivial ones, $U = $ constant). Therefore, we must take $\omega = \pm k$. In this case, U is *arbitrary*, and we recover the D'Alembert form (106). The phase speed is ± 1 independent of k; such a wave is called *nondispersive*.

Another important property of the uniform dispersive wave (236) is that it is periodic. In particular, given the *wave number* k, the wave defined by (236b) is periodic in x and t. Thus, for any fixed time t_0, u is a periodic function of x. This period, called the *wavelength*, is $L \equiv 2\pi/k$. Conversely, for a fixed $x = x_0$, U is a periodic function of time, and this period is $T \equiv 2\pi/(1 + k^2)^{1/2}$.

It is not necessary that a uniform solution be periodic. A very important class of uniform solutions that arise in nonlinear problems are the so-called solitary waves described by functions U that tend to zero as $|\theta| \to \infty$. Another important feature of the present linear problem is the absence of any amplitude dependence in the dispersion relation. This is no longer true for nonlinear waves; for example, see pp. 486–489 of [2] for a discussion of the nonlinear counterpart of (232). See also Problem 24.

Note also that the solution (236) is an exact solution of (232) for the *special initial conditions*

$$u(x, 0) = \alpha \sin(kx + \beta) \tag{238a}$$

$$u_t(x, 0) = -(1 + k^2)^{1/2}\alpha \cos(kx + \beta) \tag{238b}$$

It is important to keep in mind that for a given sinusoidal initial "deflection" (238a), we must also specify a particular initial "velocity" (238b) to produce the uniform wave (236) propagating to the right.

Because of linearity, we can add any number of such waves to obtain a "discrete wave train" propagating to the right:

$$u(x,t) = \sum_{i=1}^{N} \alpha_i \sin(k_i x - \sqrt{1 + k_i^2}\, t + \beta_i) \tag{239}$$

which also solves (232). Corresponding solutions for waves propagating to the left can also be added, and linearity ensures that these waves do not interact. Thus, if each term in the preceding series solves (232), the sum is also a solution. The corresponding statement is not true for uniform periodic waves associated with a nonlinear problem.

It is now natural to ask what role, if any, the uniform periodic waves play in the solution of a *general* initial-value problem.

3.8.2 General Initial-Value Problem

We shall show here that the waves $e^{i(kx \mp \sqrt{k^2+1}t)}$ are the fundamental building blocks in constructing the general solution of (232). Consider the solution of (232) for the general initial conditions

$$u(x, 0^+) = f(x) \tag{240a}$$

$$u_t(x, 0^+) = h(x) \tag{240b}$$

on the infinite x-axis using Fourier transforms.

If we denote the Fourier transforms of f and h by $\bar{f}(k)$ and $\bar{h}(k)$ (see Equations (35)–(36) of Chapter 1), we obtain

$$u(x,t) = \frac{1}{2\sqrt{2\pi}} \int_{-\infty}^{\infty} \left\{ \left[\bar{f}(k) - i\frac{\bar{h}(k)}{(1+k^2)^{1/2}} \right] e^{i(kx+\sqrt{1+k^2}t)} \right.$$
$$\left. + \left[\bar{f}(k) + \frac{i\bar{h}(k)}{(1+k^2)^{1/2}} \right] e^{i(kx-\sqrt{1+k^2}t)} \right\} dk \tag{241}$$

This complicated expression actually has the following simple interpretation if we regard the integral with respect to k as a continuous superposition. In this interpretation, each k defines a pair of waves of given wave number, wavelength $2\pi/k$, phase speed $\mp\sqrt{k^2 + 1}/k$, and amplitude proportional to $|\bar{f} + i\bar{h}(1 + k^2)^{-1/2}|$, propagating to the right and left. The complete solution is just the superposition of *all such waves*.

For the special case of the fundamental solution of (232), we have $f(x) = 0$, $h(x) = \delta(x)$; (241) reduces to

$$u(x,t) = \frac{1}{4\pi i} \int_{-\infty}^{\infty} \frac{e^{i(kx+\sqrt{1+k^2}t)} - e^{i(kx-\sqrt{1+k^2}t)}}{\sqrt{1+k^2}} dk \tag{242a}$$

and this can be evaluated explicitly to obtain:

$$u(x,t) = \begin{cases} \frac{1}{2}J_0(\sqrt{t^2 - x^2}); & \text{if } t > |x| \\ 0, & \text{if } t < |x| \end{cases} \tag{242b}$$

in agreement with the result (229a) calculated by similarity, since for $a = b = 0$ and $c = 1$, (227b) gives $\lambda = \frac{1}{4}$.

3.8.3 Group Velocity

In contrast to the special case (239) of discrete waves, the general initial-value problem involves a continuous superposition of waves with all values of k. This type of superposition introduces an interesting kinematic behavior, which we now study in some detail.

For simplicity, let us consider the behavior of the *wave packet* obtained by combining two unit-amplitude uniform waves (236), each with zero phase shift β but with nearly equal wave numbers k_0 and $k_0 + \Delta k$.

Thus, we take

$$u(x, t) = \sin\{k_0[x - c(k_0)t]\} + \sin\{(k_0 + \Delta k)[x - c(k_0 + \Delta k)t]\} \quad (243a)$$

where $c(k_0)$ is the phase speed for the wave number k_0, as given in (237b). For Δk small, we develop the argument of the second wave and use trigonometric identities to obtain the approximate form

$$u(x, t) = 2\cos\left\{\frac{\Delta k}{2}[x - v(k_0)t]\right\}\sin\{k_0[x - c(k_0)t]\} + O(\Delta k) \quad (243b)$$

where the speed $v(k)$ is the following function of k:

$$v(k) \equiv \frac{d}{dk}[\omega(k)] = \frac{k}{(1 + k^2)^{1/2}} \quad (244)$$

Since each wave in (243a) is an exact solution of (232), the sum given in the form (243a) or (243b) is also an exact solution. The form of (243b) exhibits the familiar phenomenon of beats, wherein the short wave defined by $\sin k_0[x - ct]$ has an amplitude $2\cos(\Delta k/2)(x - vt)$, which modulates the oscillations over the long wavelength $2\pi/\Delta k$, as shown in Figure 3.25.

Figure 3.25 illustrates the behavior in (243b) for a fixed time t over one wavelength of the amplitude modulation, and the result may be extended over all x by periodicity. In the limit $\Delta k \to dk$, we refer to the result in (243b) as the *packet of waves* in the interval $k_0 \leq k \leq k_0 + dk$ of wave space.

If we evaluate the result (243b) at time $t + \Delta t$, we see that the portion of the packet contained in the x-interval,

$$x_1(t) \equiv -\frac{\pi}{\Delta k} + vt \leq x \leq \frac{\pi}{\Delta k} + vt \equiv x_2(t)$$

will have translated *unchanged* to the right a distance $v\,\Delta t$ if we neglect terms in u having amplitude equal to $O(\Delta t)$. Thus, the packet of waves contained in the envelope $\pm 2\cos(\Delta k/2)(x - vt)$ moves with the speed $v(k_0)$ of the envelope, and this is called the *group speed* of the waves with wave number k_0.

In addition to this kinematic description of the group speed, we can, in this example, derive a dynamical description based on the propagation of the average energy contained in the packet. Because of periodicity, we need consider only the solution in the interval $x_1(t) \leq x \leq x_2(t)$.

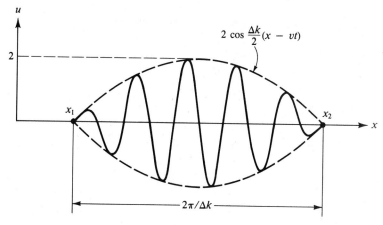

Figure 3.25

As discussed in Section 3.6.4 and Problem 18, the wave equation

$$u_{tt} - u_{xx} + F(u) = 0 \tag{245}$$

has the energy conservation law $E(t) = $ constant, where

$$E(t) \equiv \frac{1}{2} \int_{x_1}^{x_2} (u_t^2 + u_x^2 + 2G(u)) \, dx \tag{246}$$

and $dG/du \equiv F(u)$, as long as x_1 and x_2 are fixed points at which the solution of (245) vanishes.

In the case of the solution (243b), $u(x_1(t), t) = u(x_2(t), t) = 0$, but the points x_1 and x_2 move uniformly to the right with the group speed $v(k_0)$. What happens to $E(t)$ as defined by (246) in this case? We shall show next that $E(t)$ is still constant.

We have

$$E(t) \equiv \frac{1}{2} \int_{x_1(t)}^{x_2(t)} (u_t^2 + u_x^2 + u^2) \, dx \tag{247}$$

where

$$x_1(t) = -\frac{\pi}{\Delta k} + v(k_0)t$$

$$x_2(t) = \frac{\pi}{\Delta k} + v(k_0)t$$

Differentiating (247) gives

$$\frac{dE}{dt} = \int_{x_1(t)}^{x_2(t)} (u_t u_{tt} + u_x u_{xt} + uu_t) \, dx + \frac{v}{2}[u_t^2(x,t) + u_x^2(x,t) + u^2(x,t)]_{x=x_1(t)}^{x=x_2(t)} \tag{248}$$

Integrating the second term inside the integral by parts shows that:

$$\int_{x_1(t)}^{x_2(t)} u_x u_{xt}\, dx = u_x u_t \Big|_{x=x_1(t)}^{x=x_2(t)} - \int_{x_1(t)}^{x_2(t)} u_t u_{xx}\, dx$$

Therefore, (248) is equal to

$$\frac{dE}{dt} = \left[\frac{v}{2}(u_t^2 + u_x^2 + u^2) + u_x u_t \right]_{x=x_1(t)}^{x=x_2(t)} \tag{249}$$

because what remains in the integrand is identically equal to zero, since u satisfies (232).

To facilitate the evaluation of the right-hand side of (249) at the two endpoints, let us write u in the form

$$u(x,t) = 2\cos\phi(x,t)\sin\psi(x,t) + O(\Delta k) \tag{250}$$

where

$$\phi(x,t) = \frac{\Delta k}{2}(x - vt) \tag{251a}$$

$$\psi(x,t) = k_0(x - ct) \tag{251b}$$

Since $\cos\phi(x_1(t),t) = \cos\phi(x_2(t),t) = 0$, the only terms that contribute to (249) are

$$\frac{dE}{dt} = [2v(\phi_t^2 + \phi_x^2) + 4\phi_x\phi_t]\sin^2\phi\sin^2\psi \Big|_{x_1}^{x_2} = O(\Delta k) \tag{252}$$

because $\phi_t = -v\Delta k/2$ and $\phi_x = \Delta k/2$. Therefore, since the expression (250) for u is correct to $O(\Delta k)$, we conclude that $dE/dt = 0$ correct to $O(\Delta k)$ also.

The preceding calculation indicates that as the packet of waves $k_0 \le k \le k_0 + dk$ moves to the right, the average energy in this packet remains constant. These ideas also apply to the general case where the amplitude and phase shift of the two neighboring waves in (243a) are prescribed functions of k.

3.8.4 Dispersion

Next, we consider the behavior of solutions of (232) for large times. The case of a sum of discrete waves is not interesting because each wave in the sum evolves unchanged from the point of view of an observer moving with the phase speed of that wave. So, we turn our attention to a result involving a continuous super-position, as given in general by (241).

This result may be analyzed asymptotically for $t \to \infty$, $r \equiv x/t$ fixed, using the method of stationary phase (for example, see pp. 272–275 of [6]). It is equally instructive and more transparent to examine the result (242) for the fundamental solution, since this is given explicitly in terms of J_0.

We see, in fact, that result (242b) may be interpreted as the continuous superposition of the right and left propagating waves of *all wave numbers*, as defined in (242a). This latter messy expression eventually sorts itself out to give the asymptotic form [see (230a)]:

$$u = \frac{1}{[2\pi(t^2 - x^2)^{1/2}]^{1/2}} \cos\left[(t^2 - x^2)^{1/2} - \frac{\pi}{4} \right] + \cdots \tag{253a}$$

Notice that this result is not valid near the wave front $t \approx |x|$.

A little algebra shows that (253a) can also be written in the form

$$u(x,t) = -\frac{1}{[2\pi t(1-r^2)^{1/2}]^{1/2}} \sin\left[Kx - \Omega t - \frac{\pi}{4} \right] + \cdots \tag{253b}$$

where

$$K(r) \equiv \frac{r}{(1-r^2)^{1/2}}; \qquad r \equiv \frac{x}{t}; \, 0 \le r < 1 \tag{254a}$$

$$\Omega(K) \equiv [1 + K^2]^{1/2} = \frac{1}{(1-r^2)^{1/2}} \tag{254b}$$

Thus, for a fixed r, (253b) describes a *uniform wave* with constant wave number K, frequency Ω, and phase shift $-\pi/4$. The amplitude, which is given by the factor multiplying the sine function, decays like $t^{-1/2}$, and this feature is consistent with energy conservation. We can interpret the preceding behavior geometrically and say that an observer moving at the constant speed $v = r$ sees only a wave of one wave number, $K(r)$. In fact, v is *the group velocity* of the K-wave because

$$\frac{d\Omega}{dK} = \frac{K}{(1+K^2)^{1/2}} = r \tag{255}$$

We could also interpret the result (253b) to mean that for t large, wave numbers and frequencies are locally propagated with the associated group speed.

Dispersion refers to the fact that eventually waves of different wave numbers separate (disperse) as they propagate with their group velocity.

We also note that at large times, a stationary observer sees progressively longer waves arriving from the left, whereas the observer moving with speed r and focusing on a given phase sees this phase moving slowly to the right (since the local phase speed $c(K)$ is larger than r in this example).

The reason we claim that K, and hence Ω and c, change slowly with x, is because for a large fixed t, we have

$$\frac{\partial c}{\partial x} = \frac{\partial c}{\partial K}\frac{\partial K}{\partial x}; \qquad \frac{\partial \Omega}{\partial x} = \frac{\partial \Omega}{\partial K}\frac{\partial K}{\partial x}; \cdots$$

Now $\partial K/\partial x$ is small if $t \to \infty$ with r fixed because

$$\frac{\partial K}{\partial x} = \frac{r_x}{(1-r^2)^{3/2}} = \frac{1}{t}\frac{1}{(1-r^2)^{3/2}} = O(t^{-1}) \tag{256}$$

Therefore, c_x and Ω_x are $O(t^{-1})$ if $t \to \infty$ with r fixed; that is, *they vary slowly with x.*

Figure 3.26 illustrates these features for the present example.

In concluding this section, we reiterate that the preceding ideas are not restricted to the linear wave equation (232); they apply to a variety of other linear and nonlinear partial differential equations. The reader is referred to Chapter 11 of [2] for a discussion of general linear dispersive waves. Selected nonlinear

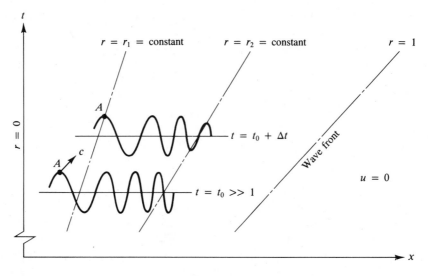

Figure 3.26

dispersive wave problems are also analyzed there. In Chapter 8, we consider uniform wave solutions of the Korteweg-deVries equation, which we derive for shallow-water waves.

3.9   The Three-Dimensional Wave Equation; Acoustics

In Section 3.3.6 we showed that the velocity potential in three-dimensional flow due to small disturbances in an initially ambient, inviscid non-heat-conducting gas obeys the dimensionless wave equation (101):

$$\Delta\phi - \phi_{tt} = 0 \tag{257}$$

Recall that the velocity is given by [see (95)]

$$\mathbf{u} = \text{grad } \phi \tag{258}$$

and the pressure and density perturbations $\tilde{p}$ and $\tilde{\rho}$ (measured from their ambient values) are related to ϕ_t according to [see (99) and (100)]

$$\tilde{p} = -\gamma\phi_t \tag{259a}$$

$$\tilde{\rho} = -\phi_t \tag{259b}$$

Thus, knowing ϕ defines all pertinent flow quantities.

3.9.1 Fundamental Solution

The fundamental solution is the solution of

$$\Delta\phi - \phi_{tt} = \delta(x)\delta(y)\delta(z)\delta(t) \tag{260a}$$

in the infinite domain with zero initial conditions:

$$\phi(x, y, z, 0^-) = \phi_t(x, y, z, 0^-) = 0 \tag{260b,c}$$

For the interpretation of (260a) as the velocity potential in acoustics, the right-hand side represents a *positive unit source of mass* [use (93) with a source term on the right-hand side to eliminate $\tilde{\rho}_t$ from the derivative of (100) with respect to t].

Since the disturbance is spherically symmetric in space, $\phi = \phi(r, t)$, and we need consider only

$$\phi_{rr} + \frac{2}{r}\phi_r - \phi_{tt} = \delta_3(r)\delta(t) \tag{261a}$$

$$\phi(r, 0^-) = \phi_t(r, 0^-) = 0 \tag{261b,c}$$

where $\delta_3(r)$ is the three-dimensional delta function defined in (34) of Chapter 2.

Consider (261a) with zero right-hand side, which can be written in the form

$$(r\phi)_{rr} - (r\phi)_{tt} = 0 \tag{262}$$

Therefore, $r\phi$ has the general D'Alembert form [see Equation (106)]

$$r\phi = v(t - r) + w(t + r) \tag{263}$$

where v and w are arbitrary functions of their respective arguments. Since w represents an incoming disturbance, it is not appropriate for the case of (261), and we discard it. (In a homogeneous unbounded domain, there is no mechanism for generating reflected disturbances traveling toward the origin when a source is turned on at the origin.)

Thus, ϕ is in the form

$$\phi(r, t) = \frac{v(t - r)}{r} \tag{264}$$

To determine v, we integrate (261a) over a sphere of radius $\varepsilon \ll 1$. This gives

$$\iiint_{r \le \varepsilon} [\Delta\phi - \phi_{tt}] \, dV = \delta(t) \tag{265}$$

Using Gauss' theorem [as in (37a) of Chapter 2] to express the volume integral of $\Delta\phi$ in terms of ϕ_r on the boundary gives

$$\iiint_{r \le \varepsilon} \Delta\phi \, dV = \iint_{r = \varepsilon} \phi_r \, dA = \iint_{r = \varepsilon} \left[-\frac{v(t - r)}{r^2} - \frac{v'(t - r)}{r} \right] dA$$

In the limit $\varepsilon \to 0$, the first term of the surface integral will contribute because $dA = O(\varepsilon^2)$, and the second term will not. So, we find

$$\lim_{\varepsilon \to 0} \iiint_{r \le \varepsilon} \Delta\phi \, dV = -\lim_{\varepsilon \to 0} \int_{\theta=0}^{\pi} \int_{\psi=0}^{2\pi} v(t - \varepsilon) \sin\theta \, d\theta \, d\psi = -4\pi v(t)$$

Also

$$\lim_{\varepsilon \to 0} \iiint_{r \le \varepsilon} \phi_{tt} \, dV = \lim_{\varepsilon \to 0} \frac{\partial^2}{\partial t^2} \iiint_{r \le \varepsilon} \phi \, dV = 0$$

Therefore, (265) reduces to

$$-4\pi v(t) = \delta(t)$$

or

$$v(t) = -\frac{\delta(t)}{4\pi}$$

and the solution of (261) is

$$\phi(r, t) = -\frac{1}{4\pi} \frac{\delta(t - r)}{r} \tag{266}$$

More generally, the solution of

$$\Delta\phi - \phi_{tt} = \delta(x - \xi)\delta(y - \eta)\delta(z - \zeta)\delta(t - \tau) \tag{267a}$$

$$\phi(x, y, z, \tau^-) = \phi_t(x, y, z, \tau^-) = 0 \tag{267b,c}$$

is

$$F(x - \xi, y - \eta, z - \zeta, t - \tau) \equiv -\frac{\delta(t - \tau - r_{PQ})}{4\pi r_{PQ}} \tag{268}$$

where $P = (x, y, z)$, $Q = (\xi, \eta, \zeta)$ and

$$r_{PQ}^2 = (x - \xi)^2 + (y - \eta)^2 + (z - \zeta)^2 \tag{269}$$

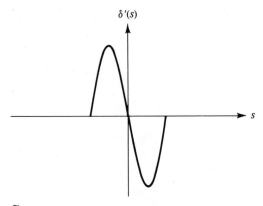

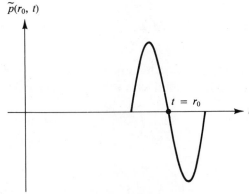

Figure 3.27

It is instructive to interpret the result in (266) in terms of the pressure disturbance $\tilde{p} = -\gamma\phi_t = \gamma\delta'(t-r)/4\pi r$, which results from turning on a unit mass source at the origin at time $t = 0$. If we regard $\delta'(s)$ qualitatively as the function sketched in the top of Figure 3.27, an observer at the fixed point $r = r_0 > 0$ receives the pressure signal indicated in the lower figure as a function of time. Thus, the pressure disturbance remains zero up until time $t = r_0$, when the disturbance emitted at the origin and $t = 0$ arrives. This disturbance is qualitatively a rapid rise in the pressure, followed by a rapid drop, followed by a return to a null value. This so-called N-wave is typical of acoustic disturbances as well as more energetic disturbances such as sonic booms or explosions.

3.9.2 Arbitrary Source Distribution

Consider the problem:

$$\Delta\phi - \phi_{tt} = S(x, y, z, t) \tag{270a}$$

$$\phi \equiv 0; \quad S \equiv 0 \quad \text{if } t < 0 \tag{270b}$$

which in acoustics corresponds to a prescribed spatial and temporal distribution of mass sources of strength S turned on at $t = 0$. Using superposition, we have

$$\phi(x, y, z, t) = \int_{\tau=0}^{t} d\tau \left[\int\int\int_{-\infty}^{\infty} S(\xi, \eta, \zeta, \tau) F(x - \xi, y - \eta, z - \zeta, t - \tau) \, d\xi \, d\eta \, d\zeta \right] \tag{271}$$

where $F(x - \xi, y - \eta, z - \zeta, t - \tau)$ is given in (268).

Changing variables from τ to $\sigma \equiv t - \tau - r_{PQ}$ and performing the $\tau(\sigma)$ integration first gives

$$\phi(x, y, z, t) = -\frac{1}{4\pi} \int\int\int_{r_{PQ} \leq t} \frac{S(\xi, \eta, \zeta, t - r_{PQ})}{r_{PQ}} \, d\xi \, d\eta \, d\zeta \tag{272}$$

Note that for a fixed $P = (x, y, z)$, the integration variables $Q = (\xi, \eta, \zeta)$ range only over the *interior* of the sphere $r_{PQ} = t$ centered at P. This result is called the *retarded potential* because of the delay effect in the time dependence in S.

As a special case, let

$$S = \delta(x)\delta(y)\delta(z)f(t) \tag{273}$$

corresponding to a *point source* at the origin having a time-varying strength f. This is an idealization in acoustics of a point source of sound, such as a speaker at the origin. Using (273) for S in (272) and performing the integrations gives:

$$\phi(x, y, z, t) = -\frac{1}{4\pi} \frac{f(t - r)}{r} \tag{274}$$

where $r^2 = x^2 + y^2 + z^2$. Thus, the given signal is received a distance r away from the source in a form that is *undistorted*, $f(t) \to f(t - r)$; rather, it is merely *attenuated* like r^{-1}. In particular, the pressure disturbance created by the source at $r = 0$ arrives undistorted but weaker some distance away. Thus, the receiver

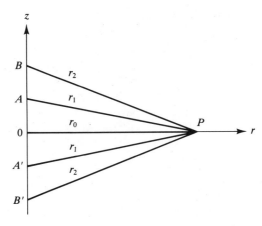

Figure 3.28

(ear) has a relatively simple task of interpreting the signal, all of which is fortunate for us. Mathematically, this feature is a consequence of the presence of the delta function in the fundamental solution, which ensures that a given disturbance propagates along a *distinct* front corresponding to the vanishing of the argument of the delta function. This situation is no longer true in two dimensions. Consider, for example, the signal due to a *line* source of variable strength $f(t)$ along the z-axis; for example,

$$S = \delta(x)\delta(y)f(t) \tag{275}$$

In this case (272) reduces to [Problem 25(a)]

$$\phi(x, y, t) = -\frac{1}{2\pi} \int_0^{t-r} \frac{f(\tau)\, d\tau}{\sqrt{(t-\tau)^2 - r^2}} \tag{276}$$

where $r^2 = x^2 + y^2$. Thus, an "integrated" version of the signal arrives at the observer location r, and this is physically obvious, since each point on the z-axis (and its mirror image along $-z$) is at a different distance from the observer, as shown in Figure 3.28. Hence, signals that arrive at $r = r_0$ at a given time $t = t_0$ were broadcast at different times. In particular, at the time $t = t_0$, the disturbance that arrives at P is made up of the signal sent from 0 at time $t = t_0 - r_0$, the signal sent from A and A' at $t = t_0 - r_1$, the signals sent from B and B' at $t = t_0 - r_2$, and so on.

As a further specialization of (276), set $f(t) = \delta(t)$, which gives the fundamental solution for the two-dimensional wave equation as

$$\phi = -\frac{1}{2\pi} \frac{H(t-r)}{\sqrt{t^2 - r^2}}; \qquad r^2 = x^2 + y^2 \tag{277}$$

where H is the Heaviside function [see also Problems 25(b), (c), (d)].

Let us return now to the result (262) and interpret the potential ϕ as a superposition of certain *averages* evaluated over different concentric spherical shells surrounding P. To do this, we define $\langle h \rangle_\rho$ to be the average value of a given function $h(\xi, \eta, \zeta)$ evaluated at the fixed point $P = (x, y, z)$ with respect to the sphere of radius ρ centered at P. Thus,

$$\langle h \rangle_\rho \equiv \frac{1}{4\pi\rho^2} \iint\limits_{r_{PQ}=\rho} h(Q)\,dA_Q \tag{278}$$

More explicitly, if we introduce the local spherical polar coordinate system at P, defined by

$$\xi - x = \rho \sin \theta \cos \psi$$
$$\eta - y = \rho \sin \theta \sin \psi \tag{279}$$
$$\zeta - z = \rho \cos \theta$$

then (278) becomes

$$\langle h \rangle_\rho = \frac{1}{4\pi\rho^2} \int_{\theta=0}^{\pi} \int_{\psi=0}^{2\pi} h(x + \rho \sin \theta \cos \psi, y + \rho \sin \theta \sin \psi, z + \rho \cos \theta) \rho^2 \sin \theta\, d\theta\, d\psi \tag{280}$$

Now, if we write (271) using τ, ρ, θ, and ψ as integration variables, we have:

$$\phi(x, y, z, t) = -\frac{1}{4\pi} \int_{\tau=0}^{t} d\tau \int_{\rho=0}^{\infty} \frac{\delta(t - \tau - \rho)}{\rho}\, d\rho$$
$$\cdot \left\{ \int_{\theta=0}^{\pi} \int_{\psi=0}^{2\pi} S(x + \rho \sin \theta \cos \psi, y + \rho \sin \theta \sin \psi, z + \rho \cos \theta, \tau) \rho^2 \sin \theta\, d\theta\, d\psi \right\} \tag{281}$$

which is just

$$\phi(x, y, z, t) = -\int_{\tau=0}^{t} \langle \rho S \rangle_{\rho=t-\tau}\, d\tau \tag{282}$$

Thus, for a fixed $P = (x, y, z)$ and time t, as τ increases from 0 to t, the contribution to ϕ consists of the *sum of the averages of* $-\rho S$ on spheres centered at P and having decreasing radii equal to $t - \tau$. The maximal radius is, of course, $\rho = t$.

3.9.3 Initial-Value Problems for the Homogeneous Equation

Consider the initial-value problem

$$\Delta\phi - \phi_{tt} = 0 \tag{283a}$$
$$\phi(x, y, z, 0^+) = 0 \tag{283b}$$
$$\phi_t(x, y, z, 0^+) = f(x, y, z) \tag{283c}$$

In acoustics, this would correspond to a prescribed initial pressure perturbation $\tilde{p}$ equal to $-\gamma f$ [see (259a)]. The preceding is equivalent to the inhomogeneous problem

$$\Delta\phi - \phi_{tt} = -\delta(t)f(x, y, z) \tag{284a}$$

with zero initial conditions

$$\phi(x, y, z, 0^-) = \phi_t(x, y, z, 0^-) = 0 \tag{284b,c}$$

Thus, we have a special case of (282) with $S = -\delta(t)f(x, y, z)$, and the solution is

$$\phi(x, y, z, t) = t\langle f(\xi, \eta, \zeta)\rangle_{\rho=t} \tag{285}$$

The other initial-value problem has

$$\Delta\phi - \phi_{tt} = 0 \tag{286a}$$

$$\phi(x, y, z, 0^+) = g(x, y, z) \tag{286b}$$

$$\phi_t(x, y, z, 0^+) = 0 \tag{286c}$$

We can reduce this to the form (283) by introducing the new dependent variable Φ defined by [see (117)]

$$\Phi(x, y, z, t) \equiv \int_0^t \phi(x, y, z, \tau)\,d\tau \tag{287}$$

Now $\partial\Phi/\partial t = \phi$, and $\partial^2\Phi/\partial t^2 = \partial\phi/\partial t$. Thus, Φ satisfies the initial conditions

$$\Phi_t(x, y, z, 0^+) = \phi(x, y, z, 0^+) = g(x, y, z) \tag{288a}$$

$$\Phi(x, y, z, 0^+) = 0 \tag{288b}$$

Also, Φ obeys the wave equation because

$$\Delta\Phi = \int_0^t \Delta\phi\,d\tau = \int_0^t \phi_{tt}\,d\tau = \phi_t(x, y, z, t) - \phi_t(x, y, z, 0^+) = \phi_t = \frac{\partial^2\Phi}{\partial t^2}$$

Hence, using (285), we have

$$\Phi(x, y, z, t) = t\langle g(\xi, \eta, \zeta)\rangle_{\rho=t}$$

or

$$\phi(x, y, z, t) = \frac{\partial}{\partial t}\{t\langle g(\xi, \eta, \zeta)\rangle_{\rho=t}\} \tag{289}$$

3.10   Examples in Acoustics and Aerodynamics

In this section we outline three specific examples of applications of the preceding results.

3.10.1 The Bursting Balloon

A spherical balloon, as shown in Figure 3.29(a), is inflated to a pressure $p = p_1 > p_0$ and radius L. At time $t = 0$, the balloon bursts. What is the pressure disturbance as a function of time that is felt at a distance $R > L$ measured from the center of the balloon? The case $R < L$ is considered in Problem 28.

If $\varepsilon \equiv (p_1 - p_0)/p_0 \ll 1$, it is appropriate to use the linear theory of Section 3.9. Since $\tilde{p} = -\gamma\phi_t$, the dimensionless formulation (using L to normalize lengths, L/a_0 to normalize time, and La_0 to normalize the velocity potential) is given by (283) with spherical symmetry—that is, $\phi = \phi(r, t)$—and with

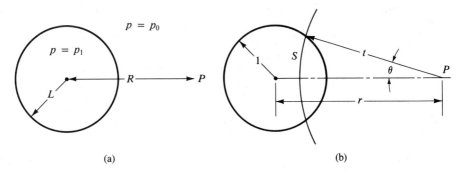

Figure 3.29

$$f(r) = \begin{cases} -\dfrac{1}{\gamma} & \text{if } r < 1 \\[2mm] 0 & \text{if } r > 1 \end{cases} \tag{290}$$

The solution for ϕ is given by (285), and we need to calculate the area of S, the portion of the sphere of radius t (centered at P) that lies inside the balloon [see Figure 3.29(b)]. Using trigonometric identities, it is easily seen that $\cos\theta = (t^2 + r^2 - 1)/2rt$. Therefore, the area of S is

$$A \equiv 2\pi t^2 \int_0^\theta \sin\theta'\, d\theta' = \frac{\pi t}{r}[1 - (r - t)^2]$$

and we find that

$$\langle f \rangle_{\rho=t} = \frac{A\left(-\dfrac{1}{\gamma}\right)}{4\pi t^2} = -\frac{1 - (r - t)^2}{4\gamma rt} \tag{291}$$

if $r - 1 < t < r + 1$ and that $\langle f \rangle_{\rho=t}$ is zero otherwise.

Equation (285) for the disturbance potential becomes

$$\phi(r, t) = \begin{cases} -\dfrac{1}{4\gamma}\left[\dfrac{1 - (r - t)^2}{r}\right], & \text{if } r - 1 < t < r + 1 \\[3mm] 0, & \text{otherwise} \end{cases} \tag{292}$$

and the overpressure $\varepsilon\tilde{p}$ is defined by

$$\tilde{p} = -\gamma\phi_t(r, t) = \begin{cases} \dfrac{r - t}{2r}, & \text{if } r - 1 < t < r + 1 \\[3mm] 0, & \text{otherwise} \end{cases} \tag{293}$$

Because of spherical symmetry, this problem can also be solved directly using the general solution (263) (see Problem 28).

Figure 3.30 shows a sketch of $\tilde{p}$ as a function of time at the *fixed position r*. We find a symmetric N-wave consisting of a sudden rise in overpressure to the value $1/2r$, followed by a linear drop to the value $-1/2r$ and a sudden rise to the zero level.

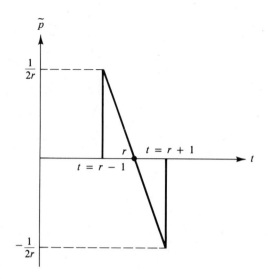

Figure 3.30

Now, suppose there is a vertical wall at a distance $r_2 > r$ from the origin. What is the reflected noise? The boundary condition to be satisfied at $r = r_2$ is $\phi_r(r_2, t) = 0$—that is, no flow normal to the wall at the wall. We can satisfy this boundary condition by introducing an image ballon at the distance $2r_2 - r$ to the right of the observer and bursting it at $t = 0$. The resulting reflected noise is then obtained by replacing r in (293) by $(2r_2 - r)$. Thus,

$$\tilde{p}_{\text{reflected}} = \begin{cases} \dfrac{2r_2 - r - t}{2r_2 - r}, & \text{if } 2r_2 - r - 1 < t < 2r_2 - r + 1 \\ 0, & \text{otherwise} \end{cases} \tag{294}$$

Of course, if t satisfies both inequalities in (293) and (294), we must also add the primary source contribution of (293).

3.10.2 Source Distribution over the Plane

Suppose we wish to solve the initial- and boundary-value problem

$$\Delta\phi - \phi_{tt} = 0 \qquad \text{on } 0 \le y \tag{295a}$$

with initial condition

$$\phi(x, y, z, 0^+) = \phi_t(x, y, z, 0^+) = 0 \tag{295b,c}$$

and boundary conditions

$$\phi_y(x, 0^+, z, t) = h(x, z, t), \qquad t > 0 \tag{295d}$$

$$\phi(x, \infty, z, t) = 0 \tag{295e}$$

It is clear from symmetry that we can replace this by the problem on $-\infty < y < \infty$ by appending the boundary conditions

$$\phi_y(x, 0^-, z, t) = -h(x, z, t), \qquad t > 0 \tag{295f}$$

$$\phi(x, -\infty, z, t) = 0 \tag{295g}$$

Let us attempt to solve (295) by introducing a source sheet of strength/unit area S equal to $2h(x, z, t)$ on the $y = 0$ plane; that is, we claim that the initial-value problem

$$\Delta\phi - \phi_{tt} = 2\delta(y)h(x, z, t); \qquad -\infty < y < \infty \tag{296a}$$

$$\phi(x, y, z, 0^+) = \phi_t(x, y, z, 0^+) = 0 \tag{296b,c}$$

with sources as indicated on the $y = 0$ plane, is equivalent to (295).

It is clear that the requirement $S = 2\delta(y)h$ is *necessary* in order that (296a) produce the correct boundary conditions at $y = 0^+$ and $y = 0^-$. This follows by integrating (296a) from $y = 0^-$ to $y = 0^+$ to obtain

$$\phi_y(x, 0^+, z, t) - \phi_y(x, 0^-, z, t) = 2h(x, z, t) \tag{297}$$

after noting that ϕ_{xx}, ϕ_{zz}, and ϕ_{tt} are continuous at $y = 0$ and hence do not contribute to (297). The above is *not enough*; we must also show that the solution of (296) tends to the *individual* limits (295d) and (295f).

The solution of (296) is a special case of (272) with $S = 2\delta h$—that is,

$$\phi(x, y, z, t) = -\frac{1}{2\pi} \iiint_{r_{PQ} \le t} \frac{\delta(\eta)h(\xi, \zeta, t - r_{PQ})}{r_{PQ}} \, d\xi \, d\eta \, d\zeta \tag{298a}$$

—and the delta function restricts the domain of integration to the intersection of the sphere $r_{PQ} \le t$ with the plane $\eta = 0$. Thus,

$$\phi(x, y, z, t) = -\frac{1}{2\pi} \iint_{r_{PQ_0} \le t} \frac{h(\xi, \zeta, t - r_{PQ_0})}{r_{PQ_0}} \, d\xi \, d\zeta \tag{298b}$$

where

$$r_{PQ_0}{}^2 = (x - \xi)^2 + y^2 + (z - \zeta)^2 \tag{298c}$$

As shown in Figure 3.31, the sphere of radius t centered at $P = (x, y, z)$ intersects with the $\eta = 0$ plane to form the circular disc D centered at $\xi = x, \eta = 0$, $\zeta = z$ and having radius $\sqrt{t^2 - y^2}$. The integral in (298b) is evaluated over D. Therefore, it is convenient to introduce the polar coordinates in the plane of D defined by

$$\xi - x = \rho \cos\theta \tag{299a}$$

$$\zeta - z = \rho \sin\theta \tag{299b}$$

The integral (298b) then becomes

$$\phi(x, y, z, t) = -\frac{1}{2\pi} \int_{\rho=0}^{\sqrt{t^2-y^2}} \int_{\theta=0}^{2\pi} \frac{h(x + \rho\cos\theta, z + \rho\sin\theta, t - \sqrt{\rho^2 + y^2})\rho \, d\rho \, d\theta}{\sqrt{\rho^2 + y^2}} \tag{300}$$

The derivative of this expression with respect to y leads to an improper integral with respect to ρ as $y \to 0^+$. This situation is entirely analogous to the one discussed in Section 2.4.4 for a surface distribution of sources for Laplace's equation. Therefore, as in (106) of Chapter 2, we split the integration with respect

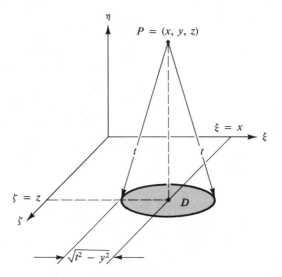

Figure 3.31

to ρ into a contribution over the small disk of radius $\rho = \varepsilon$ plus the contribution over the annulus $\varepsilon \leq \rho \leq \sqrt{t^2 - y^2}$. Denoting these contributions to ϕ as ϕ_ε and ϕ_a, we have:

$$\phi_\varepsilon(x, y, z, t; \varepsilon) = -\frac{1}{2\pi} \int_{\rho=0}^\varepsilon \int_{\theta=0}^{2\pi} \frac{h\rho \, d\rho \, d\theta}{(\rho^2 + y^2)^{1/2}} \tag{301a}$$

$$\phi_a(x, y, z, t; \varepsilon) = -\frac{1}{2\pi} \int_{\rho=\varepsilon}^{\sqrt{t^2-y^2}} \int_{\theta=0}^{2\pi} \frac{h\rho \, d\rho \, d\theta}{(\rho^2 + y^2)^{1/2}} \tag{301b}$$

where the arguments of h are as indicated in (300), and $\phi = \phi_\varepsilon + \phi_a$.

We define the limit as $y \to 0^+$ as in (99) of Chapter 2 with y and z interchanged. Now, it is easily seen that $\partial \phi_a / \partial y \to 0$ as $\varepsilon \to 0$, as long as $(y/\varepsilon) \to 0$. To evaluate $\partial \phi_\varepsilon / \partial y$ as ε and y tend to zero, we approximate ϕ_ε for ε small and then take the derivative of the result. We have

$$\phi_\varepsilon(x, y, z, t; \varepsilon) = -\frac{1}{2\pi} \int_{\rho=0}^\varepsilon \int_{\theta=0}^{2\pi} \frac{[h(x, z, t - y) + O(\rho)]\rho \, d\rho \, d\theta}{(\rho^2 + y^2)^{1/2}}$$

$$= -h(x, z, t - y) \int_{\rho=0}^\varepsilon \frac{\rho \, d\rho}{(\rho^2 + y^2)^{1/2}} + O\left(\int_0^\varepsilon \frac{\rho^2 \, d\rho}{(\rho^2 + y^2)^{1/2}} \right)$$

$$= -h(x, z, t - y)[\sqrt{\varepsilon^2 + y^2} - y] + O(\varepsilon y) \qquad \text{as } y \to 0$$

Therefore,

$$\frac{\partial \phi_\varepsilon}{\partial y}(x, y, z, t; \varepsilon) = h_t(x, z, t - y)[(\varepsilon^2 + y^2)^{1/2} - y]$$

$$- h(x, z, t - y)[y(\varepsilon^2 + y^2)^{-1/2} - 1] + O(\varepsilon)$$

In the limit $\varepsilon \to 0$, $(y/\varepsilon) \to 0^+$, we obtain

$$\lim_{\substack{\varepsilon \to 0 \\ (y/\varepsilon) \to 0^+}} \frac{\partial \phi_\varepsilon}{\partial y}(x, y, z, t; \varepsilon) = h(x, z, t)$$

Thus,

$$\phi_y(x, 0^+, z, t) = \lim_{\substack{\varepsilon \to 0 \\ (y/\varepsilon) \to 0^+}} \frac{\partial \phi_\varepsilon}{\partial y}(x, y, z, t; \varepsilon) = h(x, z, t) \tag{302}$$

Similarly, $\phi_y(x, 0^-, z, t) = -h(x, z, t)$, and this completes the proof.

3.10.3 Perturbation of a Uniform Flow

Equation (101) was obtained under the assumption that $\mathbf{u}/a_0$ is small. Suppose we wish to solve (87) and (88) for the case where $\mathbf{u}$ is a steady uniform flow (say in the x-direction) that is perturbed by a small disturbance (for example, a slender body). This is the classical linear small-disturbance theory of aerodynamics and is a special case of (101).

If we use the dimensionless variables introduced in Sections 3.3.3 and 3.9, the exact problem defined by (87) and (88) has the following dimensionless form using Cartesian tensor notation [with $p = \rho^\gamma$, according to (90)]

$$\frac{\partial \rho}{\partial t*} + \frac{\partial (\rho u_k)}{\partial x_k} = 0 \tag{303a}$$

$$\frac{\partial (\rho u_i)}{\partial t*} + \frac{\partial}{\partial x_k}(\rho u_i u_k) + \rho^{\gamma-1} \frac{\partial \rho}{\partial x_i} = 0; \qquad i = 1, 2, 3 \tag{303b}$$

Here, $x_1 = x^*$, $x_2 = y^*$, $x_3 = z^*$. Asterisks denote dimensionless space and time variables, and we sum over repeated indices. Thus, the u_i are the Cartesian velocity components with respect to the x^*, y^*, z^* frame.

If (303a) is used to eliminate $\partial \rho / \partial t$ from (303b), this equation simplifies to [see (73)]:

$$\frac{\partial u_i}{\partial t*} + u_k \frac{\partial u_i}{\partial x_k} + \rho^{\gamma-2} \frac{\partial \rho}{\partial x_i} = 0; \qquad i = 1, 2, 3 \tag{303c}$$

Now, we assume that the u_i and ρ have the following perturbation expansions:

$$u_1 = M + \varepsilon u_1^*(x^*, y^*, z^*, t^*) + \cdots \tag{304a}$$

$$u_i = \varepsilon u_i^*(x^*, y^*, z^*, t^*) + \cdots; \qquad i = 2, 3 \tag{304b}$$

$$\rho = 1 + \varepsilon \rho^*(x^*, y^*, z^*, t^*) + \cdots \tag{304c}$$

where ε is a small parameter measuring the amplitude of the disturbance and M, called the Mach number, is the dimensionless unperturbed flow speed (that is, the speed at infinity divided by the ambient speed of sound). Thus, if $\varepsilon = 0$, the flow consists of the uniform velocity $M\mathbf{i}$ in the x^* direction.

The equations corresponding to (93) and (94) for the perturbation quantities

$\mathbf{u}^* = (u_1^*, u_2^*, u_3^*)$ and ρ^* are

$$\frac{\partial \rho^*}{\partial t^*} + \operatorname{div} \mathbf{u}^* + M \frac{\partial \rho^*}{\partial x^*} = 0 \qquad (305a)$$

$$\frac{\partial \mathbf{u}^*}{\partial t^*} + M \frac{\partial \mathbf{u}^*}{\partial x^*} + \operatorname{grad} \rho^* = 0 \qquad (305b)$$

Taking the curl of (305b) gives

$$\frac{\partial}{\partial t^*}(\operatorname{curl} \mathbf{u}^*) + M \frac{\partial}{\partial x^*}(\operatorname{curl} \mathbf{u}^*) = 0 \qquad (306)$$

that is, the curl of $\mathbf{u}^*$ is constant along the paths $x^* - Mt^* = \text{constant}$. Since $\operatorname{curl} \mathbf{u}^* = 0$ at upstream infinity where $\mathbf{u}^* = 0$, we conclude that $\mathbf{u}^*$ is irrotational and defined in terms of the potential ϕ^* by

$$\mathbf{u}^* = \operatorname{grad} \phi^* \qquad (307)$$

Using this and eliminating ρ^* from (305) gives

$$\frac{\partial^2 \phi^*}{\partial t^{*2}} - (1 - M^2)\frac{\partial^2 \phi^*}{\partial x^{*2}} - \frac{\partial^2 \phi^*}{\partial y^{*2}} - \frac{\partial^2 \phi^*}{\partial z^{*2}} + 2M \frac{\partial^2 \phi^*}{\partial x^* \partial t^*} = 0 \qquad (308)$$

This equation could also have been obtained more directly from (257) using the Galilean transformation:

$$x^* = Mt + x, \qquad y^* = y, \qquad z^* = z, \qquad t^* = t, \qquad \phi^* = \phi \qquad (309)$$

This is to be expected, since the transformation $x^* = Mt + x$ corresponds to a uniform translation with velocity $-M\mathbf{i}$ of the starred frame relative to the unstarred frame. As a result, a stationary observer in the starred frame sees a uniform flow $M\mathbf{i}$, where $\mathbf{i}$ is the unit vector along x^* [see Figure 3.32].

If the disturbances *are steady in the starred frame*—that is, $\partial/\partial t^* = 0$, as would result for example if a slender rigid body were moving with constant

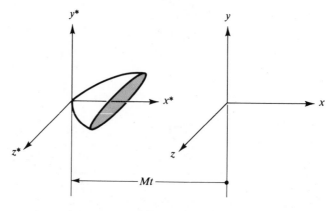

Figure 3.32

velocity $-M\mathbf{i}$ in the unstarred frame—(308) reduces to

$$\Delta_2 \phi^* - (M^2 - 1)\frac{\partial^2 \phi}{\partial x^{*2}} = 0 \tag{310}$$

where

$$\Delta_2 \equiv \partial^2/\partial y^{*2} + \partial^2/\partial z^{*2}.$$

To fix ideas, consider the flow over a winglike body, as shown in Figure 3.33. The body is slender in the y-direction, so its upper surface is defined by

$$y^* = \varepsilon F(x^*, z^*)$$

with a corresponding formula for the lower surface. The perturbation potential ϕ^* then obeys (310), and the boundary condition on the body is that the normal component of the velocity must vanish there.

The normal (not necessarily unit) to the body has components $\mathbf{n} = (-\varepsilon F_{x^*}, 1, -\varepsilon F_{z^*})$. The flow velocity has components $\mathbf{u} = (M + \varepsilon \phi^*_{x^*}, \varepsilon \phi^*_{y^*}, \varepsilon \phi^*_{z^*})$. Therefore, to $O(\varepsilon)$, the boundary condition $\mathbf{n} \cdot \mathbf{u} = 0$ on the body gives

$$\phi^*_{y^*}(x^*, 0^+, z^*) = M F_{x^*}(x^*, z^*) \tag{311}$$

with a corresponding formula for $\phi^*_{y^*}(x^*, 0^-, z^*)$. The boundary condition at infinity is that ϕ^* must vanish there [compare with (295d)–(295e)].

Notice that for $M > 1$ (supersonic flow), (310) is a two-dimensional wave equation with x^* as a timelike variable and a characteristic "speed" equal to $1/(M^2 - 1)^{1/2}$. For $M < 1$, (310) is equivalent to the three-dimensional Laplacian in the stretched coordinate system $\bar{x} = (1 - M^2)^{-1/2}x^*$, $\bar{y} = y^*$, $\bar{z} = z^*$. For steady two-dimensional flows, $\partial/\partial z^* = 0$ also, and we have the one-dimensional wave equation if $M > 1$ and the two-dimensional Laplace equation for $M < 1$. A discussion of solutions of (308), (310), or their two-dimensional versions can be found in various books on aerodynamics and is beyond the scope of this text. An idealized version is outlined in Problem 26, and the steady two-dimensional problem for a thin airfoil is outlined in Problem 6 of Chapter 4.

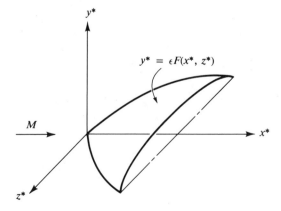

Figure 3.33

Problems

1. Substitute the expansions (10) into the nonlinear equations (7) and retain terms of order ε^2 to show that u_2 and v_2 obey

$$r(x)u_{2_{tt}} - f'(\sigma_0)u_{2_{xx}} = \tfrac{1}{2}[f'(\sigma_0) - 1](v_{1_x}^2)_x + 2f''(\sigma_0)(u_{1_x}^2)_x \tag{312a}$$

$$r(x)v_{2_{tt}} - v_{2_{xx}} = [f'(\sigma_0) - 1](u_{1_x}v_{1_x})_x + (pu_1)_x \tag{312b}$$

2. Consider shallow water flow in the X and Z directions over a variable bottom defined by the surface

$$Y_b \equiv Ab\left(\frac{X}{L_0}, \frac{Z}{L_0}, \frac{T}{T_0}\right) \tag{313a}$$

Here L_0 and T_0 are characteristic length and time scales for the bottom surface motion, and A is a characteristic amplitude. We assume, as in Section 3.2.1, that vertical motions are negligible, the density is constant, and that viscosity and surface tension may be ignored. The geometry is sketched in Figure 3.34, where $X = X_1$, $X = X_2$, $Z = Z_1$ and $Z = Z_2$ are fixed vertical planes which bound the sides of the domain of interest G. The bottom is defined by (313a) and the free surface by

$$Y_s \equiv Hh\left(\frac{X}{L_0}, \frac{Z}{L_0}, \frac{T}{T_0}\right) \tag{313b}$$

Denote the components of the flow velocity in the X and Z directions by $U(X/L_0, Z/L_0, T/T_0)$ and $W(X/L_0, Z/L_0, T/T_0)$, respectively.

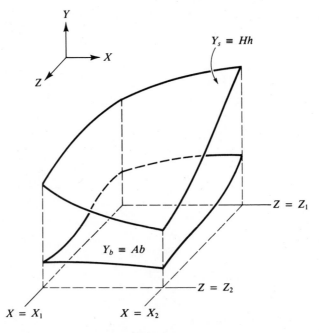

Figure 3.34

a. Use the dimensionless variables

$$x \equiv \frac{X}{L_0}; \qquad y \equiv \frac{Y}{H}; \qquad z \equiv \frac{Z}{L_0}; \qquad t \equiv \frac{T(gH)^{1/2}}{L_0}$$

$$u \equiv \frac{U}{(gH)^{1/2}}; \qquad w \equiv \frac{W}{(gH)^{1/2}}; \qquad p \equiv \frac{P}{\rho g H} \qquad (314)$$

and show that the integral law of mass conservation is

$$\frac{d}{dt} \int_{x_1}^{x_2} \int_{z_1}^{z_2} h(x, z, t)\, dx\, dz + \int_{z_1}^{z_2} [u(x_2, z, t)h(x_2, z, t) - u(x_1, z, t)h(x_1, z, t)]\, dz$$

$$+ \int_{x_1}^{x_2} [w(x, z_2, t)h(x, z_2, t) - w(x, z_1, t)h(x, z_1, t)]\, dx = 0 \qquad (315)$$

For smooth solutions, show that (315) reduces to

$$h_t + (uh)_x + (wh)_z = 0 \qquad (316)$$

b. Show that the pressure (hydrostatic) is given by

$$p(x, z, t) = \varepsilon b(x, z, t) + h(x, z, t) - y \qquad (317)$$

where $\varepsilon \equiv A/H$, and that the x and z components of the pressure force exerted by the water on that portion of the bottom surface contained in G are given by $\varepsilon \int_{x_1}^{x_2} \int_{z_1}^{z_2} h(x, z, t) b_x(x, z, t)\, dx\, dz$ and $\varepsilon \int_{x_1}^{x_2} \int_{z_1}^{z_2} h(x, z, t) \cdot b_z(x, z, t)\, dx\, dz$, respectively. Notice that in the absence of viscosity, the only stress on the bottom is the hydrostatic pressure that acts normal to the bottom.

c. Use the results in (b) to derive the following integral law of momentum conservation in the x direction:

$$\frac{d}{dt} \int_{x_1}^{x_2} \int_{z_1}^{z_2} u(x, z, t)h(x, z, t)\, dx\, dz + \varepsilon \int_{x_1}^{x_2} \int_{z_1}^{z_2} h(x, z, t)b_x(x, z, t)\, dx\, dz$$

$$+ \int_{z_1}^{z_2} \left[u^2(x, z, t)h(x, z, t) + \frac{h^2}{2}(x, z, t) \right]_{x=x_1}^{x=x_2} dz$$

$$+ \int_{x_1}^{x_2} [u(x, z, t)w(x, z, t)h(x, z, t)]_{z=z_1}^{z=z_2}\, dx = 0 \qquad (318)$$

The formula for momentum conservation in the z direction follows from (318) by replacing $u \to w$, $w \to u$, $x \to z$, and $z \to x$.

d. For smooth solutions, show that (318) and the corresponding formula for momentum conservation in the z direction reduce to

$$(uh)_t + \left(u^2 h + \frac{h^2}{2} + \varepsilon h b \right)_x + (uwh)_z = 0 \qquad (319a)$$

$$(wh)_t + \left(w^2 h + \frac{h^2}{2} + \varepsilon h b \right)_z + (uwh)_x = 0 \qquad (319b)$$

Use (316) to simplify these to the form

$$u_t + uu_x + wu_z + (h + \varepsilon b)_x = 0 \tag{320a}$$

$$w_t + uw_x + ww_z + (h + \varepsilon b)_z = 0 \tag{320b}$$

3. a. Use Laplace transforms with respect to t to solve (44) with zero initial conditions and the boundary condition (46). Distinguish carefully the two cases $x > t$ and $x < t$.
   Hint: Note that the Laplace transform of $\delta(t - x)$ is e^{-sx}, and use the convolution theorem to invert your result.

   b. A slight variation of the signaling problem discussed in Section 3.2.8 has the wavemaker speed proportional to the instantaneous local surface perturbation. The boundary condition that replaces (46) for the linearized problem is now

   $$u_1(0, t) = ch_1(0, t); \quad t > 0 \tag{321}$$

   where c is a constant. Assume that $c \neq 1$.
   Use Laplace transforms with respect to t to solve the signaling problem governed by (34), the boundary condition (321), and the initial conditions

   $$u_1(x, 0) = 0 \tag{322a}$$

   $$h_1(x, 0) = a \sin x \tag{322b}$$

   with $a = $ constant. Once you have studied Section 3.5.4, discuss why the case $c = 1$ is ill-posed.

4. Show that (72) follows from (66), (69), and (71) if we assume a smooth solution.

5. In an infinite tube, consider two different gases at rest separated by a diaphragm at $x = 0$. The gas in $x > 0$ has properties $\lambda_1, \mu_1, C_{p_1}, C_{v_1}$, all constants, and the gas in $x < 0$ has the constants $\lambda_2, \mu_2, C_{p_2}, C_{v_2}$. Initially, both gases are at rest and have the same temperature θ_0, but the pressure p_2 in $x < 0$ is slightly larger than p_1, the pressure in $x > 0$. At $t = 0$, the diaphragm is suddenly removed. Set up a perturbation expansion in terms of the small parameter $(p_2 - p_1)/p_1$ and derive the linearized equations analogous to (77) and initial conditions on the perturbation velocities, temperatures, pressures, and densities. Assume that the two gases do not mix and that the interface that was at $x = 0$ at time $t = 0$ always separates the two gases. It will be shown in Chapter 5 that the jump conditions at the interface are

$$\frac{d\xi}{dt} = u(\xi^+, t) = u(\xi^-, t) \tag{323}$$

$$p(\xi^+, t) - \tfrac{4}{3}\mu_1 u_x(\xi^+, t) = p(\xi^-, t) - \tfrac{4}{3}\mu_2 u_x(\xi^-, t) \tag{324}$$

$$\lambda_1 \theta_x(\xi^+, t) = \lambda_2 \theta_x(\xi^-, t) \tag{325}$$

where $x = \xi(t)$ is the location of the interface.
Use these results to derive what the appropriate jump conditions are for the perturbation quantities. Use Laplace transforms to show that you have a

well-posed problem and can determine u, p, θ, ρ as functions of x and t for all x and all $t \geq 0$.

6. Calculate the fundamental solution of the one-dimensional wave equation (104), using Laplace transforms with respect to t. Repeat the derivation using Fourier transforms with respect to x.

7. Derive the result (136) and show that the expressions for u_1 and h_1, given in (135) and (136), are the same as those obtained from (130) in regions ①–⑥ of the xt-plane.

8. The result that the fundamental solution of the wave equation

$$h_{1_{tt}} - h_{1_{xx}} = \delta(x)\delta(t); \qquad -\infty < x < \infty; 0 \leq t \tag{326a}$$

$$h_1(x, 0^-) = 0 \tag{326b}$$

$$h_{1_t}(x, 0^-) = 0 \tag{326c}$$

is [see (108)]

$$h_1(x, t) = \tfrac{1}{2}[H(x + t) - H(x - t)] \tag{327}$$

may appear, at first glance, to be counterintuitive if εh_1 is interpreted as the surface height perturbation for shallow-water waves [see (32b) and (38)]. We may ask how mass, momentum, or energy can be conserved if the surface above equilibrium equals $\varepsilon/2$ over the interval $-t < x < t$, which grows linearly in time.

a. Show that the problem (326) actually corresponds to the system of two first-order equations (34) with initial conditions

$$u_1(x, 0^+) = -H(x) \tag{328a}$$

$$h_1(x, 0^+) = 0 \tag{328b}$$

According to (328a) the entire body of water over $x > 0$ is initially moving to the left with speed ε, and this may provide the mechanism for a linearly expanding interval of water of above-equilibrium surface height.

b. Solve (34) subject to (328) by the method of characteristics to obtain

$$h_1(x, t) = \tfrac{1}{2}[H(x + t) - H(x - t)] \tag{329a}$$

$$u_1(x, t) = -\tfrac{1}{2}[H(x + t) + H(x - t)] \tag{329b}$$

Thus, (329a) confirms the result (327), and (329b) shows that the speed of the disturbed water is $-\varepsilon/2$ over the interval $-t < x < t$.

c. Argue physically how the semi-infinite reservoir of water of height 1 and speed $-\varepsilon$ to the right of $x = t$ "feeds" the increase in height over $-t < x < t$ because the speed of the water drops by a factor of two in this interval. Use a more careful argument and demonstrate that mass, momentum, and energy are indeed conserved over any fixed interval $x_1 \leq x \leq x_2$ for the solution (329).

9. a. For the problem discussed in Section 3.4.3, modify the initial conditions (123) so the result represents flow for $x \leq 1/2$ with a solid vertical wall located at $x = \tfrac{1}{2}$.

b. Calculate the solution for u_1 and h_1 for all $t \geq 0$ and $x \leq \frac{1}{2}$.

c. Given $\tilde{u}_1$, for what value of $\tilde{h}_1$ will there be only one wave propagating initially to the right over $x \leq \frac{1}{2}$? Summarize the solution in (b) for this case.

10. Consider the mixed boundary-value problem

$$u_{tt} - u_{xx} = 0; \qquad 0 \leq x \leq \infty, t \geq 0 \tag{330}$$

$$u_x(0,t) + bu(0,t) = c(t), \qquad t > 0 \tag{331}$$

$$u(x,0) = u_t(x,0) = 0 \tag{332}$$

where b is a constant and $c(t)$ is a prescribed analytic function of t with $c(0) = 0$, and $c(t) = 0$ if $t < 0$.

a. Transform the dependent variable $u \to v$ using

$$v(x,t) \equiv u_x(x,t) + bu(x,t) \tag{333}$$

and show that if u satisfies (330)–(332), then v is governed by

$$v_{tt} - v_{xx} = 0; \qquad 0 \leq x < \infty, t \geq 0 \tag{334}$$

$$v(0,t) = c(t) \tag{335}$$

$$v(x,0) = v_t(x,0) = 0 \tag{336}$$

Therefore, according to (155), the solution of (334)–(336) for v is

$$v(x,t) = \begin{cases} 0, & \text{if } t < x \\ c(t-x), & \text{if } t > x \end{cases} \tag{337}$$

b. Using (337) for v, solve (333) for u in the form

$$u(x,t) = \alpha(t)e^{-bx} + e^{-bx} \int_t^x c(t-\xi)e^{b\xi} \, d\xi \tag{338}$$

where α is as yet unspecified.

c. Substitute (338) into (330) and use the initial conditions (332) to prove that $\alpha(t) = 0$. Therefore, the desired solution is

$$u(x,t) = e^{-bx} \int_t^x c(t-\xi)e^{b\xi} \, d\xi \tag{339}$$

11. Consider the signaling problem on $0 \leq x \leq \infty$ for the wave equation

$$u_{tt} - c^2 u_{xx} = 0 \tag{340}$$

where $c \equiv c_1 = $ constant > 0 on $0 \leq x < 1$ and $c \equiv c_2 = $ constant $> c_1$ on $1 < x < \infty$. The initial conditions are $u(x,0) = 0$, $u_t(x,0) = 0$, and the boundary condition at $x = 0$ is the periodic signal $u(0,t) = A \sin \omega t$ for $t > 0$, where A and ω are given constants. Derive the solution in terms of appropriate primary, reflected, and transmitted waves, using the ideas in Section 3.5.5 and the interface conditions (189).

12. Use Green's function G_2, defined by (200), or symmetry to solve

$$u_{tt} - u_{xx} = p(x, t); \qquad 0 \le x; 0 \le t \tag{341}$$

$$u(x, 0) = f(x) \tag{342}$$

$$u_t(x, 0) = h(x) \tag{343}$$

$$u_x(0, t) = k(t); \qquad t > 0 \tag{344}$$

13. Calculate and sketch the remaining three Green's functions [in addition to (202)] for the wave equation on $0 \le x \le 1$. Denote, as in Problem 11 of Chapter 1:

$$G_2(x, \xi, t): \quad G_2(0, \xi, t) = G_{2_x}(1, \xi, t) = 0 \tag{345}$$

$$G_3(x, \xi, t): \quad G_{3_x}(0, \xi, t) = G_3(1, \xi, t) = 0 \tag{346}$$

$$G_4(x, \xi, t): \quad G_{4_x}(0, \xi, t) = G_{4_x}(1, \xi, t) = 0 \tag{347}$$

14. Use the appropriate Green's function and transformation to homogeneous boundary conditions to solve

$$u_{tt} - u_{xx} = 0; \qquad 0 \le x \le 1; 0 \le t \tag{348}$$

$$u(x, 0) = f(x) \tag{349}$$

$$u_t(x, 0) = 0 \tag{350}$$

$$u_x(0, t) = g(t), \qquad u(1, t) = l(t); \qquad t > 0 \tag{351}$$

15. a. Use separation of variables to derive (208), and show that (209) is just the Fourier series of G_1 as defined by (202).
    b. Solve (205) using Laplace transforms with respect to time. For $h = 1$, show that if the inversion integral is approximated for s large, one obtains the same result as (207).

16. Evaluate (212) for the case

$$f(x) = \begin{cases} 1 + x, & \text{on } 0 \le x \le \frac{1}{3} \\ 2(1 - x), & \text{on } \frac{1}{3} \le x \le 1 \end{cases} \tag{352}$$

Then show that the same result follows from (120) by extending the definition of $f(x)$ over the entire x-axis in a manner that is appropriate for the boundary conditions (210d)–(210e).

17. To what extent can we carry out the uniqueness proof of Section 3.6.4 to the case where $u(x, t)$ obeys (216a)–(216c) and the following three possible pairs of inhomogeneous boundary conditions?

$$\begin{cases} u_x(0, t) = \text{prescribed} & \text{(353a)} \\ u_x(1, t) = \text{prescribed} & \text{(353b)} \end{cases}$$

$$\begin{cases} u_x(0, t) = \text{prescribed} & \text{(354a)} \\ u(1, t) = \text{prescribed} & \text{(354b)} \end{cases}$$

$$\begin{cases} u(0, t) = \text{prescribed} & \text{(355a)} \\ u_x(1, t) = \text{prescribed} & \text{(355b)} \end{cases}$$

18. Consider the nonlinear wave equation:

$$u_{tt} - u_{xx} + F(u) = 0; \qquad 0 \le x < \pi; \qquad 0 \le t \tag{356}$$

where F is a given function with $F(0) = 0$, $F(-u) = -F(u)$, and $F > 0$ if $u > 0$. The boundary conditions are

$$u(0,t) = u(\pi, t) = 0; \qquad t > 0 \tag{357}$$

and the initial conditions are

$$u(x,0) = f(x) \tag{358a}$$

$$u_t(x,0) = g(x) \tag{358b}$$

a. Multiply (356) by u_t and integrate the result with respect to x over $0 \le x \le \pi$ to derive an energy integral analogous to (220) and involving the potential of the nonlinear force F defined by $G(u)$: $dG/du = F(u)$. Express $E(0)$ in terms of the given initial data, $f(x)$ and $g(x)$.

b. Assume a solution of (356)–(358) in the form of a series of the eigenfunctions of the linear problem, that is,

$$u(x,t) = \sum_{n=1}^{\infty} q_n(t) \sin nx \tag{359}$$

Substituting the series into the wave equation (356), show that the $q_n(t)$ obey a system of *coupled* nonlinear oscillator equations of the form

$$\frac{d^2 q_n}{dt^2} + n^2 q_n + F_n(q_1, q_2, \ldots) = 0; \qquad n = 1, 2, \ldots \tag{360}$$

Express the F_n in (360) as the Fourier coefficients of F, and show in particular that if $F = u$, $F_n = q_n$. What is F_n if $F = u + \varepsilon u^2$ and ε is a constant?

19. In this problem, we study the one-dimensional wave equation over the time-dependent domain D: $\alpha t \le x < \infty, t \ge 0$, where the constant α is restricted to $0 < \alpha < 1$.

a. Calculate Green's function of the first kind; that is, solve

$$u_{tt} - u_{xx} = \delta(x - \xi)\delta(t) \tag{361a}$$

in D for a constant $\xi > 0$ such that $\alpha \tau < \xi$. The initial conditions are

$$u(x,0^-) = u_t(x,0^-) = 0 \tag{361b,c}$$

and the boundary conditions are

$$u(\alpha t, t) = u(\infty, t) = 0 \tag{361d,e}$$

Hint: Introduce the transformation

$$\bar{x} = x - \alpha t, \qquad \bar{t} = t \tag{362a,b}$$

so that the left boundary is at the fixed point $\bar{x} = 0$ in the new frame, and solve the resulting problem by Laplace transforms with respect to $\bar{t}$. *Note:* The inversion integral can be evaluated explicitly.

b. Use the result in (a) and follow the procedure in Section 3.5.3 to solve the inhomogeneous boundary-value problem

$$u_{tt} - u_{xx} = 0 \tag{363a}$$

$$u(x,0) = u_t(x,0) = 0 \tag{363b,c}$$

$$u(\alpha t, t) = \begin{cases} 1, & \text{if } 0 < t < T \\ 0, & \text{if } T < t \end{cases} \tag{363d}$$

$$u(\infty, t) = 0 \tag{363e}$$

in D, where T is a positive constant.

c. Now calculate the solution of (363) by first transforming it to the $\bar{x}, \bar{t}$ variables defined by (362) and then using Laplace transforms to solve the resulting problem.

20. Derive (228) using similarity. Be careful to show that the constant is $\frac{1}{2}$ by requiring the result to satisfy (221) with $\delta(x)\delta(t)$ on the right-hand side.

21. Work out the results analogous to (114), (116), and (120) for the general one-dimensional wave equation (221) for the case $\lambda > 0$. Note that the fundamental solution may be written as the expression in (229) multiplied by $[H(x + t) - H(x - t)]$.

22. Consider the signaling problem for the telegraph equation:

$$u_{tt} - u_{xx} + u_t = 0 \tag{364}$$

$$u(x, 0^+) = u_t(x, 0^+) = 0 \tag{365}$$

$$u(0, t) = g(t); \qquad t > 0 \tag{366}$$

a. Take Laplace transforms with respect to t; then use the convolution theorem to derive the integral representation

$$u(x, t) = e^{-x/2}g(t - x) + \frac{x}{2} \int_x^t \frac{e^{-\tau/2}g(t - \tau)}{(\tau^2 - x^2)^{1/2}} I_1\left(\frac{1}{2}\sqrt{\tau^2 - x^2}\right) d\tau \tag{367}$$

if $t > x$ and $u = 0$ if $t < x$, where I_1 is the modified Bessel function of the first kind of order one.

b. Calculate Green's function of the first kind analogous to (140) for this case; then follow the approach in Section 3.5.3 and derive a result analogous to (154). Show that this result reduces to the expression (367).

23. In Chapter 8, we shall derive the following quasilinear third-order equation (Korteweg-deVries) in our discussion of weakly nonlinear shallow-water waves:

$$f_{0_\tau} + \frac{3}{4}f_0 f_{0_\xi} + \frac{\kappa^2}{6}f_{0_{\xi\xi\xi}} = 0 \tag{368}$$

Here $f_0(\xi, \tau)$ defines the leading approximation for part of the solution, ξ and

τ are dimensionless space and time variables, and κ is a constant.

a. Since (368) admits the trivial solution $f_0 = f_{00} = $ constant, we seek a small-disturbance approximation of the form

$$f_0(\xi, \tau) = f_{00} + f_{01}(\xi, \tau) \tag{369}$$

where $|f_{01}| \ll |f_{00}|$. Show that f_{01} obeys the linear equation

$$f_{01_\tau} + \frac{3}{4} f_{00} f_{01_\xi} + \frac{\kappa^2}{6} f_{01_{\xi\xi\xi}} = 0 \tag{370}$$

b. Assume a uniform wave solution of (370) in the form

$$f_{01}(\xi, \tau) = F(k\xi - \omega\tau) \tag{371}$$

to show that the dispersion relation is

$$\omega(k) = k\left(\frac{3}{4}f_{00} - \frac{k^2\kappa^2}{6}\right) \tag{372}$$

and that F has sinusoidal solutions.

24. Consider the nonlinear wave equation

$$u_{tt} - u_{xx} + u + \varepsilon u^3 = 0 \tag{373}$$

where ε is a positive constant.

a. Show that the assumption of a uniform wave, as in (233), leads to the nonlinear conservative oscillator problem for U:

$$(\omega^2 - k^2)U'' + U + \varepsilon U^3 = 0 \tag{374}$$

which has periodic solutions for $\omega^2 - k^2 > 0$ defined by the elliptic functions *sn* or *cn*. See Section 3.1.1 of [1].

b. To calculate the dispersion relation, note that (374) implies the integral:

$$(\omega^2 - k^2)\frac{U'^2}{2} + \frac{U^2}{2} + \frac{\varepsilon U^4}{4} = E = \text{constant} \tag{375}$$

Therefore, periodic solutions in θ correspond to closed curves in the (U, U') phase-plane for any given $E > 0$. If we normalize the period in θ to be 2π and indicate the closed contour for $E = $ constant by C, we have

$$(\omega^2 - k^2)^{1/2} \oint_C \frac{dU}{\left[2E - U^2 - \dfrac{\varepsilon U^4}{2}\right]^{1/2}} = 2\pi \tag{376}$$

where the proper sign for the square root must be used, depending on where the integration occurs. Show that for $\varepsilon \to 0$ the result reduces to the dispersion relation (237).

For $\varepsilon \neq 0$, the above result gives a relation linking ω, k *and* E. Thus, in general

$$\omega = \omega(k, E; \varepsilon) \tag{377}$$

and it is only in the limit $\varepsilon \to 0$ that the dispersion relation is independent of E (or the amplitude α).

c. For $0 < \varepsilon \ll 1$, show that the dispersion relation becomes

$$\omega(k, E; \varepsilon) = (1 + k^2)^{1/2}\left[1 + \frac{3E\varepsilon}{4(1 + k^2)} + O(\varepsilon^2)\right] \tag{378}$$

25. This problem concerns various aspects of the two-dimensional wave equation.

a. Work out the details leading from (272), in which S is given by (275), to obtain (276). Next, use similarity arguments to derive (277) directly for the two-dimensional wave equation.

b. Show that $\Phi(r, s)$, the Laplace transform of the fundamental solution of the two-dimensional wave equation, is given by

$$\Phi(r, s) = -\frac{1}{2\pi} K_0(sr) \tag{379}$$

where $r^2 = x^2 + y^2$, s is the Laplace transform variable, and K_0 is the modified Bessel function of the second kind of zero order. Use tables of Laplace transforms to show that the inversion formula for (379) gives (277).

c. Now use Fourier transforms with respect to x and y to show that $\tilde{\phi}(k_1, k_2, t)$, the transform of the fundamental solution of the two-dimensional wave equation, is given by

$$\bar{\phi}(k_1, k_2, t) = -\frac{1}{2\pi}\frac{\sin kt}{k}; \qquad k^2 = (k_1^2 + k_2^2) \tag{380}$$

Then introduce polar coordinates in the inversion integral and show that it simplifies to

$$\phi(r, t) = -\frac{1}{2\pi}\int_0^\infty J_0(kr) \sin kt \, dk \tag{381a}$$

when you use the integral representation

$$J_0(kr) \equiv \frac{1}{\pi}\int_0^\pi \cos(kr \sin \theta) \, d\theta \tag{381b}$$

for the Bessel function of the first kind of zero order. Use integral tables to show that (381a) gives (277).

d. The Hankel transform of a function $f(r)$ is denoted by $f^*(\omega)$ and is defined as

$$f^*(\omega) \equiv \int_0^\infty J_0(\omega r) f(r) r \, dr \tag{382a}$$

whenever the integral exists. Show that

$$f(r) = \int_0^\infty J_0(\omega r) f^*(\omega) \, d\omega \tag{382b}$$

is the corresponding inversion integral. Use (382a) to show that the Hankel transform of the fundamental solution of the two-dimensional wave equation is

$$\phi^*(\omega, t) = -\frac{1}{2\pi\omega} \sin \omega t \tag{383}$$

Therefore, the inversion integral (382b) directly gives the previously obtained result (381b).

e. Construct Green's function for the two-dimensional wave equation in the corner domain $x \geq 0$, $y \geq 0$ with zero boundary values; that is, solve

$$u_{tt} - u_{xx} - u_{yy} = \delta(x - \xi)\delta(y - \eta)\delta(t - \tau) \tag{384a}$$

$$u(x, y, \tau^-) = u_t(x, y, \tau^-) = 0 \tag{384b,c}$$

$$u(0, y, t) = u(x, 0, t) = 0 \tag{384d,e}$$

where ξ, η, and τ are positive constants.

Use this result to obtain an integral representation for the solution of

$$u_{tt} - u_{xx} - u_{yy} = p(x, t) \tag{385a}$$

$$u(x, y, 0) = f(x, y) \tag{385b}$$

$$u_t(x, y, 0) = g(x, y) \tag{385c}$$

$$u(x, 0, t) = u(0, y, t) = 0 \tag{385d,e}$$

on $x \geq 0$, $t \geq 0$.

Hint: Consider three separate problems as in Section 3.5.2 for the one-dimensional problem, and introduce a transformation analogous to (148) to reduce the problem for $f \neq 0$, $g = 0$ to one with $f = 0$, $g \neq 0$.

f. Now consider the case of inhomogeneous boundary conditions and solve

$$u_{tt} - u_{xx} - u_{yy} = 0 \tag{386a}$$

$$u(x, y, 0) = u_t(x, y, 0) = 0 \tag{386b,c}$$

$$u(x, 0, t) = k(x, t) \tag{386d}$$

$$u(0, y, t) = \ell(y, t) \tag{386e}$$

on $x \geq 0$, $y \geq 0$.

Hint: Transform $u(x, y, t)$ to a new dependent variable $w(x, y, t)$ defined in the form

$$u(x, y, t) \equiv w(x, y, t) + \alpha(x, y, t) \tag{387}$$

for some α such that $w(x, 0, t) = w(0, y, t) = 0$. One possible choice for α is

$$\alpha(x, y, t) = \frac{k(x, t)x}{(x^2 + y^2)^{1/2}} + \frac{\ell(x, t)y}{(x^2 + y^2)^{1/2}} \tag{388}$$

g. Use symmetry arguments to construct Green's function for the two-dimensional wave equation in the strip $-\infty < x < \infty$, $0 \le y \le 1$ with zero boundary values; that is, solve

$$u_{tt} - u_{xx} - u_{yy} = \delta(x - \xi)\delta(y - \eta)\delta(t - \tau) \tag{389a}$$

$$u(x, y, \tau^-) = u_t(x, y, \tau^-) = 0 \tag{389b,c}$$

$$u(x, 0, t) = u(x, 1, t) = 0 \tag{389d,e}$$

As in (e), use this result to calculate solutions with arbitrarily prescribed initial and boundary values.

26. Consider the potential ϕ due to a unit mass source moving with constant speed M along the x-axis. Thus, ϕ obeys

$$\Delta\phi - \phi_{tt} = \delta(x + Mt)\delta(y)\delta(z) \tag{390}$$

a. For $M < 1$, use the superposition integral (271) with the limits $-\infty$ to ∞ on τ to show that

$$\phi(x, y, z, t) = -\frac{1}{4\pi} \frac{1}{[(x + Mt)^2 + (1 - M^2)(y^2 + z^2)]^{1/2}} \tag{391}$$

Rederive this result as the fundamental solution of Laplace's equation in appropriate stretched variables.

b. For $M > 1$, introduce the transformation (309); then use (277) to solve the resulting problem and obtain

$$\phi(x, y, z, t) = \frac{-1}{2\pi} \frac{H[x + Mt - \sqrt{(M^2 - 1)(y^2 + z^2)}]}{[(x + Mt)^2 - (M^2 - 1)(y^2 + z^2)]^{1/2}} \tag{392}$$

27. Consider the signaling problem for acoustic disturbances down a semi-infinite waveguide with a square cross section. The waveguide occupies $0 \le x \le \pi$, $0 \le y \le \pi$, $0 \le z \le \infty$, and the boundary condition on the walls is the usual one of zero normal velocity there. The flow is initially at rest, and at $t = 0$ we start sending a given sinusoidal signal at the left end. So, the velocity potential satisfies

$$\Delta\phi - \phi_{tt} = 0; \qquad 0 \le x \le \pi, 0 \le y \le \pi; 0 \le z; 0 \le t \tag{393}$$

$$\phi(x, y, z, 0) = \phi_t(x, y, z, 0) = 0 \tag{394}$$

$$\phi_x(0, y, z, t) = \phi_x(\pi, y, z, t) = \phi_y(x, 0, z, t) = \phi_y(x, \pi, z, t) = 0 \tag{395}$$

$$\phi_z(x, y, 0, t) = A(x, y)\sin \omega t; \qquad t > 0 \tag{396}$$

where $A(x, y)$ is a prescribed function and ω is a constant.

a. Use separation of variables with respect to the x and y dependence, and Laplace transforms with respect to t to show that

$$\phi(x, y, z, t) = \sum_{m=0}^{\infty} \sum_{n=0}^{\infty} \psi_{mn}(z, t) \cos mx \cos ny \tag{397}$$

where

$$\psi_{mn}(z,t) = \begin{cases} -A_{mn} \displaystyle\int_z^t J_0(\sqrt{(m^2+n^2)(\tau^2-z^2)}) \sin \omega(t-\tau) \, d\tau, & \text{if } t > z \\[2mm] 0, & \text{if } t < z \end{cases} \tag{398}$$

Here J_0 is the Bessel function, and

$$A_{mn} = \frac{4}{\pi^2} \int_0^\pi \int_0^\pi A(x,y) \cos mx \cos ny \, dx \, dy \tag{399}$$

b. For $t \to \infty$, z fixed, assume that each mode ψ_{mn} has the form

$$\psi_{mn}(z,t) = \alpha_{mn} \cos \omega \left(t - \frac{z}{c_{mn}} \right) \tag{400}$$

Show by substitution into the equation governing ψ_{mn} that the constants c_{mn}, α_{mn} are given by

$$c_{mn} = \frac{\omega}{[\omega^2 - (m^2+n^2)]^{1/2}} \tag{401}$$

$$\alpha_{mn} = \frac{A_{mn}}{[\omega^2 - (m^2+n^2)]^{1/2}} \tag{402}$$

Comment on the nonexistence of such solutions if ω is smaller than the "cutoff" frequency $\omega_c \equiv \sqrt{m^2+n^2}$ associated with the given mode. Comment on the relationship of your result with the solution in (a).

28. Consider the balloon problem discussed in Section 3.10.1. The solution for ϕ is spherically symmetric. Therefore, the velocity potential depends only on r and t and has the form [see (263)]

$$\phi(r,t) = \frac{1}{r}\phi_1(r-t) + \frac{1}{r}\phi_2(r+t) \tag{403}$$

The initial conditions are

$$\phi(r,0) = 0 \tag{404a}$$

$$\phi_t(r,0) = \begin{cases} -\dfrac{1}{\gamma} & \text{if } r < 1 \\[2mm] 0 & \text{if } r > 1 \end{cases} \tag{404b}$$

Show that Equations (404) determine ϕ_1 and ϕ_2 for positive values of their arguments. Since the argument of ϕ_1 may be negative, we need one more condition. This is obtained by recalling that the homogeneous wave equation (257) corresponds to zero mass sources. In particular, since there is no source at the origin, we must have

$$\lim_{r \to 0} r^2 \phi_r(r,t) = 0, \qquad t \geq 0 \tag{405}$$

Thus, the radial velocity cannot grow at a rate faster than r^{-2} as $r \to 0$. Show that (405) determines ϕ_1 for negative values of its argument in terms of ϕ_2 for positive argument, and that ϕ_1 and ϕ_2 are given by

$$
\phi_1(z) = \begin{cases} \dfrac{1}{4\gamma}(z^2 - 1), & \text{if } -1 < z < 1 \\[2mm] 0, & \text{if } |z| > 1 \end{cases}
\tag{406a}
$$

$$
\phi_2(z) = \begin{cases} -\dfrac{1}{4\gamma}(z^2 - 1), & \text{if } 0 < z < 1 \\[2mm] 0, & \text{if } 1 < z \end{cases}
\tag{406b}
$$

Verify that using (406) in (403) gives (292) if $r > 1$. Calculate the solution for $r < 1$ and sketch the variation of $\tilde{p}$ with time for this case.

References

1. J. Kevorkian, and J. D. Cole, *Perturbation Methods in Applied Mathematics*, Springer, New York, 1981.
2. G. B. Whitham, *Linear and Nonlinear Waves*, Wiley, New York, 1974.
3. W. Lick, "Wave Propagation in Real Gases," *Advances in Applied Mechanics* 10, Fasc. 1 (1967): 1–72.
4. B. Friedman, *Principles and Techniques of Applied Mathematics*, Wiley, New York, 1956.
5. M. Abramowitz, and I. A. Stegun, *Handbook of Mathematical Functions*, National Bureau of Standards, Washington, D.C., 1964.
6. G. F. Carrier, M. Krook, and C. E. Pearson, *Functions of a Complex Variable, Theory and Technique*, McGraw-Hill, New York, 1966.

Linear Second-Order Equations with Two Independent Variables

In this chapter, we consider the general linear second-order partial differential equation in two independent variables and show that, depending upon the values of the coefficients multiplying the second derivatives, it can be reduced to one of the three canonical forms discussed in Chapters 1–3. We also derive the corresponding results for a pair of linear first-order equations. The quasilinear problem, for which the coefficients also involve the dependent variable, is discussed in Chapter 5 for a first-order equation, and in Chapter 7 for systems of first order. In this chapter, we also restrict attention to the problem in two independent variables. In general, it is not possible to reduce a second-order equation in more than two independent variables to a simple canonical form (see for example, Chapter 3, Section 2, of [1]).

4.1   A General Transformation of Variables

The general linear second-order equation in the two independent variables x, y is

$$au_{xx} + 2bu_{xy} + cu_{yy} + du_x + eu_y + fu = g \tag{1}$$

where a, b, c, d, e, f, and g are prescribed functions of x and y. We denote the linear differential operator by L and write (1) as

$$L(u) + fu = g \tag{2}$$

where

$$L \equiv a\frac{\partial^2}{\partial x^2} + 2b\frac{\partial^2}{\partial x\,\partial y} + c\frac{\partial^2}{\partial y^2} + d\frac{\partial}{\partial x} + e\frac{\partial}{\partial y} \tag{3}$$

Suppose that

$$\xi = \phi(x, y) \tag{4a}$$

$$\eta = \psi(x, y) \tag{4b}$$

is an *arbitrary* curvilinear coordinate system that we wish to use instead of the original system (x, y). A necessary condition on the functions ϕ and ψ to ensure that (4) is a coordinate transformation—that is, that for every point (x, y) there corresponds a unique point (ξ, η) and vice versa—is that the Jacobian

$$J(x, y) \equiv \phi_x \psi_y - \psi_x \phi_y \tag{5}$$

not vanish identically in the domain of interest. Let $U(\xi, \eta)$ denote the dependent variable regarded as a function of the new independent variables (ξ, η); that is,

$$u(x, y) \equiv U(\phi(x, y), \psi(x, y)) \tag{6}$$

We now calculate:

$$u_x = U_\xi \phi_x + U_\eta \psi_x; \qquad u_y = U_\xi \phi_y + U_\eta \psi_y$$

$$u_{xx} = U_{\xi\xi} \phi_x^2 + 2U_{\xi\eta} \phi_x \psi_x + U_{\eta\eta} \psi_x^2 + U_\xi \phi_{xx} + U_\eta \psi_{xx}$$

$$u_{yy} = U_{\xi\xi} \phi_y^2 + 2U_{\xi\eta} \phi_y \psi_y + U_{\eta\eta} \psi_y^2 + U_\xi \phi_{yy} + U_\eta \psi_{yy}$$

$$u_{xy} = U_{\xi\xi} \phi_x \phi_y + U_{\xi\eta}(\phi_x \psi_y + \psi_x \phi_y) + U_{\eta\eta} \psi_x \psi_y + U_\xi \phi_{xy} + U_\eta \psi_{xy}$$

Therefore, (2) transforms to

$$M(U) + FU = G \tag{7}$$

where the linear operator M has the form

$$M \equiv A \frac{\partial^2}{\partial \xi^2} + 2B \frac{\partial^2}{\partial \xi \, \partial \eta} + C \frac{\partial^2}{\partial \eta^2} + D \frac{\partial}{\partial \xi} + E \frac{\partial}{\partial \eta} \tag{8}$$

Here A, B, and C are the following quadratic forms:

$$B(\xi, \eta) \equiv a\phi_x \psi_x + b(\phi_x \psi_y + \psi_x \phi_y) + c\phi_y \psi_y \tag{9a}$$

$$A(\xi, \eta) \equiv a\phi_x^2 + 2b\phi_x \phi_y + c\phi_y^2 \tag{9b}$$

$$C(\xi, \eta) \equiv a\psi_x^2 + 2b\psi_x \psi_y + c\psi_y^2 \tag{9c}$$

The functions D and E are given by [see (3)]

$$D(\xi, \eta) = L(\phi) \tag{10a}$$

$$E(\xi, \eta) = L(\psi) \tag{10b}$$

and

$$F(\xi, \eta) = f \tag{10c}$$

$$G(\xi, \eta) = g \tag{10d}$$

Note that to compute A, B, or C we must evaluate the functions of x and y defined by the right-hand sides of (9) and then express x and y in terms of ξ, η. This latter step is always possible for a coordinate transformation, since $J \neq 0$. In deriving D and E in (10a)–(10b), we must evaluate the second-order differential operator L for the given functions $\phi(x, y)$ and $\psi(x, y)$. The resulting functions of x and y are then expressed in terms of ξ and η. We also express x and y in terms of ξ, η in the right-hand sides of (10c) and (10d) to obtain F and G.

A general property of the transformation (4) is that the sign of $b^2 - ac$ is the same as the sign of $(B^2 - AC)$ because it follows from (9) that

$$\Delta \equiv b^2 - ac = \frac{B^2 - AC}{J^2} \tag{11}$$

We shall see next that depending on whether $\Delta > 0$, $\Delta < 0$, or $\Delta = 0$, we can choose a corresponding particular transformation (4) that reduces (7) to a simple canonical form.

4.2   Classification

4.2.1 The Hyperbolic Problem, $\Delta > 0$; $A = C = 0$

Recalling the D'Alembert form of the wave equation [(225) of Chapter 3], it is natural to seek the conditions for which one is able to reduce (7) to this simple form. Clearly, if we can find functions ϕ and ψ such that $A = C = 0$ and $B \neq 0$, we can divide (7) by B to obtain the desired result.

We see from (9b) and (9c) that the equation governing ϕ that results from setting $A = 0$ is the same as the equation governing ψ in order to have $C = 0$. So, it is sufficient to consider either one of these equations, say,

$$a(x, y)\phi_x^2 + 2b(x, y)\phi_x\phi_y + c(x, y)\phi_y^2 = 0 \tag{12}$$

Now, this nonlinear first-order partial differential equation for ϕ is less formidable than it looks. Consider a level curve $\phi(x, y) = \xi_0 = \text{constant}$ on any solution surface $\xi = \phi(x, y)$ of (12) as sketched in Figure 4.1.

On this level curve, the slope $y' \equiv (dy/dx)$ is given by $y' = -\phi_x(x, y)/\phi_y(x, y)$ if $\phi_y \neq 0$. But according to (12), if $\phi_y \neq 0$, we have

$$a\frac{\phi_x^2}{\phi_y^2} + 2b\frac{\phi_x}{\phi_y} + c = 0 \tag{13a}$$

or

$$ay'^2 - 2by' + c = 0 \tag{13b}$$

Solving the quadratic expression (13b) for y' gives a real result only if $\Delta \geq 0$. We defer discussion of the case $\Delta = 0$ to Section 4.2.3. and note here that a real transformation of variables that renders $A = C = 0$ does not exist if $\Delta < 0$. In the case $\Delta > 0$, we have two real and distinct roots:

$$y' = \frac{b + \sqrt{\Delta}}{a} \tag{14a}$$

$$y' = \frac{b - \sqrt{\Delta}}{a} \tag{14b}$$

Thus, the canonical form

$$U_{\xi\eta} = \tilde{D}U_\xi + \tilde{E}U_\eta + \tilde{F}U + \tilde{G} \tag{15a}$$

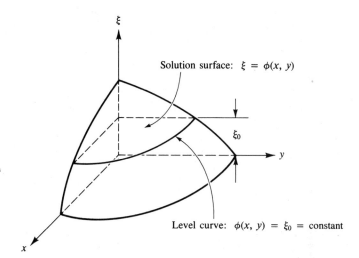

Solution surface: $\xi = \phi(x, y)$

ξ_0

Level curve: $\phi(x, y) = \xi_0 = $ constant

Figure 4.1

where

$$\tilde{D} = -\frac{D}{2B} \tag{15b}$$

$$\tilde{E} = -\frac{E}{2B} \tag{15c}$$

$$\tilde{F} = -\frac{F}{2B} \tag{15d}$$

$$\tilde{G} = \frac{G}{2B} \tag{15e}$$

is indeed possible if $\Delta > 0$. In this case, the solution of the two ordinary differential equations (14) defines two distinct families of level curves, as shown next. Since the solution of the equation $C = 0$ gives the same pair of families, we may identify the curves $\phi = $ constant with one sign for $\sqrt{\Delta}$, say the plus sign, and let the curves $\psi = $ constant correspond to the minus sign in (14).

We now verify that the two families of curves $\phi = $ constant and $\psi = $ constant so defined do indeed represent a coordinate transformation—that is, $J \neq 0$. If we divide (5) by $\phi_y \psi_y \neq 0$ (assuming temporarily that neither ϕ_y nor ψ_y vanish), we find

$$\frac{J}{\phi_y \psi_y} = \frac{\phi_x}{\phi_y} - \frac{\psi_x}{\psi_y} = -\frac{b}{a} - \frac{\sqrt{b^2 - ac}}{a} + \frac{b}{a} - \frac{\sqrt{b^2 - ac}}{a} = -\frac{2\sqrt{b^2 - ac}}{a} \neq 0$$

hence, $J \neq 0$. The two families of curves $\phi = $ constant, and $\psi = $ constant are called *characteristic curves* [see (106) and (124) of Chapter 3].

If $\phi_y \equiv 0$, then (12) implies that $\phi_x \equiv 0$ also, and we have the trivial solution $\phi = $ constant, which always satisfies (12). So, we exclude this case. If $a = 0$, we may interchange the roles of x and y so that the divisor in (14) is c. If $c = 0$ also, we need not transform (2) because it is already in canonical form. Also note that $B \equiv 0$ implies that $\Delta \equiv 0$ as can be seen by substituting (14) for ϕ_x/ϕ_y and ψ_x/ψ_y in (9a). Therefore, $B \neq 0$ and division by $2B$ is not troublesome in (15a).

We have shown that if $\Delta > 0$, we can always derive the appropriate transformation of variables (4) so that the equation governing U is in the canonical form (15). Depending upon the functions $a(x, y)$, $b(x, y)$, and $c(x, y)$, there may or may not exist a domain $\mathcal{D}$ in the (x, y) plane in which $\Delta > 0$. If $\mathcal{D}$ exists, (2) is said to be hyperbolic in $\mathcal{D}$. Notice that the existence of $\mathcal{D}$ depends only on the coefficients a, b, and c of the second-derivative terms and not on the coefficients d, e, f, or g. Moreover, this property is independent of the choice of variables used to express (2), since the sign of Δ is coordinate-invariant.

We can also transform (15a) to the alternate canonical form [see (221) of Chapter 3]

$$V_{XX} - V_{YY} = D^*V_X + E^*V_Y + F^*V + G^* \tag{16}$$

by introducing the transformation

$$X = \xi + \eta; \quad Y = \xi - \eta; \quad V(X, Y) = U\left(\frac{X + Y}{2}, \frac{X - Y}{2}\right)$$

We find

$$D^* = \tilde{D} + \tilde{E}; \quad E^* = \tilde{D} - \tilde{E}; \quad F^* = \tilde{F}; \quad G^* = \tilde{G}$$

In general, it is not possible to simplify (15) or (16) further. The exception occurs when $\tilde{D}$ and $\tilde{E}$ satisfy the condition $\tilde{D}_\xi = \tilde{E}_\eta$, in which case we can also eliminate the first-derivative terms (See Problem 2). In particular, the condition $\tilde{D}_\xi = \tilde{E}_\eta$ holds trivially if D and E are constant; we have already seen the needed transformation in Section 3.7.1.

4.2.2 Hyperbolic Examples

The general hyperbolic equation with *constant* coefficients has a, b, c, d, e, and f all constant in (1) with $\Delta = b^2 - ac > 0$. We find

$$y' = \frac{b \pm \sqrt{\Delta}}{a} = \alpha_1, \alpha_2 \tag{17}$$

where α_1 and α_2 are unequal constants. Therefore, the characteristics are the straight lines $\xi = \text{constant}$ and $\eta = \text{constant}$ defined by

$$\xi = y - \alpha_1 x; \quad \eta = y - \alpha_2 x \tag{18}$$

Using (9a), we compute

$$2B = \frac{4(ac - b^2)}{a}$$

and using (10a), we find

$$D = L(\xi) = a\xi_{xx} + 2b\xi_{xy} + c\xi_{yy} + d\xi_x + e\xi_y = -\alpha_1 d + e$$

$$E = L(\eta) = -\alpha_2 d + e$$

Therefore, U obeys

$$U_{\xi\eta} = \tilde{D}U_\xi + \tilde{E}U_\eta + \tilde{F}U + \tilde{G} \tag{19}$$

with

$$\tilde{D} = -\frac{Da}{4(ac - b^2)}; \qquad \tilde{E} = -\frac{Ea}{4(ac - b^2)}$$

$$\tilde{F} = -\frac{fa}{4(ac - b^2)}; \qquad \tilde{G} = \frac{ga}{4(ac - b^2)} \tag{20}$$

An important equation in the linearized theory of transonic aerodynamics is the *Tricomi* equation (for example, see Section 3.5 of [2]):

$$yu_{xx} - u_{yy} = 0 \tag{21}$$

Here, $a = y$, $b = 0$, $c = -1$, and $d = e = f = g = 0$. Thus, $\Delta = y$, and (21) is hyperbolic for $y > 0$. The characteristic curves satisfy

$$y' = \pm y^{-1/2}, \qquad y > 0 \tag{22}$$

We compute

$$\xi = \tfrac{2}{3}y^{3/2} - x; \qquad \eta = \tfrac{2}{3}y^{3/2} + x \tag{23}$$

The curves $\xi = $ constant and $\eta = $ constant are a pair of one-parameter families in $y > 0$ that end up with a cusp on the x-axis as shown in Figure 4.2.

4.2.3 The Parabolic Problem, $\Delta = 0$; $C = 0$

In this case, the condition $A = 0$ or $C = 0$ defines the *same* single family of characteristics satisfying

$$y' = \frac{b}{a} \tag{24}$$

Suppose that we set $C = 0$. Then the solution of (24) defines the family

$$\eta = \psi(x, y)$$

and we may choose the family $\xi = \phi(x, y)$ *arbitrarily* as long as $J \neq 0$; that is, $A \neq 0$.

Since $\Delta = 0$ also implies $B^2 - AC = 0$, we have $B = 0$, and the canonical form for U is

$$U_{\xi\xi} = \bar{D}U_\xi + \bar{E}U_\eta + \bar{F}U + \bar{G} \tag{25}$$

where

$$\bar{D} = -\frac{D}{A}; \qquad \bar{E} = -\frac{E}{A}; \qquad \bar{F} = -\frac{F}{A}; \qquad \bar{G} = \frac{G}{A}$$

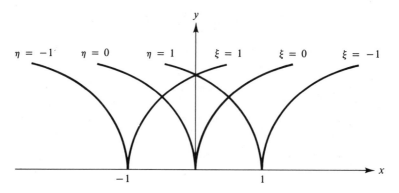

Figure 4.2

Of course, it is also possible to set $A = 0$, $C \neq 0$ and obtain the canonical form

$$U_{\eta\eta} = D^* U_\xi + E^* U_\eta + F^* U + G^* \tag{26}$$

where the starred coefficients are obtained from the coefficients in (7) by dividing by C.

For the special case where the linear equation (2) has constant coefficients, $\Delta = 0$ corresponds to $ac = b^2$; this condition does not depend on the point (x, y). We have $y' = b/a$—that is, $\eta = y - bx/a$—and we may choose $\xi = x$, as this results in $J = 1$ for all x, y. We then find $A = a$, $B = C = 0$, and the canonical form (25) is

$$U_{\xi\xi} = -\frac{d}{a} U_\xi - \frac{e}{a} U_\eta - \frac{f}{a} U + \frac{g}{a}$$

In general, for given functions $a(x, y)$, $b(x, y)$, and $c(x, y)$, the condition $\Delta(x, y) = 0$ may either be satisfied identically in some domain $\mathcal{D}$ (as for the case where $c(x, y) \equiv b^2(x, y)/a(x, y)$ in $\mathcal{D}$), or $\Delta = 0$ may be true only on some curve $\mathcal{C}$ (for example, for the Tricomi equation, the curve $\mathcal{C}$ is the x-axis), or $\Delta(x, y) = 0$ may have no real solution. If $\Delta(x, y)$ has a real solution in some domain $\mathcal{D}$ or along some curve $\mathcal{C}$, we say that (2) is parabolic in $\mathcal{D}$ or on $\mathcal{C}$.

4.2.4 The Elliptic Problem, $\Delta < 0$; $B = 0$, $A = C$

In this case, we cannot satisfy (14) anywhere for real functions $\phi(x, y)$ and $\psi(x, y)$, and therefore $A \neq 0$, $C \neq 0$. So, the only alternative for simplification is to eliminate the mixed partial derivative $U_{\xi\eta}$ in (7) by setting $B = 0$. Then requiring $A = C$ allows the remaining second-derivative terms to reduce to the Laplacian when (7) is divided by A.

Thus, we need to solve the following two coupled equations for ϕ and ψ that result from (9) when we set $B = 0$ and $A - C = 0$:

$$a(\phi_x^2 - \psi_x^2) + 2b(\phi_x\phi_y - \psi_x\psi_y) + c(\phi_y^2 - \psi_y^2) = 0 \tag{27a}$$

$$a\phi_x\psi_x + b(\phi_x\psi_y + \phi_y\psi_x) + c\phi_y\psi_y = 0 \tag{27b}$$

If we multiply (27b) by $2i$ and add this to (27a), we find the complex version of (12) in terms of the complex variable $\zeta = \xi + i\eta$; that is,

$$a\zeta_x^2 + 2b\zeta_x\zeta_y + c\zeta_y^2 = 0 \tag{28}$$

or, equivalently:

$$\frac{\phi_x + i\psi_x}{\phi_y + i\psi_y} = -\frac{b + i\sqrt{ac - b^2}}{a} \tag{29}$$

The real and imaginary parts of (29) give

$$\phi_x = \frac{c\psi_y + b\psi_x}{(ac - b^2)^{1/2}} \tag{30a}$$

$$\phi_y = -\frac{b\psi_y + a\psi_x}{(ac - b^2)^{1/2}} \tag{30b}$$

These two first-order equations for ϕ and ψ are called the *Beltrami* equations, and *any* solution $\phi(x, y)$, $\psi(x, y)$ defines a transformation to the canonical form

$$U_{\xi\xi} + U_{\eta\eta} = \text{lower-derivative terms} \tag{31}$$

Although we may eliminate one of the dependent variables in favor of the other from (30), the resulting second-order equation is in general more complicated and harder to solve than the original system (30). For example, eliminating ϕ gives

$$\left[\frac{a\psi_x + b\psi_y}{(ac - b^2)^{1/2}} \right]_x + \left[\frac{c\psi_y + b\psi_x}{(ac - b^2)^{1/2}} \right]_y = 0$$

and we note that, in general, solving this for ψ is not any easier than solving the system (30).

If a, b, and c are analytic, we can construct solutions of the Beltrami equations by solving [compare with (14)]

$$\frac{dy}{dx} = \frac{b + i\sqrt{ac - b^2}}{a} \tag{32}$$

in the complex plane. These ideas are discussed in [1] and are not pursued further here. Actually, a general solution of the Beltrami equations is not needed in order to implement the transformation to the canonical form (31); any solution that satisfies the requirement $J \neq 0$ will do. These ideas are illustrated next for the special case of constant coefficients. See also Problem 3(b).

If (1) has constant coefficients, the general solution of (30) is linear in x and y and is easily derived. To see this assume a solution of the form

$$\phi = \alpha x + \beta y; \qquad \psi = \gamma x + \delta y$$

with α, β, γ, and δ constant. Substitution into (30) gives the following two relations linking the four constants:

$$\alpha = \tilde{c}\delta + \tilde{b}\gamma; \qquad \beta = -(\tilde{b}\delta + \tilde{a}\gamma)$$

where a tilde over a constant indicates that the constant is divided by $(ac - b^2)^{1/2}$. Thus, for any choice of δ, γ, we have α, β. For simplicity, take $\delta = 0$, $\gamma = 1$ to find $\alpha = \tilde{b}$ and $\beta = -\tilde{a}$ and the transformation

$$\xi = \tilde{b}x - \tilde{a}y; \qquad \eta = x$$

This results in the canonical form

$$U_{\xi\xi} + U_{\eta\eta} + \frac{db - ea}{a(ac - b^2)^{1/2}} U_\xi + \frac{d}{a} U_\eta + \frac{f}{a} U = \frac{g}{a}$$

and division by a leads to no difficulties because $\Delta < 0$ implies that $a \neq 0$.

We can again remove the first-derivative terms U_ξ and U_η by the transformation of dependent variable $U \to W$ defined by

$$U(\xi, \eta) \equiv W(\xi, \eta) \exp - \left[\frac{db - ea}{2a(ac - b^2)^{1/2}} \xi + \frac{d}{2a} \eta \right]$$

A straightforward calculation shows that W obeys

$$W_{\xi\xi} + W_{\eta\eta} + \left[\frac{f}{a} + \frac{2dbe - e^2a - dc^2}{4a(ac - b^2)} \right] W = \frac{g}{a} \exp \left[\frac{db - ea}{2a(ac - b^2)^{1/2}} \xi + \frac{d}{2a} \eta \right]$$

4.3   The Role of Characteristics in Hyperbolic Equations

One interpretation of the characteristic curves $\phi(x, y) = \xi = $ constant and $\psi(x, y) = \eta = $ constant defined by (17) with $\Delta > 0$ is that this pair of one-parameter families of curves defines a coordinate transformation of (1) to the canonical form (15a). In this section we give two other interpretations that are equally significant for these curves.

4.3.1   Cauchy's Problem

Let $\phi(x, y) = \xi_0 = $ constant define a curve $\mathscr{C}$ in the (x, y) plane. The Cauchy problem for (1) consists of the solution of this equation subject to prescribed values of u and $\partial u/\partial n$ on $\mathscr{C}$. As usual, $\partial/\partial n$ indicates the directional derivative in the direction normal to $\mathscr{C}$. It is also understood that we have specified on which side of $\mathscr{C}$ we wish to solve (1); hence the unit normal $\mathbf{n}$ to $\mathscr{C}$ is taken to point into the domain of interest $\mathscr{D}$ (see Figure 4.3).

Now consider a curvilinear coordinate system (ξ, η) defined by

$$\xi = \phi(x, y) \tag{33a}$$

$$\eta = \psi(x, y) \tag{33b}$$

where $\phi(x, y) = \xi_0$ is the same function as the one defining $\mathscr{C}$ and the curves $\eta = $ constant are chosen to be noncollinear to the curves $\xi = $ constant; that is, $\phi_x \psi_y - \psi_x \phi_y \neq 0$ in $\mathscr{D}$.

We saw in Section 4.1 that the governing equation (1) transforms to (7):

$$A(\xi, \eta) U_{\xi\xi} + 2B(\xi, \eta) U_{\xi\eta} + C(\xi, \eta) U_{\eta\eta} + D(\xi, \eta) U_\xi + E(\xi, \eta) U_\eta + F(\xi, \eta) U = G(\xi, \eta) \tag{34}$$

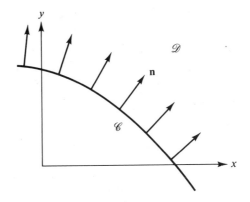

Figure 4.3

where $A, \ldots, G$ are defined in terms of the given ϕ, ψ and the original coefficients $a, \ldots, g$ in (9)–(10).

If u and $\partial u / \partial n$ are prescribed along $\mathscr{C}$, it means that we have

$$U(\xi_0, \eta) \equiv \alpha(\eta) \tag{35a}$$

$$U_\xi(\xi_0, \eta) \equiv \beta(\eta) \tag{35b}$$

for prescribed functions $\alpha(\eta)$ and $\beta(\eta)$. This information is called *Cauchy data* on $\mathscr{C}$.

We now ask whether it is possible to use the Cauchy data (35) in conjunction with the partial differential equation (34) to define $U(\xi, \eta)$ on some neighboring curve $\xi_1 = \xi_0 + \Delta\xi = $ constant.

Using a Taylor series, we find

$$U(\xi_1, \eta) = U(\xi_0, \eta) + U_\xi(\xi_0, \eta)\Delta\xi + U_{\xi\xi}(\xi_0, \eta)\frac{\Delta\xi^2}{2} + \cdots \tag{36}$$

and we know $U(\xi_0, \eta)$, $U_\xi(\xi_0, \eta)$. Therefore, to compute the Taylor series to $O(\Delta\xi^2)$, all we need is $U_{\xi\xi}(\xi_0, \eta)$.

Clearly, we can obtain $U_{\xi\xi}$ from (34) in terms of known quantities *as long as the curve $\mathscr{C}$ is not characteristic*. In fact, if $\mathscr{C}$ is not characteristic, $A \neq 0$, and we have

$$U_{\xi\xi}(\xi, \eta) = -\frac{2B}{A}U_{\xi\eta} - \frac{C}{A}U_{\eta\eta} - \frac{D}{A}U_\xi - \frac{E}{A}U_\eta - \frac{F}{A}U + \frac{G}{A} \tag{37}$$

Each of the coefficients on the right-hand side of (37) is known and can therefore be evaluated at $\xi = \xi_0$. Moreover, on $\xi = \xi_0$, U and U_ξ are the known functions of η given by the Cauchy data in the form (35). To compute U_η, $U_{\xi\eta}$, and $U_{\eta\eta}$ on $\xi = \xi_0$, we also use (35) and find:

$$U_\eta(\xi_0, \eta) = \alpha'(\eta); \qquad U_{\eta\eta}(\xi_0, \eta) = \alpha''(\eta); \qquad U_{\xi\eta}(\xi_0, \eta) = \beta'(\eta)$$

This shows that we can evaluate $U_{\xi\xi}(\xi_0, \eta)$. To evaluate each higher derivative $U_{\xi\xi\xi}$ and so on, we differentiate (37) with respect to ξ, solve for $U_{\xi\xi\xi}(\xi, \eta)$, and evaluate $U_{\xi\xi\xi}(\xi, \eta)$ along $\mathscr{C}$ in terms of known quantities there. This is always possible as long as the coefficients $A, B, \ldots$ are analytic and $A \neq 0$. The formal theorem ensuring that this construction generates a unique solution as long as $a, \ldots, g$ are analytic is attributed to Cauchy and Kowalewski. (For a proof, see Chapter 1, Section 7 of [3]). This construction is not restricted to any given type (hyperbolic, elliptic, or parabolic) of equation; however, it is not very useful as a practical solution technique [see the discussion in Section 4.4.5)]. In the next section, we outline a method based on the behavior of solutions along characteristic curves to calculate U numerically for hyperbolic equations.

Returning to the exceptional case where the curve $\mathscr{C}$ is characteristic, we see that one cannot use Cauchy data in this case to extend this data to a neighboring curve. In fact, if $\mathscr{C}$ is a characteristic $\xi = \xi_0 =$ constant, we *cannot even specify α and β arbitrarily on it* and expect the result to be part of a solution of (34). With the curves $\xi =$ constant and $\eta =$ constant as characteristics of (1), any solution must satisfy (15a) everywhere. In particular, along the characteristic $\xi = \xi_0 =$ constant, (15a) reduces to the following consistency condition governing the functions $\alpha(\eta)$ and $\beta(\eta)$:

$$2B(\xi_0, \eta)\beta'(\eta) + D(\xi_0, \eta)\beta(\eta) + E(\xi_0, \eta)\alpha'(\eta) + F(\xi_0, \eta)\alpha(\eta) = G(\xi_0, \eta) \quad (38)$$

Given $\alpha(\eta)$, this is a first-order differential equation that determines $\beta(\eta)$ to within an arbitrary constant or vice versa. Therefore, $\alpha(\eta)$ and $\beta(\eta)$ cannot be specified arbitrarily on $\xi = \xi_0$.

4.3.2 Characteristics as Carriers of Discontinuities in the Second Derivative

Regard (4) as a coordinate transformation $(x, y, u) \to (\xi, \eta, U)$, so $U(\xi, \eta)$ satisfies (34) if $u(x, y)$ satisfies (1). Let the curve $\mathscr{C}_0$ correspond to $\phi(x, y) = \xi_0$ for a fixed constant ξ_0. Assume that on either side of $\mathscr{C}_0$ we have solved (34) and that this solution has continuous second partial derivatives everywhere outside $\mathscr{C}_0$; that is, $U, U_\xi, U_\eta, U_{\xi\xi}, U_{\xi\eta}, U_{\eta\eta}$ are continuous for all $\xi \neq \xi_0$ and all η. For future reference, we denote such a solution as a *strict* solution of (1) or (34).

Now suppose that on $\mathscr{C}_0$, the functions $U, U_\xi, U_\eta, U_{\xi\eta}$, and $U_{\eta\eta}$ are also continuous but $U_{\xi\xi}$ is *not*. Thus, strictly speaking, (1) and (34) are not satisfied on $\mathscr{C}_0$ because $U_{\xi\xi}$ is not defined there. Can we choose $\mathscr{C}_0$ in such a way that (1) is also satisfied there even with $U_{\xi\xi}(\xi_0^+, \eta) \neq U_{\xi\xi}(\xi_0^-, \eta)$? To answer this question, we evaluate (34) on either side of $\xi = \xi_0$ and subtract the two resulting expressions. Since all the terms except $U_{\xi\xi}$ are continuous, we are left with

$$A(\xi_0, \eta)[U_{\xi\xi}(\xi_0^+, \eta) - U_{\xi\xi}(\xi_0^-, \eta)] = 0$$

Therefore, (34) is satisfied on $\xi = \xi_0$ either trivially if $U_{\xi\xi}$ is continuous there or if $A(\xi_0, \eta) = 0$—that is, if $\xi = \xi_0$ is the characteristic curve in the xy-plane, defined by (14a). This shows that within the framework of a *strict solution everywhere*, the characteristics $\xi = \xi_0 =$ constant are loci of possible discontinuity in $U_{\xi\xi}$. Similarly, the characteristics $\eta = \eta_0 =$ constant are loci of possible discontinuity in $U_{\eta\eta}$.

Geometrically, we may interpret this result to imply that we can join two solutions smoothly along a characteristic. For example, in Figure 4.4 we show a

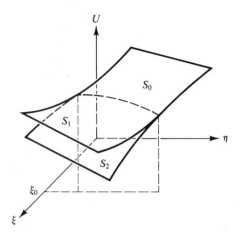

Figure 4.4

solution surface $S_0 + S_1$, which branches smoothly into the surface $S_0 + S_2$. The branching is smooth in the sense that U, U_ξ, U_η, $U_{\xi\eta}$, and $U_{\eta\eta}$ are continuous everywhere, including along $\xi = \xi_0$. On $\xi = \xi_0^+$, the value of $U_{\xi\xi}$ is different for S_1 and S_2. One or both of these values may also differ from $U_{\xi\xi}$ on $\xi = \xi_0^-$—that is, for S_0. Similarly, two solutions may be joined smoothly at a characteristic $\eta = \eta_0$ where the values of $U_{\eta\eta}$ differ.

Actually, once the jump in the value of $U_{\xi\xi}$ is specified at some point on $\mathscr{C}_0$, the propagation of this jump along $\mathscr{C}_0$ is determined by (7). To derive the equation governing the propagation of this jump, we first take the partial derivative of (7) with respect to ξ to obtain

$$AU_{\xi\xi\xi} + A_\xi U_{\xi\xi} + 2BU_{\xi\xi\eta} + 2B_\xi U_{\xi\eta} + CU_{\xi\eta\eta} + C_\xi U_{\eta\eta} + DU_{\xi\xi} + D_\xi U_\xi$$
$$+ EU_{\xi\eta} + E_\xi U_\eta + FU_\xi + F_\xi U - G_\xi = 0 \qquad (39)$$

Now, we evaluate (39) on either side of $\mathscr{C}_0$; that is, at ξ_0^+ and ξ_0^- and subtract the two resulting expressions. If we denote the jump in $U_{\xi\xi}$ by

$$\rho(\xi_0, \eta) \equiv U_{\xi\xi}(\xi_0^+, \eta) - U_{\xi\xi}(\xi_0^-, \eta)$$

and note that $A(\xi_0, \eta) = 0$ while $U_{\xi\eta}$, $U_{\eta\eta}$, U_ξ, U_η, and U are continuous on $\xi = \xi_0$, we obtain the linear first-order ordinary differential equation

$$\frac{\partial \rho}{\partial \eta} = -R(\xi_0, \eta)\rho \qquad (40a)$$

where

$$R(\xi_0, \eta) \equiv \frac{D(\xi_0, \eta) + A_\xi(\xi_0, \eta)}{2B(\xi_0, \eta)} \qquad (40b)$$

Thus, once ρ is prescribed at some point $\xi = \xi_0$, $\eta = \eta_0$ on $\mathscr{C}_0$, the solution of (40a) determines ρ everywhere along $\mathscr{C}_0$ in the form:

$$\rho(\xi_0,\eta) = \rho(\xi_0,\eta_0)\exp\left[-\int_{\eta_0}^{\eta} R(\xi_0,\sigma)\,d\sigma\right] \tag{41}$$

This result shows that ρ is identically equal to zero if it vanishes initially and that if $\rho(\xi_0,\eta_0) \neq 0$, then ρ is never equal to zero.

Similarly, discontinuities in $U_{\eta\eta}$ propagate along characteristics $\eta = $ constant according to

$$\lambda(\xi,\eta_0) = \lambda(\xi_0,\eta_0)\exp\left[-\int_{\xi_0}^{\xi} S(\sigma,\eta_0)\,d\sigma\right] \tag{42}$$

where λ now denotes the jump in $U_{\eta\eta}$:

$$\lambda(\xi,\eta_0) \equiv U_{\eta\eta}(\xi,\eta_0^+) - U_{\eta\eta}(\xi,\eta_0^-) \tag{43a}$$

and

$$S(\xi,\eta_0) \equiv [E(\xi,\eta_0) + C_\eta(\xi,\eta_0)]/2B(\xi,\eta_0) \tag{43b}$$

In summary, we have seen three alternative interpretations of the characteristic curves of (1) for the hyperbolic case:

1. They are curves that transform (1) to the canonical form (15a).
2. They are curves on which Cauchy data do not specify a unique solution in a neighborhood.
3. They are curves along which a strict solution of (1) may have a discontinuity in the second derivative normal to the curve. Thus, two solutions may be joined smoothly along a characteristic.

4.4   Solution of Hyperbolic Equations in Terms of Characteristics

We showed in Section 4.2 that the general partial differential equation (1) can be reduced to the following form if it is hyperbolic [see (15a)]

$$U_{\xi\eta} = \tilde{D}U_\xi + \tilde{E}U_\eta + \tilde{F}U + \tilde{G} \tag{44}$$

Given the coefficients d, e, f, and g, which are functions of x and y, we can explicitly derive the functions $\tilde{D}$, $\tilde{E}$, $\tilde{F}$, and $\tilde{G}$ of ξ and η from (15b)–(15e).

As pointed out earlier, in general it is not possible to simplify (44) further. If we denote

$$U_\xi(\xi,\eta) \equiv P(\xi,\eta) \tag{45a}$$

$$U_\eta(\xi,\eta) \equiv Q(\xi,\eta) \tag{45b}$$

we can interpret (44) as an equation governing the propagation of P along the $\xi = $ constant characteristics or an equation governing the propagation of Q along the $\eta = $ constant characteristics. In fact, (44) is just

$$P_\eta = \tilde{D}P + \tilde{E}Q + \tilde{F}U + \tilde{G} \equiv \tilde{H} \tag{46a}$$

or

$$Q_\xi = \tilde{D}P + \tilde{E}Q + \tilde{F}U + \tilde{G} \equiv \tilde{H} \tag{46b}$$

In this section we formulate a solution procedure based on the characteristic form (46) and discuss how we may implement this procedure numerically.

4.4.1 Cauchy Data on a Spacelike Arc

Our first task is to show that specifying Cauchy data (that is, U and $\partial U/\partial n$) on a noncharacteristic curve $\mathscr{C}_0$ is equivalent to specifying P and Q on this curve. We define a simple curve in the parametric form.

$$\xi = \xi^*(\tau) \tag{47a}$$

$$\eta = \eta^*(\tau) \tag{47b}$$

where the parameter τ varies monotonically along $\mathscr{C}_0$. We also assume that $\mathscr{C}_0$ is smooth—that is, that $d\xi^*/d\tau$ and $d\eta^*/d\tau$ are continuous.

We denote by $\mathscr{D}$ the domain on the side of $\mathscr{C}_0$ over which (44) is to be solved and let $\mathbf{n}$ denote the unit normal *into* $\mathscr{D}$ and τ the unit tangent in the direction of increasing τ. See any of the cases sketched in Figure 4.5. Here the characteristics are horizontal and vertical lines in terms of a Cartesian (ξ, η) frame, and with no loss of generality, we take the origin of this frame somewhere on the curve $\mathscr{C}_0$.

We distinguish two possible types of noncharacteristic curves, denoted by *spacelike* and *timelike*. A spacelike arc $\mathscr{S}_0$ has two characteristics, either emerging from every point on it into $\mathscr{D}$ as in Figure 4.5a, or entering every point on it from $\mathscr{D}$, as in Figure 4.5b. Thus, the two components of $\mathbf{n}$ in the ξ and η directions are either both positive or both negative for a spacelike arc. On a timelike arc $\mathscr{T}_0$, the components of $\mathbf{n}$ have different signs, and only one family of characteristics emerges from $\mathscr{T}_0$ into $\mathscr{D}$. These may be the $\eta = $ constant characteristics, as in Figure 4.5c, or the $\xi = $ constant characteristics, as in Figure 4.5d. In the preceding, the terms *entering* and *emerging* are associated with the directions of increasing ξ or η. The terms spacelike and timelike originate from the interpretation of (1) as the wave equation:

$$u_{xx} - u_{tt} = \text{lower-derivative terms} \tag{48}$$

where x is a distance and t is the time. In characteristic form, (48) becomes [see (14)]:

$$U_{\xi\eta} = \text{lower-derivative terms} \tag{49}$$

with $\xi = t + x$ and $\eta = t - x$. Now, if we want to solve (48) for $t \geq 0$, $-\infty < x < \infty$, we specify u and u_t on the x-axis, a spacelike curve. This maps to the straight line $\mathscr{S}_0: \eta = -\xi$, which may also be defined parametrically as $\xi = \tau$, $\eta = -\tau$, $-\infty < \tau < \infty$. The domain $t \geq 0$ maps to $\mathscr{D}: \xi + \eta \geq 0$ above $\mathscr{S}_0$ and $\mathbf{n} = (1/\sqrt{2}, 1/\sqrt{2})$. Therefore, the x-axis is indeed a spacelike arc according to our definition. More generally, any curve $x = f(t)$ with $|\dot{f}(t)| > 1$ is spacelike. Conversely, if $|\dot{f}| < 1$, the curve is timelike. In particular, the vertical line $x = c = \text{constant}$ in the xt-plane (which is the time axis if $c = 0$) is timelike for the solution domain on either side. In the preceding discussion, it is understood that both ξ and η increase as t increases. Hence, the choice $\eta = t - x$ rather than $\eta = x - t$ is crucial.

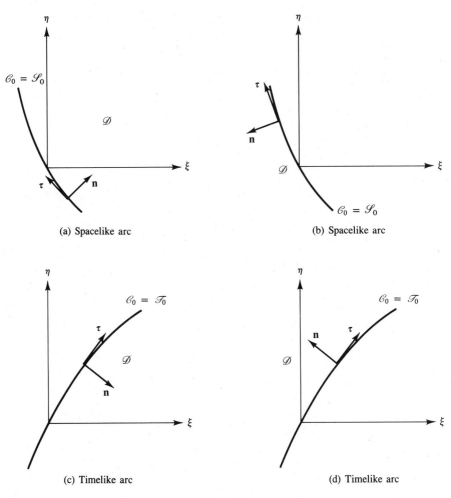

(a) Spacelike arc   (b) Spacelike arc

(c) Timelike arc   (d) Timelike arc

Figure 4.5

Consider now Cauchy data on a spacelike arc $\mathscr{S}_0$ with $\mathscr{D}$ to the right of $\mathscr{S}_0$, as in Figure 4.5a. We are given U and $\partial U/\partial n$ on $\mathscr{S}_0$ and can therefore express these parametrically in terms of τ; that is,

$$U(\xi^*(\tau), \eta^*(\tau)) \equiv U^*(\tau) = \text{given} \tag{50a}$$

$$\frac{\partial U}{\partial n}(\xi^*(\tau), \eta^*(\tau)) \equiv V^*(\tau) = \text{given} \tag{50b}$$

For the time being, let us assume that $U^*(\tau)$, $\dot{U}^*(\tau)$, and $V^*(\tau)$ are continuous on $\mathscr{S}_0$. We can then derive P and Q as continuous functions of τ on $\mathscr{S}_0$ as follows. We differentiate (50a) with respect to τ; denoting:

$$P^*(\tau) \equiv \frac{\partial U}{\partial \xi}(\xi^*(\tau), \eta^*(\tau)) \tag{51a}$$

$$Q^*(\tau) \equiv \frac{\partial U}{\partial \eta}(\xi^*(\tau), \eta^*(\tau)) \tag{51b}$$

we obtain one relation linking P^* and Q^* to the given Cauchy data and $\mathscr{S}_0$:

$$P^*\dot{\xi}^* + Q^*\dot{\eta}^* = \dot{U}^* \tag{52}$$

The second equation is just (50b), in which we express $\mathbf{n}$ in terms of $\dot{\xi}^*$ and $\dot{\eta}^*$. For the spacelike arc in Figure 4.5a, $\dot{\eta}^* > 0$ and $\dot{\xi}^* < 0$; hence

$$\mathbf{n} = \left(\frac{\dot{\eta}^*}{\dot{s}}, \frac{-\dot{\xi}^*}{\dot{s}} \right) \tag{53a}$$

where

$$\dot{s} = (\dot{\xi}^{*2} + \dot{\eta}^{*2})^{1/2} \tag{53b}$$

Therefore, (50b) becomes

$$\frac{\partial U}{\partial n} \equiv \text{grad } U \cdot \mathbf{n} \equiv P^* \frac{\dot{\eta}^*}{\dot{s}} - Q^* \frac{\dot{\xi}^*}{\dot{s}} = V^* \tag{54}$$

Equations (52) and (54) are linearly independent algebraic equations linking $\dot{U}^*$, V^* to P^*, Q^*. Solving these gives

$$P^* = \frac{\dot{\xi}^*\dot{U}^* + \dot{\eta}^*\dot{s}V^*}{\dot{s}^2} \tag{55a}$$

$$Q^* = -\frac{(\dot{\xi}^*\dot{s}V^* - \dot{\eta}^*\dot{U}^*)}{\dot{s}^2} \tag{55b}$$

Thus, knowing U^* and V^* on a given noncharacteristic curve, we compute P^* and Q^* there using (55). The formulas for the situation in Figure 4.5b are obtained from the preceding by replacing $\dot{s}$ by $-\dot{s}$.

4.4.2 Cauchy Problem; the Numerical Method of Characteristics

We are now in a position to extend the initial data U, P, Q given on $\mathscr{S}_0$ to a neighboring curve using (45) and (46) as rules for the propagation of P, Q in the characteristic directions. This is a general version of the initial-value problem for the wave equation discussed in Section 3.4. Again referring to the case of Figure 4.5a, we subdivide $\mathscr{D}$ into a rectangular grid, with a variable grid spacing (as might be dictated by the rate of change of the initial data) as shown in Figure 4.6. Denote the values of U, P, Q at each gridpoint by the associated subscript. Thus, let

$$U_{i,j} \equiv U(\xi_i, \eta_j); \qquad P_{i,j} \equiv P(\xi_i, \eta_j); \qquad Q_{i,j} \equiv Q(\xi_i, \eta_j) \tag{56}$$

Let the horizontal distance between the points (i, j) and $(i + 1, j)$ be denoted by ℓ_i; that is,

$$\xi_{i+1} - \xi_i \equiv \ell_i \tag{57a}$$

and denote the vertical distance between the (i, j) and $(i, j + 1)$ points by h_j;

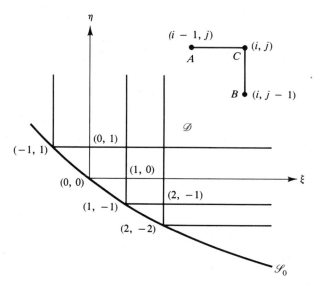

Figure 4.6

that is,

$$\eta_{j+1} - \eta_j \equiv h_j \tag{57b}$$

Assume that U, P, and Q are known at the two adjacent points: $A = (i - 1, j)$ and $B = (i, j - 1)$, and we want to compute U, P, and Q at the point $C = (i, j)$ (see Figure 4.6).

Equation (45a) gives the forward difference

$$U_{i,j} = U_{i-1,j} + \ell_{i-1} P_{i-1,j} \tag{58a}$$

and (45b) gives

$$U_{i,j} = U_{i,j-1} + h_{j-1} Q_{i,j-1} \tag{58b}$$

Each of these defines $U_{i,j}$ in terms of known quantities, and one approach is to use the average value

$$U_{i,j} = \tfrac{1}{2}(U_{i-1,j} + U_{i,j-1}) + \tfrac{1}{2}(\ell_{i-1} P_{i-1,j} + h_{j-1} Q_{i,j-1}) \tag{59}$$

To compute $P_{i,j}$, we use the forward difference of (46a):

$$P_{i,j} = P_{i,j-1} + h_{j-1} \tilde{H}_{i,j-1} \tag{60a}$$

Similarly, we obtain $Q_{i,j}$ from (46b):

$$Q_{i,j} = Q_{i-1,j} + \ell_{i-1} \tilde{H}_{i-1,j} \tag{60b}$$

Equations (59)–(60) define U, P, and Q at C in terms of known values at A and B, and this process can be repeated to generate the solution at successive points. We see that if we use Cauchy data along a finite segment of the initial curve, say between the gridpoints $(-N, N)$ and $(M, -M)$, we are able to define the solution of (44) in the triangular domain bounded by the given segment of the initial curve and the characteristics $\eta_N = \eta = $ constant and $\xi = \xi_M = $ constant. This triangular domain is the domain of dependence of the point (ξ_M, η_N) [see the discussion following (116) in Chapter 3]. For this construction, it is crucial to start with U *and a derivative of U in a direction which is not tangent to $\mathscr{C}_0$* (for example, the normal derivative) in order to be able to march the values of P and Q forward. This result confirms the argument used in Section 4.3 based on Taylor series.

4.4.3 Goursat's Problem; Boundary Conditions on a Timelike Arc

Consider a characteristic arc, say $0 \le \xi \le \xi_F$, where ξ_F may equal ∞, and a timelike arc $\mathscr{T}_0$ over the same interval, as shown in Figure 4.7. Assume that U is prescribed on both these arcs and we wish to solve (44) in the enclosed domain $\mathscr{D}_1$. This is Goursat's problem, and we demonstrate next that specifying U along these two arcs is sufficient to define U, P, and Q at all points in $\mathscr{D}_1$ using a characteristic construction.

Consider first the three adjacent points $(0,0)$, $(1,1)$, and $(1,0)$ at which the values of U are prescribed but where the values of P and Q are unknown. Let τ be a parameter along $\mathscr{T}_0$, and denote values on $\mathscr{T}_0$ by an overbar; that is, $U = \bar{U}(\tau)$, $P = \bar{P}(\tau)$, $Q = \bar{Q}(\tau)$, where the function $\bar{U}(\tau)$ is given but $\bar{P}$ and $\bar{Q}$ are unknown. The identity

$$\frac{d\bar{U}}{d\tau} = \bar{P}\frac{d\bar{\xi}}{d\tau} + \bar{Q}\frac{d\bar{\eta}}{d\tau} \tag{61}$$

for the rate of change of U on $\mathscr{T}_0$ leads to the following difference equation at $(0,0)$:

$$U_{1,1} - U_{0,0} = P_{0,0}\ell_0 + Q_{0,0}h_0 \tag{62a}$$

which provides a linear relation between $P_{0,0}$ and $Q_{0,0}$. Similarly, at $(1,1)$ we have

$$U_{2,2} - U_{1,1} = P_{1,1}\ell_1 + Q_{1,1}h_1 \tag{62b}$$

The definition of P at $(0,0)$ and $(1,0)$ gives

$$P_{0,0} = \frac{U_{1,0} - U_{0,0}}{\ell_0} \tag{63a}$$

$$P_{1,0} = \frac{U_{2,0} - U_{1,0}}{\ell_1} \tag{63b}$$

Finally, the partial differential equation (44) implies that

$$Q_{1,0} = Q_{0,0} + \ell_0\tilde{H}_{0,0} \tag{64a}$$

at $(1,0)$, and

$$P_{1,1} = P_{1,0} + h_0\tilde{H}_{1,0} \tag{64b}$$

at $(1,1)$. The $\tilde{H}_{i,j}$ in (64) are the linear functions of $P_{i,j}$, $Q_{i,j}$, $U_{i,j}$ as defined in (46).

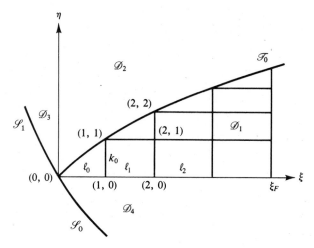

Figure 4.7

Now, (63a) defines $P_{0,0}$, and using this in (62a), we obtain $Q_{0,0}$ in the form

$$Q_{0,0} = \frac{U_{1,1} - U_{1,0}}{h_0} \tag{65}$$

Normally, we would interpret the difference expression on the right-hand side of (65) to be $Q_{1,0}$ instead of $Q_{0,0}$. However, since no information is available at the gridpoints $(-1,0)$ or $(0,-1)$, we cannot define $Q_{0,0}$ from data to the left or below the origin. In fact, (65) is the correct difference expression for sufficiently small ℓ_0, and we introduce no inconsistency in using (65), since $Q_{1,0}$ is calculated using (64a). Of course, one could also use the *exact* value for $Q_{0,0}$ that we can calculate using (61) evaluated at $\tau = 0$ and the definition $P = \partial U/\partial \xi$ at $\xi = 0$ for the data given on the ξ-axis.

Next, we obtain $P_{1,0}$ from (63b); therefore, the right-hand side of (64b) is known, and we have $P_{1,1}$. Finally, this is used in (62b) to obtain $Q_{1,1}$.

Now, consider the gridpoint $(2,0)$, at which P and Q are unknown. We define $P_{2,0}$ using

$$P_{2,0} = \frac{U_{3,0} - U_{2,0}}{\ell_2} \tag{66a}$$

and $Q_{2,0}$ using the partial differential equation; that is,

$$Q_{2,0} = Q_{1,0} + \ell_1 \tilde{H}_{1,0} \tag{66b}$$

This procedure can now be extended as follows to compute $U, P,$ and Q at all interior points in $\mathscr{D}_1$. First, we compute P at each gridpoint of the characteristic arc using [see (66a)]

$$P_{i,0} = \frac{U_{i+1,0} - U_{i,0}}{\ell_i} \tag{67a}$$

Then, we compute the $Q_{i,0}$ successively along this arc using [see (66b)]

$$Q_{i,0} = Q_{i-1,0} + \ell_{i-1}\tilde{H}_{i-1,0} \tag{67b}$$

We now have enough information to compute U, P, and Q at all gridpoints on the $\eta = \eta_1$ characteristic. In particular, using the values of U, P, and Q at the $(1,1)$ and $(2,0)$ gridpoints, we use the scheme outlined for the Cauchy problem [see (59)–(60)] to compute U, P, and Q at the $(2,1)$ gridpoint, and so on.

Similarly, once we have computed U, P, and Q at the $(2,2)$ gridpoint [using the same process as in computing the solution at $(1,1)$], we can march down the $\eta = \eta_2$ characteristic, and so on. An analogous procedure can be used to solve (44) in $\mathscr{D}_2$ if U is specified on the η-axis and the arc $\mathscr{T}_0$.

Suppose now that we wish to solve (44) in $\mathscr{D}_1 + \mathscr{D}_4$ (see Figure 4.7) subject to Cauchy data being specified on the spacelike arc $\mathscr{S}_0$ and U being specified on the timelike arc $\mathscr{T}_0$. This is the general version of the initial- and boundary-value problem for the wave equation over the semi-infinite domain discussed in Section 3.5. Clearly, this solution is defined by the union of the solution of the Cauchy problem in $\mathscr{D}_4$ (which specifies U on the ξ-axis) followed by the solution of the Goursat problem in $\mathscr{D}_1$.

4.4.4   Characteristic Boundary-Value Problem

The limiting case of the Goursat problem in $\mathscr{D}_1 + \mathscr{D}_2$ for which the arc $\mathscr{T}_0$ becomes the characteristic η-axis is denoted as the *characteristic boundary-value problem*. It is easily seen that specifying U on the ξ-axis defines P there, whereas specifying U on the η-axis defines Q there. In particular, we have both P and Q at the origin. To compute P on the η-axis, we evaluate (46a) for $\xi = 0$. The result is a linear first-order ordinary differential equation for P with respect to the independent variable η. Solving this subject to the boundary condition for P at $\eta = 0$ defines P uniquely. Equivalently, we can use the difference expression corresponding to (46a) [see (64b)]. Similarly, we compute Q on the ξ-axis by solving (46b) either exactly or in difference form subject to the boundary condition for Q at the origin [see (64a)]. Note that in this case $\bar{\xi} = 0$ on the η-axis, and (61) specializes to the definition of Q along the η-axis. The solution for all interior points can now be computed recursively using the scheme employed for the Cauchy problem.

4.4.5   Well-Posedness

Based on the examples discussed in Sections 4.4.2–4.4.4, we observe that our characteristic construction of the solution in each case provides an explicit demonstration that the type of boundary data imposed defines a "well-posed" problem in the sense that we have exactly enough information to calculate a unique solution. In particular, it was necessary to specify *both* U and $\partial U/\partial n$ on the spacelike arc $\mathscr{S}_0$ for the Cauchy problem; we would have been unable to solve the problem had we specified only U or $\partial U/\partial n$. In contrast, for Goursat's problem it would be inconsistent to specify both U and $\partial U/\partial n$ (or equivalently U, P, and Q subject to (61)) on the timelike arc $\mathscr{T}_0$. In doing so, note that the value of P at the $(1,1)$ gridpoint, for example, would depend on whether we used the partial differential equation—that is, (64b)—or the prescribed data on $\mathscr{T}_0$. The most general linear boundary condition that is allowable on $\mathcal{T}_0$ for Goursat's problem is

$$a(\tau)U(\bar{\xi}(\tau), \bar{\eta}(\tau)) + b(\tau)\frac{\partial U}{\partial n}(\bar{\xi}(\tau), \bar{\eta}(\tau)) = c(\tau) \tag{68}$$

for given functions a, b, and c. This boundary condition specializes to the case that we discussed in Section 4.4.3 if $b = 0$ and $c/a = \bar{U}$. The other limiting case $a = 0$ is also important. For example, it corresponds physically to a boundary condition on the velocity in any interpretation where (44) is a wave equation and U is a velocity potential. The derivation of the difference scheme for the general boundary condition (68) is left as an exercise.

Although we have confined our discussion to a left boundary, analogous remarks apply to a right boundary and to domains contained between two timelike arcs.

In view of the role of characteristics in propagating the values of P and Q, we note that we always need to specify two independent conditions on a spacelike arc and one condition on a timelike arc. Also, as pointed out in Section 4.4.1, we cannot specify both U and $\partial U/\partial n$ arbitrarily on a characteristic arc; the requirement that (44) holds on a characteristic arc implies that only one condition, possibly of the form (68), can be imposed there. In this case, a unique solution results only if we specify a second condition of the form (68) on an intersecting timelike arc as in the Gorusat problem; Cauchy data on an intersecting spacelike arc are inconsistent. Before concluding this discussion, the following disclaimers must be made.

The primitive numerical algorithms that we have used so far (as well as similar algorithms to be used later on in this chapter and in Chapter 7) are based on *forward differencing* along characteristic coordinates. These are not necessarily the most efficient or accurate numerical schemes for computing solutions. Rather, they provide direct and concise demonstrations that solutions may be derived in a consistent manner for certain types of initial or boundary data. A discussion of sophisticated numerical solution methods is beyond the scope of this text and is not attempted.

Our arguments concerning well-posedness have all relied on our being able to construct a unique solution in some neighborhood of a given curve with given boundary data. Needless to say, these arguments do not constitute rigorous proofs; they merely ensure that in our calculation of a solution, no inconsistencies result from the given information. A broader definition of well-posedness, which we have not addressed, requires that the solution we calculate *depends continuously on the boundary data*. In this regard, let us consider a striking counterexample first proposed by Hadamard (see the discussion in Section 4.1 of [1]). This example is designed to show that a unique extension of boundary data to some neighborhood of the boundary does not necessarily imply that the solution in the extended domain depends continuously on the boundary data.

We study Laplace's equation

$$u_{xx} + u_{yy} = 0$$

in $x \geq 0$ subject to the *two* boundary conditions on the y-axis:

$$u(0, y) = 0$$

$$u_x(0, y) = \frac{1}{n} \sin ny$$

where n is an integer. Our physical intuition suggests there must be something wrong in prescribing Cauchy data for the Laplacian. Based on our experience with Laplace's equation, we would have been more comfortable with only one of the boundary conditions at $x = 0$ and a second boundary condition at $x = \infty$. Nevertheless, the Taylor series construction we used in Section 4.3.1 can be implemented with no difficulties because the boundary data are analytic, and we can easily construct the series in powers of x. In fact, the Taylor series can be summed to give the following result that can also be obtained by separation of variables:

$$u(x, y) = \frac{1}{n^2} \sinh nx \sin ny$$

For any fixed $x > 0$, this result predicts oscillations in y with unbounded amplitude [since $(\sinh nx)/n^2 \to \infty$ as $n \to \infty$] and unbounded wave number n. But the Cauchy data tend to zero uniformly as $n \to \infty$. Thus, the solution does not depend continuously on the boundary data. Also note that even if n is a fixed finite number, this solution is unstable in the sense that $u \to \infty$ as $x \to \infty$ for any fixed y for which $\sin ny \neq 0$. There is indeed something very wrong with prescribing Cauchy data for the Laplacian.

In the absence of an exact solution, we must rely on further information—for example, physical reasoning—to argue that specifying Cauchy data on the boundary for an elliptic equation leads to an ill-posed problem. In Section 4.4.7 we demonstrate another type of ill-posedness for a wave equation where disturbances grow in the far-field.

4.4.6 The General Solution of Cauchy's Problem; the Riemann Function

In Section 3.4.2 we showed that the general solution of the inhomogeneous wave equation with arbitrarily prescribed initial conditions on $-\infty < x < \infty$ could be written explicitly once the fundamental solution was known. Here we derive an analogous result for the Cauchy problem for the general hyperbolic equation (44) in terms of the so-called Riemann function.

We begin with (44), written in operator form

$$\mathscr{L}(U) \equiv U_{\xi\eta} - \tilde{D}U_\xi - \tilde{E}U_\eta - \tilde{F}U = \tilde{G} \tag{69}$$

for given functions $\tilde{D}$, $\tilde{E}$, $\tilde{F}$, and $\tilde{G}$ of the characteristic variables ξ, η. We now define the adjoint operator $\mathscr{M}$ of $\mathscr{L}$ by the requirement that for any given pair of functions, U and V, the condition

$$V\mathscr{L}(U) - U\mathscr{M}(V) = J_\xi + K_\eta \tag{70}$$

holds for certain functions J, K of U, V and their first derivatives. To evaluate

$V\mathscr{L}(U)$ we write

$$VU_{\xi\eta} = (VU_{\xi})_{\eta} - V_{\eta}U_{\xi} = (VU_{\xi})_{\eta} - (V_{\eta}U)_{\xi} + V_{\xi\eta}U \qquad (71a)$$

$$-V\tilde{D}U_{\xi} = -(\tilde{D}VU)_{\xi} + (\tilde{D}V)_{\xi}U \qquad (71b)$$

$$-V\tilde{E}U_{\eta} = -(\tilde{E}VU)_{\eta} + (\tilde{E}V)_{\eta}U \qquad (71c)$$

and conclude by comparing (71) with (70) that the adjoint operator is

$$\mathscr{M}(V) \equiv V_{\xi\eta} + (\tilde{D}V)_{\xi} + (\tilde{E}V)_{\eta} - \tilde{F}V$$
$$\equiv V_{\xi\eta} + \tilde{D}V_{\xi} + \tilde{E}V_{\eta} + (\tilde{D}_{\xi} + \tilde{E}_{\eta} - \tilde{F})V \qquad (72)$$

and that

$$V\mathscr{L}(U) - U\mathscr{M}(V) = -(\tilde{D}VU + UV_{\eta})_{\xi} + (-\tilde{E}VU + VU_{\xi})_{\eta} \qquad (73)$$

Note incidentally that a necessary and sufficient condition for $\mathscr{L}$ to be *self-adjoint*—that is, $\mathscr{M} = \mathscr{L}$—is that $\tilde{D} = \tilde{E} = 0$. We also note that the functions J and K, which we have exhibited in (73) as

$$J \equiv -(\tilde{D}VU + UV_{\eta}) \qquad (74a)$$

$$K \equiv -\tilde{E}VU + VU_{\xi} \qquad (74b)$$

are not unique for a given $\mathscr{L}$; we can always add a function J^* to J and a function K^* to K with the property $J_{\xi}^* + K_{\eta}^* = 0$ without altering the form of the right-hand side of (70). Thus, for example, we can also satisfy the right-hand side of (70) with $\tilde{J}, \tilde{K}$ where

$$\tilde{J} = J + J^* \equiv -\tilde{D}VU + \tfrac{1}{2}VU_{\eta} - \tfrac{1}{2}V_{\eta}U \qquad (75a)$$

$$\tilde{K} = K + K^* \equiv -\tilde{E}VU + \tfrac{1}{2}VU_{\xi} - \tfrac{1}{2}V_{\xi}U \qquad (75b)$$

since the added functions

$$J^* = \tfrac{1}{2}(VU_{\eta} + V_{\eta}U) \qquad (76a)$$

$$K^* = -\tfrac{1}{2}(VU_{\xi} + V_{\xi}U) \qquad (76b)$$

satisfy $J_{\xi}^* + K_{\eta}^* = 0$.

The reason that we seek an expression of the form (70) hinges on exploiting the two-dimensional Gauss theorem (for example, see (44) of Chapter 2) for the vector with components $\tilde{J}$ and $\tilde{K}$ which are functions of ξ, η:

$$\iint_{\mathscr{D}} (\tilde{J}_{\xi} + \tilde{K}_{\eta}) \, d\xi \, d\eta = \oint_{\Gamma} (\tilde{J} \, d\eta - \tilde{K} \, d\xi) \qquad (77)$$

where Γ is the boundary of the domain $\mathscr{D}$. For our purposes, we choose $\mathscr{D}$ to be the domain bounded by a spacelike arc $\mathscr{S}_0$ extending from the two points ① $= (\xi_1, \eta_0)$ to ② $= (\xi_0, \eta_2)$ and the two characteristics C_1, C_2 emerging from these two points and intersecting at the point ⓪ $= (\xi_0, \eta_0)$, as shown in Figure 4.8. The arrows indicate the direction in which Γ is to be traversed in the line integral in (77). In view of (73), (77) gives

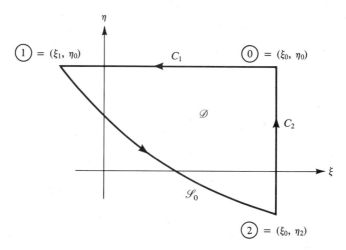

Figure 4.8

$$\iint_{\mathscr{D}} [V\mathscr{L}(U) - U\mathscr{M}(V)]\, d\xi\, d\eta = \oint_{\Gamma} [(-\tilde{D}VU + \tfrac{1}{2}VU_\eta - \tfrac{1}{2}V_\eta U)\, d\eta$$

$$+ (\tilde{E}VU - \tfrac{1}{2}VU_\xi + \tfrac{1}{2}V_\xi U)\, d\xi] \qquad (78)$$

After the following integrations by parts,

$$\tfrac{1}{2}\int_{C_2} VU_\eta\, d\eta = \tfrac{1}{2}VU\Big|_{\xi_0,\eta_2}^{\xi_0,\eta_0} - \tfrac{1}{2}\int_{C_2} UV_\eta\, d\eta \qquad (79a)$$

$$\tfrac{1}{2}\int_{C_1} VU_\xi\, d\xi = \tfrac{1}{2}VU\Big|_{\xi_0,\eta_0}^{\xi_1,\eta_0} - \tfrac{1}{2}\int_{C_1} V_\xi U\, d\xi \qquad (79b)$$

the result in (78) simplifies to

$$\iint_{\mathscr{D}} [V\mathscr{L}(U) - U\mathscr{M}(V)]\, d\xi\, d\eta$$

$$= \int_{C_1} (\tilde{E}V + V_\xi)U\, d\xi - \int_{C_2} (\tilde{D}V + V_\eta)U\, d\eta + V(\xi_0,\eta_0)U(\xi_0,\eta_0)$$

$$- \tfrac{1}{2}V(\xi_1,\eta_0)U(\xi_1,\eta_0) - \tfrac{1}{2}V(\xi_0,\eta_2)U(\xi_0,\eta_2) + \int_{\mathscr{S}_0} \Lambda(U,V) \qquad (80)$$

Here we have introduced the notation

$$\Lambda(U,V) \equiv (-\tilde{D}VU + \tfrac{1}{2}VU_\eta - \tfrac{1}{2}V_\eta U)\, d\eta + (\tilde{E}VU - \tfrac{1}{2}VU_\xi + \tfrac{1}{2}V_\xi U)\, d\xi \quad (81)$$

and the line integrals in (80)–(81) are to be evaluated in the directions indicated in Figure 4.8. Equation (80) is true for arbitrary functions U and V. Now, if we choose U to satisfy $\mathscr{L}(U) = \tilde{G}$ and choose V to satisfy $\mathscr{M}(V) = 0$, the left-hand side of (80) simplifies to $\iint_{\mathscr{D}} V\tilde{G}\, d\xi\, d\eta$. To eliminate the line integrals along C_1

and C_2, we set

$$V_\xi = -\tilde{E}V \quad \text{on} \quad C_1 \tag{82a}$$

and

$$V_\eta = -\tilde{D}V \quad \text{on} \quad C_2 \tag{82b}$$

Solving these equations subject to the boundary condition $V(\xi_0, \eta_0) = 1$ gives

$$V(\xi_0, \eta) = \exp\left[-\int_{\eta_0}^{\eta} \tilde{D}(\xi_0, \sigma)\, d\sigma \right] \tag{83a}$$

$$V(\xi, \eta_0) = \exp\left[-\int_{\xi_0}^{\xi} \tilde{E}(\sigma, \eta_0)\, d\sigma \right] \tag{83b}$$

The partial differential equation $\mathcal{M}(V) = 0$ is hyperbolic, and we saw in Section 4.4.4 that the solution of this equation subject to the characteristic boundary conditions (83) is well-posed and has a unique solution. We denote this solution as

$$V = R(\xi, \eta, \xi_0, \eta_0) \tag{84}$$

the Riemann function associated with the operator $\mathcal{L}$ at the point ξ_0, η_0. The result (80) now becomes

$$U(\xi_0, \eta_0) = \tfrac{1}{2} R(\xi_1, \eta_0, \xi_0, \eta_0) U(\xi_1, \eta_0) + \tfrac{1}{2} R(\xi_0, \eta_2, \xi_0, \eta_0) U(\xi_0, \eta_2)$$

$$- \int_{\mathcal{S}_0} \Lambda[U(\xi, \eta), R(\xi, \eta, \xi_0, \eta_0)]$$

$$+ \iint_{\mathcal{D}} \tilde{G}(\xi, \eta) R(\xi, \eta, \xi_0, \eta_0)\, d\xi\, d\eta \tag{85}$$

If we are given Cauchy data on $\mathcal{S}_0$, we know U, P, and Q there [see(55)], and if we have calculated the Riemann function R explicitly, then Λ is a known function of ξ, η, ξ_0, η_0; hence (85) gives an explicit result for U. This result is analogous to D'Alembert's solution for the inhomogeneous wave equation in that U consists of a double integral involving $\tilde{G}$ over the domain of dependence $\mathcal{D}$ of ξ_0, η_0, plus an integral over that portion of the initial arc contained in the domain of dependence, and contributions form the endpoints of the initial arc. Of course, the solution for R is not easy to obtain for a general $\mathcal{L}$, but once this is available, the Cauchy problem for the inhomogeneous equation (69) can be written explicitly. The result (85) is useful for deriving estimates for U even in cases where R is not available explicitly.

Consider now the relatively simpler problem that results when we remove the first derivative terms in $\mathcal{L}(U) = 0$ for the case of constant coefficients. This problem was discussed in Section 3.7, and we have

$$\mathcal{L}(U) \equiv U_{\xi\eta} + \lambda U = 0 \tag{86a}$$

for $\lambda = $ constant, say, $\lambda > 0$. This operator is self-adjoint, and the Riemann function obeys (86a) subject to

$$U(\xi_0, \eta) = U(\xi, \eta_0) = 1 \tag{86b}$$

Using similarity arguments, it is easy to show that the solution is

$$R(\xi, \eta, \xi_0, \eta_0) = J_0(2\sqrt{\lambda(\xi - \xi_0)(\eta - \eta_0)}) \tag{87}$$

in the domain $\xi \leq \xi_0, \eta \leq \eta_0$ or $\xi \geq \xi_0, \eta \geq \eta_0$. This result is exactly twice the fundamental solution [see (228a) of Chapter 3]. It is left as an exercise [Problem 4(f)–(g)] to show that the superposition results based on the fundamental solution give precisely the same expression as (85) for this example.

4.4.7 Weak Solutions; Propagation of Discontinuities in *P* and *Q*; Stability

In many physical applications, the Cauchy data or boundary data are discontinuous. For example, in the acoustic approximation of the bursting balloon discussed in Section 3.10.1, the time derivative of the velocity potential (representing the pressure perturbation) is discontinuous at the surface of the balloon, say at $r = 1$, on the axis $0 \leq r < \infty$. A discontinuity in u itself occurs naturally in a signaling problem, say over $0 \leq x < \infty$ whenever the boundary and initial values of u do not agree at $x = 0$, $t = 0$. Examples of this type were routinely handled in Chapter 3 using the method of characteristics. How do such discontinuities propagate into the solution domain in the general case?

Before starting this discussion, it is important to acknowledge that if a discontinuity in U, U_ξ, or U_η exists along some curve $\mathscr{C}$ in the "solution" domain of (44), then (44) is not satisfied on $\mathscr{C}$; we have a strict solution on either side of $\mathscr{C}$ but not on $\mathscr{C}$ itself. In contrast, recall that discontinuities in $U_{\xi\xi}$ across a $\xi = $ constant characteristic or discontinuities in $U_{\eta\eta}$ across an $\eta = $ constant characteristic are perfectly allowable within the context of a strict solution.

A solution for which U, U_ξ, or U_η becomes discontinuous on a curve $\mathscr{C}$ is denoted as a *weak solution*; we study such solutions in detail for nonlinear problems in Chapters 5 and 7. As we have already observed from our results in Chapter 3, discontinuities in the solution occur in linear problems only if the initial or boundary data are discontinuous, and in such cases these discontinuities propagate along characteristics. We next confirm that this observation remains true for the general linear problem (44), and we work out the details for discontinuities in the first derivative. The discussion of discontinuities in U is deferred until Section 4.5 because this case is best treated in terms of a system of two first-order equations. In fact, the simple examples from water waves studied in Sections 3.4.3 and 3.5.4 clearly indicate the efficiency of calculations in terms of characteristics using the system of two first-order equations.

Consider now the general problem (44), where Cauchy data have a discontinuity in P and Q at some point (ξ_0, η_0) on a spacelike arc $\mathscr{S}_0$, whereas U is continuous everywhere on $\mathscr{S}_0$. As shown in Figure 4.9, we subdivide the domain of interest into the three parts $\mathscr{D}_1$, $\mathscr{D}_0$, and $\mathscr{D}_2$ and subdivide $\mathscr{S}_0$ into $\mathscr{S}_1$ and $\mathscr{S}_2$. The solution of the Cauchy problems in $\mathscr{D}_1$ and $\mathscr{D}_2$ are strict solutions because the U, P, Q are continuous on $\mathscr{S}_1$ and $\mathscr{S}_2$ separately. In particular, the values $U(\xi_0^-, \eta)$ and $U(\xi, \eta_0^-)$ are provided by these solutions.

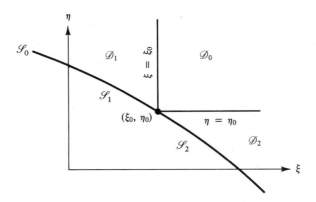

Figure 4.9

This is as far as we can proceed within the framework of a strict solution; the solution in $\mathcal{D}_0$ cannot be calculated unless we assume that P is continuous across $\eta = \eta_0$ and Q is continuous across $\xi = \xi_0$. These are certainly plausible assumptions; in fact, they are implicit in our interpretation of (44) as a rule for the propagation of P and Q along $\xi = $ constant and $\eta = $ constant characteristics, respectively. It then follows that U is continuous across both these characteristics; that is,

$$U(\xi_0^+, \eta) = U(\xi_0^-, \eta) \tag{88a}$$

$$U(\xi, \eta_0^+) = U(\xi, \eta_0^-) \tag{88b}$$

The values of $U(\xi_0^+, \eta)$ and $U(\xi, \eta_0^+)$ can then be used as boundary values to solve the characteristic boundary-value problem in $\mathcal{D}_0$. Notice that the values of $U(\xi_0^+, \eta)$ and $U(\xi, \eta_0^+)$ *must not* be calculated using the averages indicated in (59) because this would introduce a spurious discontinuity in U along the ξ_0, η_0 characteristics. This difficulty does not arise if the solution in $\mathcal{D}_0$ is calculated separately starting with the given characteristic boundary values (88).

The given initial discontinuities in P and Q at the point ξ_0, η_0 thus propagate along the $\xi = \xi_0$ and $\eta = \eta_0$ characteristics, respectively. These propagation rules can be derived explicitly by an analogous but more direct approach than that used in Section 4.3.2 for computing how jumps in second derivatives propagate. Consider first how the initial jump in Q propagates. We evaluate (44) in either side of $\eta = \eta_0$ and subtract the result, noting that P and U are continuous across $\eta = \eta_0$. This leads to the following first-order ordinary differential equation along $\eta = \eta_0$:

$$\frac{\partial \kappa}{\partial \xi} = \tilde{E}(\xi, \eta_0)\kappa \tag{89a}$$

where κ denotes the jump in Q; that is,

$$\kappa(\xi, \eta_0) \equiv Q(\xi, \eta_0^+) - Q(\xi, \eta_0^-) \tag{89b}$$

Thus, we have the explicit propagation rule

$$\kappa(\xi, \eta_0) = \kappa(\xi_0, \eta_0) \exp\left[\int_{\xi_0}^{\xi} \tilde{E}(\sigma, \eta_0)\, d\sigma \right] \tag{90}$$

which implies, in particular, that a discontinuity $\kappa(\xi, \eta_0)$ occurs if and only if $\kappa(\xi_0, \eta_0) \neq 0$.

Similarly, the discontinuity in P,

$$\mu(\xi_0, \eta) \equiv P(\xi_0^+, \eta) - P(\xi_0^-, \eta) \tag{91}$$

propagates along the $\xi = \xi_0$ characteristic according to

$$\mu(\xi_0, \eta) = \mu(\xi_0, \eta_0) \exp\left[\int_{\eta_0}^{\eta} \tilde{D}(\xi_0, \sigma)\, d\sigma \right] \tag{92}$$

Equations (90) and (92) may be used to characterize the behavior of the solution for large ξ and η. We argue that if an initial discontinuity in the values of P or Q at some point tends to grow as $\xi \to \infty$ or $\eta \to \infty$, then the solution is unstable. We see from (90) and (92) that a necessary condition for stability is that $\tilde{D} < 0$ for sufficiently large η and $\tilde{E} < 0$ for sufficiently large ξ. (For the situation sketched in Figure 4.5b, we reverse the signs.) Notice that if $\tilde{D}$ and $\tilde{E}$ are constant, this stability condition is identical to the one derived in Section 3.7.2 using the asymptotic behavior of the fundamental solution.

An unstable problem for (44) is ill-posed in a sense similar to the case discussed in Section 4.4.5. Here, it is the solution in the far-field (that is, $\xi \to \infty, \eta \to \infty$) that does not depend continuously on the Cauchy data. In physical applications, a linearized problem such as (44) is derived by making a small perturbation approximation in a nonlinear problem. For example, in Chapter 3, we derived the linearized acoustic approximation from the nonlinear conservation laws assuming small disturbances. If linearization leads to an unstable problem for (44), then either the basic nonlinear model is flawed, or the small perturbation assumption is not valid in the far-field. See the example discussed in Problem 7.

4.5 Hyperbolic Systems of Two First-Order Equations

In our discussion so far in this chapter, we have concentrated on the general second-order equation (1), which is the primary mathematical model in a number of physical applications—for example, the diffusion equation for one-dimensional heat conduction, Laplace's equation for steady heat conduction in two dimensions, and the one-dimensional wave equation for the small-amplitude vibrations of a string. In other applications, the second-order form (1) is the result of eliminating one of the two dependent variables from a model that occurs naturally as a system of two first-order equations. This is the case, for example, for steady incompressible irrotational flow in two dimensions. As discussed in Chapter 2, Laplace's equation is a consequence of the two first-order equations div $\mathbf{v} = 0$, curl $\mathbf{v} = 0$, where $\mathbf{v}$ is the velocity vector. Also, in our study of shallow-water waves of small amplitude in Chapter 3, we derived the one-dimensional wave equation for the velocity or free-surface perturbations after eliminating one of these variables from the basic first-order equations for mass and momentum conservation.

In those applications where the mathematical model of the basic physical problem is in the form of a system of first-order equations, it is important to reconsider some of the results in Sections 4.1–4.4, particularly for the hyperbolic case.

4.5.1 The Perburbation of a Quasilinear System Near a Known Solution

In many physical applications, (1) governs the perturbations to a known solution of a quasilinear system of equations. We have already seen some examples of this in our discussion of acoustics, shallow water waves, and the like, where we perturbed about the trivial solution of no flow. In Chapters 5 and 7 we study in what sense such quasilinear systems represent physical conservation laws. For the purposes of our discussion here, we start with a given pair of quasilinear equations written in vector form,

$$\mathbf{v}_t + A\mathbf{v}_x = \mathbf{f} \tag{93}$$

where $\mathbf{v}$ and $\mathbf{f}$ are two-dimensional vectors with components

$$\mathbf{v} = (v_1, v_2) \tag{94a}$$

$$\mathbf{f} = (f_1, f_2) \tag{94b}$$

and A is a linear operator with (2×2) matrix components A_{ij}. We assume that the components of $\mathbf{f}$ and A are functions of v_1, v_2, x, t:

$$A_{ij} = A_{ij}(v_1, v_2, x, t) \tag{95a}$$

$$f_i = f_i(v_1, v_2, x, t) \tag{95b}$$

where the indices i, j may equal 1 or 2. Since the A_{ij} do not depend on the $\partial v_i/\partial x$ or $\partial v_i/\partial t$, the two components of (93) represent a quasilinear pair of first-order equations. The vector equation (93) may result from multiplying the more general equation $C\mathbf{v}_t + D\mathbf{v}_x = \mathbf{g}$ by the inverse of C, which we have tacitly assumed to be nonsingular in the solution domain of interest.

Let

$$\mathbf{v}^{(0)}(x, t) = (v_1^{(0)}(x, t), v_2^{(0)}(x, t)) \tag{96}$$

be a known solution of (93). If $\mathbf{f} = 0$, an important special case has $\mathbf{v}^{(0)}$ equal to a constant vector. A number of examples of this case were discussed in Chapter 3. A more general case when $\mathbf{f}$ is independent of t has $\mathbf{v}^{(0)}$ independent of t also—that is, a steady-state solution of the pair of ordinary differential equations that result from (93) if $\partial/\partial t = 0$. In the slightly more general case with $\mathbf{f}$ depending on $x - ct$, $c = $ constant, we look for a solution where $\mathbf{v}^{(0)}$ is a function of $x - ct$. This is the nonlinear version of the "uniform wave" solution considered in Section 3.8.1. In the most general case, $\mathbf{v}^{(0)}(x, t)$ may represent some particular *exact* solution of (93) [see the example of Section 4.5.6] that we wish to perturb in the following sense.

Assume that we want to solve an initial-value problem for (93) in $-\infty < x < \infty$, $0 \le t$ where the given initial value $\mathbf{v}(x,0)$ is close to the one corresponding to the nominal solution (96). More precisely, assume that the initial value of $\mathbf{v}$ is given in the form

$$\mathbf{v}(x,0;\varepsilon) = \mathbf{a}(x) + \varepsilon\mathbf{b}(x) \tag{97}$$

for a specified vector function $\mathbf{b}(x)$ and a small parameter ε. In (97) $\mathbf{a}(x) = \mathbf{v}^{(0)}(x,0)$, the special solution (96) evaluated at $t = 0$. The perturbation idea is that the solution $\mathbf{v}(x,t;\varepsilon)$ of the initial-value problem (93), (97) is close to the special solution $\mathbf{v}^{(0)}(x,t)$, and we assume that it has the form:

$$\mathbf{v}(x,t;\varepsilon) = \mathbf{v}^{(0)}(x,t) + \varepsilon\mathbf{u}(x,t) + O(\varepsilon^2) \tag{98}$$

In Chapter 8 we explore the conditions under which this "regular perturbation" idea is correct.

Assuming that the expansion (98) is correct for sufficiently small ε, we need to compute the expressions that result for A and $\mathbf{f}$ when (98) is substituted for $\mathbf{v}$. In most physical applications A and $\mathbf{f}$ are analytic functions of v_1 and v_2, and therefore they have the following expansions in the neighborhood of the nominal solution $\mathbf{v}^{(0)}(x,t)$:

$$\begin{aligned}
A_{ij}(v_1,v_2,x,t;\varepsilon) &= A_{ij}(v_1^{(0)} + \varepsilon u_1, v_2^{(0)} + \varepsilon u_2, x, t; \varepsilon) + O(\varepsilon^2) \\
&= A_{ij}(v_1^{(0)}(x,t), v_2^{(0)}(x,t), x, t; 0) \\
&\quad + \varepsilon\left(\frac{\partial A_{ij}}{\partial v_1}u_1 + \frac{\partial A_{ij}}{\partial v_2}u_2 + \frac{\partial A_{ij}}{\partial \varepsilon}\right) + O(\varepsilon^2)
\end{aligned} \tag{99a}$$

$$\begin{aligned}
f_i(v_1,v_2,x,t;\varepsilon) &= f_i(v_1^{(0)} + \varepsilon u_1, v_2^{(0)} + \varepsilon u_2, x, t; \varepsilon) + O(\varepsilon^2) \\
&= f_i(v_1^{(0)}(x,t), v_2^{(0)}(x,t), x, t; 0) \\
&\quad + \varepsilon\left(\frac{\partial f_i}{\partial v_1}u_1 + \frac{\partial f_i}{\partial v_2}u_2 + \frac{\partial f_i}{\partial \varepsilon}\right) + O(\varepsilon^2)
\end{aligned} \tag{99b}$$

where we evaluate the arguments of the partial derivatives of A_{ij} and f_i for $v_1 = v_1^{(0)}(x,t)$, $v_2 = v_2^{(0)}(x,t)$ and $\varepsilon = 0$. In (99) we have generalized (95) to allow the $A_{ij}f_i$ to also depend on ε. Thus, (96) is an exact solution of (93), for $\varepsilon = 0$.

Substituting the expansions (98)–(99) into (93), ignoring terms of $O(\varepsilon^2)$, and taking into account that $\mathbf{v}^{(0)}(x,t)$ is a solution of (93) results in the following *linear* vector partial differential equation for $\mathbf{u}(x,t)$:

$$\mathbf{u}_t + A^{(0)}\mathbf{u}_x + B\mathbf{u} = \mathbf{f}^{(1)} \tag{100a}$$

and initial condition

$$\mathbf{u}(x,0) = \mathbf{b}(x) \tag{100b}$$

Here we have introduced the notation $A^{(0)}$ for the linear operator with components that are functions of x, t defined by the $O(1)$ term in (99a); that is,

$$A_{ij}^{(0)}(x,t) \equiv A_{ij}(v_1^{(0)}(x,t), v_2^{(0)}(x,t), x, t; 0) \tag{101a}$$

The components of the operator B and vector $\mathbf{f}^{(1)}$ are also functions of x, t defined by

$$B_{ij}(x, t) \equiv -\frac{\partial f_i}{\partial v_j} + \sum_{\ell=1}^{2} \frac{\partial A_{i\ell}}{\partial v_j} \frac{\partial v_\ell^{(0)}}{\partial x} \tag{101b}$$

$$f_i^{(1)} \equiv \frac{\partial f_i}{\partial \varepsilon} - \sum_{\ell=1}^{2} \frac{\partial A_{i\ell}}{\partial \varepsilon} \frac{\partial v_\ell^{(0)}}{\partial x} \tag{101c}$$

where again we have set $v_1 = v_1^{(0)}(x, t)$, $v_2 = v_2^{(0)}(x, t)$, and $\varepsilon = 0$ in the arguments of f_i and the partial derivatives of $A_{i\ell}$. In component form, (100a) reads:

$$\frac{\partial u_i}{\partial t} + \sum_{j=1}^{2} A_{ij}^{(0)}(x, t) \frac{\partial u_j}{\partial x} + \sum_{j=1}^{2} B_{ij}(x, t) u_j = f_i^{(1)}(x, t); \qquad i = 1, 2 \tag{102}$$

We note that if $\mathbf{v}^{(0)}$ is a constant vector (this can be true only if $\mathbf{f} = 0$), then $B = 0$ and $\mathbf{f}^{(1)} = 0$. In this case $A^{(0)}$ is a constant operator only if the operator A in the quasilinear problem (93) does not depend on x and t. The linearized equations for shallow-water waves [see (34) of Chapter 3] and the linearized equations for one-dimensional acoustics [see (84) of Chapter 3] are examples of this special case. In both these examples, the nominal solution $\mathbf{v}^{(0)}$ is a constant vector.

The extension of the preceding results to higher dimensions is straight-forward; in effect (102) generalizes to a system of N equations for the N components u_i. This case is discussed for the quasilinear problem (93) in Section 7.2.

The case of more independent variables is also important. For example, in three-dimensional acoustics, we have the system of four equations [see (93), (94) and (99) of Chapter 3] for the perturbations in the density and the three components of velocity. This is a special case that reduces to the simple canonical form for the velocity potential [see (101) of Chapter 3]. In general, such a reduction is not possible, and we discuss only the case of two independent variables.

4.5.2 Characteristics

Here again we define a characteristic curve for (102) based on whether it is possible to extend given Cauchy data on a curve $\mathscr{C}$ to a neighboring curve.

Let $\mathscr{C}$ be defined in the implicit form

$$\phi(x, t) = \xi_0 = \text{constant} \tag{103}$$

and assume that u_1 and u_2 are given on $\mathscr{C}$. Now (ϕ_x, ϕ_t) are the components of a normal to $\mathscr{C}$ and $(\phi_t, -\phi_x)$ are tangential components. Therefore, knowing the u_i on $\mathscr{C}$ means that we also know the tangential derivative $(u_{i_x}\phi_t - u_{i_t}\phi_x)$ there. It then follows that we can express u_{i_t}, say, in terms of u_{i_x} on $\mathscr{C}$ in the form

$$u_{i_t} = -\lambda u_{i_x} + \text{known terms} \tag{104a}$$

where we have assumed $\phi_x \neq 0$ and have denoted

$$\lambda \equiv -\frac{\phi_t}{\phi_x} \tag{104b}$$

If we use (104) in (102), this reduces to the following pair of linear algebraic equations for u_{1_x} and u_{2_x} on $\mathscr{C}$:

$$\sum_{j=1}^{2} (A_{ij}^{(0)} - \delta_{ij}\lambda)u_{j_x} = \text{known terms}; \qquad i = 1, 2 \tag{105}$$

where δ_{ij} is the Kronecker delta ($\delta_{ij} = 1, i = j$; $\delta_{ij} = 0, i \neq j$). System (105) has a unique solution if the determinant of coefficients does not vanish. In this case, the given Cauchy data can be extended to a neighborhood of $\mathscr{C}$.

The vanishing of the determinant of coefficients in (105) corresponds exactly to the quadratic expression defining the eigenvalues of the operator $A^{(0)}$. To see this, let $\mathbf{w}$ be an eigenvector of $A^{(0)}$ belonging to the eigenvalue λ; that is,

$$A^{(0)}\mathbf{w} = \lambda\mathbf{w}; \qquad \mathbf{w} \neq \mathbf{0} \tag{106a}$$

or

$$(A^{(0)} - \lambda I)\mathbf{w} = \mathbf{0}; \qquad \mathbf{w} \neq \mathbf{0} \tag{106b}$$

where I is the identity operator. Since (106b) holds for a nonzero vector $\mathbf{w}$, the determinant of $(A^{(0)} - \lambda I)$, which is just the determinant of coefficients in (105), must vanish. Setting $\det\{A^{(0)} - \lambda I\} = 0$ gives the quadratic expression

$$\lambda^2 - (A_{11}^{(0)} + A_{22}^{(0)})\lambda + A_{11}^{(0)}A_{22}^{(0)} - A_{12}^{(0)}A_{21}^{(0)} = 0 \tag{107a}$$

which has the two solutions

$$\lambda_{1,2}(x,t) = \tfrac{1}{2}\{A_{11}^{(0)} + A_{22}^{(0)} \pm [(A_{11}^{(0)} - A_{22}^{(0)})^2 + 4A_{12}^{(0)}A_{21}^{(0)}]^{1/2}\} \tag{107b}$$

in which henceforth we shall use the plus sign for λ_1 and the minus sign for λ_2.

The eigenvalues λ_i are real and distinct if

$$(A_{11}^{(0)} - A_{22}^{(0)})^2 + 4A_{12}A_{21} > 0$$

and we concentrate on this case, which corresponds exactly to the hyperbolic problem for the second-order equation (1).

The characteristic curves $\phi_1(x,t) = \xi_1 = \text{constant}$ and $\phi_2(x,t) = \xi_2 = \text{constant}$ are defined by the solutions of $dx/dt = \lambda_1(x,t)$ and $dx/dt = \lambda_2(x,t)$, respectively, and since λ_1 and λ_2 are real and distinct, these two families of curves define a coordinate system in terms of the ξ_1, ξ_2 variables. Again, we note that Cauchy data along a characteristic curve do not define the solution uniquely near this curve, and that for a given solution of (102), u_1 and u_2 are not independent; they must satisfy (105) with the left-hand side equal to zero.

The following definitions and results from linear algebra will be helpful in the next section as well as in Section 7.2, where we consider the corresponding quasilinear problem.

Given an $n \times n$ matrix with elements A_{ij}, we may regard these as the *components of a linear operator A* with respect to some unspecified basis $\mathbf{b}_1, \ldots, \mathbf{b}_n$ according to the definition

$$A\mathbf{b}_j = \sum_{k=1}^{n} A_{kj}\mathbf{b}_k$$

for each $j = 1, \ldots, n$.

If the eigenvalues of this $\{A_{ij}\}$ matrix are *real and distinct*, the operator A is *symmetric* (or self-adjoint).

A symmetric operator A has the property

$$A\mathbf{u} \cdot \mathbf{v} = \mathbf{u} \cdot A\mathbf{v} \tag{108}$$

for any pair of vectors $\mathbf{u}$ and $\mathbf{v}$. Note that the given matrix $\{A_{ij}\}$ need not be symmetric (a matrix $\{\bar{A}_{ij}\}$ is *symmetric* if $\bar{A}_{ij} = \bar{A}_{ji}$ for any i and j). However, we can prove that there exists an orthogonal basis with respect to which the components of A form a symmetric matrix.

More importantly, the eigenvectors $\mathbf{w}_1, \ldots, \mathbf{w}_n$ of A associated with $\lambda_1, \ldots, \lambda_n$ are *mutually orthogonal* ($\mathbf{w}_i \cdot \mathbf{w}_j = 0$ if $i \neq j$). This result follows from the fact that if $\mathbf{w}_i$ and $\mathbf{w}_j$ (with $i \neq j$) are two eigenvectors of A associated with the eigenvalues λ_i and λ_j, respectively, (108) reduces to $\lambda_i(\mathbf{w}_i \cdot \mathbf{w}_j) = \lambda_j(\mathbf{w}_i \cdot \mathbf{w}_j)$. But, $\lambda_i \neq \lambda_j$; therefore $\mathbf{w}_i \cdot \mathbf{w}_j = 0$. Hence we can use $\mathbf{w}_1, \ldots, \mathbf{w}_n$ to form a basis. We can also show [see (113)] that the components of A with respect to this orthogonal basis of eigenvectors is the *diagonal matrix* $(\lambda_i \delta_{ij})$, where δ_{ij} is the Kronecker delta.

4.5.3 Transformation to Characteristic Variables

In this section we shall proceed from the component form (102) of the governing equations and verify explicitly that a transformation to a basis of eigenvectors diagonalizes the matrix $\{A_{ij}^{(0)}\}$. In Section 7.2.1, we give an equivalent but less direct derivation proceeding from (100a) and based on the property (108) of A and the orthogonality of its eigenvectors.

Recall [or observe directly from (106)] that any scalar times an eigenvector is still an eigenvector. Therefore, one can specify only the ratio of components of a two-dimensional eigenvector. In our case, the ratio of the components of $\mathbf{w}_i$, the eigenvector belonging to λ_i, must be $-A_{12}^{(0)}/(A_{11}^{(0)} - \lambda_i)$, and we choose

$$\mathbf{w}_i = (-A_{12}^{(0)}, A_{11}^{(0)} - \lambda_i); \qquad i = 1, 2 \tag{109}$$

Note that we may assume $A_{12}^{(0)} \neq 0$ because if $A_{12}^{(0)} = 0$ in (102), we can interchange components so $A_{21}^{(0)}$ occurs in (109). If $A_{21}^{(0)} = 0$ also, we need not proceed further because (102) is already in characteristic form [see (113)].

Since the λ_i are distinct, the $\mathbf{w}_i$ are linearly independent and may be chosen as a new basis. Let the components of $\mathbf{u}$ with respect to the $\mathbf{w}_1, \mathbf{w}_2$ basis be (U_1, U_2). Using (109), we obtain the following linear transformation linking the u_i to the U_i:

$$u_i = \sum_{j=1}^{2} W_{ij} U_j; \qquad i = 1, 2 \tag{110a}$$

where the transformation matrix $\{W_{ij}\}$ is given by

$$\{W_{ij}\} \equiv \begin{pmatrix} -A_{12}^{(0)} & -A_{12}^{(0)} \\ A_{11}^{(0)} - \lambda_1 & A_{11}^{(0)} - \lambda_2 \end{pmatrix} \tag{110b}$$

To calculate the system of equations governing the U_i, we first use (110a) to express the u_i in terms of the U_i in (102) to obtain

$$\sum_{k=1}^{2} \left(W_{\ell k} \frac{\partial U_k}{\partial t} + \frac{\partial W_{\ell k}}{\partial t} U_k \right) + \sum_{j=1}^{2} \sum_{k=1}^{2} A_{\ell j} \left(W_{jk} \frac{\partial U_k}{\partial x} + \frac{\partial W_{jk}}{\partial x} U_k \right)$$

$$+ \sum_{j=1}^{2} \sum_{k=1}^{2} B_{\ell j} W_{jk} U_k = f_\ell^{(1)}; \qquad \ell = 1, 2 \tag{111}$$

Now, we multiply (111) by $V_{i\ell}$ and sum over ℓ, where $\{V_{ij}\}$ is the inverse matrix of $\{W_{ij}\}$; that is,

$$\{V_{ij}\} \equiv \frac{1}{A_{12}^{(0)}(\lambda_2 - \lambda_1)} \begin{pmatrix} A_{11}^{(0)} - \lambda_2 & A_{12}^{(0)} \\ \lambda_1 - A_{11}^{(0)} & -A_{12}^{(0)} \end{pmatrix} \tag{112}$$

The final result takes the form:

$$\frac{\partial U_i}{\partial t} + \lambda_i(x, t) \frac{\partial U_i}{\partial x} + \sum_{k=1}^{2} C_{ik}(x, t) U_k = F_i(x, t); \qquad i = 1, 2 \tag{113}$$

Here the C_{ik} are given by

$$C_{ik} \equiv \sum_{\ell=1}^{2} V_{i\ell} \frac{\partial W_{\ell k}}{\partial t} + \sum_{\ell=1}^{2} \sum_{j=1}^{2} V_{i\ell} A_{\ell j} \frac{\partial W_{jk}}{\partial x} + \sum_{\ell=1}^{2} \sum_{j=1}^{2} V_{i\ell} B_{\ell j} W_{jk} \tag{114a}$$

which can also be written in matrix form as the three products

$$\{C\} = \{V\} \left\{ \frac{\partial W}{\partial t} \right\} + \{V\}\{A\} \left\{ \frac{\partial W}{\partial x} \right\} + \{V\}\{B\}\{W\} \tag{114b}$$

The F_i are given by

$$F_i \equiv \sum_{\ell=1}^{2} V_{i\ell} f_\ell^{(1)} \tag{114c}$$

The inverse relation to (110a) gives the U_i in terms of the u_i:

$$U_i = \sum_{j=1}^{2} V_{ij} u_j \tag{114d}$$

The characteristic diagonal form for the differential operator in (113) becomes even more transparent once we change independent variables to ξ_1 and ξ_2. Regard the U_i, λ_i, C_{ij}, F_i in (113) as functions of ξ_1, ξ_2 without changing the notation for the sake of simplicity. We compute

$$\frac{\partial U_1}{\partial t} = \frac{\partial U_1}{\partial \xi_1} \phi_{1_t} + \frac{\partial U_1}{\partial \xi_2} \phi_{2_t} \tag{115a}$$

$$\frac{\partial U_1}{\partial x} = \frac{\partial U_1}{\partial \xi_1} \phi_{1_x} + \frac{\partial U_1}{\partial \xi_2} \phi_{2_x} \tag{115b}$$

Therefore,

$$\frac{\partial U_1}{\partial t} + \lambda_1 \frac{\partial U_1}{\partial x} = (\phi_{1_t} + \lambda_1 \phi_{1_x}) \frac{\partial U_1}{\partial \xi_1} + (\phi_{2_t} + \lambda_1 \phi_{2_x}) \frac{\partial U_1}{\partial \xi_2}$$

$$= (\lambda_1 - \lambda_2) \phi_{2_x} \frac{\partial U_1}{\partial \xi_2} \tag{116a}$$

since $\lambda_i = -\phi_{i_t}/\phi_{i_x}$. Similarly, interchanging indices gives

$$\frac{\partial U_2}{\partial t} + \lambda_2 \frac{\partial U_2}{\partial x} = (\lambda_2 - \lambda_1)\phi_{1_x}\frac{\partial U_2}{\partial \xi_1} \tag{116b}$$

Thus, (113) simply reduces to a definition of the directional derivative of U_1 along the $\phi_1 = \xi_1 = $ constant characteristics and of U_2 along the $\phi_2 = \xi_2 = $ constant characteristics; that is,

$$\frac{\partial U_1}{\partial \xi_2} + \sum_{k=1}^{2} \tilde{C}_{1k}(\xi_1,\xi_2)U_k = \tilde{F}_1(\xi_1,\xi_2) \tag{117a}$$

$$\frac{\partial U_2}{\partial \xi_1} + \sum_{k=1}^{2} \tilde{C}_{2k}(\xi_1,\xi_2)U_k = \tilde{F}_2(\xi_1,\xi_2) \tag{117b}$$

where

$$\tilde{C}_{1k} \equiv \frac{C_{1k}}{(\lambda_1 - \lambda_2)\phi_{2_x}}; \qquad \tilde{C}_{2k} \equiv \frac{C_{2k}}{(\lambda_2 - \lambda_1)\phi_{1_x}}$$

$$\tilde{F}_1 \equiv \frac{F_1}{(\lambda_1 - \lambda_2)\phi_{2_x}}; \qquad \tilde{F}_2 \equiv \frac{F_2}{(\lambda_2 - \lambda_1)\phi_{1_x}}$$

Eliminating U_1 in favor of U_2 (or vice versa) in (117) gives the second-order equation (69), which can be solved explicitly if the Reimann function is known (as is the case, for example, if the $\tilde{C}_{ij}$ are constants). Another special case for which (117) is explicitly solvable has the $\tilde{C}_{ii}$ and $\tilde{F}_i$ independent of ξ_i and $\tilde{C}_{ik} = 0, i \neq k$, because the system then decouples into the following pair of linear ordinary differential equations for U_1 and U_2:

$$\frac{\partial U_1}{\partial \xi_2} + \tilde{C}_{11}(\xi_2)U_1 = \tilde{F}_1(\xi_2); \qquad \frac{\partial U_2}{\partial \xi_1} + \tilde{C}_{22}(\xi_1)U_2 = \tilde{F}_2(\xi_1)$$

The solution has the form

$$U_i(\xi_1,\xi_2) = \frac{1}{R_i(\xi_i)}\left[K_i(\xi_i) + \int^{\xi_j} \tilde{F}_i(\sigma)R_i(\sigma)\,d\sigma \right]; \qquad i = 1, 2; i \neq j$$

where the K_i are arbitrary functions and

$$R_i(\xi_j) \equiv \exp\left\{ \int^{\xi_j} \tilde{C}_{ii}(\sigma)\,d\sigma \right\}; \qquad i = 1, 2; i \neq j$$

We can also solve (117) exactly if $\tilde{C}_{11} = \tilde{C}_{12} = 0$ and $\tilde{F}_i$ is independent of ξ_i either for $i = 1$ or for $i = 2$ [see the example of Section 4.5.6]. In the case $\tilde{C}_{11} = \tilde{C}_{12} = 0$ and $F_1(\xi_2)$, for instance, U_1 is a function of ξ_2 alone according to (117a). This result reduces (117b) to a solvable linear ordinary differential equation for U_2.

In general (117) cannot be solved explicitly [see (184) of Problem 7], and we have to use a numerical solution, as discussed next.

4.5.4 Numerical Solutions; Propagation of Discontinuities

Consider the Cauchy problem where we assume that U_1 and U_2 are prescribed on a spacelike arc $\mathcal{S}_0$ in the $\xi_1\xi_2$-plane, as sketched in Figure 4.10. In particular, consider the two adjacent points $\alpha = (\xi_1^{(0)}, \xi_2^{(0)} + \Delta\xi_2)$, $\beta = (\xi_1^{(0)} + \Delta\xi_1, \xi_2^{(0)})$, on $\mathcal{S}_0$ and the point $\gamma = (\xi_1^{(0)} + \Delta\xi_1, \xi_2^{(0)} + \Delta\xi_2)$ a small distance away. To compute U_1 at γ, we use the known values of U_1 and U_2 at β in (117a) and obtain

$$U_1(\gamma) = U_1(\beta) + \Delta\xi_2 \left[-\sum_{k=1}^{2} \tilde{C}_{1k}(\beta)U_k(\beta) + \tilde{F}_1(\beta) \right] \tag{118a}$$

and $U_2(\gamma)$ follows from (117b):

$$U_2(\gamma) = U_2(\alpha) + \Delta\xi_1 \left[-\sum_{k=1}^{2} \tilde{C}_{2k}(\alpha)U_k(\alpha) + \tilde{F}_2(\alpha) \right] \tag{118b}$$

This construction is easy to implement and remains valid in the case where the initial data are discontinuous. In fact, we see that a finite discontinuity in the Cauchy data for U_2 at some point propagates along the $\xi_2 = $ constant characteristic passing through that point—that is, in the ξ_1 direction from that point and vice versa. Such discontinuities arise naturally in many physical applications and, as in Section 4.4.7, we can calculate explicitly how they propagate. For example, let the Cauchy data for U_1 and U_2 have a finite discontinuity at the point $\xi_1 = \bar{\xi}_1$, $\xi_2 = \bar{\xi}_2$ on $\mathcal{S}_0$, as shown in Figure 4.10. Let us denote the jump in U_1 by μ and the jump in U_2 by κ; that is,

$$\mu(\bar{\xi}_1, \xi_2) \equiv U_1(\bar{\xi}_1^+, \xi_2) - U_1(\bar{\xi}_1^-, \xi_2)$$

$$\kappa(\xi_1, \bar{\xi}_2) \equiv U_2(\xi_1, \bar{\xi}_2^+) - U_2(\xi_1, \bar{\xi}_2^-)$$
$$\tag{119}$$

Now, if we evaluate (117a) on either side of the $\xi_1 = \bar{\xi}_1$ characteristic and subtract the result, we have

$$\frac{\partial\mu}{\partial\xi_2} + \tilde{C}_{11}\mu = 0 \tag{120a}$$

because U_2 is continuous on the $\xi_1 = \bar{\xi}_1$ characteristic. Similarly, the difference of (117b) on either side of the $\xi_2 = \bar{\xi}_2$ characteristic gives

$$\frac{\partial\kappa}{\partial\xi_1} + \tilde{C}_{22}\kappa = 0 \tag{120b}$$

Therefore, μ and κ propagate according to

$$\mu(\bar{\xi}_1, \xi_2) = \mu(\bar{\xi}_1, \bar{\xi}_2)\exp\left\{ -\int_{\bar{\xi}_2}^{\xi_2} \tilde{C}_{11}(\bar{\xi}_1, \sigma)\,d\sigma \right\} \tag{121a}$$

$$\kappa(\xi_1, \bar{\xi}_2) = \kappa(\bar{\xi}_1, \bar{\xi}_2)\exp\left\{ -\int_{\bar{\xi}_1}^{\xi_1} \tilde{C}_{22}(\sigma, \bar{\xi}_2)\,d\sigma \right\} \tag{121b}$$

This result can be used to directly determine the stability of solutions of (117) [see Problem 7(c)].

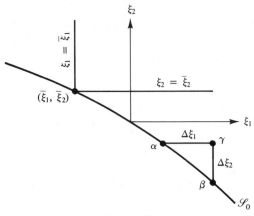

Figure 4.10

Consider now the solution of (117) in a domain bounded by a timelike arc $\mathcal{T}_0$ and a spacelike arc $\mathcal{S}_0$. The geometry is identical to the situation discussed in Figure 4.7 and is repeated in Figure 4.11. If we have U_1 and U_2 prescribed on the spacelike arc $\mathcal{S}_0$ and we wish to solve the system (117) in the domain $\mathcal{D}_1 + \mathcal{D}_4$, we need also to specify U_2 on the timelike arc $\mathcal{T}_0$. Similarly, with Cauchy data on $\mathcal{S}_1$, we need to specify U_1 on $\mathcal{T}_0$ to solve (117) in $\mathcal{D}_2 + \mathcal{D}_3$.

To fix ideas, consider the three adjacent points $\alpha = (\Delta\xi_1, \Delta\xi_2)$ on $\mathcal{T}_0$, $\beta = (\Delta\xi_1, 0)$ on the ξ_1-axis, and the origin O. Having solved the Cauchy problem in $\mathcal{D}_4$, we have U_1 and U_2 on the ξ_1-axis. If the limiting boundary value of U_2 as $\xi_2 \to 0^+$ on $\mathcal{T}_0$ agrees with the limiting value of U_2 obtained from the Cauchy data as $\xi_1 \to 0^+$ on $\mathcal{S}_0$, then U_2 is continuous across β, and we use the values of U_1 and U_2 at β obtained from the solution in $\mathcal{D}_4$ to continue the solution into $\mathcal{D}_1$. In particular, we use (117a) to compute U_1 at α as follows:

$$U_1(\alpha) = U_1(\beta) + \Delta\xi_2[-\tilde{C}_{11}(\beta)U_1(\beta) - \tilde{C}_{12}(\beta)U_2(\beta) + \tilde{F}_1(\beta)] \tag{122}$$

If, however, the two limiting values of U_2 at the origin do not agree, then we cannot use the value of U_2 at β predicted by the solution in $\mathcal{D}_4$; we must calculate all the values of U_2 on $\xi_2 = 0^+$ by solving (117b) subject to the boundary condition for U_2 at the origin obtained from the data on $\mathcal{T}_0$ as $\xi_2 \to 0^+$. In particular, at β we find

$$U_2(\beta) = U_2(0) + \Delta\xi_1[-\tilde{C}_{21}(0)U_1(0) - \tilde{C}_{22}(0)U_2(0) + \tilde{F}_2(0)] \tag{123}$$

where $U_2(0)$ is the limiting value of the boundary data on $\mathcal{T}_0$ as $\xi_2 \to 0^+$. Now, we use this value of $U_2(\beta)$ in (122) to calculate $U_1(\alpha)$. In this case, a discontinuity in U_2 propagates along the ξ_1-axis.

The solution can now be continued into $\mathcal{D}_1$. We use the values of U_1 and U_2 at α in conjunction with (117b) to calculate U_2 at a neighboring point to the right, and so on.

In the characteristic boundary-value problem, we wish to solve (117) in $\mathcal{D}_1 + \mathcal{D}_2$, and we proceed as before with given boundary values for U_1 on the ξ_1-axis and for U_2 on the ξ_2-axis.

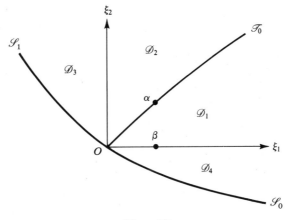

Figure 4.11

4.5.5 Connection with the Second-Order Equation

Here we explore the relation between system (113) and the second-order equation (1) for the hyperbolic case. To reduce (113) to a second-order equation, let us eliminate one of the variables U_1 or U_2. We exclude the case $C_{ij} = 0$, $i \neq j$, because we have seen that this corresponds to an uncoupled pair of equations for U_1 and U_2. If $C_{21} \neq 0$, say, we can solve (113) with $i = 2$ for U_1 and substitute the result into (113) with $i = 1$ to obtain the following second-order equation for U_2:

$$\frac{\partial^2 U_2}{\partial t^2} + (\lambda_1 + \lambda_2)\frac{\partial^2 U_2}{\partial x\,\partial t} + \lambda_1 \lambda_2 \frac{\partial^2 U_2}{\partial x^2} + d^* \frac{\partial U_2}{\partial t} + e^* \frac{\partial U_2}{\partial x} + f^* U_2 = g^* \quad (124)$$

where

$$d^* \equiv C_{11} + C_{22} + C_{21}\left[\left(\frac{1}{C_{21}}\right)_t + \lambda_1\left(\frac{1}{C_{21}}\right)_x\right] \qquad (125a)$$

$$e^* \equiv \lambda_1 C_{22} + \lambda_2 C_{11} + C_{21}\left[\left(\frac{\lambda_2}{C_{21}}\right)_t + \lambda_1\left(\frac{\lambda_2}{C_{21}}\right)_x\right] \qquad (125b)$$

$$f^* \equiv C_{11}C_{22} - C_{12}C_{21} + C_{21}\left[\left(\frac{C_{22}}{C_{21}}\right)_t + \lambda_1\left(\frac{C_{22}}{C_{21}}\right)_x\right] \qquad (125c)$$

$$g^* \equiv F_2 C_{11} - F_1 C_{12} + C_{21}\left[\left(\frac{F_2}{C_{21}}\right)_t + \lambda_1\left(\frac{F_2}{C_{21}}\right)_x\right] \qquad (125d)$$

In view of the definition (14) for the characteristic slopes, we can immediately identify (124) with (1) if we divide (1) by a, replace $x \to t$, $y \to x$, $d/a \to d^*$, $e/a \to e^*$, $f/a \to f^*$, and $g/a \to g^*$. Thus, given a hyperbolic system of two first-order equations, (124)–(125) define a unique second-order hyperbolic equation for U_2. A corresponding second-order hyperbolic equation for U_1 can also be uniquely defined.

The converse problem—that is, that of associating a pair of first-order equations for a given second-order hyperbolic equation—while possible, is not unique. To see this, it suffices to consider the constant coefficient problem

$$u_{tt} - u_{xx} + ku = p(x, t) \tag{126}$$

which we have shown to be equivalent to the most general hyperbolic equation with constant coefficients. For the purposes of this discussion, assume $k > 0$ and p is a given function of x, t.

We ask whether we can find a linear transformation of the dependent variables u, u_t, u_x such that (126) is equivalent to (113). The most general linear transformation is

$$U_i = \alpha_i u_t + \beta_i u_x + \gamma_i u; \qquad i = 1, 2 \tag{127}$$

for as yet unspecified constants $\alpha_i, \beta_i, \gamma_i$. Substituting (127) into (113) shows that the following pair of equations must be satisfied identically:

$$\alpha_i u_{tt} + (\beta_i + \lambda_i \alpha_i) u_{xt} + \lambda_i \beta_i u_{xx}$$

$$+ \left(\gamma_i + \sum_{j=1}^{2} C_{ij} \alpha_j \right) u_t + \left(\lambda_i \gamma_i + \sum_{j=1}^{2} C_{ij} \beta_j \right) u_x$$

$$+ \left(\sum_{j=1}^{2} C_{ij} \gamma_j \right) u = F_i; \qquad i = 1, 2 \tag{128}$$

where $\lambda_1 = 1$ and $\lambda_2 = -1$.

Suppose that we require (128) for $i = 1$ to vanish identically and to reduce to (126) if $i = 2$. It is easily seen that we must set $\alpha_1 = 0, \beta_1 = 0, \gamma_1 = -C_{12}, C_{11} = 0$, $F_1 = 0, \alpha_2 = 1, \beta_2 = 1, \gamma_2 = 0, C_{21} = -k/C_{12}, C_{22} = 0, F_2 = p$, and C_{12} is arbitrary. One choice has $C_{12} = -1$, in which case the transformation (127) becomes

$$U_1 = u \tag{129a}$$

$$U_2 = u_t + u_x \tag{129b}$$

and the system (113) appears in the form

$$U_{1_t} + U_{1_x} - U_2 = 0 \tag{130a}$$

$$U_{2_t} - U_{2_x} + kU_1 = p \tag{130b}$$

Other choices of C_{12} lead to different forms for (130). A second class of transformations results from requiring both equations (128) reduce to (126). Again the choice of constants is not unique; one selection that accomplishes this is

$$\alpha_1 = 1, \quad \beta_1 = -1, \quad \gamma_1 = -\sqrt{k}, \quad C_{11} = 0, \qquad C_{12} = \sqrt{k}, \quad F_1 = p$$

and

$$\alpha_2 = 1, \quad \beta_2 = 1, \qquad \gamma_2 = \sqrt{k}, \qquad C_{21} = -\sqrt{k}, \quad C_{22} = 0, \qquad F_2 = p$$

The transformation (127) is now

$$U_1 = u_t - u_x - \sqrt{k}u \tag{131a}$$

$$U_2 = u_t + u_x + \sqrt{k}u \tag{131b}$$

and system (113) is

$$U_{1_t} + U_{1_x} + \sqrt{k}U_2 = p \tag{132a}$$

$$U_{2_t} - U_{2_x} - \sqrt{k}U_1 = p \tag{132b}$$

Other choices of the constants are also possible, so there is no unique pair of first-order equations that we can associate with a given second-order hyperbolic equation. However, regardless of the particular choice of decomposition, the final solution for u is the same (See Problem 8).

4.5.6 Perturbation of the Dam-Breaking Problem

We illustrate the ideas in Sections 4.5.1–4.5.4 using a classical exact solution of the shallow-water equations for the dam-breaking problem. At time $t = 0$, we assume that one suddenly removes a "dam" at $x = 0$ retaining water of nearly unit height and nearly zero velocity over the negative x-axis, as sketched in Figure 4.12. We also assume that the shallow-water equations [see (22) of Chapter 3] correctly model the flow. Thus, using the notation in Section 4.5.1, we wish to solve

$$v_{1_t} + v_1 v_{1_x} + v_{2_x} = 0 \tag{133a}$$

$$v_{2_t} + (v_1 v_2)_x = 0 \tag{133b}$$

where $v_1(x, t)$ denotes the vertically averaged horizontal speed and $v_2(x, t)$ denotes the height of the free surface measured from the bottom.

Initial conditions corresponding to nearly quiescent flow are

$$v_1(x, 0^-) = \varepsilon H(-x)b_1(x) \tag{134a}$$

$$v_2(x, 0^-) = H(-x)[1 + \varepsilon b_2(x)] \tag{134b}$$

where H is the Heaviside function and $b_1(x)$, $b_2(x)$ are arbitrary prescribed

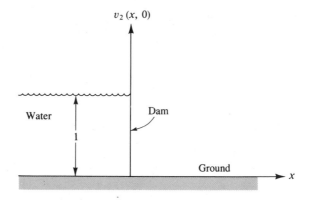

Figure 4.12

functions. The boundary condition of no flow at the dam boundary requires that $b_1(0) = 0$, and by an appropriate choice of the vertical scale, we can also set $b_2(0) = 0$ with no loss of generality. The small parameter ε measures the depar-ture of the initial conditions from the quiescent state $v_1^{(0)} = 0$, $v_2^{(0)} = H(-x)$.

In Section 7.4.2 we derive the following *exact* solution for the $\varepsilon = 0$ problem

$$v_1^{(0)}(x,t) = \begin{cases} \dfrac{2}{3}\left(\dfrac{x}{t}+1\right), & -t < x < 2t \\ 0, & x \le -t \quad \text{or} \quad x \ge 2t \end{cases} \tag{135a}$$

$$v_2^{(0)}(x,t) = \begin{cases} \dfrac{1}{9}\left(2-\dfrac{x}{t}\right)^2, & -t < x < 2t \\ 1, & x \le -t \\ 0, & x \ge 2t \end{cases} \tag{135b}$$

which is easy to verify by direct substitution into (133). As shown in Figure 4.13a, the initial discontinuity in the height develops into the moving parabolic profile

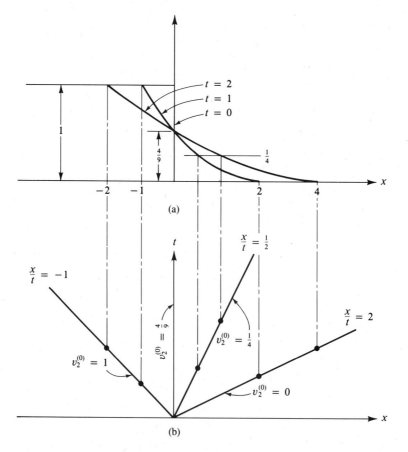

(a)

(b)

Figure 4.13

given in (135b), which spreads to the right with speed 2 and to the left into the quiescent water with speed -1. On rays $x/t = $ constant in the xt-plane, both the speed $v_1^{(0)}$ and height $v_2^{(0)}$ remain constant. This is shown in Figure 4.13b. Since the solution differs from the initial state only in the domain $-\infty < x < 2t$, $t > 0$, we shall concentrate on the perturbation solution over this domain. Consider first the domain $-t < x < 2t$, $t > 0$.

Using the result (135) in (101), we find the following expressions for the matrix components of $A^{(0)}$ and B ($\mathbf{f}^{(1)} = 0$):

$$\{A_{ij}^{(0)}(x,t)\} = \begin{pmatrix} v_1^{(0)} & 1 \\ v_2^{(0)} & v_1^{(0)} \end{pmatrix} \tag{136a}$$

$$\{B_{ij}(x,t)\} = \begin{pmatrix} v_{1_x}^{(0)} & 0 \\ v_{2_x}^{(0)} & v_{1_x}^{(0)} \end{pmatrix} \tag{136b}$$

It follows from (136a) and (107b) that the two characteristic speeds λ_1 and λ_2 are given by

$$\lambda_1 = v_1^{(0)}(x,t) + \sqrt{v_2^{(0)}(x,t)} = \frac{1}{3}\left(\frac{x}{t} + 4\right) \tag{137a}$$

$$\lambda_2 = v_1^{(0)}(x,t) - \sqrt{v_2^{(0)}(x,t)} = \frac{x}{t} \tag{137b}$$

Therefore, the eigenvectors (109) are given by

$$\mathbf{w}_1 = (-1, -\sqrt{v_2^{(0)}(x,t)}) \tag{138a}$$

$$\mathbf{w}_2 = (-1, \sqrt{v_2^{(0)}(x,t)}) \tag{138b}$$

and the transformation matrix $\{W\}$ is

$$\{W_{ij}\} = \begin{pmatrix} -1 & -1 \\ -\sqrt{v_2^{(0)}} & \sqrt{v_2^{(0)}} \end{pmatrix} \tag{139a}$$

with inverse

$$\{V_{ij}\} = \begin{pmatrix} -\dfrac{1}{2} & -\dfrac{1}{2\sqrt{v_2^{(0)}}} \\[2mm] -\dfrac{1}{2} & \dfrac{1}{2\sqrt{v_2^{(0)}}} \end{pmatrix} \tag{139b}$$

We also compute the following matrix components for $\{C\}$ using (114a):

$$\{C_{ij}\} = \begin{pmatrix} 0 & 0 \\ \dfrac{1}{3t} & \dfrac{1}{t} \end{pmatrix} \tag{140}$$

Therefore, the characteristic form (113) is

$$U_{1_t} + \frac{1}{3}\left(\frac{x}{t} + 4\right)U_{1_x} = 0 \tag{141a}$$

$$U_{2_t} + \frac{x}{t} U_{2_x} + \frac{1}{3t} U_1 + \frac{1}{t} U_2 = 0 \tag{141b}$$

for the transformed variables

$$U_1 = -\frac{1}{2} u_1 - \frac{1}{2\sqrt{v_2^{(0)}}} u_2 \tag{142a}$$

$$U_2 = -\frac{1}{2} u_1 + \frac{1}{2\sqrt{v_2^{(0)}}} u_2 \tag{142b}$$

Notice that the equation for U_1 is decoupled from that for U_2. Thus, we can first solve (141a) for U_1 and then substitute the result in (141b) to obtain an inhomogeneous first-order equation for U_2 that can also be solved. This will become more transparent after we introduce the characteristic independent variables ξ_1 and ξ_2 instead of x and t. In preparation for this transformation, we first solve the characteristic differential equations

$$\frac{dx}{dt} = \frac{1}{3}\left(\frac{x}{t} + 4\right) \tag{143a}$$

$$\frac{dx}{dt} = \frac{x}{t} \tag{143b}$$

This gives the two families of characteristic curves:

$$\phi_1(x, t) \equiv -xt^{-1/3} + 2t^{2/3} = \xi_1 = \text{constant} \tag{144a}$$

$$\phi_2(x, t) \equiv \frac{x}{t} = \xi_2 = \text{constant} \tag{144b}$$

In Figure 4.14 we show the two families of characteristics $\xi_1 = \text{constant}$ and $\xi_2 = \text{constant}$. The domain of interest is the triangular region ② bounded by $\xi_2 = -1$ (that is, $x = -t$) and $\xi_1 = 0$ (that is, $x = 2t$). In this domain, only the curves $\xi_1 = \text{constant} > 0$, which lie to the right of the ray $\xi_2 = -1$, are of interest. We note that the ray $x = 2t$ coincides with *both* characteristics $\xi_2 = 2$ and $\xi_1 = 0$. Thus, the coordinate transformation (144) breaks down along this ray. This does not pose any serious difficulties, as we shall see later on.

The solution of the linearized problem in region ③, to the right of $\xi_2 = 2$, is the trivial solution $u_1(x, t) = u_2(x, t) = 0$. In region ①, to the left of the ray $\xi_2 = -1$, we need to solve the system

$$\mathbf{u}_t + A \mathbf{u}_x = 0 \tag{145a}$$

with

$$\{A_{ij}\} = \begin{pmatrix} 0 & 1 \\ 1 & 0 \end{pmatrix} \tag{145b}$$

subject to the initial condition

$$u_1(x, 0) = b_1(x) \tag{146a}$$

$$u_2(x, 0) = b_2(x) \tag{146b}$$

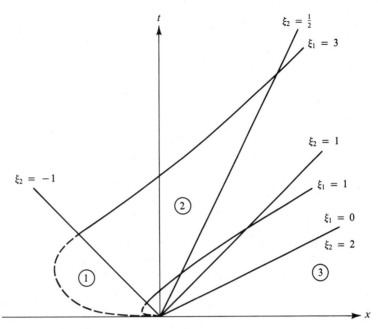

Figure 4.14

This defines $u_1(x, t)$ and $u_2(x, t)$ in ① (see Problem 10b), including the ray $\xi_2 = -1$. Thus, we have a characteristic boundary-value problem to solve in ②, where we know the value of U_1 on the ray $\xi_2 = -1$ and $U_2 = 0$ along the ray $\xi_1 = 0$.

To derive the equations in terms of the ξ_1 and ξ_2 variables, we solve (144) for x and t and use the result

$$x = \xi_2 \left(\frac{\xi_1}{2 - \xi_2} \right)^{3/2} \tag{147a}$$

$$t = \left(\frac{\xi_1}{2 - \xi_2} \right)^{3/2} \tag{147b}$$

in (117) to obtain

$$\frac{\partial U_1}{\partial \xi_2} = 0 \tag{148a}$$

$$\xi_1 \frac{\partial U_2}{\partial \xi_1} + \frac{3}{2} U_2 + \frac{1}{2} U_1 = 0 \tag{148b}$$

This confirms our earlier observation that U_1 can be solved first in the form

$$U_1 = \Gamma(\xi_1) \tag{149a}$$

which substituted into (148b) leads to the solution for U_2:

$$U_2 = \xi_1^{-3/2} \left[\Omega(\xi_2) - \frac{1}{2} \int_0^{\xi_1} s^{1/2} \Gamma(s) \, ds \right] \tag{149b}$$

Here Γ and Ω are arbitrary functions of their arguments that must be determined in terms of the boundary conditions on the bounding characteristics for region ②. The solution in region ①, when evaluated on $\xi_2 = -1$, defines $\Gamma(\xi)$. It also follows that we must set $\Omega(\xi_2) \equiv 0$ because U_2 must tend to zero as $\xi_1 \to 0$ along *any* ray $\xi_2 = $ constant with $-1 \le \xi_2 \le 2$. Thus,

$$U_2(\xi_1) = -\frac{1}{2\xi_1^{3/2}} \int_0^{\xi_1} s^{1/2} \Gamma(s)\, ds \tag{149c}$$

In effect, (149c) is the solution of the ordinary differential equation (148b) along the rays $\xi_2 = $ constant with the boundary condition $U_2(0) = 0$. In this regard, note that $U_2(0)$ is well behaved as long as $\Gamma(0)$ is a finite constant; in fact, $\Gamma(0) = 0$ [see Problem 10(c)], which means $U_2(0) = 0$.

Problems

1. Classify (1) for the case $a = \alpha x^m y^n$, $b = 0$, $c = \beta x^r y^s$, with integer values of m, n, r, s and constants α, β. Derive the characteristics and the canonical form (15a) for the hyperbolic case.

2. Consider the transformation of dependent variable $U \to W$ in (15a) defined by

$$U(\xi, \eta) = W(\xi, \eta) R(\xi, \eta) \tag{150}$$

for an arbitrary function $R(\xi, \eta)$. Show that the partial differential equation that results from (15a) for W will be free of W_ξ and W_η if R satisfies

$$R_\eta - \tilde{D}R = 0 \tag{151a}$$
$$R_\xi - \tilde{E}R = 0 \tag{151b}$$

Prove that a necessary and sufficient condition for this system to have a solution is that

$$\tilde{D}_\xi = \tilde{E}_\eta \tag{152}$$

in which case

$$R(\xi, \eta) = \exp\left\{ \int_{\xi_0}^{\xi} \tilde{E}(\sigma, \eta_0)\, d\sigma + \int_{\eta_0}^{\eta} \tilde{D}(\xi, s)\, ds \right\} \tag{153}$$

where we have normalized the solution so that $R(\xi_0, \eta_0) = 1$. Show also that W satisfies

$$W_{\xi\eta} + (\tilde{D}_\xi - \tilde{E}\tilde{D} - \tilde{F})W = \frac{\tilde{G}}{R} \tag{154}$$

3. a. Derive the canonical form (15a) for the Tricomi equation (21) when $y > 0$.
   b. Now consider (21) in the $y < 0$ half-plane. Find an appropriate change of variables $y \to z$ where z is real so that the Beltrami equations (30) reduce to the Cauchy-Riemann equations

$$\phi_x = \psi_z \tag{155a}$$

$$\psi_x = -\phi_z \tag{155b}$$

One solution of (155) has $\phi = z$ and $\psi = -x$. Use this result to transform (21) to the canonical form (31) in which you exhibit any lower-derivative terms explicitly. Notice that the preceding rather trivial solution sufficed, and we do not need to solve (155) or Laplace's equation for ϕ or ψ that result from (155), in general.

4. Consider the following Cauchy problem for the inhomogeneous dispersive wave equation ($k = $ constant > 0)

$$u_{tt} - u_{xx} + ku = p(x,t), \qquad -\infty < x < \infty; 0 \le t \tag{156a}$$

$$u(x,0) = f(x) \tag{156b}$$

$$u_t(x,0) = g(x) \tag{156c}$$

a. Introduce the characteristic coordinates

$$\xi \equiv t + x \tag{157a}$$

$$\eta \equiv t - x \tag{157b}$$

and transform (156a) to the form

$$U_{\xi\eta} = \tilde{F}U + \tilde{G} \tag{158}$$

where $\tilde{F} \equiv -k/4$ and

$$U(\xi,\eta) \equiv u\left(\frac{\xi - \eta}{2}, \frac{\xi + \eta}{2}\right) \tag{159a}$$

$$\tilde{G}(\xi,\eta) \equiv \frac{1}{4}p\left(\frac{\xi - \eta}{2}, \frac{\xi + \eta}{2}\right) \tag{159b}$$

b. Now, parameterize the initial curve $t = 0$ in the form

$$\xi = \tau \tag{160a}$$

$$\eta = -\tau \tag{160b}$$

—that is, $\xi^* = \tau, \eta^* = -\tau$ in (47)—and show that the initial conditions (156b) and (156c) imply that U^* of (50a) and P^*, Q^* of (55) are given by

$$U^*(\tau) = f(\tau) \tag{161a}$$

$$P^*(\tau) = \frac{g(\tau) + \dot{f}(\tau)}{2} \tag{161b}$$

$$Q^*(\tau) = \frac{g(\tau) - \dot{f}(\tau)}{2} \tag{161c}$$

c. Solve (158) subject to (161) numerically in the rectangle $|x| \le 1, 0 \le t \le 1$. Let $p(x,t) = (1 - x^2)\cos \pi t/2$ in this rectangle and $p = 0$ outside. Let

$f(x) = \sin \pi x$, $g(x) = e^{-x^2}$ on $|x| \le 1$, and $f = g = 0$ for $|x| > 1$. Use a uniform mesh spacing equal to 0.1 in ξ and η.

d. Now, assume that $f(x)$ is continuous for all x and that $g(x)$ has a finite discontinuity at $x = 0$—that is, $g(0^+) - g(0^-) \equiv \rho \ne 0$—but is continuous everywhere else. What are the initial discontinuities in P and Q and how do these propagate?

e. Suppose you wish to define the initial data at $t = 0$ in such a way that U^* and P^* are continuous everywhere on $t = 0$ but Q^* has a discontinuity at $x = 0$. How must you define f, $\dot{f}$, and g? For this case, how does this initial discontinuity propagate?

f. Use the Riemann function (87) in (85) to define an integral representation of the solution. Then express this result in terms of the x, t variables.

g. Use the fundamental solution of (156a) derived in Section 3.7.2 to solve this initial-value problem and show that your results agree with those in part (f).

5. In Section 3.10 we derived an expression for the pressure perturbation due to a bursting balloon. In particular, we found the jump in pressure along the characteristic $r - t = 1$ to be $1/2r$ and the jump in pressure along the characteristic $r - t = -1$ to be $-1/2r$. These results are valid as long as $r > 1$. Reconsider this problem as the solution of

$$\frac{\partial^2 \phi}{\partial r^2} + \frac{2}{r} \frac{\partial \phi}{\partial r} - \frac{\partial^2 \phi}{\partial t^2} = 0; \qquad 0 \le r, 0 \le t \tag{162a}$$

with initial condition

$$\phi(r, 0) = 0 \tag{162b}$$

$$\phi_t(r, 0) = \begin{cases} -\dfrac{1}{\gamma}, & \text{if } r < 1 \\[2mm] 0, & \text{if } r > 1 \end{cases} \tag{162c}$$

Since we have invoked spherical symmetry and reduced the governing equation to the one-dimensional form (162a) with r restricted to $0 \le r$, we must specify a boundary condition at the origin. Because of the absence of a source at the origin, we must have [see (405) of Problem 28 in Chapter 3]

$$\lim_{r \to 0} r^2 \phi_r(r, t) = 0, \qquad t \ge 0 \tag{163}$$

In light of our discussion of Section 4.4.7, derive the expression for the propagation of the initial discontinuity at $r = 1$ along the outgoing characteristic $r - t = 1$ and compare your result with the value derived in Chapter 3. Next, consider the incoming characteristic $r + t = 1$, which reflects from the t-axis and becomes the outgoing characteristic $r - t = -1$. How does the initial discontinuity propagate inward and then reflect? Again, compare your result with the expression derived in Chapter 3.

6. The nonlinear equation governing the dimensionless velocity potential Φ for steady two-dimensional supersonic flow of an inviscid perfect gas is

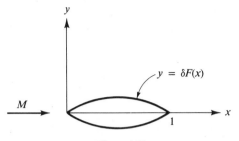

Figure 4.15

$$(a^2 - \Phi_x^2)\Phi_{xx} - 2\Phi_x\Phi_y\Phi_{xy} + (a^2 - \Phi_y^2)\Phi_{yy} = 0 \tag{164a}$$

where a is the dimensionless local speed of sound and is related to Φ by

$$a^2 + \frac{\gamma - 1}{2}(\Phi_x^2 + \Phi_y^2) = 1 + \frac{M^2(\gamma - 1)}{2} = \text{constant} \tag{164b}$$

Here $M > 1$ is the constant Mach number (dimensionless velocity) at $x = -\infty$, and γ is the constant ratio of specific heats. A derivation of (164) from equations (87)–(90) of Chapter 3 can be found in Section 2.4 of [2]. The details of the calculations outlined next are also given there. The boundary conditions for flow over a symmetric body, as sketched in Figure 4.15, are as follows:

1. *The tangency condition:*
    $\Phi_y(x, \pm \delta F(x); M, \delta) = \pm \delta F'(x)\Phi_x(x, \pm \delta F(x); M, \delta)$ where $F(x)$ is the shape of the body (airfoil) and the dimensionless parameter δ is the ratio of half the maximum thickness to the length.
2. *The upstream condition:*

$$\Phi_x(-\infty, y; M, \delta) = M \tag{165a}$$

$$\Phi_y(-\infty, y; M, \delta) = 0 \tag{165b}$$

Thus, the velocity potential depends on the two dimensionless parameters M and δ that characterize the flow speed and airfoil thickness, respectively. Φ also depends on γ, but we consider this to be a fixed constant for all flows.

a. For $y \geq 0$, assume an expansion for Φ in the form

$$\Phi(x, y; M, \delta) = M\{x + \delta\phi_1(x, y; M) + \delta^2\phi_2(x, y; M) + O(\delta^3)\} \tag{166}$$

valid for thin airfoils; that is, $0 < \delta \ll 1$. Show that the assumed expansion for Φ implies that a^2 has the expansion

$$a^2 = 1 - (\gamma - 1)M^2\delta\phi_{1_x} + O(\delta^2) \tag{167}$$

and that substitution of (166) and (167) into (164a) gives the following wave equations governing ϕ_1 and ϕ_2:

$$(M^2 - 1)\phi_{1_{xx}} - \phi_{1_{yy}} = 0 \tag{168a}$$

$$(M^2 - 1)\phi_{2_{xx}} - \phi_{2_{yy}} = -M^2[(\gamma - 1)M^2 + 2]\phi_{1_x}\phi_{1_{xx}} - 2M^2\phi_{1_y}\phi_{1_{xy}} \tag{168b}$$

Show also that the tangency boundary condition requires

$$\phi_{1_y}(x, 0; M) = F'(x) \tag{169a}$$

$$\phi_{2_y}(x, 0; M) = F'(x)\phi_{1_x}(x, 0; M) - F(x)\phi_{1_{yy}}(x, 0; M) \tag{169b}$$

and that the upsteam condition implies

$$\phi_1(-\infty, y; M) = \phi_2(-\infty, y; M) = 0 \tag{170}$$

b. Show that the solution for ϕ_1 is

$$\phi_1(x, y; M)$$

$$= \begin{cases} -\dfrac{1}{\sqrt{M^2 - 1}} F(x - \sqrt{M^2 - 1}\,y), & \text{if } 0 \le x - \sqrt{M^2 - 1}\,y \le 1 \\ \\ 0, & \text{otherwise} \end{cases} \tag{171}$$

Thus, the assumed expansion (166) breaks down as $M \to 1$. A discussion of the correct expansion for $M \approx 1$ is given in [2]. Henceforth, assume that $M - 1$ is not small.

c. Introduce the characteristic coordinates

$$\xi \equiv x - \sqrt{M^2 - 1}\,y \tag{172a}$$

$$\eta \equiv x + \sqrt{M^2 - 1}\,y \tag{172b}$$

and denote

$$\tilde{\phi}_i(\xi, \eta) \equiv \phi_i\left(\frac{\xi + \eta}{2}, \frac{\eta - \xi}{2\sqrt{M^2 - 1}}\right); \qquad i = 1, 2 \tag{173}$$

Show that (168b) transforms to

$$\tilde{\phi}_{2_{\xi\eta}} = -\frac{(\gamma + 1)}{4} \frac{M^4}{(M^2 - 1)^2} F'(\xi)F''(\xi) \tag{174}$$

d. Solve (174) in the form

$$\tilde{\phi}_2(\xi, \eta) = \frac{(\gamma + 1)^2}{8} \frac{M^4}{(M^2 - 1)^2} \eta[F'^2(0) - F'^2(\xi)] + \tilde{f}_2(\xi); \quad 0 < \xi < 1 \tag{175}$$

and $\tilde{\phi}_2 = 0$ if $\xi > 1$ or $\xi < 0$; then determine $\tilde{f}_2(\xi)$ using the boundary condition (169b).

Notice that because of the term proportional to η in (175), $\delta^2\phi_2$ is of order δ^2, as implied by the expansion (166), *only as long as* $\eta = O(1)$; if η is as large as $O(\delta^{-1})$, then $\delta^2\phi_2$ is $O(\delta)$ in violation of the ordering of terms in (166). Now, $\eta \to \infty$ with ξ fixed implies that $y \to \infty$ along a ray $\xi =$ constant. Thus, the assumed form of the perturbation expansion (166) is correct only for $y = O(1)$; it breaks down to $O(\delta^2)$ when y is $O(\delta^{-1})$. Such a "nonuniformity" in the second-order term of a perturbation expansion was also encountered in our study of shallow-water waves of small amplitude and in acoustics [see (178) of Chapter 3]. We distinguish the preceding nonuniformity in the second-order term of a perturbation expansion from the more serious breakdown associated with an *unstable first-order* solution, as mentioned at the conclusion of Section 4.4 and illustrated in Problem 7(c). In our problem, the solution to $O(\delta)$ as given by ϕ_1 in (171) is well behaved; the difficulty occurs only to higher order and in the far-field. In Chapter 8 we discuss this question in detail for a related problem and derive a so-called multiple-scale expansion that, unlike (166), remains valid in the far-field.

7. The conservation equations for shallow-water flow over a flat inclined bottom with friction are

$$H_T + (UH)_X = 0: \quad \text{mass conservation} \qquad (176a)$$

$$HU_T + HUU_X + gHH_X - gsH + cU^2 = 0: \quad \text{momentum conservation} \quad (176b)$$

As shown in Figure 4.16, X is measured along the bottom and Y is normal to it, $Y = H(X, T)$ locates the free surface, s is the small inclination angle of the bottom, U is the water speed (averaged over $0 \le Y \le H$) in the X-direction, c is a dimensionless constant friction coefficient, and g is the constant acceleration of gravity. For a detailed derivation see Section 5.1.1.

a. The uniform flow solution $H = H_0 = \text{constant}$ and $U = U_0 = \text{constant}$ satisfies (176a) identically and (176b) if

$$H_0 = cU_0^2/gs \qquad (177)$$

or

$$U_0 = (gsH_0/c)^{1/2} \qquad (178)$$

—that is, if the friction and gravitational forces are exactly in balance. Introduce the dimensionless variables

$$x \equiv \frac{X}{L_0}; \quad t \equiv \frac{T}{L_0/V_0}; \quad h \equiv \frac{H}{H_0}; \quad u \equiv \frac{U}{V_0} \qquad (179)$$

where L_0 is a characteristic length and V_0 is a characteristic speed. Show that the choice

$$V_0 \equiv (gH_0)^{1/2}; \quad L_0 \equiv H_0/s$$

results in the simple dimensionless form

$$h_t + (uh)_x = 0 \qquad (180a)$$

$$hu_t + huu_x + hh_x - h + \frac{u^2}{F^2} = 0 \qquad (180b)$$

where F is the dimensionless uniform speed

$$F \equiv \frac{U_0}{(gH_0)^{1/2}} \qquad (181)$$

called the Froude number.

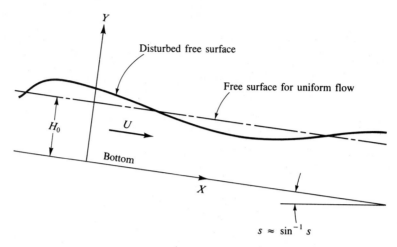

Figure 4.16

b. Now show that if one assumes a perturbation expansion [see (98)]

$$u = F + \varepsilon u_1(x, t) + O(\varepsilon^2) \tag{182a}$$

$$h = 1 + \varepsilon u_2(x, t) + O(\varepsilon^2) \tag{182b}$$

for a given small parameter ε, then $\mathbf{u} = (u_1, u_2)$ obeys (100a) with $\mathbf{f}^{(1)} = 0$ and

$$A^{(0)} = \begin{pmatrix} F & 1 \\ 1 & F \end{pmatrix} \tag{183a}$$

$$B = \begin{pmatrix} \dfrac{2}{F} & -1 \\ 0 & 0 \end{pmatrix} \tag{183b}$$

For this case, show that $\lambda_1 = F + 1$, $\lambda_2 = F - 1$, $\xi_1 = x - (F + 1)t$, $\xi_2 = x - (F - 1)t$, $U_1 = -(u_1 + u_2)/2$, $U_2 = -(u_1 - u_2)/2$, and that (117) becomes

$$\frac{\partial U_1}{\partial \xi_2} + \left(\frac{1}{2F} - \frac{1}{4}\right) U_1 + \left(\frac{1}{2F} + \frac{1}{4}\right) U_2 = 0 \tag{184a}$$

$$\frac{\partial U_2}{\partial \xi_1} + \left(-\frac{1}{2F} + \frac{1}{4}\right) U_1 + \left(-\frac{1}{2F} - \frac{1}{4}\right) U_2 = 0 \tag{184b}$$

c. Consider the propagation of an initial discontinuity in U_1 and U_2 along $\xi_1 = $ constant and $\xi_2 = $ constant characteristics, respectively. Show that the discontinuity in U_1 decays exponentially with time if $F < 2$ and that the discontinuity in U_2 always decays (because F is positive). For the unstable case $F > 2$, the linear theory fails and we must study the non-linear equations (180). For a discussion, see Section 3.2 of [4].

d. Eliminate U_1 from (113) for this example and show that U_2 obeys

$$\left[\frac{\partial}{\partial t} + (F+1)\frac{\partial}{\partial x}\right]\left[\frac{\partial}{\partial t} + (F-1)\frac{\partial}{\partial x}\right]U_2 + \frac{2}{F}\left(\frac{\partial}{\partial t} + \frac{3F}{2}\frac{\partial}{\partial x}\right)U_2 = 0$$

(185)

This is a special case of the general hyperbolic equation

$$\left(\frac{\partial}{\partial t} + \lambda_1\frac{\partial}{\partial x}\right)\left(\frac{\partial}{\partial t} + \lambda_2\frac{\partial}{\partial x}\right)u + \delta\left(\frac{\partial}{\partial t} + a\frac{\partial}{\partial x}\right)u = 0$$

(186)

with the constant characteristic speeds $\lambda_2 < \lambda_1$ and $\delta = $ constant, $a = $ constant. Show that (186) is stable as $t \to \infty$ if $\delta > 0$ and $\lambda_2 < a < \lambda_1$. Verify that for (185) the stability conditions are satisfied if $0 < F < 2$.

8. Consider the wave equation (126) with $k = 0$ on $-\infty < x < \infty$ subject to the initial conditions

$$u(x, 0) = f(x)$$

(187a)

$$u_t(x, 0) = h(x)$$

(187b)

The D'Alembert solution was calculated in Chapter 3 in the form

$$u(x, t) = \frac{1}{2}[f(x+t) + f(x-t)] + \frac{1}{2}\int_{x-t}^{x+t} h(s)\,ds$$

$$+ \frac{1}{2}\int_0^t d\tau \int_{x-(t-\tau)}^{x+(t-\tau)} p(s, \tau)\,ds$$

(188)

a. Use the decomposition (129) with $k = 0$ to show that the initial-value problem reduces to the following pair of equations for $U_1(\xi_1, \xi_2)$, $U_2(\xi_1, \xi_2)$:

$$\frac{\partial U_1}{\partial \xi_2} = \frac{1}{2}U_2$$

(189a)

$$\frac{\partial U_2}{\partial \xi_1} = \frac{1}{2}p\left(\frac{\xi_2 - \xi_1}{2}, \frac{\xi_2 + \xi_1}{2}\right)$$

(189b)

where $\xi_1 = t - x$, $\xi_2 = t + x$, and that the conditions (187) imply

$$U_1(\xi_1, -\xi_1) = f(-\xi_1)$$

(190a)

$$U_2(-\xi_2, \xi_2) = h(\xi_2) + f'(\xi_2)$$

(190b)

b. Solve (189) subject to (190) and show that your results agree with (188).

9. In Problem 2 of Chapter 3, we derived the shallow-water equations for flow over a variable bottom. Specializing these results to the one-dimensional case and the problem of an isolated bump of unit length moving uniformly with dimensionless speed (Froude number) $F = $ constant > 0 to the left, we have (see Figure 4.17a)

$$\bar{h}_t + (\bar{u}\bar{h})_{\bar{x}} = 0$$

(191a)

$$\bar{u}_{\bar{t}} + \bar{u}\bar{u}_{\bar{x}} + (\bar{h} + \varepsilon\bar{b}_{\bar{x}}) = 0$$

(191b)

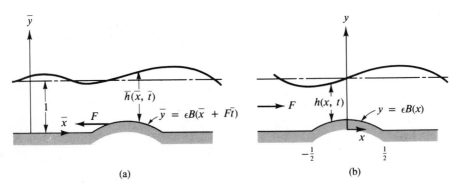

Figure 4.17

where the bump height is $\varepsilon \bar{b}$ with

$$\bar{b}(\bar{x}, \bar{t}) \equiv B(\bar{x} + F\bar{t}) \tag{192}$$

For an isolated bump, the function B vanishes when $|\bar{x} + F\bar{t}| > \frac{1}{2}$.

a. Introduce the Galilean transformation (see Figure 4.17b)

$$x = \bar{x} + F\bar{t} \tag{193a}$$

$$t = \bar{t} \tag{193b}$$

$$\bar{u}(\bar{x}, \bar{t}) = u(x, t) - F \tag{193c}$$

$$\bar{h}(\bar{x}, \bar{t}) = h(x, t) \tag{193d}$$

to a coordinate system attached to and moving with the bump. Thus, $\bar{b}(\bar{x}, \bar{t}) = B(x)$. Show that (191) transforms to

$$h_t + (uh)_x = 0 \tag{194a}$$

$$u_t + uu_x + (h + \varepsilon B)_x = 0 \tag{194b}$$

b. Consider the initial-value problem for (191), where the bump is at rest in quiescent water for $t < 0$ and is started impulsively at $t = 0$. We thus have

$$\bar{u}(\bar{x}, 0) = 0 \tag{195a}$$

$$h(\bar{x}, 0) = 1 - \varepsilon \bar{b}(\bar{x}, 0) \tag{195b}$$

The corresponding initial conditions for (194) are

$$u(x, 0) = F \tag{196a}$$

$$h(x, 0) = 1 - \varepsilon B(x) \tag{196b}$$

Assume that u and h have the following expansions valid for small ε:

$$u(x, t; \varepsilon, F) = F + \varepsilon u_1(x, t; F) + \varepsilon^2 u_2(x, t; F) + O(\varepsilon^3) \tag{197a}$$

$$h(x, t; \varepsilon, F) = 1 + \varepsilon h_1(x, t; F) + \varepsilon^2 h_2(x, t; F) + O(\varepsilon^3) \tag{197b}$$

and derive the equations governing u_1, h_1 and u_2, h_2.

c. Express the systems you derived in part (b) for u_1, h_1 and u_2, h_2 in characteristic form, then solve these for the initial conditions implied by

(196). Observe the breakdown of the solution for u_1 and h_1 if $F \approx 1$, and the breakdown of the solution for u_2 and h_2 in the far-field. These questions are discussed in Problem 16 of Chapter 8.

10. a. Work out the details of the results in Section 4.5.6 up to (144).
    b. Consider the initial-value problem for which

    $$b_1(x) = 0 \tag{198a}$$

    $$b_2(x) = \sin x \tag{198b}$$

    Calculate the perturbation solution $u_1(x, t), u_2(x, t)$ in region ①. What are the values of U_1 and U_2 that result along the ray $x = -t$?
    c. For the case solved in part (b), evaluate $\Gamma(\xi_1)$ and show that setting $\Omega(\xi_2) = 0$ leads to a well-behaved expression for $U_2(\xi_1)$ with $U_2(0) = 0$. Discuss where discontinuities in u_1 and u_2 occur.

11. Consider the linear system of n first-order equations defined in vector form by

    $$\bar{A}\frac{\partial \mathbf{u}}{\partial t} + \bar{B}\frac{\partial \mathbf{u}}{\partial x} + \bar{C}\mathbf{u} = \mathbf{f} \tag{199}$$

    where $\mathbf{u} = (u_1, \ldots, u_n)$, $\mathbf{f} = (f_1(x, t), f_2(x, t), \ldots, f_n(x, t))$, and $\bar{A}, \bar{B}, \bar{C}$ are $n \times n$ matrices with components that are given functions of x, t. Show that the condition under which $\phi(x, t) = \xi_0 = $ constant is a characteristic curve is that the determinant of the matrix $\bar{D} = \phi_x \bar{B} + \phi_t \bar{A}$ vanishes. Equation (106b) is the special case of this for $n = 2$ and under the assumption that $\bar{A}^{-1}$ exists.

12. In Section 4.3.1, we saw that the following is one of the ways to define a characteristic curve $\phi(x, y) = \xi = $ constant for a hyperbolic equation in two independent variables: Given (1) and Cauchy data on $\phi(x, y) = \xi$, we cannot compute $\partial^2 u / \partial \xi^2$ on this curve. In this problem, we study how this idea generalizes to the three-dimensional wave equation with a space-dependent signal speed; that is.

    $$\sum_{j=1}^{3} \frac{\partial^2 v}{\partial x_j^2} - \frac{1}{c^2(x_1, x_2, x_3)}\frac{\partial^2 v}{\partial x_4^2} = 0 \tag{200}$$

    Here x_1, x_2, x_3 are Cartesian coordinates, and x_4 is the time t. Assume that instead of $x_1, \ldots, x_4$, we introduce new variables $\xi_1, \ldots, \xi_4$ defined in general by the four functions

    $$\phi_j(x_1, \ldots, x_4) = \xi_j = \text{constant}, \qquad j = 1, \ldots, 4 \tag{201}$$

    a. Denote

    $$V(\phi_1(x_1, \ldots, x_4), \ldots, \phi_4(x_1, \ldots, x_4)) \equiv v(x_1, \ldots, x_4) \tag{202}$$

    and show that (200) transforms to

    $$\sum_{k=1}^{4}\sum_{\ell=1}^{4}\frac{\partial^2 V}{\partial \xi_k \partial \xi_\ell}\left(\sum_{j=1}^{3}\frac{\partial \phi_k}{\partial x_j}\frac{\partial \phi_\ell}{\partial x_j} - \frac{1}{c^2}\frac{\partial \phi_k}{\partial x_4}\frac{\partial \phi_\ell}{\partial x_4}\right)$$
    $$+ \sum_{k=1}^{4}\frac{\partial V}{\partial \xi_k}\left(\sum_{j=1}^{3}\frac{\partial^2 \phi_k}{\partial x_j^2} - \frac{1}{c^2}\frac{\partial^2 \phi_k}{\partial x_4^2}\right) = 0 \tag{203}$$

b. Now assume that we are given Cauchy data on $\phi_4(x_1,\ldots,x_4) = 0$; that is, we are given the values of $V(\xi_1,\xi_2,\xi_3,0)$ and $(\partial V/\partial \xi_4)(\xi_1,\xi_2,\xi_3,0)$. It then follows that we also know $\partial V/\partial \xi_1$, $\partial V/\partial \xi_2$, $\partial V/\partial \xi_3$ as well as $\partial^2 V/\partial \xi_1 \partial \xi_4$, $\partial^2 V/\partial \xi_2 \partial \xi_4$, and $\partial^2 V/\partial \xi_3 \partial \xi_4$ on $\xi_4 = 0$. But, we do not know $\partial^2 V/\partial \xi_4^2$ on $\xi_4 = 0$. Show that on $\xi_4 = 0$, (203) takes the form

$$\frac{\partial^2 V}{\partial \xi_4^2}\left[\sum_{j=1}^{3}\left(\frac{\partial \phi_4}{\partial x_j}\right)^2 - \frac{1}{c^2(x_1,x_2,x_3)}\left(\frac{\partial \phi_4}{\partial x_4}\right)^2\right] = \text{known terms on } \xi_4 = 0 \tag{204}$$

Thus, either the quadratic form

$$Q \equiv \sum_{j=1}^{3}\left(\frac{\partial \phi_4}{\partial x_j}\right)^2 - \frac{1}{c^2}\left(\frac{\partial \phi_4}{\partial x_4}\right)^2 \tag{205}$$

is not equal to zero and we can solve for $\partial^2 V/\partial \xi_4^2$, or $Q = 0$ and the Cauchy data cannot be specified arbitrarily on $\xi_4 = 0$. The four-dimensional manifold $\phi_4(x_1,\ldots,x_4) = 0$ that satisfies $Q = 0$ is called a characteristic manifold. If we assume a solution of $Q = 0$ in the form

$$x_4 = u(x_1,x_2,x_3) \tag{206}$$

we see that u obeys the eikonal equation [see Section 6.1.1.]

$$\left(\frac{\partial u}{\partial x_1}\right)^2 + \left(\frac{\partial u}{\partial x_2}\right)^2 + \left(\frac{\partial u}{\partial x_3}\right)^2 = \frac{1}{c^2(x_1,x_2,x_3)} \tag{207}$$

c. Parallel the development in Section 4.3.2 to show that $Q = 0$ also represents the manifold on which the first derivatives of a solution of (203) as well as all second derivatives except $\partial^2 V/\partial \xi_4^2$ are continuous. But $\partial^2 V/\partial \xi_4^2$ may be discontinuous. A more general discussion of these concepts can be found in Sections 1–2, Chapter 6, of [3].

References

1. P. Garabedian, *Partial Differential Equations*, Wiley, New York, 1964.
2. J. D. Cole, and L. P. Cook, *Transonic Aerodynamics*, North-Holland, New York, 1986.
3. R. Courant, and D. Hilbert, *Methods of Mathematical Physics*, Vol. *II: Partial Differential Equations* by R. Courant, Wiley-Interscience, New York 1962.
4. G. B. Whitham, *Linear and Nonlinear Waves*, Wiley-Interscience, New York, 1974.

C H A P T E R   5

Quasilinear First-Order Equations

The first four chapters have been almost exclusively concerned with linear equations. In this chapter we start our discussion of nonlinear problems by studying the quasilinear partial differential equation of first order for a scalar dependent variable. As in previous chapters, we begin by considering in some detail a number of physical examples that lead to first-order equations.

5.1 The Scalar Conservation Law; Quasilinear Equations

In Chapter 3 we derived the *coupled pair* of integral conservation laws for mass and momentum [see (16) and (20) of Chapter 3] governing the speed u and height h of shallow-water flow. These conservation laws led to the coupled pair of quasilinear equations (22) of Chapter 3 under the assumption of smooth solutions.

Here we are looking at an even simpler situation, where only one conservation law suffices to define the evolution of a quantity. This level of description usually results from a relationship based either on a physical limiting case or an empirical assumption that links the two dependent variables.

Consider the basic law of "mass" conservation for a one-dimensional flow over the fixed interval $x_1 \leq x \leq x_2$ in the absence of sources

$$\frac{d}{dt} \int_{x_1}^{x_2} \rho(x, t)\, dx = \phi(x_1, t) - \phi(x_2, t) \tag{1a}$$

Here ρ is a "density" and ϕ is a "flux." In general, the flux is a given function of the density, a second dependent variable λ, and the independent variables x, t—that is,

$$\phi(x, t) \equiv \Phi(\rho(x, t), \lambda(x, t), x, t) \tag{1b}$$

For example, (1a) gives the law of mass conservation for shallow-water flow with $\rho = h$, $\lambda = u$, and $\Phi = uh$ [see (16) of Chapter 3]. In this example we used the second conservation law, for momentum, to complete the problem formulation [see (20) of Chapter 3]. This leads to a pair of equations for the two dependent variables u and h.

In some examples, it is possible to bypass the second conservation law and obtain an explicit algebraic relation

$$\phi(x, t) = \Phi(\rho(x, t), x, t) \tag{1c}$$

linking ϕ to ρ, and most of the discussion in Sections 5.1, 5.2, and 5.3 concerns such problems. Equation (1c) may result from a physically consistent simplifying assumption concerning the second conservation law, as illustrated by the example in Section 5.1.1. Sometimes, the second conservation law is not available, and one invokes an empirical or experimentally derived expression for (1c). This is the case, for example, when (1a) describes traffic flow, as discussed in Section 5.1.2.

In any given physical interpretation of (1a), we can define a *flow speed* $v(x, t)$ by

$$v(x, t) = \frac{\phi}{\rho} \tag{2a}$$

A second quantity with units of speed is

$$c(\rho, x, t) = \frac{\partial \Phi}{\partial \rho} \tag{2b}$$

and we shall see that c defines a local *signal speed* for the flow.

For smooth solutions—that is, if ρ and ϕ are continuously differentiable functions of x and t—(1a) with (1c) implies

$$\rho_t + \phi_x = 0 \tag{3a}$$

Using (2b), this gives the following equation for $\rho(x, t)$:

$$\rho_t + c(\rho, x, t)\rho_x + \Phi_x(\rho, x, t) = 0 \tag{3b}$$

5.1.1 Flow of Water in a Conduit with Friction

We consider the flow of water down a conduit having a rectangular cross section and inclined at a constant angle s to the horizontal, as shown in Figure 5.1. Let G denote the domain bounded by the free surface, the bottom, and the two vertical planes $X = X_1$ and $X = X_2$.

Equation (1a) with $\rho \equiv \rho_0 B_0 H(X, T)$ and $\Phi \equiv \rho_0 B_0 H(X, T)U(X, T)$ becomes

$$\frac{d}{dT} \int_{X_1}^{X_2} \rho_0 B_0 H(X, T)\,dX = -\rho_0 B_0 H(X, T)U(X, T)\Big|_{X=X_1}^{X=X_2} \tag{4}$$

We are using dimensional T, X, Y, H, U variables, where T is the time, X is the distance measured along the bottom, and Y is the distance normal to it. Thus, H is the height of the free surface in the Y direction and U is the X-component of the flow velocity averaged over $0 \leq Y \leq H$. The constant density of water is ρ_0, and B_0 is the constant breadth of the conduit.

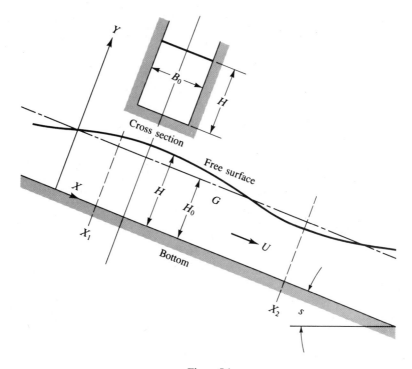

Figure 5.1

To complete the problem formulation, we must relate U to H, and we use momentum conservation. As in Section 3.2, we assume hydrostatic balance in the Y-direction and ignore surface tension, but now we do account for viscous effects approximately through the introduction of a friction force exerted on the fluid by the solid boundaries of the conduit.

We identify the following contributions to the momentum balance for G in the direction parallel to the bottom:

$$\frac{d}{dT} \int_{X_1}^{X_2} \rho_0 B_0 H(X, T) U(X, T)\, dX = \text{rate of change of momentum of } G \quad (5a)$$

$$-\rho_0 B_0 U^2(X, T) H(X, T) \Big|_{X=X_1}^{X=X_2} = \text{net inflow of momentum into } G \quad (5b)$$

$$-\int_0^{H(X_2, T)} \rho_0 B_0 g(H - Y) \cos s\, dY = \left\{ \begin{array}{l} \text{streamwise component of} \\ \text{the pressure force on} \\ G \text{ at } X = X_2 \end{array} \right. \quad (5c)$$

$$+\int_0^{H(X_1, T)} \rho_0 B_0 g(H - Y) \cos s\, dY = \left\{ \begin{array}{l} \text{streamwise component of} \\ \text{the pressure force on} \\ G \text{ at } X = X_1 \end{array} \right. \quad (5d)$$

$$+\int_{X_1}^{X_2} \rho_0 B_0 g H(X, T) \sin s\, dX = \left\{ \begin{array}{l} \text{streamwise component of the} \\ \text{gravity force on } G \end{array} \right. \quad (5e)$$

$$-\int_{X_1}^{X_2} C\rho_0[B_0 + 2H(X,T)]U^2(X,T)\,dX = \left\{\begin{array}{l}\text{friction force acting}\\\text{on } G \text{ due to solid}\\\text{boundaries.}\end{array}\right. \tag{5f}$$

In (5f), C is a dimensionless friction coefficient, which we assume to be a constant independent of Reynolds number, and we have used the standard formula

Friction force $= C\rho_0 AU^2$

for the friction force over a surface of wetted area A. For a rectangular conduit, the wetted area of an element of length dX is $dA = (B_0 + 2H)\,dX$.

After integrating (5c) and (5d) and dividing by the factor $\rho_0 B_0$, we obtain the following integral conservation law:

$$\frac{d}{dT}\int_{X_1}^{X_2} H(X,T)U(X,T)\,dX = -\left\{U^2(X,T)H(X,T) + \frac{1}{2}gH^2(X,T)\cos s\right\}_{X=X_1}^{X=X_2}$$
$$+ \int_{X_1}^{X_2}\left\{gH(X,T)\sin s - C\left(1 + \frac{2H}{B_0}\right)U^2(X,T)\right\}dX \tag{6}$$

For a broad conduit, $H/B_0 \ll 1$ and may be ignored. Also, if s is small, we may set $\sin s \approx s$, $\cos s \approx 1$. Then (4) combined with (6) reduce to (176) of Chapter 4 for smooth solutions.

A special solution of (4) and (6) corresponds to uniform flow ($U \equiv U_0 =$ constant, $H \equiv H_0 =$ constant) with the friction and gravitational forces in perfect balance; that is,

$$gH_0 \sin s = C(1 + 2\sigma)U_0^2$$

where σ is the dimensionless parameter $\sigma \equiv H_0/B_0$.

For any given H_0, this defines a unique U_0, or vice versa:

$$U_0 = \sqrt{\frac{gH_0 \sin s}{C(1 + 2\sigma)}} \tag{7a}$$

$$H_0 = \frac{C(1 + 2\sigma)U_0^2}{g \sin s} \tag{7b}$$

We introduce the following dimensionless variables in terms of a length scale (L_0) and speed (V_0) that are to be specified

$$x \equiv \frac{X}{L_0}; \qquad t = \frac{T}{L_0/V_0}; \qquad h = \frac{H}{H_0}; \qquad u = \frac{U}{V_0} \tag{8}$$

It is easy to see that the choice

$$L_0 = \frac{H_0}{\tan s} \tag{9a}$$

$$V_0 = (gH_0 \cos s)^{1/2} \tag{9b}$$

simplifies the notation and leads to

$$\frac{d}{dt}\int_{x_1}^{x_2} h(x,t)\,dx = -\{h(x,t)u(x,t)\}_{x=x_1}^{x=x_2} \tag{10a}$$

$$\frac{d}{dt} \int_{x_1}^{x_2} u(x,t)h(x,t)\,dx = -\left\{ u^2(x,t)h(x,t) + \frac{1}{2}h^2(x,t) \right\}_{x=x_1}^{x=x_2}$$

$$+ \int_{x_1}^{x_2} \left\{ h(x,t) - \frac{1 + 2\sigma h(x,t)}{F^2} u^2(x,t) \right\} dx \qquad (10b)$$

Here, F is the dimensionless speed (Froude number) defined by

$$F^2 = \frac{U_0^2(1 + 2\sigma)}{gH_0 \cos s} \qquad (11a)$$

and (7) reduces to the condition

$$F^2 = \frac{\tan s}{C} \qquad (11b)$$

relating F, C, and s.

For smooth solutions, (10a) and (10b) give the pair of quasilinear equations

$$h_t + (hu)_x = 0 \qquad (12a)$$

$$(uh)_t + \left(u^2 h + \frac{1}{2}h^2 \right)_x = h - \frac{1 + 2\sigma h}{F^2} u^2 \qquad (12b)$$

and using (12a) to eliminate h_t from (12b) gives

$$h_t + uh_x + hu_x = 0 \qquad (13a)$$

$$u_t + uu_x + h_x = 1 - \frac{1 + 2\sigma h}{F^2} \frac{u^2}{h} \qquad (13b)$$

If σ and s are both small, (13b) reduces to

$$u_t + uu_x + h_x = 1 - \frac{u^2}{F^2 h} \qquad (13c)$$

and we shall use (13c) henceforth for simplicity. Since $\cos s \approx 1$, we may regard h as the height of the free surface in the vertical direction.

The crucial assumption that is now made to reduce (13) to a single equation for h is that the left-hand side of (13c) equals zero as a first approximation; that is, even though u and h are no longer constant, they evolve in such a way that $u_t + uu_x + h_x \approx 0$. For a discussion of the validity of this assumption, see Section 3.2 of [1]. Thus, we set $u = F\sqrt{h}$ in (10a) and obtain

$$\frac{d}{dt} \int_{x_1}^{x_2} h(x,t)\,dx = -\{Fh^{3/2}\}_{x=x_1}^{x=x_2}. \qquad (14a)$$

or the scalar first-order quasilinear equation

$$h_t + \frac{3F}{2} h^{1/2} h_x = 0 \qquad (14b)$$

for smooth solutions.

In this example, the simplifying physical assumption $u = Fh^{1/2}$ defines the flux in the form $\Phi = Fh^{3/2}$ and reduces the coupled pair of equations (10) or (13a) and (13c) to the single equation (14a) or (14b). In certain cases, one must resort to an empirical formula for Φ to make progress, as in the case of traffic flow discussed next.

5.1.2 Traffic Flow

Consider traffic on a one-lane road with no on and off ramps. Let ρ denote the traffic density—that is, the number of cars per unit distance X of road. In this application, it is reasonable to set

$$\phi(X, T) = \Phi(\rho(X, T), X, T) \tag{15}$$

where the dependence of Φ on ρ has the qualitative behavior shown in Figure 5.2.

This graph depicts Φ at a given point X on the road and at a given time T. It shows the obvious fact that if there are no cars ($\rho = 0$), then $\Phi = 0$. Conversely, there is a maximum density ρ_{max} of cars for which $\Phi = 0$ also, because the cars are stacked bumper to bumper and cannot move. At some intermediate density ρ_0, the flux has a maximum value Φ_{max}. The flux also depends in general on X and T to reflect driving conditions at each location along the road and time of day. One could derive these empirical relationships by making a large number of observations covering all traffic conditions at different locations and times. One could improve the model in (15) by allowing ϕ to depend also on ρ_X, to reflect the fact that drivers tend to slow down when moving into a region of increasing traffic density and vice versa. Thus, a plausible refinement would rephrase (15) as

$$\phi(X, T) = \Phi(\rho(X, T), X, T) - k\rho_X(X, T) \tag{16}$$

where k is a positive constant. The simplest expression (16) that retains all the essential features has a quadratic dependence of Φ on ρ, i.e.,

$$\phi(X, T) = R\rho(\rho_{max} - \rho) - k\rho_X \tag{17}$$

with R, ρ_{max} and k constant.

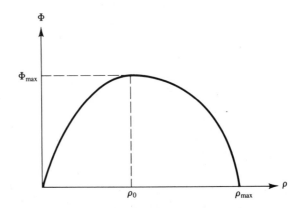

Figure 5.2

In this case (1a) gives

$$\rho_T + [R\rho(\rho_{max} - \rho)]_X = k\rho_{XX} \tag{18}$$

for smooth solutions.

Let us introduce the dimensionless variables

$$u \equiv \frac{\rho_{max}/2 - \rho}{\rho_{max}/2}; \qquad x \equiv \frac{X}{L_0}; \qquad t \equiv \frac{T}{L_0/R\rho_{max}} \tag{19}$$

where L_0 is a characteristic length scale associated with the initial condition for (18). We then find

$$u_t + [\tfrac{1}{2}(u^2 - 1)]_x = \varepsilon u_{xx} \tag{20a}$$

or

$$u_t + uu_x = \varepsilon u_{xx}; \qquad \varepsilon \equiv \frac{k}{R\rho_{max}L_0} \tag{20b}$$

where ε is a dimensionless parameter that is the ratio of the *diffusion length* $k/R\rho_{max}$ to L_0. Equation (20) is Burgers' equation, discussed in Chapter 1. Of course, in the present interpretation, solutions of (20) are valid only for $0 \le \rho \le \rho_{max}$—that is, for $|u| \le 1$.

A number of other physical problems may be modeled by (1); the reader is referred to Chapter 3 of [1] for a detailed discussion and references to original sources.

5.2   Continuously Differentiable Solution of the Quasilinear Equation in Two Independent Variables

In Section 5.1 we showed that a typical integral conservation law of the form (1) corresponds to a quasilinear equation if the solution is *smooth*—that is, if it is continuous and has continuous first partial derivatives. In this section, we study the geometric properties of such solutions as a guide in deriving their analytic form. We then point out how the smoothness requirements may break down in physically realistic situations. This motivates the idea of *weak solutions* discussed in Section 5.3. These weak solutions are based on the integral conservation law formulation and admit discontinuities.

5.2.1 Geometrical Aspects of Solutions

The general quasilinear equation for u as a function of the two independent variables x, y is

$$a(x, y, u)u_x + b(x, y, u)u_y = c(x, y, u) \tag{21}$$

where a, b, and c are prescribed functions of x, y, u. We also assume that in some domain of interest, a, b, and c are continuous and have continuous first derivatives and that a and b do not vanish simultaneously in this domain along a given solution. The problem is quasilinear because a, b, and c depend on u but do not involve u_x or u_y. The fact that u is involved in these coefficients means that we must take into consideration the solution $u(x, y)$ in any geometric interpretation of (21).

Therefore, let us examine (21) in the three-dimensional Cartesian space x, y, u. Let $P = (x_0, y_0, u_0)$ be a given point on a solution $u = \phi(x, y)$ of (21); we write this solution in the form

$$\Phi(x, y, u) \equiv \phi(x, y) - u = 0 \tag{22}$$

A normal vector $\mathbf{n}$ to the solution surface $\Phi = 0$ has components $(u_x, u_y, -1)$. Neither the magnitude of this normal, nor the fact that the normal points in the negative u direction is relevant; these are consequences of an arbitrary constant multiplier that does not alter (22). At the point P, the coefficients a, b, and c each equal a number, which we denote by $a_0 \equiv a(x_0, y_0, u_0)$, $b_0 \equiv b(x_0, y_0, u_0)$, and $c_0 \equiv c(x_0, y_0, u_0)$. We note that at P, (21) is just a statement of the vanishing of the dot product between the normal vector $\mathbf{n}_0 \equiv (u_x(x_0, y_0), u_y(x_0, y_0), -1)$ and the constant vector $\boldsymbol{\sigma}_0 \equiv (a_0, b_0, c_0)$. Since (21) is a *linear* algebraic relation linking u_x and u_y at P, it follows that the normals to all the possible solution surfaces through P must lie *in a plane perpendicular to* $\boldsymbol{\sigma}_0$. In other words, $\boldsymbol{\sigma}_0$ *is a tangent vector to all the possible solution surfaces through* P. One may also interpret $\boldsymbol{\sigma}_0$ as the intersection of the one-parameter family of possible solution surfaces of (21) at P, as sketched in Figure 5.3a. Again, the magnitude and sense of $\boldsymbol{\sigma}_0$ are irrelevant.

The curve generated in x, y, u space by following the local $\boldsymbol{\sigma}$ direction from a given initial point is called a *characteristic curve*; its projection on the $u = 0$ plane is called a *characteristic ground curve*.

Let $\mathscr{S}$ be a reference surface on which the characteristic direction at every point is not tangent to $\mathscr{S}$. Therefore, the characteristic curves that pass through $\mathscr{S}$ can be labeled using two parameters (corresponding, for example, to the two coordinates needed to specify each point on $\mathscr{S}$). This two-parameter family of curves fills the x, y, u space, at least in some neighborhood of $\mathscr{S}$. In order to isolate a specific solution surface $\Phi = 0$ generated by a one-parameter subfamily of characteristic curves, let us pick some curve $\mathscr{C}_0$ on $\mathscr{S}$ and consider only those characteristics that pass through $\mathscr{C}_0$. These characteristics generate a surface, on every point of which (21) is satisfied. This solution surface may be geometrically visualized by regarding it as an infinitely dense set of characteristic curves one layer thick.

We see that in order to specify a solution surface, we must require that it contain a prescribed *initial curve* $\mathscr{C}_0$. The local construction of the solution surface near $\mathscr{C}_0$ may be visualized as follows: We introduce the parameter τ, which varies along $\mathscr{C}_0$, and the parameter s (this may be an arc length), which varies along each characteristic curve emerging from $\mathscr{C}_0$. By extending the characteristic curves a short distance Δs, we generate a thin strip Δs wide to one side of $\mathscr{C}_0$. The forward edge of this strip defines a new curve $\mathscr{C}_1$, which is distinct from $\mathscr{C}_0$ because none of the characteristic directions are tangent to $\mathscr{C}_0$ (see Figure 5.3b).

We now repeat this process and extend the solution over a new thin strip to $\mathscr{C}_2$, and so on. This construction generates a solution surface as long as the successive curves $\mathscr{C}_1$, $\mathscr{C}_2$, ..., do not become characteristic over a finite arc.

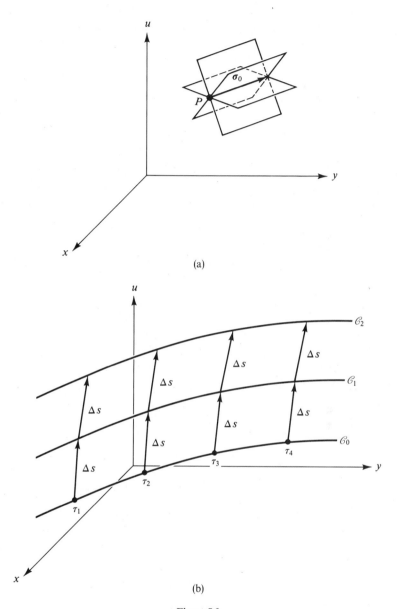

(a)

(b)

Figure 5.3

5.2.2 Characteristic Curves; the Solution Surface

We have seen that at each point $P = (x_0, y_0, u_0)$, there is a unique characteristic vector $\sigma_0 = (a_0, b_0, c_0)$ that is tangent to the characteristic curve passing through P. If s is a parameter that varies monotonically along a characteristic curve, the coordinates $Q = (x_0 + \Delta x, y_0 + \Delta y, u_0 + \Delta u)$ of a point located at a small distance from P in the characteristic direction of increasing s are obtained by setting $\Delta x = a_0 \Delta s$, $\Delta y = b_0 \Delta s$, $\Delta u = c_0 \Delta s$. Thus, in the limit as $\Delta s \to 0$, the characteristic curves that are so generated obey the system

$$\frac{dx}{ds} = a(x, y, u) \tag{23a}$$

$$\frac{dy}{ds} = b(x, y, u) \tag{23b}$$

$$\frac{du}{ds} = c(x, y, u) \tag{23c}$$

Since the system (23) is autonomous, its solution can be expressed in terms of $s - s_0$, where s_0 is a constant that fixes the origin of the independent variable, and two other arbitrary constants c_1 and c_2:

$$x = \bar{X}(s - s_0, c_1, c_2) \tag{24a}$$

$$y = \bar{Y}(s - s_0, c_1, c_2) \tag{24b}$$

$$u = \bar{U}(s - s_0, c_1, c_2) \tag{24c}$$

We can choose s_0, c_1, and c_2 so that (24) passes through a given point x_0, y_0, z_0. The form of (24) also confirms the claim made earlier that the characteristics are a two-parameter family of curves, since we can regard $s - s_0 = \sigma$ to be a new variable along each curve identified by the two constants c_1, c_2.

We can isolate any desired one-parameter subfamily of (24) by regarding c_1, s_0, and c_2 as functions of a single parameter τ. This gives

$$x = \bar{X}(s - s_0(\tau), c_1(\tau), c_2(\tau)) \equiv X(s, \tau) \tag{25a}$$

$$y = \bar{Y}(s - s_0(\tau), c_1(\tau), c_2(\tau)) \equiv Y(s, \tau) \tag{25b}$$

$$u = \bar{U}(s - s_0(\tau), c_1(\tau), c_2(\tau)) \equiv U(s, \tau) \tag{25c}$$

The particular one-parameter subfamily that passes through the initial curve $\mathscr{C}_0$ is obtained as follows. We specify $\mathscr{C}_0$ in the parametric form:

$$x = x_0(\tau) \tag{26a}$$

$$y = y_0(\tau) \tag{26b}$$

$$u = u_0(\tau) \tag{26c}$$

where x_0, y_0, and u_0 are continuously differentiable functions of τ, and we assume that the ground curve $x_0(\tau)$, $y_0(\tau)$ does not intersect with itself. Then we require the functions $X(s, \tau)$, $Y(s, \tau)$, and $U(s, \tau)$ in (25) to satisfy the initial conditions

$$X(0, \tau) = x_0(\tau) \tag{27a}$$

$$Y(0, \tau) = y_0(\tau) \tag{27b}$$

$$U(0, \tau) = u_0(\tau) \tag{27c}$$

Thus, we have set $s = 0$ on the initial curve. The conditions (27) specify the three functions $s_0(\tau)$, $c_1(\tau)$, $c_2(\tau)$, and the expressions $X(s, \tau)$, $Y(s, \tau)$, $U(s, \tau)$, which result in (25), define a one-parameter family of characteristic curves that pass through $\mathscr{C}_0$. In practice, we shall derive the solution of (23) directly in the form (25).

To exhibit the solution surface $u = \phi(x, y)$ that passes through $\mathscr{C}_0$, we solve (25a) and (25b) for s and τ in terms of x and y. This is always possible as long as the Jacobian

$$\Delta(s, \tau) \equiv X_s Y_\tau - Y_s X_\tau \tag{28}$$

does not vanish. We know that if the characteristics are nowhere tangent to $\mathscr{C}_0$, then the directions of increasing s and τ along $\mathscr{C}_0$ are not collinear. Therefore, at least in some neighborhood of $s = 0$, we have $\Delta \neq 0$, and we express the solutions of (25a) and (25b) for s and τ in the form

$$s = S(x, y) \tag{29a}$$

$$\tau = T(x, y) \tag{29b}$$

Substituting these into (25c) defines the solution surface in the form

$$u = U(S(x, y), T(x, y)) \equiv \phi(x, y) \tag{30}$$

The result (30) is available whenever $\Delta \neq 0$, and we can prove the following theorem (for instance, see Chapter 3, Section 1.2, of [2]). Every integral surface is generated (in the sense just discussed) by a one-parameter family of characteristic curves. Conversely, every one-parameter family of characteristic curves generates an integral surface.

If $\Delta(0, \tau) = 0$, we have two possibilities: Either the initial curve is a characteristic curve, in which case an infinite number of solutions pass through $\mathscr{C}_0$, or $\mathscr{C}_0$ is not characteristic, and it is not possible to calculate a continuously differentiable solution. We now illustrate these features by studying different initial conditions for Burgers' equation with $\varepsilon = 0$ and a constant source term [consider (20b) with $t \to y$]

$$u u_x + u_y = 1 \tag{31}$$

This equation is also discussed in [2]. It is ideal for purposes of illustration, as one can readily construct explicit solutions that exhibit the various features of quasilinear equations.

The characteristic differential equations for (31) are

$$\frac{dx}{ds} = u \tag{32a}$$

$$\frac{dy}{ds} = 1 \tag{32b}$$

$$\frac{du}{ds} = 1 \tag{32c}$$

Solving these subject to an unspecified initial curve $\mathscr{C}_0$: $x_0(\tau)$, $y_0(\tau)$, $u_0(\tau)$, we find

$$x = \frac{s^2}{2} + su_0(\tau) + x_0(\tau) \tag{33a}$$

$$y = s + y_0(\tau) \tag{33b}$$

$$u = s + u_0(\tau) \tag{33c}$$

The Jacobian of the transformation (33a)–(33b) is

$$\Delta(s,\tau) \equiv [s + u_0(\tau)]y_0'(\tau) - [su_0'(\tau) + x_0'(\tau)] \tag{34}$$

and this may vanish at $s = 0$ or along some other curve, depending on the choice of initial data.

$\Delta(0,\tau) \neq 0$

To illustrate a case where a unique continuously differentiable solution exists near $\mathscr{C}_0$, let

$$x_0(\tau) = \tau, \qquad y_0(\tau) = \tau, \qquad u_0(\tau) = 2$$

We then calculate $\Delta(s,\tau) = s + 1$ from (34), and this does not vanish at $s = 0$. We therefore expect a continuously differentiable solution to exist for $s > -1$. The difficulty at $s = -1$ will become clear when we derive the details of the solution.

The one-parameter family of characteristic curves (32) for this case is given by

$$x = \frac{s^2}{2} + 2s + \tau \tag{35a}$$

$$y = s + \tau \tag{35b}$$

$$u = s + 2 \tag{35c}$$

Consider first the characteristic ground curves (35a)–(35b). Eliminating s gives the one-parameter family of parabolas defined by

$$x + 2 - \tau - \tfrac{1}{2}(y + 2 - \tau)^2 = 0 \tag{36}$$

which are sketched in Figure 5.4. To generate the family, we just translate the parabola $x = y^2/2$ corresponding to $\tau = 2$ parallel to itself along the straight line $y = x$. We also notice that these parabolas envelope along the straight line $y = x + \tfrac{1}{2}$, which, in fact, corresponds to $\Delta(s,\tau) = 0$—that is, $s = -1$.

To see what happens at $s = -1$, we eliminate τ from (35a)–(35b). This gives the quadratic expression $s^2 + 2s + 2(y - x) = 0$ for s. The root that corresponds to $y = x$ at $s = 0$ is $s = -1 + \sqrt{1 + 2(x - y)}$. Therefore, the solution is given by

$$u = 1 + \sqrt{1 + 2(x - y)} \tag{37}$$

Thus, u is a constant along the lines $x - y = $ constant. The radical in (37) vanishes along the line $y = x + \tfrac{1}{2}$—that is, $s = -1$—and the solution does not exist to the left of this line. Moreover, u_x and u_y become infinite along this line, which represents an edge of regression. In fact, we might be tempted to claim the existence of a second branch for u corresponding to the negative sign in front of the radical. This branch gives the mirror image of the surface $u(x,y)$ relative to

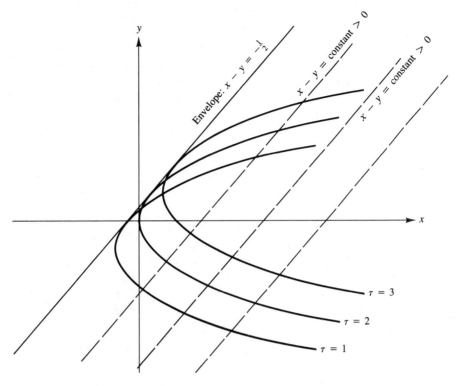

Figure 5.4

the plane $u = 1$. This claim is, however, entirely inconsistent; it would result in a two-valued solution with the lower branch not satisfying the correct initial condition.

Notice that the curve defined by $\Delta(s, \tau) = 0$, $u = 1$—that is, the straight line $y = x + \frac{1}{2}$, $u = 1$—is *not a characteristic curve*; it is the envelope of characteristic ground curves.

$\Delta(0, \tau) = 0$, $\mathscr{C}_0$ is not a Characteristic

The difficulty encountered at $s = -1$ in the previous example occurs on the initial curve for the case

$$x_0 = \tau^2 \tag{38a}$$

$$y_0 = 2\tau \tag{38b}$$

$$u_0 = \tau \tag{38c}$$

It is easily seen that $\Delta(0, \tau) = 0$. In order for the initial curve to be characteristic, it must satisfy $dx_0/d\tau = u_0(\tau)$, $dy_0/d\tau = 1$, and $du_0/d\tau = 1$, which it does not. We consider the characteristic curves

$$x = \frac{s^2}{2} + s\tau + \tau^2 \tag{39a}$$

$$y = s + 2\tau \tag{39b}$$

$$u = s + \tau \tag{39c}$$

and eliminate s from (39a)–(39b) to obtain the one-parameter family of parabolas defined by

$$F(x, y, \tau) \equiv x - \frac{\tau^2}{2} - \frac{1}{2}(y - \tau)^2 = 0 \tag{40}$$

Again, each member of this family is obtained by translating the parabola $x - y^2/2 = 0$, as indicated in (40). The family (40) has the envelope $x - y^2/4 = 0$ (obtained by eliminating τ from (40) and from the expression for $F_\tau = 0$). Thus, the projection of the initial curve on the xy-plane is now the envelope of characteristic ground curves.

If we calculate a formal expression for $u(x, y)$ from (39), we find $u = y/2 \pm (x - y^2/4)^{1/2}$. Therefore, the initial curve is an edge of regression from which the two branches of the radical define two surfaces over the domain $x > y^2/4$. Again, *this is not a solution* in the strict sense as u_x and u_y are infinite on the initial curve and u is two-valued for $x > y^2/4$. For the linear equation (21) where the coefficients a, b, c do not depend on u, the case $\Delta = 0$ on a noncharacteristic curve $\mathscr{C}$ means that the characteristic equations (23) define a vertical surface through $\mathscr{C}$ (see Problem 5).

$\Delta(0, \tau) = 0$, $\mathscr{C}_0$ *Is a Characteristic*

Now we consider the initial curve

$$x_0 = \tau^2/2, \qquad y_0 = \tau, \qquad u_0 = \tau$$

and find $\Delta(0, \tau) = 0$. We also verify that $dx_0/d\tau = \tau = u_0$, $dy_0/d\tau = 1$, and $du_0/d\tau = 1$. Therefore, the initial curve is a characteristic. In fact, the one-parameter family (33) now has the degenerate form

$$x = \frac{(s + \tau)^2}{2} \tag{41a}$$

$$y = s + \tau \tag{41b}$$

$$u = s + \tau \tag{41c}$$

in terms of the *single* parameter $\sigma \equiv s + \tau$, which implies that no unique surface $u(x, y)$ is possible. For example, any solvable algebraic relation $w(u - y) + u^2/2 - x = 0$ defines a solution surface $u(x, y)$ implicitly, where w is an arbitrary function of a single variable with $w(0) = 0$.

Steepening of a Wave

In Section 5.1.2 we showed that (20b) describes the evolution of traffic density for an idealized model. Consider the limiting case $\varepsilon = 0$ (drivers do not react to density gradients), for which we have

$$u_t + uu_x = 0 \tag{42a}$$

and let the initial condition be

$$u(x,0) = \begin{cases} \cos 2\pi x, & \text{if } 0 \leq x \leq 1 \\ 1, & \text{if } x \leq 0 \text{ or } x \geq 1 \end{cases} \tag{42b}$$

Note that (42b) defines a continuously differentiable function on $-\infty < x < \infty$. According to (19), $u = 1$ corresponds to zero density, whereas $u = -1$ has density equal to the maximum value at which there is zero flux. Therefore, the initial condition (42b) represents an isolated initial distribution of traffic over the interval with a peak density ($u = -1$) at $x = 1/2$ and no traffic initially outside the unit interval.

We parametrize (42b) to read

$$x_0 = \tau \tag{43a}$$

$$t_0 = 0 \tag{43b}$$

$$u_0 = \begin{cases} \cos 2\pi\tau, & \text{if } 0 \leq \tau \leq 1 \\ 1, & \text{if } \tau \leq 0 \text{ or } \tau \geq 1 \end{cases} \tag{43c}$$

and obtain the characteristic curves

$$x = u_0(\tau)s + \tau \tag{44a}$$

$$t = s \tag{44b}$$

$$u = u_0(\tau) \tag{44c}$$

The solution for u can be expressed in implicit form as

$$u = u_0[x - u_0 t] \tag{45}$$

where u_0 is the function defined by (43c). The characteristic ground curves are the one-parameter family of straight lines $x = u_0(\tau)t + \tau$, on which u remains a constant equal to its initial value $u_0(\tau)$. The slope dx/dt of each characteristic ground curve equals u_0, and this varies over the unit interval. The resulting pattern is shown in Figure 5.5a, where an envelope is clearly defined for sufficiently large t.

We can calculate the envelope of characteristic ground curves either by setting $\Delta(s,t) = 0$ or by eliminating τ between $F(x,t,\tau) \equiv u_0(\tau)t + \tau - x = 0$ and $F_\tau = 0$. Let us take the second approach: Setting $F_\tau = 0$ gives $\tau = (1/2\pi)\sin^{-1}(1/2\pi t)$. Therefore, the envelope first forms at $t = 1/2\pi$. Setting $\tau = (1/2\pi)\sin^{-1}(1/2\pi t)$ in $F = 0$ gives two branches. One branch has

$$x = x_R(t) \equiv \left(t^2 - \frac{1}{4\pi^2}\right)^{1/2} + \frac{1}{2\pi}\sin^{-1}\left(\frac{1}{2\pi t}\right); \quad \frac{1}{4} \leq x < \infty, t \geq \frac{1}{2\pi} \tag{46a}$$

with $0 < \sin^{-1}(1/2\pi t) \leq \pi/2$, and defines the envelope of the characteristic ground curves emerging from $0 \leq x \leq 1/4$, $t = 0$. The second branch is defined by

$$x = x_L(t) \equiv -\left(t^2 - \frac{1}{4\pi^2}\right)^{1/2} + \frac{1}{2\pi}\sin^{-1}\left(\frac{1}{2\pi t}\right); \quad -\infty < x \leq \frac{1}{4}, t \geq \frac{1}{2\pi} \tag{46b}$$

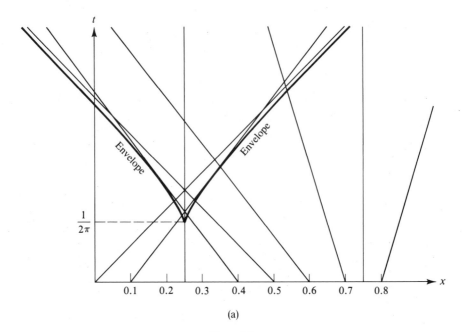

(a)

Figure 5.5a

with $\pi/2 \leq \sin^{-1}(1/2\pi t) < \pi$, and this branch defines the envelope of the characteristic ground curves emerging from $\frac{1}{4} \leq x \leq \frac{1}{2}$, $t = 0$. The second branch is just the reflection of the first branch with respect to the line $x = \frac{1}{4}$. We note that the two branches form a cusp at $x = \frac{1}{4}$, $t = 1/2\pi$; that is, $(dx_R/dt) = (dx_L/dt) = 0$ there. The first branch approaches the characteristic ground curve $\tau = 0$—that is, $x = t$ as $t \to \infty$—whereas the second branch approaches the characteristic ground curve $\tau = \frac{1}{2}$—that is, $x = \frac{1}{2} - t$ as $t \to \infty$. The domain below the two envelope curves has u well defined, whereas above these curves u is three-valued.

The solution (45) for u as a function of x is sketched in Figure 5.5b for the times $t = 0$, 0.1, $1/2\pi$, 0.4. The points A, B, C, D, E on the initial profile locate the values of u at $x = 0$, $\frac{1}{4}$, $\frac{1}{2}$, $\frac{3}{4}$, and 1. These same phases are located by primes for $t = 0.1$, double primes for $t = 1/2\pi$, and triple primes for $t = 0.4$. Since the characteristic ground curves that originate over the interval $0 \leq x \leq \frac{1}{2}$ converge, the initial profile ABC over this subinterval steepens, and at $t = 1/2\pi$, u_x has an infinite slope at B, that is, $x = \frac{1}{4}$. For values of $t > 1/2\pi$, the "solution" is triple-valued in the subinterval $x_L < x < x_R$ spanned by the envelope and is therefore undefined. In the next section, we shall see that a *weak solution*, where a discontinuity in u is allowed, can be constructed for all $t \geq 1/2\pi$. Such a solution will have a stationary discontinuity at the point $x = \frac{1}{4}$, preventing the crossing of the characteristics originating from $x > \frac{1}{4}$ with those originating from $x < \frac{1}{4}$. Thus, the weak solution will consist of the profile to the left of S_- (if $u > 0$) and to the right of S_+ (if $u < 0$) in Figure 5.5b.

The steepening of the profile ABC and the flattening of the profile CDE with time is physically consistent with the traffic behavior we have postulated. In

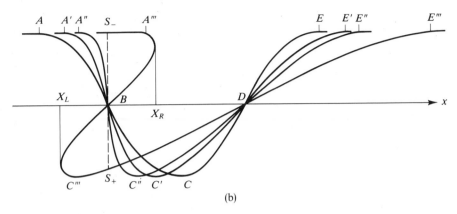

(b)

Figure 5.5b

particular, whenever the density is less than ρ_{max}, there is a flow of vehicles to the right. This causes the cars behind the point of maximum initial density to tend to pile up, so the point at which $\rho = \rho_{max}$ moves to the left. Conversely, the cars initially located in the subinterval $\frac{1}{2} \leq x \leq 1$ gradually spread out over a larger and larger interval, so the initial profile flattens out. We shall comment further on the behavior of the solution for $t > 1/2\pi$ after we have discussed shocks in the next section.

The preceding phenomena are not restricted to (42a) and carry over to the general quasilinear equation (3b) resulting from the conservation law (1). The signal speed c given by (2b) is the characteristic speed at which the initial data are locally propagated. Also, since $d\rho/ds = -\Phi_x$ along a given characteristic, we see that the initial value of ρ remains constant if $\Phi_x = 0$; it increases if $\Phi_x < 0$ and decreases if $\Phi_x > 0$. Now, if $\partial c/\partial \rho > 0$, characteristics locally tend to bunch up (hence waves steepen) if $\rho_x < 0$; they spread out (and waves flatten) if $\rho_x > 0$. The opposite behavior occurs if $\partial c/\partial \rho < 0$.

5.3   Weak Solutions: Shocks, Fans, and Interfaces

In many physical applications, smooth solutions, as postulated in the derivations in the previous sections, are not appropriate; discontinuities may arise either in the initial data or because the coefficients of the governing equations are discontinuous as, for example, at the interface between two different media. We have also seen that even if the differential equation and initial data are smooth, solutions may steepen and "break," and we should look for a description involving discontinuities when this occurs.

It is evident from the derivations of all the integral conservation laws we have studied so far in this book that these remain valid if the dependent variable or its partial derivatives with respect to x and t are discontinuous at some point in the interval $x_1 \leq x \leq x_2$. For example, the law of mass conservation (1a) still holds if ρ is discontinuous. In this section, we shall study how such discontinuities propagate in a solution framework that does not require smoothness.

5.3.1 Shock Speed for a System of Integral Conservation Laws

Here and in the next three subsections, we broaden our scope to *systems* of integral conservation laws because the derivations are not altered significantly. We consider the following general system of n *integral conservation laws* for the n independent variables $u_1, \ldots, u_n$.

$$
\frac{d}{dt} \int_{x_1}^{x_2} \Psi_j(u_i(x,t), x, t)\, dx = \Phi_j(u_i(x_1,t), x_1, t) - \Phi_j(u_i(x_2,t), x_2, t)
$$

$$
+ \int_{x_1}^{x_2} \Lambda_j(u_i(x,t), x, t)\, dx, \quad j = 1, \ldots, n \qquad (47a)
$$

Here, x_1 and x_2 are arbitrary fixed points with $x_1 < x_2$, and the i-subscript indicates the dependence on all the components. Henceforth, we shall also omit stating in every case that $j = 1, \ldots, n$ in our results. We assume that the Ψ_i, Φ_i, Λ_i are well-behaved functions of their arguments. (For the time being, we regard these functions to be continuous and to have continuous first partial derivatives with respect to the u_i, x, t for all $t \geq 0$ and all x: $x_1 \leq x \leq x_2$. This assumption will be relaxed when we consider the behavior of solutions at an interface.)

If the u_i are smooth functions of x and t—that is, they are continuous and have continuous first-partial derivatives with respect to x and t—(47a) implies that they obey the system of *divergence relations*

$$
\frac{\partial \Psi_j}{\partial t} + \frac{\partial \Phi_j}{\partial x} - \Lambda_j = 0 \qquad (47b)
$$

For example, in the problem discussed in Section 5.1.1, we have [see (10)]: $n = 2$, $u_1 = h$, $u_2 = u$, $\Psi_1 = u_1$, $\Psi_2 = u_1 u_2$, $\Phi_1 = u_1 u_2$, $\Phi_2 = u_1 u_2^2 + \frac{1}{2} u_1^2$, $\Lambda_1 = 0$, and $\Lambda_2 = u_1 - (1 + 2\sigma u_1)u_2^2/F^2$. The divergence relations corresponding to (47b) are given by (12), and these in turn simplify to (13).

We have argued that typical integral conservation laws (47a) remain valid—that is, physically consistent—even if the u_i, or $\partial u_i/\partial t$, $\partial u_i/\partial x$ have discontinuities. In fact, it is the system of integral conservation laws (47a) that gives the basic problem description; the system of partial differential equations that results from simplifying the divergence relations (47b) is valid only for smooth solutions. The implications of this point are considered in detail in Section 5.3.3. Here, we wish to explore what restrictions, if any, are implied by (47a) on possible discontinuities in the solution.

To be more specific, assume that a curve Γ_s, defined by $x = \xi(t)$ in the xt-plane, divides the domain of interest D: $x_1 \leq x \leq x_2$, $t \geq 0$, into two subdomains, D_1: $x_1 \leq x < \xi(t)$ and D_2: $\xi(t) < x \leq x_2(t)$, and that the u_i, $\partial u_i/\partial x$, $\partial u_i/\partial t$ are continuous in D_1 and D_2. Thus, Γ_s is a locus of possible discontinuities in the u_i, $\partial u_i/\partial x$, $\partial u_i/\partial t$, and we shall also refer to Γ_s as a *shock* for the case where the u_i are discontinuous. The question is: What information concerning $\xi(t)$ can we draw from (47a)?

To this end we approximate (47a) in the limit as x_1 and x_2 are taken sufficiently close to ξ. We split the interval of integration for the term on the left-hand side of (47a) into the two subintervals (x_1, ξ) and (ξ, x_2). If we then regard $(x_2 - \xi)$ and $(\xi - x_1)$ as being small, we find

$$\int_{x_1}^{x_2} \Psi_j(u_i, x, t)\, dx = \int_{x_1}^{\xi(t)} \Psi_j(u_i, x, t)\, dx + \int_{\xi(t)}^{x_2} \Psi_j(u_i, x, t)\, dx$$
$$= \Psi_j^-(t)(\xi(t) - x_1) + \Psi_j^+(t)(x_2 - \xi(t))$$
$$+ O((\xi - x_1)^2) + O((x_2 - \xi)^2) \tag{48}$$

where we have used the notation

$$\Psi_j^-(t) \equiv \Psi_j\{u_i(\xi^-(t), t), \xi^-(t), t\} \tag{49a}$$
$$\Psi_j^+(t) \equiv \Psi_j\{u_i(\xi^+(t), t), \xi^+(t), t\} \tag{49b}$$

Differentiating (48) with respect to t gives

$$\frac{d}{dt} \int_{x_1}^{x_2} \Psi_j(u_i, x, t)\, dx = -\dot{\xi}(t)[\Psi_j] + O(\xi - x_1) + O(x_2 - \xi) \tag{50}$$

and the [ ] notation means

$$[\Psi_j] \equiv \Psi_j^+(t) - \Psi_j^-(t) \tag{51}$$

The right-hand side of (47a) contributes only $-[\Phi_j]$ if terms of order $(\xi - x_1)$ and $(x_2 - \xi)$ are neglected. Therefore, in the limit as $x_1 \uparrow \xi$ and $x_2 \downarrow \xi$, we have

$$\dot{\xi}(t)[\Psi_j] = [\Phi_j] \tag{52}$$

This condition defines the shock speed $\dot{\xi}$ in terms of the values of the solution on either side. Illustrative examples are discussed later.

For the special case where the Ψ_j and Φ_j are linear in the u_i, let us denote

$$\Psi_j = \sum_{k=1}^{n} \bar{A}_{jk}(x, t)u_k \tag{53a}$$

$$\Phi_j = \sum_{k=1}^{n} \bar{B}_{jk}(x, t)u_k \tag{53b}$$

where the elements of the matrices $\{\bar{A}_{jk}\}$ and $\{\bar{B}_{jk}\}$ are continuously differentiable functions of x and t. Equation (52) implies that

$$[u_j] \sum_{k=1}^{n} \{\dot{\xi}(t)\bar{A}_{jk}(x, t) - \bar{B}_{jk}(x, t)\} = 0 \tag{54}$$

In order that (54) hold, either (1) solutions are continuous—that is, $[u_j] = 0$ for each $j = 1, \ldots, n$—or (2) if the vector with components $[u_1], [u_2], \ldots, [u_n]$ is nonzero, the coefficient matrix in (54) has a zero determinant. This latter case gives precisely the condition that $x = \xi(t)$ is a characteristic curve for the linear system

$$\sum_{k=1}^{n} \left\{ \bar{A}_{jk} \frac{\partial u_k}{\partial t} + \bar{B}_{jk} \frac{\partial u_k}{\partial x} \right\} + \cdots = 0 \tag{55}$$

associated with (47b) for this case [see Problem 11 of Chapter 4].

The preceding discussion confirms our observation, based on the numerical solution of (55) by the method of characteristics in Chapter 4, that discontinuities propagate only along characteristics for the linear problem. Specializing further to the scalar linear equation

$$a(x,t)u_t + b(x,t)u_x = c(x,t)u \tag{56}$$

we see that for $[u] \neq 0$, (54) simply defines the characteristic ground curves as loci of discontinuity in u. Moreover, in the linear case, such discontinuities occur only if the initial data are prescribed to be discontinuous; smooth initial data cannot lead to discontinuous solutions as in the quasilinear problem.

5.3.2 Formal Definition of a Weak Solution

We now present a formal definition of weak solutions of the system of divergence relations (47b).

To begin with, consider only smooth solutions of (47b); let the solution domain in the xt-plane be denoted by D and let its boundary be Γ. Let $\zeta(x,t)$ be any continuously differentiable function of x and t that vanishes on Γ. Now, since (47b) holds *everywhere* in D, we have

$$\iint_D \zeta \left(\frac{\partial \Psi_j}{\partial t} + \frac{\partial \Phi_j}{\partial x} - \Lambda_j \right) dx\, dt = 0 \tag{57a}$$

We can write (57a) as

$$\iint_D \{ (\zeta \Psi_j)_t + (\zeta \Phi_j)_x \}\, dx\, dt = \iint_D (\zeta_t \Psi_j + \zeta_x \Phi_j + \zeta \Lambda_j)\, dx\, dt \tag{57b}$$

and since ζ, Ψ_j, and Φ_j are continuously differentiable functions of x and t, the left-hand side of (57b) can be expressed as a contour integral over Γ, using the two-dimensional Gauss theorem. This contour integral vanishes identically because the integrand has ζ as a factor and $\zeta = 0$ on Γ. Thus, we conclude that for smooth solutions of (47), the following integral [the right-hand side of (57b)] must vanish for all smooth functions ζ such that $\zeta = 0$ on Γ

$$\iint_D (\zeta_t \Psi_j + \zeta_x \Phi_j + \zeta \Lambda_j)\, dx\, dt = 0 \tag{58}$$

Notice that the integrand in (58) does not involve any derivatives of Ψ_j or Φ_j; the only derivatives that occur are for ζ.

We take advantage of this feature to define a *weak solution* of (47) as one that satisfies (58) in D for any smooth ζ that vanishes on Γ. This definition makes sense even if the u_i, $\partial u_i/\partial t$, and $\partial u_i/\partial x$ have finite discontinuities in D because the integral (58) remains well defined. If the u_i, $\partial u_i/\partial t$, and $\partial u_i/\partial x$ are continuous in D, the statement that (58) holds immediately implies (47b). Therefore, we have produced a more general definition of a solution which does not require the

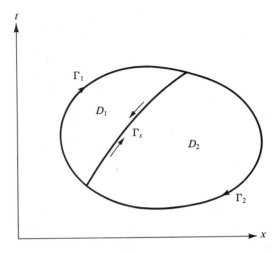

Figure 5.6

smoothness of the u_i. Let us now see whether this definition results in a shock speed formula which is consistent with (52).

As in Section 5.3.1, we consider a domain D which is divided into two parts D_1 and D_2 by a shock curve Γ_s: $x = \xi(t)$. Assume that in each of the domains D_1 and D_2, the solutions of (47b) are smooth, and denote the boundary of D_1 by $\Gamma_1 + \Gamma_s$ and the boundary of D_2 by $\Gamma_2 + \Gamma_s$, as indicated in Figure 5.6.

Since the solution for the u_i is smooth in each subdomain D_1 and D_2, we have [according to (57b) and the two-dimensional Gauss theorem]:

$$\iint_{D_1} (\zeta_t \Psi_j + \zeta_x \Phi_j + \zeta \Lambda_j)\, dx\, dt = \iint_{D_1} \{(\zeta\Psi_j)_t + (\zeta\Phi_j)_x\}\, dx\, dt$$

$$= \int_{\Gamma_s} \{\zeta\Psi_j\, dx - \zeta\Phi_j\, dt\} \tag{59a}$$

$$\iint_{D_2} (\zeta_t \Psi_j + \zeta_x \Phi_j + \zeta \Lambda_j)\, dx\, dt = \iint_{D_2} \{(\zeta\Psi_j)_t + (\zeta\Phi_j)_x\}\, dx\, dt$$

$$= -\int_{\Gamma_s} \{\zeta\Psi_j\, dx - \zeta\Phi_j\, dt\} \tag{59b}$$

where now the integrals over Γ_s, on which $\zeta \neq 0$, remain. Note that in the right-hand side of (59b), we have inserted a minus sign in front of the integral over Γ_s. Therefore, the contour integral in (59b) (before it is multiplied by -1) is evaluated in the same direction along Γ_s as the contour integral in (59a). Adding (59a) and (59b) and using the definition of a weak solution to set equal to zero the sum of the integrals over D_1 and D_2 that results on the left-hand side, we find

$$0 = \int_{\Gamma_s} \zeta\{[\Psi_j]\, dx - [\Phi_j]\, dt\} = 0 \tag{60}$$

where [ ] again denotes the difference in values on either side of the shock curve. Since (60) is true for arbitrary ζ, the integrand must vanish, and we obtain the shock relation (52) derived earlier.

Up until now, we have assumed that the Ψ_j, Φ_j, and Λ_j are smooth functions of their arguments everywhere in D; discontinuities in these functions arose only because the u_i were discontinuous on the shock curve $x = \xi(t)$. This is not true if $x = \xi(t)$ is an *interface* separating the domains D_1 and D_2, where the Ψ_j, Φ_j, and Λ_j may have a different functional dependence on the u_i, x, or t. However, our results remain valid as long as the Ψ_j, Φ_j, and Λ_j are smooth functions of their arguments in the individual domains D_1 and D_2. This allows us to use (52) when $x = \xi(t)$ is such an interface, as is illustrated in the next section.

5.3.3 The Correct Shock and Interface Conditions

We have seen in Sections 5.3.1 and 5.3.2 that the physically relevant integral conservation law (47a) is the starting point for the definition of the shock condition (52). In this section, we shall illustrate the fact that the system of partial differential equations we obtain from (47b) for smooth solutions may be associated with *different divergence relations*. Therefore, without knowledge of the underlying physical principles, if we were given only a system of partial differential equations governing smooth solutions, we would be unable to deduce from these the correct conservation laws or shock conditions.

A Scalar Problem

Consider the scalar problem [see (20)] for traffic flow with $\varepsilon = 0$. The integral conservation law is

$$\frac{d}{dt} \int_{x_1}^{x_2} u(x,t)\,dx = \tfrac{1}{2}\{u^2(x_1,t) - u^2(x_2,t)\} \tag{61a}$$

and this implies the divergence relation

$$u_t + \left(\frac{u^2}{2}\right)_x = 0 \tag{61b}$$

for smooth solutions. This simplifies to

$$u_t + uu_x = 0 \tag{61c}$$

Equations (61a)–(61c) give three levels of description: The most general, valid for discontinuous solutions, is (61a). Equation (61b) is a direct consequence of (61a) for smooth solutions. Equation (61c) is a consequence of (61b). The correct shock condition (52) for (61a) is

$$\dot{\xi}(t)[u] = \left[\frac{u^2}{2}\right]$$

or

$$\dot{\xi}(u^+ - u^-) = \left\{\frac{(u^+)^2}{2} - \frac{(u^-)^2}{2}\right\}$$

or

$$\dot{\xi} = \tfrac{1}{2}(u^+ + u^-) \tag{62}$$

We shall illustrate how this result can be used to derive a weak solution in later sections.

Now, suppose we don't know (61a) and start from (61c). The transformation of dependent variable

$$u(x,t) = e^{v(x,t)} \tag{63}$$

implies that *if u is a smooth solution of* (61b), *then v(x, t) is a smooth solution of*

$$v_t + e^v v_x = 0 \tag{64}$$

for *u* and *v* related by (63). This is as far as we can go. If we were to claim that (64) is a consequence of the divergence relation

$$v_t + (e^v)_x = 0 \tag{65}$$

we would obtain the shock condition

$$\dot{\xi}[v] = [e^v]$$

or

$$\dot{\xi} = \frac{u^+ - u^-}{\log u^+ - \log u^-} \tag{66}$$

which does not agree with (62). In fact, we could also set $u = v^2$ to obtain $v_t + v^2 v_x = 0$ or $v_t + (v^3/3)_x = 0$ and claim

$$\dot{\xi} = \frac{1}{3} \frac{[v^3]}{[v]} = \frac{1}{3} \frac{(u^+)^{3/2} - (u^-)^{3/2}}{(u^+)^{1/2} - (u^-)^{1/2}}$$

and so on. We cannot tell what is the correct shock condition if the only information available is (61c). We need the basic integral conservation law (61a) in order to select the correct shock condition.

Shallow-Water Waves, the Bore Conditions

To illustrate this further, consider the integral conservation laws for shallow-water flow [see (16) and (20) of Chapter 3]. Mass conservation requires

$$\frac{d}{dt} \int_{x_1}^{x_2} h(x,t)\, dx = u(x,t)h(x,t)\Big|_{x=x_2}^{x=x_1} \tag{67a}$$

and momentum conservation requires

$$\frac{d}{dt} \int_{x_1}^{x_2} u(x,t)h(x,t)\, dx = \{u^2(x,t)h(x,t) + \tfrac{1}{2}h^2(x,t)\}\Big|_{x=x_2}^{x=x_1} \tag{67b}$$

The divergence relations associated with (67) are

$$h_t + (uh)_x = 0 \tag{68a}$$

$$(uh)_t + \left(u^2h + \frac{h^2}{2}\right)_x = 0 \tag{68b}$$

Therefore, the correct shock conditions are

$$\dot{\xi}[h] = [uh] \tag{69a}$$

$$\dot{\xi}[uh] = \left[u^2h + \frac{h^2}{2}\right] \tag{69b}$$

In this context a shock is called a *bore*.

Observe that for smooth solutions, (68a)–(68b) are equivalent to [see (22) of Chapter 3]

$$h_t + uh_x + hu_x = 0 \tag{70a}$$

$$u_t + uu_x + h_x = 0 \tag{70b}$$

Now, if we were to interpret (70b) as being the result of the divergence relation

$$u_t + \left(\frac{u^2}{2} + h\right)_x = 0 \tag{71}$$

we would obtain the physically inconsistent second shock condition

$$\dot{\xi}[u] = \left[\frac{u^2}{2} + h\right]$$

instead of (69b).

A second observation regarding (70) is in order. We pointed out earlier [see the remarks following (27) of Chapter 3] that for smooth solutions, the integral conservation law (67b) for momentum and the integral conservation law of energy, given by (26) of Chapter 3, both lead to (70b). If we adopt conservation of momentum as the basic law governing discontinuous solutions, then energy *will not be conserved across a bore*. In fact, we shall show in Section 5.3.4 that the shock relations (69) admit two types of discontinuities characterized by the relative water levels on either side of the discontinuity; these are bores that propagate into regions of either *lower* or *higher* water than found behind the bore. We shall show that in the former case—that is, if the water behind the bore is higher than the water in front—the total energy in some interval $x_1 \le x \le x_2$ containing the bore *will decrease with time*. This behavior is selected as being physically realistic because, in a dissipative model, the turbulence generated in the bore would tend to decrease the energy. The case of a bore propagating into a region of higher water will be excluded because it implies the physically unrealistic result that the energy increases in the interval $x_1 \le x \le x_2$.

Inviscid Non-Heat-Conducting Gas, the Rankine-Hugoniot Conditions, and the Interface Conditions

A somewhat more involved problem occurs in gas dynamics, and we refer to the dimensional integral conservation laws of mass, momentum, and energy [see (55) of Chapter 3 with $\mu = 0$, $\lambda = 0$]:

$$\frac{d}{dt}\int_{x_1}^{x_2}\rho(x,t)\,dx = \rho(x,t)u(x,t)\Big|_{x=x_2}^{x=x_1} \tag{72a}$$

$$\frac{d}{dt}\int_{x_1}^{x_2}\rho(x,t)u(x,t)\,dx = \{\rho(x,t)u^2(x,t)+p(x,t)\}\Big|_{x=x_2}^{x=x_1} \tag{72b}$$

$$\frac{d}{dt}\int_{x_1}^{x_2}\left\{\rho(x,t)\frac{u^2}{2}(x,t)+\frac{p(x,t)}{\gamma-1}\right\}dx$$
$$= \left\{u(x,t)\left(\rho(x,t)\frac{u^2}{2}(x,t)+\frac{\gamma}{\gamma-1}p(x,t)\right)\right\}\Big|_{x=x_2}^{x=x_1} \tag{72c}$$

In (72c), we have used the equation of state [(54a) of Chapter 3] to express the temperature in terms of the pressure p and density ρ. The constant γ is the ratio of specific heats [(62) of Chapter 3].

Now, we have three integral conservation laws for the three dependent variables ρ, u, p. For smooth solutions, (72) imply the divergence relations

$$\rho_t + (u\rho)_x = 0 \tag{73a}$$

$$(\rho u)_t + (\rho u^2 + p)_x = 0 \tag{73b}$$

$$\left\{\rho\frac{u^2}{2}+\frac{p}{\gamma-1}\right\}_t + \left\{u\left(\rho\frac{u^2}{2}+\frac{\gamma}{\gamma-1}p\right)\right\}_x = 0 \tag{73c}$$

These imply the shock conditions

$$\dot\xi[\rho] = [u\rho] \tag{74a}$$

$$\dot\xi[\rho u] = [\rho u^2 + p] \tag{74b}$$

$$\dot\xi\left[\rho\frac{u^2}{2}+\frac{p}{\gamma-1}\right] = \left[u\left(\rho\frac{u^2}{2}+\frac{\gamma}{\gamma-1}p\right)\right] \tag{74c}$$

An alternate way of writing (74) is

$$[\rho v] = 0 \tag{75a}$$

$$[\rho v^2 + p] = 0 \tag{75b}$$

$$\left[\rho\frac{v^2}{2}+\frac{\gamma}{\gamma-1}p\right] = 0 \tag{75c}$$

where $v(t)$ is the flow speed relative to the shock; that is,

$$v^+(t) \equiv u(\xi^+(t),t) - \dot\xi(t) \tag{76a}$$

$$v^-(t) \equiv u(\xi^-(t),t) - \dot\xi(t) \tag{76b}$$

Equations (75) are the *Rankine-Hugoniot* relations.

For smooth solutions, (73a)–(73c) simplify to yield

$$\rho_t + (\rho u)_x = 0 \tag{77a}$$

$$u_t + uu_x + \frac{p_x}{\rho} = 0 \tag{77b}$$

$$\left(\frac{p}{\rho^\gamma}\right)_t + u\left(\frac{p}{\rho^\gamma}\right)_x = 0 \tag{77c}$$

Equation (77c) states that (p/ρ^γ) remains constant along particle paths—that is, on curves $dx/dt = u$. Now, since the entropy is a function of (p/ρ^γ), this implies that the entropy remains constant along particle paths, a result that is physically consistent only if the particle path does not cross a shock.

We can establish how the entropy behaves across a shock from the vantage of the more accurate flow description provided by the conservation equations in which μ and λ are retained [see (55) of Chapter 3]. In this description, *all flows in a gas having constant ambient properties are smooth*; an initial discontinuity immediately evolves into a thin region (shock layer) across which the flow changes rapidly. One can then show that the entropy must *increase* downstream of such a shock layer; for example, see Section 6.15 of [1]. We shall explore this question in more detail for Burgers' equation in Section 5.3.6.

Suppose now that we consider two different gasses on either side of a diaphragm that is suddenly removed. In order to exhibit a distinct interface $\xi(t)$, we make an assumption that is physically somewhat unrealistic for gasses; namely, that the two gasses do not mix across the interface. What are the consequences of the conditions (74) in this context?

If ξ is such an interface, then $u(\xi(t), t)$ must be continuous for all t, and (74a) gives $\dot{\xi}[\rho] = u(\xi, t)[\rho]$—that is, the intuitively obvious result that the interface moves with the speed $\dot{\xi} = u(\xi, t)$ of the gas on either side. Since $u(\xi, t)$ is continuous and equals $\dot{\xi}$, (74b) reduces to $[p] = 0$; that is, the interface cannot sustain a pressure difference. These results, when used in (74c), give an identity. In summary, for two inviscid non-heat-conducting gases at an interface, we must require the speed to be continuous and equal to the interface speed, and we must require the pressure to be continuous. The density (and hence the temperature) and the gas constant γ may be discontinuous.

Interface between Two Different Viscous Heat-Conducting Gases

Consider now the more accurate description where the two gases on either side of the interface are regarded as viscous and heat-conducting. The integral conservation laws of mass momentum and energy are given by (55) of Chapter 3, and these lead to the following interface conditions:

$$\dot{\xi}[\rho] = [\rho u] \tag{78a}$$

$$\dot{\xi}[\rho u] = \left[\rho u^2 + p - \frac{4}{3}\mu u_x\right] \tag{78b}$$

$$\dot{\xi}\left[\rho\left(\frac{u^2}{2} + C_v\theta\right)\right] = \left[\rho u\left(\frac{u^2}{2} + C_v\theta\right) + pu - \frac{4}{3}\mu u u_x - \lambda\theta_x\right] \tag{78c}$$

The pressure p, temperature θ, and density ρ are related according to the equation of state (54a) of Chapter 3. The constants μ, λ, C_p, and C_v are all, in general, different for the two gases.

Again, we must have u continuous at ξ, and (78a) gives the result

$$\dot{\xi} = u(\xi^+, t) = u(\xi^-, t) \tag{79a}$$

that the interface moves with the speed of the gas on either side. When this result is used in (78b), we conclude that

$$[p - \tfrac{4}{3}\mu u_x] = 0 \tag{79b}$$

which gives a balance between the pressure and the viscous stress across the interface. Equations (79a) and (79b) simplify (78c) to the physically obvious result

$$[\lambda \theta_x] = 0 \tag{79c}$$

That is, the heat flux is continuous at the interface.

   To calculate a flow with an interface, we must solve the governing equations on either side; these are (54a), (56a), (57a), and (57b) of Chapter 3. This calculation provides a solution with certain unknown constants, which are determined when the interface conditions (79) are imposed. Problem 5 of Chapter 3 illustrates the ideas for a small disturbance theory for which $|u| \ll 1$; hence the interface is stationary in the first approximation.

5.3.4 Constant Speed Shocks; Nonuniqueness of Weak Solutions

Solutions for which the dependent variables remain constant on either side of a shock are interesting and simple special cases that provide much insight into the behavior of more complicated situations. We see immediately from (52) that if the u_i on either side of the shock are constant and the Ψ_j, Φ_j do not depend on x or t, then the $[\Psi_j]$, $[\Phi_j]$ and hence $\dot{\xi}$ are also all constant. Such special solutions occur either if the initial data are constant on either side of a point on the x-axis or if they exhibit an appropriate symmetry.

The Scalar Problem: Shocks and Fans

Both shocks and fans occur for the scalar example (61), which we consider next.

   First, assume that the initial condition is the piecewise constant function

$$u(x, 0) = \begin{cases} u_1 = \text{constant}, & \text{if } x < x_0 \\ u_2 = \text{constant} < u_1 & \text{if } x > x_0 \end{cases} \tag{80}$$

Thus, the characteristics emerging from the $x < x_0$ portion of the x-axis all have the same speed u_1, which is *greater* than the speed u_2 of the characteristics emerging from $x > x_0$. These two families of characteristics immediately intersect, and the solution is not defined in the triangular region $x_0 + u_2 t < x < x_0 + u_1 t, t > 0$. This is shown in Figure 5.7a, where the arrows indicate the direction of increasing t. For $x > x_0 + u_1 t$, we have $u = u_2 = \text{constant}$, whereas for $x < x_0 + u_2 t, u = u_1 = \text{constant}$.

   It is clear that a shock must be introduced at the point $x = x_0, t = 0$, and the initial speed of this shock follows immediately from the shock condition (62). We have $\dot{\xi}(0^+) = \tfrac{1}{2}(u_1 + u_2)$. But, for this problem, the shock propagates into a region where the values of u on either side of the shock remain constant, and we conclude that $\dot{\xi}(t) = \text{constant} = (u_1 + u_2)/2$; that is, $\xi(t) = x_0 + (u_1 + u_2)t/2$. Thus, the shock has a speed equal to the average of the speeds of the characteristics on either side, and the characteristics converge (as t increases) toward the

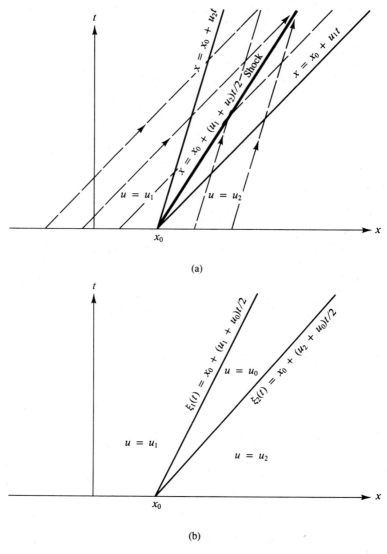

Figure 5.7

shock. In the limit of a weak discontinuity—that is, $u_1 \approx u_2$—we see that $\dot{\xi} \to u$, the characteristic speed.

We also observe that for this initial-value problem, it is not possible to have more than one shock emerging from $(x_0, 0)$. For instance, let us assume the situation depicted in Figure 5.7b, where the two shocks with constant speeds, $\dot{\xi}_1 = (u_1 + u_0)/2$ and $\dot{\xi}_2 = (u_2 + u_0)/2$, bound the triangular domain $\xi_1 < x < \xi_2$, in which $u = u_0 = $ constant. In order to have $\dot{\xi}_2 > \dot{\xi}_1$, we must have $u_2 > u_1$, but this contradicts the original premise that $u_2 < u_1$. It is easily seen that no choice of u_0 leads to a consistent picture. So we conclude that the weak solution

$$u(x,t) = \begin{cases} u_1 = \text{constant}, & \text{if } x < x_0 + \dfrac{(u_1 + u_0)t}{2} \\[4mm] u_2 = \text{constant} < u_1 & \text{if } x > x_0 + \dfrac{(u_1 + u_2)t}{2} \end{cases} \tag{81}$$

is unique if $u_2 < u_1$.

Now suppose that we reverse the inequality relating u_1 and u_2 and consider the initial-value problem

$$u(x,0) = \begin{cases} u_1 = \text{constant}, & \text{if } x < x_0 \\ u_2 = \text{constant} > u_1, & \text{if } x > x_0 \end{cases} \tag{82}$$

The picture in the xt-plane is shown in Figure 5.8a, where no characteristics enter the triangular region

$$T: \quad x_0 + u_1 t < x < x_0 + u_2 t, \qquad t > 0$$

In this case, there are *infinitely* many possible weak solutions that satisfy the integral conservation law (61a) and jump condition (62). We can insert N shocks in T, where $N = 1, 2, \ldots$, without violating (62). For example, with $N = 1$, we have the situation sketched in Figure 5.8b, where

$$u(x,t) = \begin{cases} u_1 = \text{constant}, & \text{if } x < x_0 + \dfrac{(u_1 + u_2)t}{2} \\[4mm] u_2 = \text{constant} < u_1, & \text{if } x > x_0 + \dfrac{(u_1 + u_2)t}{2} \end{cases} \tag{83a}$$

This is formally identical to (79), except now $u_2 > u_1$. As a result, the characteristics diverge from the shock as t increases. In Figure 5.8c, we show the case of $N = 2$. We have the two shocks with speeds $\dot{\xi}_1 = (u_1 + u_0)/2$ and $\dot{\xi}_2 = (u_2 + u_0)/2$ dictated by the *arbitrary* choice of u_0 in the interval $u_1 < u_0 < u_2$. The solution is

$$u(x,t) = \begin{cases} u_1 = \text{constant}, & \text{if } x < x_0 + \dfrac{(u_1 + u_0)t}{2} \\[4mm] u_0 = \text{constant}, & \text{if } x_0 + \dfrac{(u_1 + u_0)t}{2} < x < x_0 + \dfrac{(u_2 + u_0)t}{2} \\[4mm] u_2 = \text{constant}, & \text{if } x > x_0 + \dfrac{(u_2 + u_0)t}{2} \end{cases} \tag{83b}$$

which is perfectly consistent with (62). We can continue this process for any N.

We conclude that for this case the integral conservation law *does not specify a unique solution*; we must invoke some further information to decide what is an *admissible weak solution*.

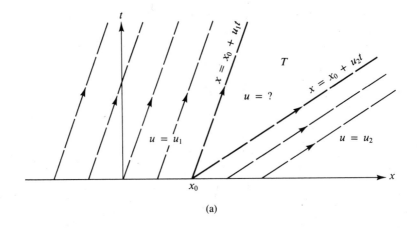

(a)

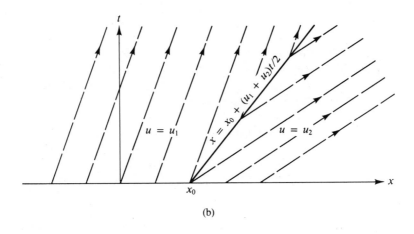

(b)

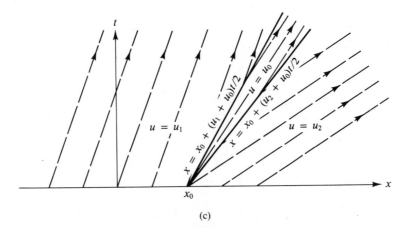

(c)

Figure 5.8

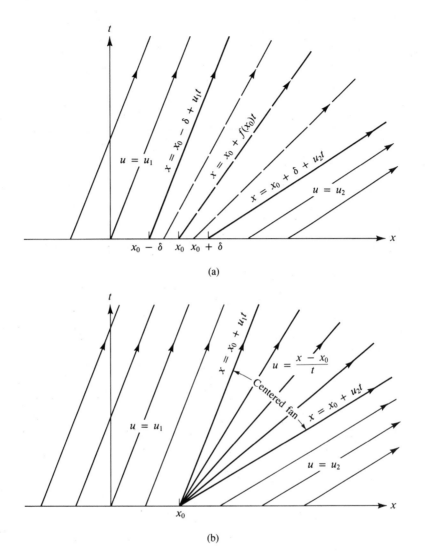

Figure 5.9

One point of view is to regard the initial condition (80) as the limiting case as $\delta \to 0$ of the *smooth initial condition*

$$u(x,0) = \begin{cases} u_1 = \text{constant}, & \text{if } x \leq x_0 - \delta \\ f(x), & \text{if } x_0 - \delta \leq x \leq x_0 + \delta \\ u_2 = \text{constant} > u_1, & \text{if } x \geq x_0 + \delta \end{cases} \tag{84}$$

for some monotone increasing function $f(x)$ with $f(x_0 - \delta) = u_1, f'(x_0 - \delta) = 0$, $f(x_0 + \delta) = u_2$, and $f'(x_0 + \delta) = 0$. The characteristics now fan out of the interval $x_0 - \delta \leq x \leq x_0 + \delta$, as shown in Figure 5.9a. The solution is smooth everywhere and has the parametric form

$$x = \tau + f(\tau)t; \qquad u = f(\tau)$$

for $x_0 - \delta \le \tau \le x_0 + \delta$. For $x \le x_0 - \delta + u_1 t$, $u(x,t) = u_1$, and for $x \ge x_0 + \delta + u_2 t$, $u(x,t) = u_2$. In the limit as $\delta \to 0$, we obtain the *centered fan* shown in Figure 5.9b, where now $u(x,t) = (x - x_0)/t$ in T. Therefore, we have obtained a weak solution for the initial-value problem (82) in the form

$$u(x,t) = \begin{cases} u_1 = \text{constant}, & \text{if } x \le x_0 + u_1 t \\[2mm] \dfrac{x - x_0}{t}, & \text{if } x_0 + u_1 t \le x \le x_0 + u_2 t \\[2mm] u_2 = \text{constant}, & \text{if } x > x_0 + u_2 t \end{cases} \tag{85}$$

Note that u is continuous along the rays $t > 0$, $x = x_0 + u_1 t$, and $x = x_0 + u_2 t$, but u_x and u_t are not.

The solution (85) may be regarded as the limiting case as $N \to \infty$ of the sequence of solutions (83) with N finite discontinuities. Thus, each member of the one-parameter family of rays emerging from x_0 may be regarded as an infinitesimally weak discontinuity.

Since (61c) has straight characteristics on which u is constant, we conclude that a centered fan is appropriate at $x = x_0$, $t = 0$ whenever we have discontinuous initial data such that $u(x_0^+, 0) > u(x_0^-, 0)$; the initial data need not be piecewise constant. It is interesting to note that a centered fan is also a similarity solution of (61c) in T with boundary conditions $u = u_1$ on $x = x_0 + u_1 t$ and $u = u_2$ on $x = x_0 + u_2 t$.

An alternative view of the solution of (61c), subject to any initial condition, is to regard it as the limiting case as $\varepsilon \to 0$ of the corresponding initial-value problem for Burgers' equation (20b). This quasilinear second-order equation is parabolic and has smooth solutions for $t > 0$ even if the initial data are discontinuous. In Section 5.3.6 we shall construct the exact solution for the two initial conditions (80) and (82) and show that as $\varepsilon \to 0$, these solutions indeed tend to the weak solutions, (81) and (85), respectively, that we have derived.

A concise condition, which excludes solutions such as (83a) and (83b), is to require

$$u^+ \le \dot{\xi} \le u^- \tag{86}$$

for admissible weak solutions. Equation (86) is referred to as an *entropy condition* because in the context of gas dynamics, the corresponding condition excludes discontinuities across which the entropy does not increase. One can prove (see [3]) that condition (86) is sufficient to isolate a unique weak solution of (61) in all cases. For the more general scalar divergence relation

$$u_t + \{\phi(u)\}_x = 0 \tag{87}$$

the corresponding entropy condition is

$$\phi'(u^+) \le \dot{\xi} \le \phi'(u^-) \tag{88}$$

Finally, we note that shocks are not associated only with discontinuous initial data; the problem discussed in Section 5.2.2 gives an example of smooth initial data for which the solution breaks down at some subsequent time. In this

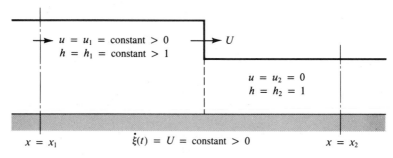

Figure 5.10

example, we derive a weak solution for $t \geq 1/2\pi$ by inserting a stationary shock at $x = \frac{1}{4}$, $t \geq 1/2\pi$. Because of the symmetry of the initial data relative to the point $x = \frac{1}{4}$, the values of u on either side of the line $x = \frac{1}{4}$, $t \geq 1/2\pi$ are equal in magnitude but differ in sign. Therefore, the shock condition (62) remains valid for all $t \geq 1/2\pi$ for this stationary shock. The weak solution for $t \geq 1/2\pi$ is the waveform to the left of S^- and to the right of S^+, with a stationary discontinuity at the point B, as shown in Figure 5.5b.

The Uniformly Propagating Bore

Because $u =$ constant and $h =$ constant solve the shallow-water equations (70), it is natural to ask whether there exist weak solutions of these equations that correspond to a bore propagating with uniform speed. This would require u and h to have different constant values ahead of and behind the bore. Using an appropriate coordinate frame (Galilean transformation) and dimensionless variables, we can regard the water ahead of the bore to be at rest and of unit depth, as shown in Figure 5.10.

Such a bore is an idealization of the steady flow produced by an incoming high tide into an estuary. The coordinate system moves with the speed of the estuary outflow. In the laboratory, we could generate this flow by impulsively setting a wavemaker into motion with constant speed u_1 relative to water at rest (see Figure 3.7). In the first case, the height h_1 of the incoming tide is prescribed, whereas in the laboratory model, the speed u_1 corresponds to the wavemaker speed and is prescribed.

Inserting the values $u^+ = 0$, $h^+ = 1$, $u^- = u_1$, and $h^- = h_1$ into the bore conditions (69) and denoting the bore speed by $U =$ constant gives

$$U = \frac{u_1 h_1}{h_1 - 1} \tag{89a}$$

$$U = \frac{u_1^2 h_1 + h_1^2/2 - \frac{1}{2}}{u_1 h_1} \tag{89b}$$

Thus, we have two equations for the two unknowns (U, h_1) if u_1 is specified or the unknowns (U, u_1) if h_1 is specified. If the unknowns are U and h_1, one obtains a cubic for h_1 after eliminating U from (89a)–(89b) (see Problem 9a). Here, we consider the case where (U, u_1) are the unknowns, and we can solve

(89a)–(89b) for these as explicit functions of h_1 in the form

$$u_1 = \pm(h_1 - 1)\left(\frac{h_1 + 1}{2h_1}\right)^{1/2} \tag{90a}$$

$$U = \pm\left[\frac{h_1(h_1 + 1)}{2}\right]^{1/2} \tag{90b}$$

Given h_1, there are two possible bores that satisfy the shock relations (89). The upper sign corresponds to bores propagating to the right, and the speed of the water behind these bores is positive or negative depending on the sign of $h_1 - 1$. The reverse is true for the lower sign. Of course, we must keep in mind that we are viewing the problem from a coordinate frame with respect to which the water ahead of the bore is at rest.

To decide which of the two possible solutions (90) is physically realistic, let us calculate $\dot{E}(t)$, the time rate of change of energy less the net influx of energy and work done in some fixed interval $x_1 \leq x \leq x_2$ containing a uniformly propagating bore. If no energy is added or dissipated in this interval, $\dot{E}(t) = 0$.

Referring to the expression (26) of Chapter 3, we have

$$2\dot{E}(t) = \frac{d}{dt}\int_{x_1}^{x_2} (u^2 h + h^2)\,dx + \{u^3 h + 2uh^2\}_{x=x_1}^{x=x_2} \tag{91a}$$

Using (50) with $\dot{\xi} = U = \text{constant}$, we find

$$2\dot{E}(t) = -U[u^2 h + h^2] + [u^3 h + 2uh^2] \tag{91b}$$

or

$$2\dot{E}(t) = -U(1 - u_1^2 h_1 - h_1^2) - u_1^3 h_1 - 2u_1 h_1^2 \tag{91c}$$

If we now insert the solution (90) for u_1 and U in (91c) and simplify the result, we find, after some algebra, that

$$2\dot{E} = \mp\left(\frac{h_1 - 1}{2}\right)\left(\frac{h_1 + 1}{2h_1}\right)^{1/2}(5h_1^2 + 2h_1 + 1) \tag{92}$$

where the upper and lower signs in (92) correspond to the upper and lower signs in (90).

For a physically realistic bore that dissipates energy, we must have $\dot{E} < 0$; therefore, we must pick the upper sign if $h_1 > 1$ and the lower sign if $h_1 < 1$. Figure 5.10 corresponds to the case $h_1 > 1$. For the second alternative, $h_1 < 1$, $u_1 > 0$ and $U < 0$. It is easily seen that a Galilean transformation that results in $u_1 = 0$ and a reversing of the flow direction, merely reproduces Figure 5.10. Therefore, for flows to the right into quiescent water, Figure 5.10 is the only physically realistic bore. It has $h_1 > 1$ and $U > u_1 > 0$. In general, an admissible bore must propagate into water of lower height. The mathematically allowable solution shown in Figure 5.11 for the choice $h_1 > 1$ and the lower signs in (90) results in a physically unrealistic flow for which $\dot{E} > 0$.

It is also useful to contrast the behavior of solutions resulting from piecewise constant initial data for the scalar problem $u_t + uu_x = 0$ and the present two-

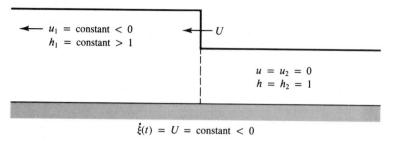

Figure 5.11

component model. In the scalar case, *any* piecewise constant initial condition of the form (80) results in a uniformly propagating shock as long as $u_1 > u_2$. In contrast, for the problem governed by (70), the initial condition

$$h(x,0) = \begin{cases} 1, & \text{if } x > x_0 \\ h_1 > 1, & \text{if } x < x_0 \end{cases}; \qquad u(x,0) = \begin{cases} 0, & \text{if } x > x_0 \\ u_1 > 0, & \text{if } x < x_0 \end{cases}$$

gives a uniformly propagating shock only if u_1 and h_1 are related by (90a); for other choices of u_1 and h_1, the solution will also involve a centered fan; this is discussed in Chapter 7.

We also confirm that for weak bores, (90a)–(90b) reduce to the results calculated in Chapter 3 for the linearized theory. If the water level and speed behind the bore do not differ much from those of the quiescent state in front, we have $h_1 = 1 + \varepsilon \tilde{h}_1$ and $u_1 = \varepsilon \tilde{u}_1$, where ε is a small parameter and $\tilde{h}_1$ and $\tilde{u}_1$ are $O(1)$ constants. Substituting this expression for h_1 into (90b) shows that the bore speed to $O(1)$ is just the unit characteristic speed. Equation (90a) gives $\tilde{u}_1 = \pm \tilde{h}_1$. For the linear theory, both solutions are possible, and in fact these uniform solutions were essentially calculated in Chapter 3 (see Figure 3.17).

The Uniformly Propagating Shock in Gas Dynamics

Here, the system of conservation laws (72) governs the evolution of the three variables ρ, u, p, and we wish to study the problem of a shock propagating into a quiescent gas (density $\rho_2 \equiv \rho_0 = $ constant, $u_2 = 0$, $p_2 \equiv p_0 = $ constant). It is convenient to adopt the dimensionless variables used in (59)–(60) of Chapter 3, where pressures are normalized using p_0, densities using ρ_0, and speeds using the ambient speed of sound $a_0 \equiv \sqrt{\gamma p_0/\rho_0}$. We then normalize the time by T_0, some characteristic time, and normalize distances by $a_0 T_0$. The dimensionless form of (73) is then

$$\rho_t + (u\rho)_x = 0; \qquad \text{mass} \tag{93a}$$

$$(\rho u)_t + \left(\rho u^2 + \frac{p}{\gamma} \right)_x = 0; \qquad \text{momentum} \tag{93b}$$

$$\left(\frac{\rho u^2}{2} + \frac{p}{\gamma(\gamma - 1)} \right)_t + \left(\frac{\rho u^3}{2} + \frac{pu}{\gamma - 1} \right)_x = 0; \qquad \text{energy} \tag{93c}$$

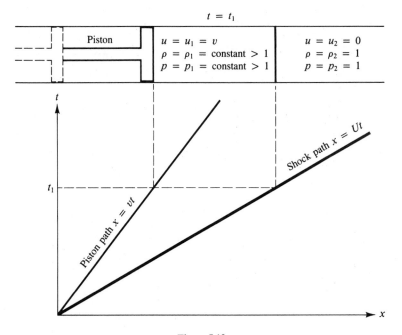

$t = t_1$

Piston	$u = u_1 = v$	$u = u_2 = 0$
	$\rho = \rho_1 = \text{constant} > 1$	$\rho = \rho_2 = 1$
	$p = p_1 = \text{constant} > 1$	$p = p_2 = 1$

Figure 5.12

where, for simplicity, we use the same notation as in (73) for the dimensionless variables. The extra factor $1/\gamma$ multiplying the dimensionless p in (93b)–(93c) is a direct result of our choice of a_0 rather than $\sqrt{p_0/\rho_0}$ for a velocity scale.

Let us again focus our attention on the problem of a piston that is pushed impulsively with constant speed v into a gas at rest. The picture is analogous to the one discussed in Problem 9 for a wavemaker being pushed into quiescent water, except that we now have three unknowns instead of two. These are the density and pressure behind the shock and the shock speed. Again, we assume that the gas speed behind the shock equals the piston speed v and is known. Thus, we have $\rho_2 = 1, u_2 = 0, p_2 = 1$, and $u_1 = v =$ prescribed, and we wish to calculate ρ_1, p_1, and the shock speed U (see Fig. 5.12). The more general problem of a shock propagating into a gas moving with constant speed u_2 can be obtained by replacing $u_1 \to u_1 - u_2$, $U \to U - u_2$ in our results.

If we set $\rho^+ = 1, \rho^- = \rho_1, u^+ = 0, u^- = v = u_1, p^+ = 1, p^- = p_1$, and $\dot{\xi} = U$ in the shock conditions associated with (93), we obtain

$$U(\rho_1 - 1) = u_1 \rho_1 \tag{94a}$$

$$U\rho_1 u_1 = \rho_1 u_1^2 + \frac{1}{\gamma}(p_1 - 1) \tag{94b}$$

$$U\left(\frac{\rho_1 u_1^2}{2} + \frac{p_1}{\gamma(\gamma - 1)} - \frac{1}{\gamma(\gamma - 1)}\right) = \frac{\rho_1 u_1^3}{2} + \frac{p_1 u_1}{\gamma - 1} \tag{94c}$$

We can solve (94a) for ρ_1 to obtain

$$\rho_1 = \frac{U}{U - u_1} \tag{95a}$$

Solving (94b) for p_1 and using (95a) gives

$$p_1 = 1 + \gamma u_1 U \tag{95b}$$

Finally, using (95a) and (95b) to eliminate ρ_1 and p_1 from (94c) gives

$$U^2 - \frac{(\gamma + 1)}{2} U u_1 - 1 = 0 \tag{95c}$$

Since u_1 is known, we can solve this quadratic for U and find the two roots

$$U^{(1)} \equiv \frac{\gamma + 1}{4} u_1 + \left[\left(\frac{\gamma + 1}{4} u_1 \right)^2 + 1 \right]^{1/2} \tag{95d}$$

$$U^{(2)} \equiv \frac{\gamma + 1}{4} u_1 - \left[\left(\frac{\gamma + 1}{4} u_1 \right)^2 + 1 \right]^{1/2} \tag{95e}$$

Each of these roots, when substituted into (95a) and (95b), gives a solution for ρ_1 and a solution for p_1. The following table lists the intervals over which U, ρ_1, and p_1 range as u_1 varies over all possible values.

	u_1	U	ρ_1	p_1
1	$0 < u_1 < \infty$	$1 < U^{(1)} < \infty$	$1 < \rho_1 < \frac{\gamma + 1}{\gamma - 1}$	$1 < p_1 < \infty$
2	$0 < u_1 < \infty$	$-1 < U^{(2)} < 0$	$1 > \rho_1 > 0$	$1 > p_1 > -\frac{\gamma - 1}{\gamma + 1}$
3	$-\infty < u_1 < 0$	$0 < U^{(1)} < 1$	$0 < \rho_1 < 1$	$-\frac{\gamma - 1}{\gamma + 1} < p_1 < 1$
4	$-\infty < u_1 < 0$	$-\infty < U^{(2)} < -1$	$\frac{\gamma + 1}{\gamma - 1} > \rho_1 > 1$	$\infty > p_1 > 1$

It is easily seen that if we change the sign of u_1 in case 4, we simply recover case 1, and if we change the sign of u_1 in case 3, we recover case 2. Therefore, cases 3 and 4 are equivalent to cases 2 and 1, respectively, for flow to the left. Our task is now to decide which of cases 1 or 2 corresponds to a physically consistent shock. The answer hinges on how the entropy behaves across the shock in each case.

We can show (See Problem 10) by direct computation that in case 1 the entropy behind the shock is higher than its upstream value. The reverse is true for case 2, where the entropy decreases downstream of the shock. As it is physically inconsistent to have the entropy decrease, we discard case 2. Thus, shocks behind which the density and pressure drop are ruled out. Actually, case

2 corresponds to an impulsive piston motion with constant speed $u_1 < 0$ *out of* a gas at rest. We shall discuss the solution of this problem in Chapter 7 when we study simple waves.

Therefore, for a shock propagating to the right into a gas at rest, there is only one physically consistent solution given by case 1. The values of U, ρ_1, and p_1 are given by

$$U = \frac{\gamma + 1}{4} u_1 + \left[\left(\frac{\gamma + 1}{4} u_1 \right)^2 + 1 \right]^{1/2} \rightarrow \begin{cases} 1, & \text{as } u_1 \rightarrow 0 \\ \dfrac{\gamma + 1}{2} u_1, & \text{as } u_1 \rightarrow \infty \end{cases} \tag{96a}$$

$$\rho_1 = \frac{4 + 4u_1 \left[\left(\dfrac{\gamma + 1}{4} u_1 \right)^2 + 1 \right]^{1/2} + (\gamma + 1)u_1^2}{4 + 2(\gamma - 1)u_1^2} \rightarrow \begin{cases} 1, & \text{as } u_1 \rightarrow 0 \\ \dfrac{\gamma + 1}{\gamma - 1}, & \text{as } u_1 \rightarrow \infty \end{cases} \tag{96b}$$

$$p_1 = 1 + \frac{\gamma(\gamma + 1)}{4} u_1^2 + \gamma u_1 \left[\left(\frac{\gamma + 1}{4} u_1 \right)^2 + 1 \right]^{1/2} \rightarrow \begin{cases} 1, & \text{as } u_1 \rightarrow 0 \\ \dfrac{\gamma(\gamma + 1)}{2} u_1, & \text{as } u_1 \rightarrow \infty \end{cases} \tag{96c}$$

These formulas generalize to the case of a shock propagating into a gas that is moving with constant speed u_2 by replacing u_1 everywhere with $u_1 - u_2$ and replacing U with $U - u_2$. It is interesting to note that for an infinitely strong shock, that is, $u_1 \rightarrow \infty$, the density ratio across the shock tends to a finite value, $\rho_1 \rightarrow (\gamma + 1)/(\gamma - 1)$, whereas the pressure ratio tends to infinity: $p_1 \rightarrow [\gamma(\gamma + 1)/2]u_1 \rightarrow \infty$ as $u_1 \rightarrow \infty$.

The properties of weak shocks—that is, $u_1 \ll 1$—are also interesting and will be important in later discussions. We expand each of (96) in powers of u_1 and retain terms up to $O(u_1^3)$ to find, after some algebra, that

$$U = 1 + \frac{\gamma + 1}{4} u_1 + \frac{(\gamma + 1)^2}{32} u_1^2 + O(u_1^4) \tag{97a}$$

$$\rho = 1 + u_1 + \frac{3 - \gamma}{4} u_1^2 + \frac{\gamma^2 - 14\gamma + 17}{32} u_1^3 + O(u_1^4) \tag{97b}$$

$$p = 1 + \gamma u_1 + \frac{\gamma(\gamma + 1)}{4} u_1^2 + \frac{\gamma(\gamma + 1)^2}{32} u_1^3 + O(u_1^4) \tag{97c}$$

Note that U tends to the characteristic speed $U \rightarrow 1$ as $u_1 \rightarrow 0$ and that $\rho - 1$ and $p - 1$ are both of order u_1. However, the change in entropy across a weak shock is extremely small, of order u_1^3 to be precise. To show this, we introduce the dimensionless entropy change s across the shock by (see Problem 10)

$$s \equiv \frac{S_1 - S_2}{C_v} = \log \frac{p_1}{\rho_1^\gamma}$$

and compute the following value for s using (97b)–(97c):

$$s = \frac{\gamma(\gamma - 1)}{12} u_1^3 + O(u_1^4) \tag{97d}$$

For example, consider a moderate shock where $u_1 = 0.3$. We find $p - 1 = 0.50235$ and $\rho - 1 = 0.33749$, but $s = 0.00275$, a very small number indeed. This fact provides a significant simplification in the analysis of problems with weak shocks. In particular, in any perturbation problem where a weak shock propagates into a domain of constant entropy, we can use the isentropic flow equations for computing the flow up to second order in the disturbances. This question is discussed in more detail in Chapter 7.

Finally, let us reexamine the analogy between shallow-water flow and the flow of a compressible gas. As pointed out in Section 3.3.4, there is an *exact* analogy between *smooth flows* for these two problems if $\gamma = 2$ and if we identify u in both cases and h (or $\sqrt{h}$) with ρ (or a). This analogy *does not carry over exactly* to discontinuous solutions because of the fact that p/ρ^γ does not remain constant across a shock. To see this, compare (93a)–(93b), in which $\gamma = 2$, $\rho = h$, with (68a)–(68b). These equations correspond if, in addition, we set $p = h^2$; that is, $p = \rho^\gamma$ with $\gamma = 2$. But, although $p/\rho^\gamma = 1$ in front of the shock, we have just shown that this quantity increases across the shock. Therefore, the analogy between the two discontinuous flows is only qualitative.

It is easy to exhibit the numerical extent of this discrepancy between the shock and bore speeds and the flow speeds behind these discontinuities for various values of ρ (or h). In the following table, we give the values of u_1 and U (behind the bore) predicted by (90) (with the upper signs) for four values of h_1.

h_1	u_1	U
1	0	1
2	0.866	1.732
3	1.633	2.449
4	2.372	3.162

We now use the values of u_1 obtained from the preceding table in (96a)–(96b) with $\gamma = 2$ to compute U and ρ_1:

u_1	U	ρ_1
0	1	1
0.866	1.842	1.887
1.633	2.806	2.393
2.372	3.820	2.639

The discrepancies between the bore and shock speeds on the one hand and h_1 and ρ_1 on the other increase as u_1 increases, as expected. In fact, $\rho_1 \to 3$ while

$h_1 \rightarrow \infty$ as $u_1 \rightarrow \infty$. Thus, even for moderate shocks, the hydraulic analogy gives only a qualitative description. Moreover, $\gamma = \frac{7}{5}$ for a diatomic gas such as air, and the requirement $\gamma = 2$ introduces further discrepancies in this case.

5.3.5 An Example of Shock Fitting for the Scalar Problem

In all the examples discussed in Section 5.3.4, the dependent variables are constant on either side of the shock. This is why the shock moves with constant speed. In this section, we give an example that illustrates the basic idea of how to *fit* a curved shock into the xt-plane in order to prevent characteristics having varying values of u from crossing.

Consider the initial-value problem

$$u(x,0) = \begin{cases} 1, & \text{if } |x| > 1 \\ -1 + |x|, & \text{if } |x| < 1 \end{cases} \tag{98}$$

for the scalar problem (61). If we interpret u as the dimensionless traffic density defined in (19), (98) corresponds to a discrete, piecewise-linear initial distribution of cars over the interval $-1 \leq x \leq 1$ with a maximum $(u = -1)$ at $x = 0$. Outside this interval there are no cars $(u = 1)$.

We parametrize the initial curve $(t = 0)$ to be

$$x_0(\tau) = \tau \tag{99a}$$

$$t_0(\tau) = 0 \tag{99b}$$

$$u_0(\tau) = \begin{cases} 1, & \text{if } |\tau| > 1 \\ -1 + |\tau|, & \text{if } |\tau| < 1 \end{cases} \tag{99c}$$

The characteristic equations

$$\frac{dt}{ds} = 1 \tag{100a}$$

$$\frac{dx}{ds} = u \tag{100b}$$

$$\frac{du}{ds} = 0 \tag{100c}$$

are now solved subject to (99), and we find

$$t = s \tag{101a}$$

$$x = u_0(\tau)s + \tau \tag{101b}$$

$$u = u_0(\tau) \tag{101c}$$

or

$$x = u_0(\tau)t + \tau \tag{102a}$$

$$u = u_0(\tau) \tag{102b}$$

Let us examine the patterns of characteristic ground curves and associated

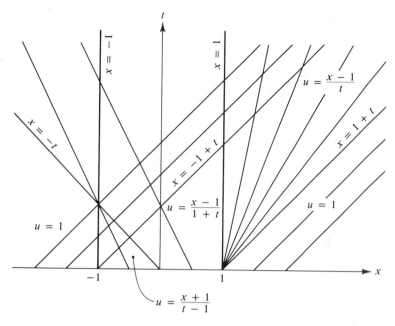

Figure 5.13

values of u that emerge from all portions of the x-axis according to (102). Starting with $x < -1$, we have

$$u = 1 \tag{103a}$$

on the family of straight lines (See Fig. 5.13).

$$x = t + \tau, \qquad -\infty < \tau < -1 \tag{103b}$$

The next segment of the x-axis has

$$u = -(1 + \tau) \tag{104a}$$

$$x = -(1 + \tau)t + \tau \tag{104b}$$

for $-1 < \tau < 0$; that is,

$$u = \frac{x + 1}{t - 1} \tag{105}$$

when we solve (104b) for τ and use the result in (104a).

Members of the family (103b) intersect with members of (104b) for $t > 0$ starting at $x = -1$. Thus, in the domain covered by both families, the solution is ambiguous. We prevent the crossing of characteristics by inserting a shock that starts at the point A: $t = 0$, $x = -1$ (see Figure 5.14).

The shock speed obeys (62) with $u_1 = 1$ and $u_2 = (x + 1)/(t - 1)$; that is,

$$\frac{d\xi}{dt} = \frac{1}{2}\left(1 + \frac{\xi + 1}{t - 1}\right) \tag{106a}$$

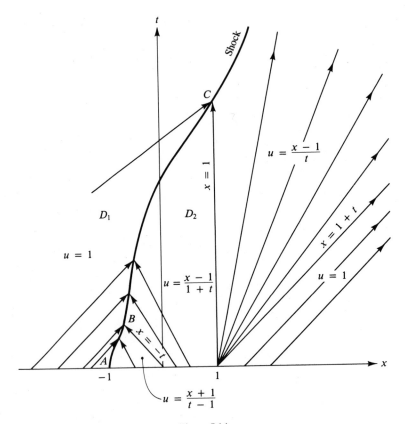

Figure 5.14

Simplifying (106a), we obtain

$$\frac{d\xi}{dt} + \frac{1}{2(1-t)}\xi = -\frac{t}{2(1-t)} \tag{106b}$$

This linear equation, subject to the initial condition $\xi(0) = -1$, can be easily solved to give

$$\xi = -2 + t + \sqrt{1-t}; \qquad 0 \le t \le \tfrac{3}{4} \tag{107}$$

As noted, the shock (107) is appropriate only as long as it continues to have $u_1 = 1$ to the left and $u_2 = (x+1)/(t-1)$ to the right. The solution for u ceases to be given by (105) at the point B, where the shock crosses the characteristic $x = -t$ emerging from the origin. Setting $\xi = -t$ in (107) gives $t = \tfrac{3}{4}$. Therefore, the shock (107) is valid up to the point B: $x = -\tfrac{3}{4}, t = \tfrac{3}{4}$. This point occurs before the singular point $x = -1, t = 1$ shown in Figure 5.13, where all the lines (104b) meet, so that this singular behavior is not present in the weak solution as seen in Figure 5.14.

To continue the shock beyond B, we must use the solution u_2 associated with the characteristics that emerge from $0 < x < 1, t = 0$. This solution is given by:

$$u = \tau - 1 \tag{108a}$$

$$x = (\tau - 1)t + \tau \tag{108b}$$

for $0 < \tau < 1$, or

$$u(x,t) = \frac{x-1}{1+t} \tag{109}$$

Therefore, the shock speed formula beyond B is

$$\frac{d\xi}{dt} = \frac{1}{2}\left(1 + \frac{\xi - 1}{1+t}\right) \tag{110}$$

This is again a linear equation, which we solve subject to the initial condition $\xi(\frac{3}{4}) = -\frac{3}{4}$ and find

$$\xi = 2 + t - \sqrt{7(1+t)}; \quad \frac{3}{4} \le t \le 6 \tag{111}$$

Formula (111) is correct as long as (109) is true—that is, up to the characteristic $x = 1$. The intersection of (111) with $x = 1$ occurs at $C: x = 1, t = 6$.

The continuation of the solution past $x = 1$ requires that we insert a centered fan at $x = 1, t = 0$, since $u(1^+, 0) > u(1^-, 0)$ [see (82)]. This is the solution

$$u(x,t) = \frac{x-1}{t}; \quad 1 \le x \le 1 + t, t > 0 \tag{112}$$

Therefore, the third segment of the shock satisfies

$$\frac{d\xi}{dt} = \frac{1}{2}\left(1 + \frac{\xi - 1}{t}\right); \quad \xi(6) = 1 \tag{113}$$

This has the solution

$$\xi = 1 + t - \sqrt{6t}; \quad 6 \le t < \infty \tag{114}$$

and it is easily seen that the curve defined by (114) remains to the left of the characteristic $x = 1 + t$. In fact, the distance between the shock and $x = 1 + t$ is $\sqrt{6t}$.

The three segments of the shock curve join smoothly (that is, have equal slopes) at B and C, and the resulting curve divides and xt-plane into two regions: (1) $D_1: x < \xi(t)$, where the solution is $u = 1$, and (2) $D_2: x > \xi(t)$, where the solution is defined by (103), (105), (109) and $u = 1$ for $x \ge 1 + t$. The solution in D_2 is smooth everywhere except along the boundaries $x = -t, x = 1$, and $x = 1 + t$, where u_x and u_t are discontinuous. This completes the weak solution of the initial-value problem (98). Additional examples are given in Problems 6–8.

5.3.6 Exact Solution of Burgers' Equation: Shock Layer, Corner Layer

In Chapter 1 we showed that the initial-value problem for Burgers' equation (20b) could be solved exactly for initial data on $-\infty < x < \infty$. In this section we construct these exact solutions for the two special initial-value problems (80) and

(82) in order to study the limiting behavior as we let $\varepsilon \to 0$.

We consider the equation

$$\bar{u}_t + \bar{u}\bar{u}_{\bar{x}} = \varepsilon \bar{u}_{\bar{x}\bar{x}}; \qquad -\infty < \bar{x} < \infty \qquad (115)$$

for $\bar{u}(\bar{x}, \bar{t}; \varepsilon)$; with no loss of generality, we adopt the simpler initial condition:

$$\bar{u}(\bar{x}, 0; \varepsilon) = \begin{cases} 1, & \text{if } \bar{x} < 0 \\ -1, & \text{if } \bar{x} > 0 \end{cases} \qquad (116)$$

instead of (80). It is easily seen that the transformation

$$\bar{x} = \frac{x - x_0 - (u_1 + u_2)t/2}{2/(u_1 - u_2)} \qquad (117a)$$

$$\bar{t} = \frac{t}{4/(u_1 - u_2)^2} \qquad (117b)$$

$$\bar{u} = \frac{2u - (u_1 + u_2)}{u_1 - u_2} \qquad (117c)$$

reduces (20b) to (115) (that is, it leaves Burgers' equation invariant) and takes the initial condition (80) to (116). Similarly, the transformation

$$\bar{x} = \frac{x - x_0 - (u_1 + u_2)t/2}{2/(u_2 - u_1)} \qquad (118a)$$

$$\bar{t} = \frac{t}{4/(u_2 - u_1)^2} \qquad (118b)$$

$$\bar{u} = \frac{2u - (u_1 + u_2)}{u_2 - u_1} \qquad (118c)$$

reduces (20b) to (115) and the initial condition (82) to

$$\bar{u}(\bar{x}, 0; \varepsilon) = \begin{cases} -1, & \text{if } \bar{x} < 0 \\ +1, & \text{if } \bar{x} > 0 \end{cases} \qquad (119)$$

Thus, we need study only the initial conditions corresponding to a stationary shock and a symmetric fan on $-\bar{t} \le \bar{x} \le \bar{t}$ for the reduced problem where $\varepsilon = 0$.

The Shock Layer

We start with (115) and (116) and omit the overbars for simplicity. According to (118) of Chapter 1, the exact solution for the initial-value problem on $-\infty < x < \infty$ is

$$u(x, t; \varepsilon) = \frac{\displaystyle\int_{-\infty}^{\infty} \frac{x - \xi}{t} g(\xi; \varepsilon) \exp\left[-\frac{(x - \xi)^2}{4\varepsilon t} \right] d\xi}{\displaystyle\int_{-\infty}^{\infty} g(\xi; \varepsilon) \exp\left[-\frac{(x - \xi)^2}{4\varepsilon t} \right] d\xi} \qquad (120)$$

where $g(x; \varepsilon)$ is defined in terms of the initial condition (116) to be

$$g(x; \varepsilon) \equiv \exp\left[-\frac{1}{2\varepsilon} \int_0^x u(s, 0; \varepsilon)\, ds\right] = \begin{cases} e^{-x/2\varepsilon}, & \text{if } x < 0 \\ e^{x/2\varepsilon}, & \text{if } x > 0 \end{cases} \tag{121}$$

If we use (121) in (120) and break up the intervals of integration over the negative and positive axes, we find

$$u(x, t; \varepsilon) = \frac{\bar{F}(x, t; \varepsilon) + \bar{G}(x, t; \varepsilon)}{F(x, t; \varepsilon) + G(x, t; \varepsilon)} \tag{122}$$

where

$$F(x, t; \varepsilon) \equiv \int_0^\infty \exp\left[\frac{\xi}{2\varepsilon} - \frac{(x + \xi)^2}{4\varepsilon t}\right] d\xi \tag{123a}$$

$$\bar{F}(x, t; \varepsilon) \equiv \int_0^\infty \left(\frac{x + \xi}{t}\right) \exp\left[\frac{\xi}{2\varepsilon} - \frac{(x + \xi)^2}{4\varepsilon t}\right] d\xi \tag{123b}$$

$$G(x, t; \varepsilon) \equiv \int_0^\infty \exp\left[\frac{\xi}{2\varepsilon} - \frac{(x - \xi)^2}{4\varepsilon t}\right] d\xi \tag{123c}$$

$$\bar{G}(x, t; \varepsilon) \equiv \int_0^\infty \left(\frac{x - \xi}{t}\right) \exp\left[\frac{\xi}{2\varepsilon} - \frac{(x - \xi)^2}{4\varepsilon t}\right] d\xi \tag{123d}$$

Integration by parts shows that

$$\bar{F}(x, t; \varepsilon) = 2\varepsilon \exp\left(-\frac{x^2}{4\varepsilon t}\right) + F(x, t; \varepsilon) \tag{124a}$$

$$\bar{G}(x, t; \varepsilon) = 2\varepsilon \exp\left(-\frac{x^2}{4\varepsilon t}\right) + G(x, t; \varepsilon) \tag{124b}$$

Therefore, (122) simplifies to

$$u(x, t; \varepsilon) = \frac{F(x, t; \varepsilon) - G(x, t; \varepsilon)}{F(x, t; \varepsilon) + G(x, t; \varepsilon)} \tag{125}$$

In fact, the integrals defining F and G can be evaluated explicitly. To evaluate F, we "complete the square" in the argument of the exponential in (123a). This gives the identity

$$\frac{\xi}{2\varepsilon} - \frac{(x + \xi)^2}{4\varepsilon t} = -\left[\frac{\xi + x + t}{2\sqrt{\varepsilon t}}\right]^2 + \frac{t - 2x}{4\varepsilon}$$

so that F has the form

$$F(x, t\varepsilon) = \exp\left(\frac{t - 2x}{4\varepsilon}\right) \int_0^\infty \exp\left[-\left(\frac{\xi + x - t}{2\sqrt{\varepsilon t}}\right)^2\right] d\xi \tag{126a}$$

The change of variable $(\xi + x - t)/2\sqrt{\varepsilon t} = s$ transforms (126a) to

$$F(x, t; \varepsilon) = 2\sqrt{\varepsilon t} \exp\left(\frac{t - 2x}{4\varepsilon}\right) \int_{(x-t)/2\sqrt{\varepsilon t}}^\infty e^{-s^2}\, ds$$

and using the definition of the complementary error function [see (62) of

Chapter 1], we find

$$F(x, t; \varepsilon) = \sqrt{\pi \varepsilon t} \, \exp\left(\frac{t - 2x}{4\varepsilon}\right) \mathrm{erfc}\left(\frac{x - t}{2\sqrt{\varepsilon t}}\right) \tag{126b}$$

This result is correct regardless of the sign of $(x - t)$.

Similar calculations, which are omitted for brevity, show that

$$G(x, t; \varepsilon) = \sqrt{\pi \varepsilon t} \, \exp\left(\frac{2x + t}{4\varepsilon}\right) \mathrm{erfc}\left(-\frac{x + t}{2\sqrt{\varepsilon t}}\right). \tag{127}$$

Substituting (126b) and (127) into (125) and canceling the common factors in the numerator and denominator give the *explicit exact solution*

$$u(x, t; \varepsilon) = \frac{e^{-x/\varepsilon} \mathrm{erfc}\left(\dfrac{x - t}{2\sqrt{\varepsilon t}}\right) - \mathrm{erfc}\left(-\dfrac{x + t}{2\sqrt{\varepsilon t}}\right)}{e^{-x/\varepsilon} \mathrm{erfc}\left(\dfrac{x - t}{2\sqrt{\varepsilon t}}\right) + \mathrm{erfc}\left(-\dfrac{x + t}{2\sqrt{\varepsilon t}}\right)} \tag{128}$$

The limiting behavior of (128) as $\varepsilon \to 0$ depends on the values of x and t. In particular, we note that for $t > 0$, the arguments of the complementary error functions are positive or negative depending on whether $(x - t)$ and $-(x + t)$ are positive or negative, respectively. Moreover, if $(x - t)/2t^{1/2} \neq 0$ and $-(x + t)/2t^{1/2} \neq 0$ and if these expressions are held fixed as $\varepsilon \to 0$, the arguments of the error functions tend to $\pm\infty$ depending on the signs of $(x - t)$ and $-(x + t)$. Therefore, we shall need the asymptotic expansion for $\mathrm{erfc}(y)$ for real y as $y \to \pm\infty$. Integration by parts of the defining integral for $\mathrm{erfc}(y)$ written in the form

$$\mathrm{erfc}(y) = \frac{1}{\sqrt{\pi}} \int_{y^2}^{\infty} e^{-s} s^{-1/2} \, ds$$

gives (see Problem 4b of Chapter 8)

$$\mathrm{erfc}(y) = \frac{e^{-y^2}}{\pi^{1/2} y} [1 + O(y^{-2})], \qquad \text{as } y \to \infty \tag{129a}$$

To calculate the behavior as $y \to -\infty$, we write $\mathrm{erfc}(y) = 1 - \mathrm{erf}(y)$ and use the fact that $\mathrm{erf}(y)$ is odd to find $\mathrm{erfc}(y) = 1 + \mathrm{erf}(-y) = 2 - \mathrm{erfc}(-y)$. Therefore,

$$\mathrm{erfc}(y) = 2 + \frac{e^{-y^2}}{\pi^{1/2} y} [1 + O(y^{-2})], \qquad \text{as } y \to -\infty \tag{129b}$$

We subdivide the xt-plane into the three domains:

①: $t > 0, x - t > 0, x + t > 0$

②: $t > 0, x - t < 0, x + t > 0$

③: $t > 0, x - t < 0, x + t < 0$

We use (129) to evaluate the leading contribution of the two terms that make up

(128) in ①, ②, and ③ as well as the boundaries between ① and ②, and ② and ③. We then use these expressions to compute the leading term for u as $\varepsilon \to 0$. The results are summarized in the following table.

Region	$e^{-x/\varepsilon}\operatorname{erfc}\left(\dfrac{x-t}{2\sqrt{\varepsilon t}}\right)$	$\operatorname{erfc}\left(\dfrac{x+t}{2\sqrt{\varepsilon t}}\right)$	u
① $t>0,\ x-t>0$ $x+t>0$	$2\left(\dfrac{\varepsilon t}{\pi}\right)^{1/2}\dfrac{1}{(x-t)}\exp\left[-\dfrac{(x+t)^2}{4\varepsilon t}\right]$	2	-1
$t>0,\ x=t$	$\exp\left(-\dfrac{x}{\varepsilon}\right)$	2	-1
② $t>0,\ x-t<0$ $x+t>0$	$2\exp\left(-\dfrac{x}{\varepsilon}\right)$	2	$-\tanh\dfrac{x}{2\varepsilon}$
$t>0,\ x=-t$	$2\exp\left(\dfrac{t}{\varepsilon}\right)$	1	$+1$
③ $t>0,\ x-t<0$ $x+t<0$	$2\exp\left(-\dfrac{x}{\varepsilon}\right)$	$2\left(\dfrac{\varepsilon t}{\pi}\right)^{1/2}\dfrac{1}{(x+t)}\exp\left[-\dfrac{(x+t)^2}{4\varepsilon t}\right]$	$+1$

The result that we have in region ②,

$$u \to -\tanh\frac{x}{2\varepsilon} \qquad (130)$$

is significant only along the $x = 0$ axis, since in the limit $\varepsilon \to 0$ with $x > 0$ (130) gives $u \to -1$, as in ①, and with $x < 0$, we have $u \to +1$, as in ③. In fact, we see that the behavior of the hyperbolic tangent is relevant only in a *thin layer of $O(\varepsilon)$* around $x = 0$. This statement can be formalized as follows: We introduce a rescaled variable $x^* = x/\varepsilon$ and define

$$u^*(x^*, t; \varepsilon) \equiv u(\varepsilon x^*, t; \varepsilon) \qquad (131)$$

We then find

$$\lim_{\substack{\varepsilon \to 0 \\ x^* \text{ fixed} \\ t > 0}} u(\varepsilon x^*, t; \varepsilon) = -\tanh\frac{x^*}{2} \qquad (132)$$

Equation (132) defines the *shock structure* for Burgers' equation and can be derived independently of the exact solution by applying the limit process in (132) to the differential equation (115). This idea will be fully explored in Section 8.3 when we study matched asymptotic expansions.

For the time being, we note that the exact solution does indeed tend to the weak solution (81) with a shock obeying (62). We see that the term εu_{xx} is important only to $O(1)$ in a thin layer of thickness $O(\varepsilon)$ centered at the shock location; this term serves to smooth out the discontinuity in the weak solution of the $\varepsilon = 0$ problem.

The Centered Fan, Corner Layer

We now consider the initial condition (119) and compute

$$g(x; \varepsilon) \equiv \exp\left[-\frac{1}{2\varepsilon}\int_0^x u(s, 0; \varepsilon)\, ds\right] = \begin{cases} e^{x/2\varepsilon}, & \text{if } x < 0 \\ e^{-x/2\varepsilon}, & \text{if } x > 0 \end{cases} \tag{133}$$

for which the exact solution (120) takes the form [see (125)]

$$u(x, t; \varepsilon) = \frac{-H(x, t; \varepsilon) + I(x, t; \varepsilon)}{H(x, t; \varepsilon) + I(x, t; \varepsilon)} \tag{134}$$

after we simplify the integrals in the numerator by integrating by parts. The functions H and I are defined by

$$H(x, t; \varepsilon) \equiv \int_0^\infty \exp\left[-\frac{\xi}{2\varepsilon} - \frac{(x + \xi)^2}{4\varepsilon t}\right] d\xi \tag{135a}$$

$$I(x, t; \varepsilon) \equiv \int_0^\infty \exp\left[-\frac{\xi}{2\varepsilon} - \frac{(x - \xi)^2}{4\varepsilon t}\right] d\xi \tag{135b}$$

Again, we can evaluate H and I explicitly in terms of error functions, and we find

$$H(x, t; \varepsilon) = \sqrt{\pi \varepsilon t}\, \exp\left(\frac{2x + t}{4\varepsilon}\right) \operatorname{erfc}\left(\frac{x + t}{2\sqrt{\varepsilon t}}\right) \tag{136a}$$

$$I(x, t; \varepsilon) = \sqrt{\pi \varepsilon t}\, \exp\left(\frac{t - 2x}{4\varepsilon}\right) \operatorname{erfc}\left(\frac{t - x}{2\sqrt{\varepsilon t}}\right) \tag{136b}$$

Using these in (134), we obtain the following explicit result for $u(x, t; \varepsilon)$:

$$u(x, t; \varepsilon) = \frac{-\operatorname{erfc}\left(\dfrac{x + t}{2\sqrt{\varepsilon t}}\right) + e^{-x/\varepsilon}\operatorname{erfc}\left(\dfrac{t - x}{2\sqrt{\varepsilon t}}\right)}{\operatorname{erfc}\left(\dfrac{x + t}{2\sqrt{\varepsilon t}}\right) + e^{-x/\varepsilon}\operatorname{erfc}\left(\dfrac{t - x}{2\sqrt{\varepsilon t}}\right)} \tag{137}$$

If we examine the leading contribution as $\varepsilon \to 0$ for each of the two terms occurring in (137), we obtain the expressions listed in the following table in the three regions.

Region	$\operatorname{erfc}\left(\dfrac{x + t}{2\sqrt{\varepsilon t}}\right)$	$e^{-x/\varepsilon}\operatorname{erfc}\left(\dfrac{t - x}{2\sqrt{\varepsilon t}}\right)$	u
① $t > 0,\ x - t > 0$ $x + t > 0$	$2\left(\dfrac{\varepsilon t}{\pi}\right)^{1/2}\dfrac{1}{(x + t)}\exp\left[-\dfrac{(x + t)^2}{4\varepsilon t}\right]$	$2e^{-x/\varepsilon}$	$+1$
② $t > 0,\ x - t < 0$ $x + t > 0$	$2\left(\dfrac{\varepsilon t}{\pi}\right)^{1/2}\dfrac{1}{(x + t)}\exp\left[-\dfrac{(x + t)^2}{4\varepsilon t}\right]$	$2\left(\dfrac{\varepsilon t}{\pi}\right)^{1/2}\dfrac{1}{(x - t)}\exp\left[-\dfrac{(x + t)^2}{4\varepsilon t}\right]$	$\dfrac{x}{t}$
③ $t > 0,\ x - t < 0$ $x + t < 0$	2	$2\left(\dfrac{\varepsilon t}{\pi}\right)^{1/2}\dfrac{1}{(x - t)}\exp\left[-\dfrac{(x + t)^2}{4\varepsilon t}\right]$	-1

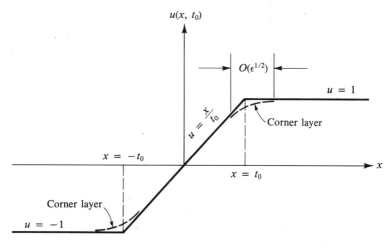

Figure 5.15

We see that in the limit $\varepsilon \to 0$, the exact solution indeed tends to the weak solution (85). As pointed out earlier, the weak solution (85) has discontinuous derivatives u_x and u_t along the rays $x = \pm t$. For example, the weak solution for u as a function of x at a fixed time $t = t_0 > 0$ is the piecewise linear profile shown in Figure 5.15.

It is interesting to show that in the neighborhood of the rays $x = \pm t$, where the weak solution has a *corner*, the asymptotic behavior of the exact solution consists of a *corner layer*, which smooths out this corner. It suffices to consider the corner at $x = t$, since the behavior at the other corner follows by symmetry. We note that the arguments of the error functions in (137) involve $(x - t)/\varepsilon^{1/2}$ and $(x + t)/\varepsilon^{1/2}$. This suggests that near $x = t$, we must hold the variables $x_c \equiv (x - t)/\varepsilon^{1/2}$ and $t_c \equiv t$ fixed as $\varepsilon \to 0$. Thus, we have

$$\operatorname{erfc}\left(\frac{x+t}{2\sqrt{\varepsilon t}}\right) = \operatorname{erfc}\left[\left(\frac{t_c}{\varepsilon}\right)^{1/2}\left(1 + \frac{x_c}{2t_c^{1/2}}\right)\right]$$

or

$$\operatorname{erfc}\left(\frac{x+t}{2\sqrt{\varepsilon t}}\right) = \left(\frac{\varepsilon}{\pi t_c}\right)^{1/2}\exp\left(-\frac{t_c}{\varepsilon} - \frac{x_c}{\varepsilon^{1/2}} - \frac{x_c^2}{4t}\right)[1 + O(\varepsilon^{1/2})] \qquad (138a)$$

when we use (129) and regard $(t_c/\varepsilon)^{1/2} \to \infty$ and $(1 + x_c/2t_c^{1/2}) = O(1)$. We also find

$$e^{-x/\varepsilon}\operatorname{erfc}\left(\frac{t-x}{2\sqrt{\varepsilon t}}\right) = \exp\left(\frac{t_c}{\varepsilon} - \frac{x_c}{\varepsilon^{1/2}}\right)\operatorname{erfc}\left(-\frac{x_c}{2t_c^{1/2}}\right) \qquad (138b)$$

Therefore, the dominant behavior for u near this corner is given by

$$u = \frac{-\left(\dfrac{\varepsilon}{\pi t_c}\right)^{1/2}\exp\left(-\dfrac{x_c^2}{4t_c}\right) + \operatorname{erfc}\left(-\dfrac{x_c}{2t_c^{1/2}}\right)}{\left(\dfrac{\varepsilon}{\pi t_c}\right)^{1/2}\exp\left(-\dfrac{x_c^2}{4t_c}\right) + \operatorname{erfc}\left(-\dfrac{x_c}{2t_c^{1/2}}\right)} + O(\varepsilon) \qquad (139)$$

when we substitute (138a)–(138b) into (137) and cancel the common factor $\exp(t_c/\varepsilon - x_c/\varepsilon^{1/2})$. Expanding (137) further for $\varepsilon^{1/2}$ small gives the *corner layer approximation*

$$u = 1 - 2\left(\frac{\varepsilon}{\pi t_c}\right)^{1/2} \frac{\exp\left(-\dfrac{x_c^2}{4t_c}\right)}{\operatorname{erfc}\left(-\dfrac{x_c}{2t_c^{1/2}}\right)} + O(\varepsilon) \tag{140}$$

It is easily seen that as $x_c \to -\infty$, the corner layer limit (140) gives $u \to x/t$ and that $u \to 1$ as $x_c \to \infty$. Thus, (140) smoothly joins the two linear profiles within a layer of thickness $O(\varepsilon^{1/2})$ centered around $x = t$. This is the dotted curve in Figure 5.15. In Chapter 8 we shall see that (140) may also be regarded as the limiting solution of Burgers' equation as $\varepsilon \to 0$ with $u_c = (u - 1)/\varepsilon^{1/2}$, x_c, and t_c fixed.

In the preceding discussion, we have confined our attention to piecewise constant initial values. One can also show, with considerably more effort and the use of asymptotic expansions, that the exact solution of Burgers' equation for more general initial data does, in fact, tend to the admissible weak solution in the limit $\varepsilon \to 0$.

Therefore, for the initial-value problem on $-\infty < x < \infty$, use of the correct integral conservation law (61a) combined with the appropriate entropy condition (86) provides a very useful approximation everywhere except in $O(\varepsilon)$ layers centered around shocks and $O(\varepsilon^{1/2})$ layers centered around the two boundary characteristics of centered fans. The solution of Burgers' equation with boundary conditions leads to $O(\varepsilon)$ boundary layers and $O(\varepsilon^{1/2})$ transition layers in addition to the shock and corner layers that we found here. A detailed discussion of the various possible approximations is given in Section 4.1.3. of Chapter 4.

5.4   The Quasilinear Equation in n Independent Variables

The general quasilinear equation in n independent variables $x_1, \ldots, x_n$ has the form [see (21)]

$$\sum_{j=1}^{n} a_j(x_i, u) \frac{\partial u}{\partial x_j} = a(x_i, u) \tag{141}$$

where the a_j and a are given functions of the x_i and u. It is assumed that in some solution domain these functions are continuous and have continuous first partial derivatives and that the a_j do not vanish simultaneously.

5.4.1   The Initial-Value Problem

The geometrical ideas developed in Section 5.2 generalize in a straightforward way to the present $(n + 1)$-dimensional problem. In particular, (141) implies that through each point (x_i, u), we have a characteristic curve defined by the solution of the system

$$\frac{dx_j}{ds} = a_j(x_i, u) \tag{142a}$$

$$\frac{du}{ds} = a(x_i, u) \tag{142b}$$

Again, the subscript i indicates that all the components are present in general, and we omit the explicit reminder that $j = 1, \ldots, n$. The characteristic curves can be expressed in the parametric form

$$x_j = \bar{X}_j(s - s_0, c_i) \tag{143a}$$

$$u = \bar{U}(s - s_0, c_i) \tag{143b}$$

involving the n arbitrary constants c_i and the additive constant s_0. Thus, they define an n-parameter family of curves that fills some portion of the space of x_i, u.

The solution of an initial-value problem for (141) consists of finding the n-dimensional manifold $u = \phi(x_1, \ldots, x_n)$ that satisfies (141) and passes through a given smooth $(n - 1)$-dimensional manifold $\mathscr{C}_0$. We may prescribe $\mathscr{C}_0$ in parametric form as follows:

$$x_j = x_j^{(0)}(\tau_1, \ldots, \tau_{n-1}) \tag{144a}$$

$$u = u^{(0)}(\tau_1, \ldots, \tau_{n-1}) \tag{144b}$$

in terms of the $(n - 1)$ parameters $\tau_1, \ldots, \tau_{n-1}$. Here the functions $x_i^{(0)}$, $u^{(0)}$ are continuous and have continuous first partial derivatives with respect to $\tau_1, \ldots, \tau_{n-1}$.

By regarding s_0 and the c_i to be functions of $\tau_1, \ldots, \tau_{n-1}$, we can generate an $(n - 1)$-parameter subfamily of (143) in the form:

$$x_j = \bar{X}_j(s - s_0(\tau_1, \ldots, \tau_{n-1}), c_i(\tau_1, \ldots, \tau_{n-1})) \equiv X_j(s, \tau_1, \ldots, \tau_{n-1}) \tag{145a}$$

$$u = \bar{U}(s - s_0(\tau_1, \ldots, \tau_{n-1}), c_i(\tau_1, \ldots, \tau_{n-1})) \equiv U(s, \tau_1, \ldots, \tau_{n-1}) \tag{145b}$$

For fixed $\tau_1, \ldots, \tau_{n-1}$, the functions X_j and U of s also define a characteristic curve in the space of x_i, u.

We now specify the family (145) by requiring it to pass through the initial manifold $\mathscr{C}_0$; that is, we set

$$X_j(0, \tau_1, \ldots, \tau_{n-1}) = x_j^{(0)}(\tau_1, \ldots, \tau_{n-1}) \tag{146a}$$

$$U(0, \tau_1, \ldots, \tau_{n-1}) = u^{(0)}(\tau_1, \ldots, \tau_{n-1}) \tag{146b}$$

where we have chosen $s = 0$ on $\mathscr{C}_0$. For a given manifold $\mathscr{C}_0$, the conditions (146) fix the $(n + 1)$ functions s_0, c_i of $\tau_1, \ldots, \tau_{n-1}$ and define an $(n - 1)$-parameter family of characteristic curves that pass through $\mathscr{C}_0$ in the form

$$x_j = X_j(s, \tau_1, \ldots, \tau_{n-1}) \tag{147a}$$

$$u = U(s, \tau_1, \ldots, \tau_{n-1}) \tag{147b}$$

In practice, we shall solve the system (142) directly in the form (147).

The solution manifold $u = \phi(x_i)$ is obtained by first solving the system (147a) for s and $\tau_1, \ldots, \tau_{n-1}$ as functions of the x_i and then substituting these expressions into (147b). We can invert (147a) as long as the Jacobian

$$\Delta(s, \tau_1, \ldots, \tau_{n-1}) \equiv \frac{\partial(X_1, \ldots, X_n)}{\partial(s, \tau_1, \ldots, \tau_{n-1})} \tag{148}$$

does not vanish.

5.4.2 The Characteristic Manifold: Existence and Uniqueness of Solutions

In preparation for dealing with the case $\Delta = 0$, we introduce the idea of a characteristic manifold.

$\mathscr{C}$ is a characteristic $(n - 1)$-dimensional manifold in the $(n + 1)$-dimensional space of x_i, u if at every point (x_i, u) on $\mathscr{C}$, the characteristic vector

$$\boldsymbol{\sigma} \equiv (a_1, \ldots, a_n, a) \tag{149}$$

is tangent to $\mathscr{C}$. In order to obtain an analytic description of a characteristic manifold based on this geometric statement, we define the manifold $\mathscr{C}$ by the $(n + 1)$ functions

$$x_j = \tilde{X}_j(\tau_1, \ldots, \tau_{n-1}) \tag{150a}$$

$$u = \tilde{U}(\tau_1, \ldots, \tau_{n-1}) \tag{150b}$$

Now the $(n - 1)$ vectors

$$\mathbf{T}_m = \left(\frac{\partial \tilde{X}_1}{\partial \tau_m}, \ldots, \frac{\partial \tilde{X}_n}{\partial \tau_m}, \frac{\partial \tilde{U}}{\partial \tau_m}\right); \qquad m = 1, \ldots, n - 1 \tag{151}$$

are linearly independent tangent vectors to $\mathscr{C}$ (see (212) of Chapter 2). The characteristic vector $\boldsymbol{\sigma}$ is tangent to $\mathscr{C}$ at some point if $\boldsymbol{\sigma}$ can be expressed as a linear combination of the $\mathbf{T}_m$ at that point—that is, if there exist $(n - 1)$ constants $\lambda_1, \ldots, \lambda_{n-1}$ such that

$$\boldsymbol{\sigma} = \sum_{m=1}^{n-1} \lambda_m \mathbf{T}_m \tag{152}$$

The characteristic vector $\boldsymbol{\sigma}$ is *everywhere* tangent to $\mathscr{C}$ if we can find $(n - 1)$ *functions* $\lambda_1, \ldots, \lambda_{n-1}$ of $\tau_1, \ldots, \tau_{n-1}$ such that (152) holds everywhere on $\mathscr{C}$. In component form, (152) implies that we must have the $(n + 1)$ conditions

$$a_j = \sum_{m=1}^{n-1} \lambda_m \frac{\partial \mathbf{X}_j}{\partial \tau_m} \tag{153a}$$

$$a = \sum_{m=1}^{n-1} \lambda_m \frac{\partial \mathbf{U}}{\partial \tau_m} \tag{153b}$$

hold everywhere on $\mathscr{C}$ for $(n - 1)$ functions $\lambda_1, \ldots, \lambda_{n-1}$ in order that $\mathscr{C}$ be a characteristic manifold.

It is easy to prove that if $\Delta = 0$, the n conditions (153a) are automatically satisfied. To see this, note that Δ is the determinant of the $n \times n$ matrix

$$\Delta \equiv \det \begin{pmatrix} \dfrac{\partial X_1}{\partial s} & \cdots & \dfrac{\partial X_n}{\partial s} \\[2ex] \dfrac{\partial X_1}{\partial \tau_1} & \cdots & \dfrac{\partial X_n}{\partial \tau_1} \\[1ex] \vdots & & \vdots \\[1ex] \dfrac{\partial X_1}{\partial \tau_{n-1}} & \cdots & \dfrac{\partial X_n}{\partial \tau_{n-1}} \end{pmatrix} = \det \begin{pmatrix} a_1 & \cdots & a_n \\[2ex] \dfrac{\partial X_1}{\partial \tau_1} & \cdots & \dfrac{\partial X_n}{\partial \tau_1} \\[1ex] \vdots & & \vdots \\[1ex] \dfrac{\partial X_1}{\partial \tau_{n-1}} & \cdots & \dfrac{\partial X_n}{\partial \tau_{n-1}} \end{pmatrix}$$

Therefore, $\Delta = 0$ implies that the first row vector $(a_1, \ldots, a_n)$ is linearly dependent on the $(n-1)$ row vectors $(\partial X_1/\partial\tau_1, \ldots, \partial X_n/\partial\tau_1), \ldots, (\partial X_1/\partial\tau_{n-1}, \ldots, \partial X_n/\partial\tau_{n-1})$; that is, there exist $(n-1)$ functions $\lambda_1, \ldots, \lambda_{n-1}$ of $\tau_1, \ldots, \tau_{n-1}$ such that

$$(a_1, \ldots, a_n) = \lambda_1\left(\frac{\partial X_1}{\partial \tau_1}, \ldots, \frac{\partial X_n}{\partial \tau_1}\right) + \cdots + \lambda_{n-1}\left(\frac{\partial X_1}{\partial \tau_{n-1}}, \ldots, \frac{\partial X_n}{\partial \tau_{n-1}}\right) \tag{154}$$

Identifying components on each side of (154) gives (153a). The converse is also true; that is, if there exist $(n-1)$ functions $\lambda_1, \ldots, \lambda_{n-1}$ of $\tau_1, \ldots, \tau_{n-1}$ such that (153a) holds everywhere on some manifold $\mathscr{C}$, then $\Delta = 0$ on $\mathscr{C}$.

Thus, a necessary condition for $\mathscr{C}$ to be a characteristic manifold is $\Delta = 0$. But this is not sufficient; one must also be able to show that (153b) holds for the functions $\lambda_1, \ldots, \lambda_{n-1}$ used to satisfy (153a). Some examples will be worked out later on to illustrate these ideas.

One can also prove the following theorem relating a characteristic manifold to a family of characteristic curves (see Section 2, Chapter 2 of [2]). Every characteristic manifold is generated by an $(n-2)$-parameter family of characteristic curves. Conversely, every $(n-2)$-parameter family of characteristic curves generates a characteristic manifold. We shall also illustrate this result for a specific example later on. Note, incidentally, that for $n = 2$, a characteristic manifold is just a characteristic curve; it is only for $n \geq 3$ that the characteristic manifold has a dimension higher than one and differs from a characteristic curve.

The theorem that concerns the existence and uniqueness of the solution of a given initial-value problem is now stated without proof (see [2]): The solution of (141), subject to the initial condition (144), exists and is unique in some neighborhood of $\mathscr{C}_0$ if $\Delta(0, \tau_1, \ldots, \tau_{n-1}) \neq 0$. In the event $\Delta = 0$ everywhere on $\mathscr{C}_0$, nonunique solutions exist only if $\mathscr{C}_0$ is a characteristic manifold; if $\Delta = 0$ but $\mathscr{C}_0$ is not a characteristic manifold, one cannot derive a solution of (141) passing through $\mathscr{C}_0$.

5.4.3 A Linear Example

We study the linear problem

$$x_2 \frac{\partial u}{\partial x_1} - x_1 \frac{\partial u}{\partial x_2} + \frac{\partial u}{\partial x_3} = 1 \tag{155}$$

and consider first the initial-value problem $u = 0$ on the conical surface $x_3 = x_1^2 + x_2^2$. A parametric form for $\mathscr{C}_0$ is

$$x_1^{(0)} = \tau_1; \qquad x_2^{(0)} = \tau_2; \qquad x_3^{(0)} = \tau_1^2 + \tau_2^2; \qquad u^{(0)} = 0 \tag{156}$$

The characteristic equations (142) specialize to

$$\frac{dx_1}{ds} = x_2; \qquad \frac{dx_2}{ds} = -x_1; \qquad \frac{dx_3}{ds} = 1; \qquad \frac{du}{ds} = 1 \tag{157}$$

The general solution of this system is easy to compute because the first two equations do not involve x_3 and u, and the last two are trivially solved. We find [see (147)] the two-parameter family of characteristic curves:

$$x_1 = x_2^{(0)}(\tau_1, \tau_2)\sin s + x_1^{(0)}(\tau_1, \tau_2)\cos s \equiv X_1(s, \tau_1, \tau_2) \tag{158a}$$

$$x_2 = -x_1^{(0)}(\tau_1, \tau_2)\sin s + x_2^{(0)}(\tau_1, \tau_2)\cos s \equiv X_2(s, \tau_1, \tau_2) \tag{158b}$$

$$x_3 = s + x_3^{(0)}(\tau_1, \tau_2) \equiv X_3(s, \tau_1, \tau_2) \tag{158c}$$

$$u = s + u^{(0)}(\tau_1, \tau_2) \equiv U(s, \tau_1, \tau_2) \tag{158d}$$

The particular two-parameter family that passes through the initial manifold (156) is

$$x_1 = \tau_2\sin s + \tau_1\cos s \tag{159a}$$

$$x_2 = \tau_2\cos s - \tau_1\sin s \tag{159b}$$

$$x_3 = s + \tau_1^2 + \tau_2^2 \tag{159c}$$

$$u = s \tag{159d}$$

Using (158) in the definition of Δ gives

$$\Delta(s, \tau_1, \tau_2) = x_1^{(0)}\left(\frac{\partial x_3^{(0)}}{\partial \tau_2}\frac{\partial x_1^{(0)}}{\partial \tau_1} - \frac{\partial x_3^{(0)}}{\partial \tau_1}\frac{\partial x_1^{(0)}}{\partial \tau_2}\right)$$
$$+ x_2^{(0)}\left(\frac{\partial x_3^{(0)}}{\partial \tau_2}\frac{\partial x_2^{(0)}}{\partial \tau_1} - \frac{\partial x_3^{(0)}}{\partial \tau_1}\frac{\partial x_2^{(0)}}{\partial \tau_2}\right) + \left(\frac{\partial x_1^{(0)}}{\partial \tau_1}\frac{\partial x_2^{(0)}}{\partial \tau_2} - \frac{\partial x_1^{(0)}}{\partial \tau_2}\frac{\partial x_2^{(0)}}{\partial \tau_1}\right) \tag{160}$$

and for the special case (156), we have

$$\Delta(s, \tau_1, \tau_2) = \Delta(0, \tau_1, \tau_2) = 1 \tag{161}$$

Therefore, we expect a unique solution manifold to result from (159). It is easily verified that this manifold is

$$u = x_3 - (x_1^2 + x_2^2) \tag{162}$$

Let us now demonstrate that the two-dimensional manifold generated by an arbitrary one-parameter family of characteristic curves is a characteristic manifold. One way to define a general one-parameter family of characteristic curves is to set $\tau_2 = r(\tau_1)$ (for an arbitrary function r) in (158). To generate a two-dimensional manifold from the resulting one-parameter family of curves, we replace $s \to \tau_2$ and $\tau_1 \to \tau_1$ and regard the new τ_1, τ_2 as the two variables on the manifold. Then $x_2^{(0)}$, $x_1^{(0)}$, $x_3^{(0)}$, and $u^{(0)}$ may be regarded as arbitrary functions

f, g, h, and k of τ_1, respectively, and we obtain a two-dimensional manifold in the form:

$$x_1 = f(\tau_1)\sin\tau_2 + g(\tau_1)\cos\tau_2 \equiv \tilde{X}_1(\tau_1,\tau_2) \tag{163a}$$

$$x_2 = -g(\tau_1)\sin\tau_2 + f(\tau_1)\cos\tau_2 \equiv \tilde{X}_2(\tau_1,\tau_2) \tag{163b}$$

$$x_3 = \tau_2 + h(\tau_1) \equiv \tilde{X}_3(\tau_1,\tau_2) \tag{163c}$$

$$u = \tau_2 + k(\tau_1) \equiv \tilde{U}(\tau_1,\tau_2) \tag{163d}$$

In order to prove that the manifold (163) is characteristic, we must find λ_1 and λ_2 such that the four equations (153) are satisfied. For our case, these are

$$-g(\tau_1)\sin\tau_2 + f(\tau_1)\cos\tau_2 = \lambda_1[f'(\tau_1)\sin\tau_2 + g'(\tau_1)\cos\tau_2]$$
$$+ \lambda_2[f(\tau_1)\cos\tau_2 - g(\tau_1)\sin\tau_2] \tag{164a}$$

$$-f(\tau_1)\sin\tau_2 - g(\tau_1)\cos\tau_2 = \lambda_1[-g'(\tau_1)\sin\tau_2 + f'(\tau_1)\cos\tau_2]$$
$$+ \lambda_2[-f(\tau_1)\sin\tau_2 - g(\tau_1)\sin\tau_2] \tag{164b}$$

$$1 = \lambda_1 h'(\tau_1) + \lambda_2 \tag{164c}$$

$$1 = \lambda_1 k'(\tau_1) + \lambda_2 \tag{164d}$$

Solving (164a)–(164b) for λ_1 and λ_2 gives $\lambda_1 = 0$ and $\lambda_2 = 1$, and these values indeed also satisfy (164c)–(164d). Therefore, the two-dimensional manifold (163) is a characteristic manifold. The converse is also true—any characteristic manifold can be generated by a one-parameter family of characteristic curves.

For the particular initial-value problem (156), we see that we may interpret the solution (162) to be generated either by the two-parameter family of characteristic curves (159) or the following one-parameter (ξ) family of characteristic manifolds:

$$x_1 = \xi\sin\tau_2 + \tau_1\cos\tau_2 \equiv \tilde{X}_1(\tau_1,\tau_2;\xi) \tag{165a}$$

$$x_2 = \xi\cos\tau_2 - \tau_1\sin\tau_2 \equiv \tilde{X}_2(\tau_1,\tau_2;\xi) \tag{165b}$$

$$x_3 = \tau_2 + \tau_1^2 + \xi^2 \quad\ \equiv \tilde{X}_3(\tau_1,\tau_2;\xi) \tag{165c}$$

$$u = \tau_2 \quad\quad\quad\quad\ \equiv \tilde{U}(\tau_1,\tau_2;\xi) \tag{165d}$$

It is easily seen that for any fixed ξ, (165) defines a characteristic manifold, and we have already derived the solution (162) from just such a set of equations (albeit, before we had labeled $\xi \to \tau_2$, $\tau_1 \to \tau_1$, $\tau_2 \to s$).

Consider now the initial manifold $\mathscr{C}_0$ with

$$x_1^{(0)} = \tau_1\cos\tau_2; \quad x_2^{(0)} = -\tau_1\sin\tau_2; \quad x_3^{(0)} = \tau_2 + \tau_1^2; \quad u^{(0)} = \tau_2 \tag{166}$$

It is easily seen that (160) gives $\Delta(0,\tau_1,\tau_2) = 0$ in this case. The family of characteristic curves (158) that results for this choice is the degenerate one,

$$x_1 = \tau_1\cos(s + \tau_2); \quad\quad x_2 = -\tau_1\sin(s + \tau_2)$$

$$x_3 = (s + \tau_2) + \tau_1^2; \quad\quad u = (s + \tau_2) \tag{167}$$

in which s and τ_2 occur only in the combination $(s + \tau_2)$. This implies that (167) actually defines just the characteristic manifold $\mathscr{C}_0$ (instead of a one-parameter family of characteristic manifolds or a two-parameter family of characteristic curves). To verify this statement, note that (153a) gives the four conditions

$$-\tau_1 \sin(s + \tau_2) = \lambda_1 \cos(s + \tau_2) - \lambda_2 \tau_1 \sin(s + \tau_2)$$

$$\tau_1 \cos(s + \tau_2) = -\lambda_1 \sin(s + \tau_2) - \lambda_2 \tau_1 \cos(s + \tau_2)$$

$$1 = 2\lambda_1 \tau_1 + \lambda_2; \qquad 1 = \lambda_2$$

which are satisfied with $\lambda_1 = 0$, $\lambda_2 = 1$. In this case, the solution manifold is not unique. We can exhibit this nonuniqueness by noting that the implicit formula

$$u = x_3 - (x_1^2 + x_2^2) + F\left(u + \tan^{-1}\frac{x_2}{x_1}\right) \tag{168}$$

defines a solution u for any function F as long as $F(0) = 0$ and (168) can be solved for u.

Finally, if we define $x_1^{(0)}$, $x_2^{(0)}$, $x_3^{(0)}$ as in (166) but choose $u^{(0)} \neq \tau_2$, we cannot solve the equations that correspond to (167). In this case, $\Delta = 0$, but $\mathscr{C}_0$ is not a characteristic manifold.

5.4.4 A Quasilinear Example

Consider the generalization of (61c) to three independent variables—that is,

$$\frac{\partial u}{\partial x_1} + u\left(\frac{\partial u}{\partial x_2} + \frac{\partial u}{\partial x_3}\right) = 0 \tag{169}$$

The characteristics satisfy the equations

$$\frac{dx_1}{ds} = 1; \qquad \frac{dx_2}{ds} = u; \qquad \frac{dx_3}{ds} = u; \qquad \frac{du}{ds} = 0$$

which can be solved in the form

$$x_1 = s + x_1^{(0)}(\tau_1, \tau_2) \tag{170a}$$

$$x_2 = u^{(0)}(\tau_1, \tau_2)s + x_2^{(0)}(\tau_1, \tau_2) \tag{170b}$$

$$x_3 = u^{(0)}(\tau_1, \tau_2)s + x_3^{(0)}(\tau_1, \tau_2) \tag{170c}$$

$$u = u^{(0)}(\tau_1, \tau_2) \tag{170d}$$

Let us restrict attention to solutions that pass through the initial manifold

$$x_1^{(0)} = 0; \qquad x_2^{(0)} = \tau_1; \qquad x_3^{(0)} = \tau_2; \qquad u^{(0)} = \sin \tau_1 \sin \tau_2$$

for which we compute

$$x_1 = s \tag{171a}$$

$$x_2 = s \sin \tau_1 \sin \tau_2 + \tau_1 \tag{171b}$$

$$x_3 = s \sin \tau_1 \sin \tau_2 + \tau_2 \tag{171c}$$

$$u = \sin \tau_1 \sin \tau_2 \tag{171d}$$

and

$$\Delta(s,\tau_1,\tau_2) = 1 + s\sin(\tau_1 + \tau_2) \tag{172}$$

Thus, $\Delta(0,\tau_1,\tau_2) = 1$, and a unique solution exists near the initial manifold. Since we cannot solve for τ_1 and τ_2 in terms of x_1, x_2, and x_3 in closed form, we write the solution in the implicit form

$$u = \sin(x_2 - x_1 u)\sin(x_3 - x_1 u) \tag{173}$$

This solution first breaks down when $x_1 = 1$ and cannot be extended to $x_1 > 1$ for values of x_1, x_2, x_3 that lie on the surface

$$\tau_1(x_1,x_2,x_3) + \tau_2(x_1,x_2,x_3) = -\sin^{-1}(1/x_1) \tag{174}$$

where $\tau_1(x_1,x_2,x_3)$ and $\tau_2(x_1,x_2,x_3)$ are the solutions of (171b) and (171c) in which $s = x_1$.

Although some aspects of the theory of weak solutions, as discussed in Section 5.3, can be extended to higher dimensions, we shall not present these results here. Certainly, the geometry of shock manifolds for dimensions greater than 1 becomes more complicated. But the difficulties are not confined just to questions of geometry. For example, it is no longer possible to derive an exact solution of the two-dimensional Burgers' equation

$$\frac{\partial u}{\partial x_1} + u\left(\frac{\partial u}{\partial x_2} + \frac{\partial u}{\partial x_3}\right) = \varepsilon\left(\frac{\partial^2 u}{\partial x_2^2} + \frac{\partial^2 u}{\partial x_3^2}\right) \tag{175}$$

Therefore, it is more difficult to establish what is an admissible weak solution of the integral conservation law that led to (169).

Problems

1. Consider the two-parameter family of ellipses

$$\frac{(x - x_0)^2}{4} + (y - y_0)^2 = 1 \tag{176}$$

which may be written in the form

$$x = x_0 + 2\cos\tau \tag{177a}$$

$$y = y_0 + \sin\tau \tag{177b}$$

Show that in order for these ellipses to envelope on the unit circle $x^2 + y^2 = 1$, we must have

$$x_0 = \pm\frac{4\tan\tau}{\sqrt{1 + 4\tan^2\tau}} - 2\cos\tau \tag{178a}$$

$$y_0 = \pm\frac{1}{\sqrt{1 + 4\tan^2\tau}} - \sin\tau \tag{178b}$$

Describe these one-parameter subfamilies geometrically.

2. Calculate the solution of

$$u_x + u_y = u^2 \tag{179}$$

passing through the curve $u = x$ on $y = -x$ and show that it becomes infinite along the hyperbola $x^2 - y^2 = 4$. What is the significance of this hyperbola?

3. Show that for any given one-parameter family of smooth curves

$$x = X(s, \tau); \qquad y = Y(s, \tau); \qquad u = U(s, \tau) \tag{180}$$

for which (28) does not vanish in some region, we may associate a *linear* first-order partial differential equation

$$a(x, y)u_x + b(x, y)u_y = c(x, y) \tag{181}$$

such that the function $u(x, y)$ obtained from (180) solves (181). Thus, given the solution $u(x, y)$ of a *quasilinear* equation, we can always interpret this as the solution of another linear equation.

Specialize your results to the example (37) and show that one can interpret this as the solution of

$$(1 + \sqrt{1 + 2(x - y)})u_x + u_y = 1 \tag{182}$$

4. Solve (31) for the following initial-value problems.

   a. $u(x, 0) = x$ \hfill (183)

   b. $u(x, 0) = x^2$ \hfill (184)

   In each case, discuss where the solution breaks down and the nature of the singularity there.

5. Consider the special case of (21)

$$\tilde{a}(x, y)u_x + \tilde{b}(x, y)u_y = c(x, y, u) \tag{185}$$

   with given continuously differentiable coefficients $\tilde{a}$, $\tilde{b}$, c, where $\tilde{a}$ and $\tilde{b}$ do not depend on u. Assume also that $\tilde{a}$ and $\tilde{b}$ do not vanish simultaneously in the domain of interest.

   a. Show that the characteristic ground curves are independent of the initial data and dependent only on $\tilde{a}$ and $\tilde{b}$. Therefore, they define a one-parameter family of nonintersecting curves with no singular points in the domain of interest.

   b. Specialize your results in part (a) to the case $\tilde{a} = -y$, $\tilde{b} = x$, $c = -u + 1$, and assume that the projection of the initial curve on the xy-plane is the positive x-axis. What is the largest domain in the xy-plane over which a solution of (185) can be found? Calculate this solution for the initial curve $u(x, 0) = \sin x$ for $x > 0$.

   c. Now assume that the solution of the characteristic equations (23) for (185) have been found in the form (25) and that these functions satisfy (27) for a certain initial curve $\mathscr{C}_0$: $x_0(\tau)$, $y_0(\tau)$, $u_0(\tau)$ for which $\Delta(0, \tau) = 0$. Assume also that $Y_s U_\tau - U_s Y_\tau \neq 0$ on $\mathscr{C}_0$ so that one can solve the system (25) for

x as a function of y and u in the form

$$x = f(y, u) \tag{186}$$

We wish to prove that f is actually independent of u—that is, that u is a *vertical* surface through $\mathscr{C}_0$ in this case. Show first that

$$\frac{\partial \Delta(s, \tau)}{\partial s} = [\tilde{a}_x(X, Y) + \tilde{b}_y(X, Y)]\Delta(s, \tau) \tag{187}$$

Therefore, $\Delta(s, \tau) \equiv 0$ if $\Delta(0, \tau) = 0$. Next, show that

$$\Delta(s, \tau) = (U_s Y_\tau - Y_s U_\tau)\frac{\partial f}{\partial u}(Y, U) \tag{188}$$

when (186) is used. Therefore, $\partial f/\partial u = 0$; that is, f depends only on y.

d. Give a particular example of the situation described in part (c).

6. Consider the initial-value problem

$$u_t + uu_x = 0 \tag{189a}$$

$$u(x, 0) = \begin{cases} 0, & \text{if } x \geq 0 \\ f(x), & \text{if } x < 0 \end{cases} \tag{189b}$$

a. What conditions must be imposed on the function $f(x)$ in order that the solution of (189) be continuously differentiable in the half-plane $-\infty < x < \infty, 0 \leq t < \infty$?

b. Now consider the inverse problem where $f(x)$ is unknown. Instead, we are told that a shock, defined by the prescribed function

$$x = \phi(t); \qquad 0 \leq t < \infty \tag{190}$$

separates the half-plane into two domains, in each of which the solution of (189) is continuously differentiable. The shock curve (190) is a consequence of the divergence form (61b) of (189a). Show that $\phi(t)$ cannot be prescribed arbitrarily. Derive the conditions that must be imposed on $\phi(t)$ in order to ensure that (190) is a consistent shock.

c. For functions $\phi(t)$ satisfying the conditions you derived in part (b), calculate $f(x)$ in terms of $\phi(t)$.

d. Show that the special case

$$\phi(t) = (1 + t)^{1/2} - 1 \tag{191}$$

is a consistent shock according to part (b) and leads to

$$f(x) = 1 + x \tag{192}$$

when you use your result in part (c).

7. Calculate the weak solution of (61a) for the following initial conditions:

a. $u(x, 0) = \begin{cases} 1, & \text{if } |x| > 1 \\ 0, & \text{if } |x| < 1 \end{cases}$ \hfill (193)

b. $u(x,0) = \begin{cases} 0, & \text{if } |x| > 1 \\ 1-x, & \text{if } 0 < x < 1 \\ -1-x, & \text{if } -1 < x < 0 \end{cases}$ 　　　　　(194)

c. $u(x,0) = \begin{cases} 1, & \text{if } x < -1 \\ 0 & \text{if } -1 < x < 1 \\ -1, & \text{if } 1 < x \end{cases}$ 　　　　　(195)

d. $u(x,0) = \begin{cases} 1, & \text{if } |x| \geq 1 \\ x^2, & \text{if } |x| \leq 1 \end{cases}$ 　　　　　(196)

8. Consider the integral conservation law with a source term

$$\frac{d}{dt} \int_{x_1}^{x_2} u(x,t)\, dx + \frac{1}{2}\{u^2(x_2,t) - u^2(x_1,t)\} = \int_{x_1}^{x_2} \Lambda(x)\, dx \qquad (197)$$

where

$$\Lambda(x) = \begin{cases} 1, & \text{if } |x| > \frac{1}{2} \\ \alpha = \text{constant}, & \text{if } |x| < \frac{1}{2} \end{cases} \qquad (198)$$

Calculate the weak solution for the initial-value problem

$$u(x,0) = C = \text{constant} \qquad (199)$$

for all ranges of values of the constants α and C. In particular, show that the interface condition at $x = \pm\frac{1}{2}$ is that u is continuous there. Use this condition to connect solutions across the two interfaces, and fit shocks and fans where appropriate.

9. A wavemaker at the origin is impulsively pushed with constant speed v into quiescent water of unit height over $0 \leq x < \infty$. A bore with speed $U = $ constant > 0 starts propagating to the right into the water at rest. The speed of the water behind the bore is u_1 and the height is h_1. Clearly, we must have $u_1 = v$ in order to satisfy the boundary condition at the wavemaker. Therefore, u_1 is known and U and h_1 are unknown.

a. Show that eliminating U from (89) gives the cubic equation

$$F_i(h_1) \equiv h_1^3 - h_1^2 - (1 + 2v^2)h_1 + 1 = 0 \qquad (200)$$

for h_1. Show that for any given $v > 0$, this equation has one negative root (which we discard) and two positive roots $h_1^{(1)} < 1$ and $h_1^{(2)} > 1$. Using (89a), we see that the root $h_1^{(1)}$ results in a negative bore speed. Therefore, the appropriate solution is $h_1^{(2)}$. Specialize your results to the numerical example $v = \sqrt{3/2}$ for which $h_1^{(2)} = 2$ and $U = \sqrt{3}$.

b. Now assume that there is a vertical wall at some sufficiently large distance $x = x_0$, so the incoming bore will reflect from this wall and propagate back to the left. Equivalently, we may regard the flow for $0 \leq x \leq x_0$ as resulting from the given wavemaker and an image wavemaker starting at $t = 0$ from $x = 2x_0$ and moving impulsively with speed $-v$. Show that the height h_3 of the water behind the reflected bore is given by the *larger* of the two positive roots of the cubic

$$F_r(h_3) \equiv h_3^3 - h_1 h_3^2 - (h_1^3 - h_1 + 1)h_3 + h_1^3 = 0 \qquad (201)$$

for a known h_1 and that this root gives $h_3 > h_1$.

10. Consider the uniformly propagating shocks corresponding to cases (1) and (2) in the table of Section 5.3.4 (p. 297). The dimensionless change in the entropy S across a shock is given by

$$\frac{S_1 - S_2}{C_v} \equiv s = \log \frac{p_1}{\rho_1^\gamma} \qquad (202)$$

where the subscript 1 denotes values behind the shock and 2 denotes values ahead. Show that s is positive for case (1) and negative for case (2).

11. Consider the equation

$$uu_x + u_y + yu_z = 1 \qquad (203)$$

a. Calculate the two-parameter family of characteristic curves.
b. Solve (203) for the initial-value problem $u = 0$ on $y = x^2 + z^2$.
c. Given an example of a noncharacteristic initial manifold on which $\Delta(0, \tau_1, \tau_2)$, as defined by (148), vanishes.

References

1. G. B. Whitham, *Linear and Nonlinear Waves*, Wiley-Interscience, New York, 1974.
2. R. Courant and D. Hilbert, *Methods of Mathematical Physics*, Vol. II: *Partial Differential Equations* by R. Courant, Wiley-Interscience, New York, 1962.
3. P. D. Lax, "Hyperbolic Systems of Conservation Laws and the Mathematical Theory of Shock Waves," Regional Conference Series in Applied Mathematics, Vol. II, S.I.A.M., Philadelphia, 1973.

Nonlinear First-Order Equations

Nonlinear first-order partial differential equations arise in geometrical optics, in the description of dynamical systems by Hamilton-Jacobi theory, and other applications. In this chapter we begin with a discussion of the underlying physical principles and then study the mathematical theory that provides a unifying description of a number of different problems.

6.1   Geometrical Optics: A Nonlinear Equation

In this section, we shall derive a nonlinear equation that is the basic mathematical model in geometrical optics, dynamics, and variational calculus. These links are established in later sections; here our discussion is based on the problem in optics. The results that we shall derive are also valid in acoustics or any process involving the propagation of a disturbance in an isotropic medium with a given space-dependent signal speed. Here discussion proceeds from physical principles, and the results are shown in Section 6.3 to be consequences of the general theory for the nonlinear first-order equations.

6.1.1 Huyghens' Construction; the Eikonal Equation

In geometrical optics, we study the propagation boundary of an optical disturbance (wave front) without regard to such factors as the intensity, frequency, or phase of the light wave. In fact, we only distinguish between domains through which a disturbance has passed and undisturbed ones and keep track of the boundary separating these two domains at any given time t. Moreover, we assume that a disturbance at some time $t = t_0$ at the point $P_0 = (x_0, y_0, z_0)$ propagates locally in an isotropic manner with speed $c_0 = c(x_0, y_0, z_0)$; that is, at time $t_0 + \Delta t$, the disturbance that originated at t_0 and P_0 has spread along a *spherical* surface of radius $c_0 \Delta T$ and center at P_0. Every point on the disturbance surface or wave front is also regarded as a continuous emitter of disturbances, consequently advancing the wave front into the medium.

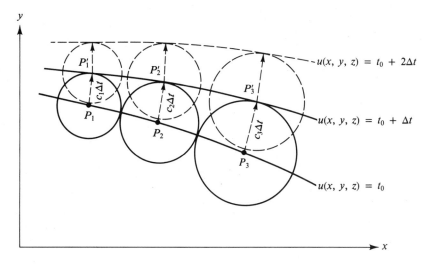

Figure 6.1

Suppose that we have a medium with a given signal speed $c(x, y, z)$ at every point and assume that at time $t = t_0$, the wave front is defined by the surface

$$u(x, y, z) = t_0 = \text{constant} \tag{1}$$

as shown in cross section in Figure 6.1. To fix ideas, let the region toward increasing y be the undisturbed zone at time $t = t_0$.

Now, each point on the surface defined by (1) emits a disturbance with a propagation speed c, which depends on location. In Figure 6.1, we illustrate the situation where c increases with x. Consider a sequence of points $P_1, P_2, P_3, \ldots$ lying on the surface (1). At time $t = t_0 + \Delta t$, the disturbances emitted from P_1, P_2, P_3 will be located along the spheres centered at P_1, P_2, P_3 and having radii equal to $c_1 \Delta t, c_2 \Delta t, c_3 \Delta t, \ldots$. Therefore, the wave front at time $t = t_0 + \Delta t$ will be the *envelope to all these spheres*. This geometrical construction, which is attributed to Huyghens, can be translated into an analytical description of the surface u once we recognize that *light rays are orthogonal to wave points*. In fact, the light rays emanating from a point P are a one-parameter family of radial vectors centered at P, and the particular rays that connect P_1 to P_1', P_2 to $P_2', \ldots$ are each orthogonal to the new front $P_1' P_2' P_3' \cdots$ at time $t = t_0 + \Delta t$. Let us denote the infinitesimal displacement vector along a light ray by $d\boldsymbol{\sigma}$. In Cartesian form, $d\boldsymbol{\sigma} \equiv dx\mathbf{i} + dy\mathbf{j} + dz\mathbf{k}$, where $\mathbf{i}, \mathbf{j}, \mathbf{k}$ are unit vectors in the x, y, and z directions, respectively. It then follows from the definition of the gradient of a scalar function that

$$|\text{grad } u| \cdot |d\boldsymbol{\sigma}| = dt \tag{2}$$

because $d\boldsymbol{\sigma}$ is orthogonal to grad u. But along a light ray we have

$$dt = \frac{|d\boldsymbol{\sigma}|}{c} \tag{3}$$

Therefore, eliminating dt from (2) and (3) gives

$$|\text{grad } u|^2 = \frac{1}{c^2} \tag{4a}$$

or

$$u_x^2 + u_y^2 + u_z^2 = \frac{1}{c^2(x, y, z)} \tag{4b}$$

in terms of the Cartesian coordinates x, y, z. Equation (4) is called the *eikonal* equation. It has a number of other interpretations besides the one just discussed. For example, see (71), (81), and (117) of this chapter. In Problem 12 of Chapter 4 we showed that (4b) governs characteristic manifolds of the wave equation. Thus, Huyghens' construction reconfirms our original interpretation of characteristics as wave fronts along which discontinuities propagate.

Consider the special case of (4b) where $c = c_0 = $ constant and $\partial/\partial z \equiv 0$—that is, disturbances do not vary in the z-direction, and introduce a point disturbance initially, say at $t = 0$, $x = y = 0$. Actually, this corresponds to a line of disturbances along the z-axis. Since c is constant, this point disturbance in the xy-plane must propagate along the front $x^2 + y^2 = c_0^2 t^2$, which is a circle of radius $c_0 t$ centered at the origin. In x, y, u space, this front is the surface of the right circular cone

$$u = \frac{\sqrt{x^2 + y^2}}{c_0} \tag{5}$$

which is easily seen to be a solution of (4b) for this special case. We shall rederive this result in Section 6.3 from the general theory (see also Problem 5 for an example with variable c).

6.1.2 The Equation for Light Rays

To simplify the derivation, we consider the two-dimensional case and denote $u_x = p$ and $u_y = q$. We shall show that along a light ray, the following system of five first-order equations is satisfied:

$$\frac{dx}{d\sigma} = cp \tag{6a}$$

$$\frac{dy}{d\sigma} = cq \tag{6b}$$

$$\frac{du}{d\sigma} = \frac{1}{c} \tag{6c}$$

$$\frac{dp}{d\sigma} = -\frac{c_x}{c^2} \tag{6d}$$

$$\frac{dq}{d\sigma} = -\frac{c_y}{c^2} \tag{6e}$$

where $d\sigma \equiv (dx^2 + dy^2)^{1/2}$ is the infinitesimal arc length along a light ray.

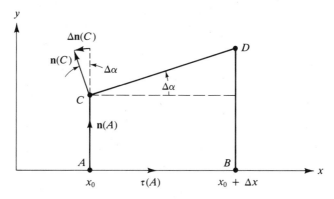

Figure 6.2

To prove (6a)–(6b), note that the vector with components $(dx/d\sigma, dy/d\sigma)$ is a unit tangent along a light ray. The unit normal to a wave front, defined by the surface $u(x, y) = $ constant, is by definition the vector with components (cp, cq). Therefore, (6a)–(6b) just states the already observed fact that light rays are normal to wave fronts. Equation (6c) is simply a restatement of (3), and we show next that (6d)–(6e) define the curvature of light rays.

To simplify this derivation further, let the wave front at $t = 0$ be tangent to the x-axis at some point $A = (x_0, 0)$, as shown in Figure 6.2.

The sketch shows that $c(A) \equiv c(x_0, 0)$ is smaller than $c(B) \equiv c(x_0 + \Delta x, 0)$ because after a time Δu, the light ray AC emerging from A has traveled a shorter distance than the ray BD.

The vectors **AB**, **AC**, and **BD** have the following components:

$$\mathbf{AB} = (\Delta x, 0); \qquad \mathbf{AC} = (0, c(A)\Delta u)$$

$$\mathbf{BD} = (0, c(B)\Delta u) = (0, c(A)\Delta u + c_x(A)\Delta x \Delta u)$$

Therefore, the infinitesimal turning angle $d\alpha$ of the wave front is

$$d\alpha \equiv \lim_{\Delta x \to 0} \frac{|\mathbf{BD} - \mathbf{AC}|}{|\mathbf{AB}|} = c_x(A)\,du$$

In general, for an arbitrary initial wave front orientation, we would have

$$\frac{d\alpha}{du} = \text{grad } c \cdot \tau \tag{7}$$

where τ is the unit tangent to the wave front—that is,

$$\tau \equiv (cq, -cp)$$

since the unit normal is

$$\mathbf{n} \equiv (cp, cq)$$

Also, we note from Figure 6.2 that $\mathbf{n}(C) = -(\sin \Delta\alpha)\mathbf{T}(A) + (\cos \Delta\alpha)\mathbf{n}(A) = -\Delta\alpha\tau(A) + \mathbf{n}(A) + O(\Delta\alpha^2)$. Therefore, $\quad \Delta\mathbf{n}(C) \equiv \mathbf{n}(C) - \mathbf{n}(A) = -\Delta\alpha\mathbf{T}(A) +$

$O(\Delta\alpha^2)$, and in the limit $\Delta\alpha \to 0$, the infinitesimal change $d\mathbf{n}$ in the unit normal is given by

$$d\mathbf{n} = -d\alpha\boldsymbol{\tau}$$

or

$$\frac{d\mathbf{n}}{d\sigma} = -\frac{d\alpha}{d\sigma}\boldsymbol{\tau} \tag{8}$$

Now

$$\frac{d\alpha}{d\sigma} = \frac{d\alpha}{du}\frac{du}{d\sigma} = \frac{1}{c}\,\mathrm{grad}\,c\cdot\boldsymbol{\tau} = qc_x - pc_y$$

where we have used (6c) for $du/d\sigma$.

Therefore, the first component of (8) is

$$\frac{d}{d\sigma}(cp) = -cq(qc_x - pc_y) \tag{9a}$$

and the second component is

$$\frac{d}{d\sigma}(cq) = cp(qc_x - pc_y) \tag{9b}$$

We develop the left-hand side of (9a) to obtain

$$c\frac{dp}{d\sigma} + p\left(c_x\frac{dx}{d\sigma} + c_y\frac{dy}{d\sigma}\right) = -cc_x q^2 + cc_y pq$$

Using (6a)–(6b) for $dx/d\sigma$ and $dy/d\sigma$ in the preceding, canceling $cc_y pq$ from both sides of the equation, and then dividing by c gives (6d). Similarly, (6e) follows from (9b). In Section 6.3, we shall show that the system (6) defines certain *characteristic strips* associated with the eikonal equation $p^2 + q^2 = 1/c^2$.

In the three-dimensional problem, with $x = x_1$, $y = x_2$, $z = x_3$, $\partial u/\partial x_i = p_i$, we have the system of seven first-order equations

$$\frac{dx_i}{d\sigma} = cp_i; \qquad i = 1, 2, 3 \tag{10a}$$

$$\frac{du}{d\sigma} = \frac{1}{c} \tag{10b}$$

$$\frac{dp_i}{d\sigma} = -\frac{1}{c^2}\frac{\partial c}{\partial x_i}; \qquad i = 1, 2, 3 \tag{10c}$$

An alternative description of the light rays in the two-dimensional problem is to eliminate u (which merely specifies the time along a ray) and σ (which specifies the length of a ray) and to derive the equation governing the ray trajectories in the xy-plane. Let us express y as a function of x along a ray and let $' \equiv d/dx$. We can combine (6a)–(6b) to write $y'^2 = q^2/p^2$, or

$$y'^2 \triangleq \frac{1 - c^2 p^2}{c^2 p^2}$$

when we use the eikonal equation $p^2 + q^2 = 1/c^2$. Solving this for p^2 gives

$$p^2 = \frac{1}{c^2(1 + y'^2)} \tag{11}$$

Dividing (6d) by (6a) gives $p' = -c_x/c^3 p$, from which it follows that

$$(p^2)' = -2\frac{c_x}{c^3} = \left(\frac{1}{c^2}\right)_x \tag{12}$$

We now differentiate (11) with respect to x and use (12) for $(p^2)'$ to find (after some algebra)

$$cy'' + (c_y - y'c_x)(1 + y'^2) = 0 \tag{13}$$

This second-order nonlinear equation defines a ray trajectory in a given medium with specified $c(x, y)$ once we prescribe $y(x_0)$ and $y'(x_0)$. In particular, note that if $c = $ constant, (13) reduces to $y'' = 0$, or $y(x)$ is a straight line, as expected.

A more fundamental interpretation of (13) is that if a light ray passes through two fixed points (x_0, y_0) and (x_1, y_1), the path that it takes (as defined by a solution of (13), subject to the two boundary conditions $y(x_0) = y_0$, $y(x_1) = y_1$), is a path of minimum time. This is *Fermat's principle*, discussed next.

6.1.3 Fermat's Principle

Consider two fixed points $P = (x_0, y_0)$ and $Q = (x_1, y_1)$ in the plane and assume that the speed of light $c(x, y)$ is prescribed everywhere. The time elapsed for light to travel from P to Q is given by the line integral

$$u = \int_P^Q \frac{d\sigma}{c} = \int_{x_0}^{x_1} \frac{(1 + y'^2)^{1/2}}{c(x, y)} dx \tag{14}$$

Ostensibly, u depends on the path $y(x)$ connecting the two points P and Q, and the minimum-time path is obtained by requiring the variation of u to vanish. This will be worked out in general in Section 6.2.2. Here, we anticipate and restate the result (32), which is basic in the calculus of variations: The variation of the functional

$$u = \int_{x_0}^{x_1} L(x, y, y') dx \tag{15}$$

vanishes for a given L, fixed endpoints (x_0 and x_1), and fixed values of y at these endpoints $[y(x_0) = y_0 = $ fixed, $y(x_1) = y_1 = $ fixed$]$ if the following equation holds:

$$\left(\frac{\partial L}{\partial y'}\right)' - \left(\frac{\partial L}{\partial y}\right) = 0 \tag{16}$$

In our case $L = s(y')/c(x, y)$, where $s \equiv (1 + y'^2)^{1/2}$. Therefore, we calculate

$$\left(\frac{\partial L}{\partial y'}\right)' = \left(\frac{y'}{cs}\right)'$$

$$= \frac{-c_x y' - c_y y''}{c^2 s} + \frac{y''}{sc} - \frac{y' y''}{cs^3} \tag{17}$$

and

$$\frac{\partial L}{\partial y} = -\frac{c_y s}{c^2} \tag{18}$$

Substituting (17) and (18) into (16) and simplifying gives (13).

The formula (13) with x as independent variable becomes awkward if rays have a vertical tangent. The corresponding formula with the roles of x and y reversed fails when rays have a horizontal tangent. In general, a result valid for arbitrary ray paths and in three dimensions can be derived by combining (10a) and (10c) to obtain a system of three second-order equations for x_1, x_2, x_3 with the arc length σ along a ray as the independent variable. These equations are not independent because the condition $(dx/d\sigma)^2 + (dy/d\sigma)^2 + (dz/d\sigma)^2 = 1$ must hold along a ray (see Problem 1).

6.2   Applications Leading to the Hamilton-Jacobi Equation

In Section 6.1, we studied the basic problem for geometrical optics from three points of view. Looking for surfaces of $t = $ constant along which disturbances propagate, we obtained the eikonal equation. The orthogonal trajectories to these surfaces define the light rays, which in turn could be regarded as paths of minimum time.

This same multiplicity of interpretations can be found in a number of applications governed by the Hamilton-Jacobi equation. This equation is a general version of the eikonal equation, and in this section we study how it arises in the calculus of variations, geometry, and dynamics.

6.2.1 The Variation of a Functional

Let the n continuously differentiable functions $q_1(s)$, $q_2(s)$, $\ldots$, $q_n(s)$ be the components of a vector $\mathbf{q}(s)$. We shall borrow the terminology used in dynamics (which, as we shall show later on, provides one interpretation of our results) and refer to the $q_1, \ldots, q_n$ as *coordinates*. In this case s will be the time, but for the present purposes s is an unspecified indepndent variable. Let L, the *Lagrangian*, be a given function of the $(2n + 1)$ variables $s, q_1, \ldots, q_n, \dot{q}_1, \ldots, \dot{q}_n$, where a dot denotes a derivative with respect to s. We shall use the abbreviated notation $L(s, q_i, \dot{q}_i)$ where the subscript i indicates that all n components of a vector quantity occur. We say that a *motion* is given if the q_i are prescribed functions of s. Thus, along a given motion, L is a scalar function of s.

Consider now the following functional:

$$J \equiv \int_{s_I}^{s_F} L(s, q_i, \dot{q}_i) \, ds \tag{19}$$

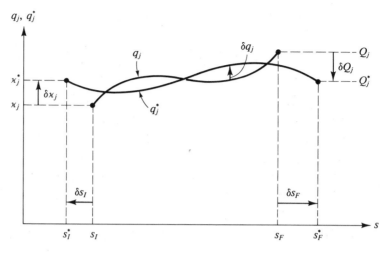

Figure 6.3

For a given motion, and if the integral exists over the interval $s_I \le s \le s_F$, J is a number. This number depends on the functional form of L, the given motion, and the values of s_I and s_F. Suppose that we now vary the motion and the endpoints slightly and evaluate J; that is, we compute

$$J^* \equiv \int_{s_I^*}^{s_F^*} L(s, q_i^*, \dot{q}_i^*)\, ds \tag{20}$$

where

$$
\begin{aligned}
s_I^* &\equiv s_I + \delta s_I; & s_F^* &\equiv s_F + \delta s_F \\
q_j^*(s) &\equiv q_j(s) + \delta q_j(s); & \dot{q}_j^*(s) &\equiv \dot{q}_j(s) + \delta \dot{q}_j(s), & j = 1, \ldots, n
\end{aligned}
\tag{21}
$$

Henceforth, for simplicitly, we shall omit pointing out that $j = 1, \ldots, n$. The δ-notation is somewhat awkward and is adopted only for the sake of tradition; it indicates that the quantity in question is a small perturbation of the associated variable. Note carefully that $\delta q_j(s)$ does not indicate that $q_j(s)$ is multiplied by δ. Rather, $\delta q_j(s)$ is a completely independent function of s that merely introduces a small perturbation to $q_j(s)$. Also, $\delta \dot{q}_j(s) \equiv (d/ds)[\delta q_j(s)]$. We shall refer to δq_j as the variation of q_j; a typical curve for $q_j(s)$ and $q_j^*(s)$ is sketched in Figure 6.3.

As seen in this figure, we have used the notation

$$
\begin{aligned}
\kappa_j &\equiv q_j(s_I); & Q_j &\equiv q_j(s_F) \\
\kappa_j^* &\equiv q_j^*(s_I^*); & Q_j^* &\equiv q_j^*(s_F^*)
\end{aligned}
\tag{22}
$$

to indicate the values of q_j and q_j^* at the initial and final points. It then follows that

$$\kappa_j^* = \kappa_j + \delta \kappa_j = q_j^*(s_I + \delta s_I) = q_j(s_I + \delta s_I) + \delta q_j(s_I + \delta s_I) \tag{23}$$

Strictly speaking, q_j is not defined outside the interval $s_I \le s \le s_F$ and $q_j(s_I + \delta s_I)$ may be ambiguous if $\delta s_I < 0$. However, regardless of the sign of δs_I, we extend the definition of q_j by linear extrapolation and set

$$q_j(s_I + \delta s_I) = q_j(s_I) + \dot{q}_j(s_I)\delta s_I$$
$$= \kappa_j + \dot{q}_j(s_I)\delta s_I \tag{24a}$$

We also have

$$\delta q_j(s_I + \delta s_I) = \delta q_j(s_I) \tag{24b}$$

and we have neglected quadratic terms in small quantities in both equations (24). If we use (24) in the right-hand side of (23) and equate perturbation quantities, we find

$$\delta \kappa_j = \delta q_j(s_I) + \dot{q}_j(s_I)\delta s_I$$

The notation

$$\dot{\kappa}_j \equiv \dot{q}_j(s_I); \qquad \dot{Q}_j \equiv \dot{q}_j(s_F) \tag{25}$$

then leads to the formula

$$\delta \kappa_j = \delta q_j(s_I) + \dot{\kappa}_j \delta s_I \tag{26a}$$

and the corresponding expression

$$\delta Q_j = \delta q_j(s_F) + \dot{Q}_j \delta s_F \tag{26b}$$

for the endpoint.

We are now ready to compute $\delta J \equiv J^* - J$. First, we split the interval of integration for J^* in (20) and write this as

$$J^* = \int_{s_I}^{s_F} L(s, q_i^*, \dot{q}_i^*)\, ds - \int_{s_I}^{s_I + \delta s_I} L(s, q_i^*, \dot{q}_i^*)\, ds + \int_{s_F}^{s_F + \delta s_F} L(s, q_i^*, \dot{q}_i^*)\, ds \tag{27}$$

If we ignore quadratic terms in perturbation quantities, the three integrals in (27) can be approximated as follows:

$$\int_{s_I}^{s_F} L(s, q_i^*, \dot{q}_i^*)\, ds = \int_{s_I}^{s_F} \left\{ L(s, q_i, \dot{q}_i) + \sum_{j=1}^{n} \frac{\partial L}{\partial q_j} \delta q_j + \sum_{j=1}^{n} \frac{\partial L}{\partial \dot{q}_j} \delta \dot{q}_j \right\} ds$$

$$\int_{s_I}^{s_I + \delta s_I} L(s, q_i^*, \dot{q}_i^*)\, ds = L(s_I, \kappa_i, \dot{\kappa}_i)\, \delta s_I$$

$$\int_{s_F}^{s_F + \delta s_F} L(s, q_i^*, \dot{q}_i^*)\, ds = L(s_F, Q_i, \dot{Q}_i)\, \delta s_F$$

Therefore,

$$\delta J = L(s_F, Q_i, \dot{Q}_i)\, \delta s_F - L(s_I, \kappa_i, \dot{\kappa}_i)\, \delta s_I + \int_{s_I}^{s_F} \sum_{j=1}^{n} \left\{ \frac{\partial L}{\partial q_j} \delta q_j + \frac{\partial L}{\partial \dot{q}_j} \delta \dot{q}_j \right\} ds \tag{28}$$

Let us denote

$$\frac{\partial L}{\partial \dot{q}_j} \equiv p_j \tag{29}$$

and refer to the p_j as the *momenta*, using the terminology of dynamics.

Along a given motion, the p_j are know functions of s, and we denote the endpoint values of the p_j by

$$p_j(s_I) \equiv \phi_j; \qquad p_j(s_F) \equiv P_j$$

Now, if we integrate the second term in the integrand in (28) by parts and collect the coefficients of δs_I and δs_F, we find on using (26) that

$$\delta J = \left[L(s_F, Q_i, \dot{Q}_i) - \sum_{j=1}^{n} P_j \dot{Q}_j \right] \delta s_F - \left[L(s_I, \kappa_i, \dot{\kappa}_i) - \sum_{j=1}^{n} \phi_j \dot{\kappa}_j \right] \delta s_I$$
$$+ \sum_{j=1}^{n} (P_j \delta Q_j - \phi_j \delta \kappa_j) + \int_{s_I}^{s_F} \sum_{j=1}^{n} \left\{ \frac{\partial L}{\partial q_j}(s, q_i, \dot{q}_i) - \frac{d}{ds} \left[\frac{\partial L}{\partial \dot{q}_j}(s, q_i, \dot{q}_i) \right] \right\} \delta q_j \, ds$$
$$\tag{30}$$

This defines the variations of the functional J in (19) for arbitrary variations of the endpoints, the values of the q_i at the endpoints, and the values of the functions $q_i(s)$ in the interval $s_I \le s \le s_F$.

6.2.2 A Variational Principle; the Euler-Lagrange Equations

Suppose that we require

$$\delta J = 0 \tag{31a}$$

subject to

$$\delta \kappa_i = \delta Q_i = 0 \tag{31b}$$

and

$$\delta s_I = \delta s_F = 0 \tag{31c}$$

This means that we allow the δq_i to be arbitrary (small) functions of s in the fixed interval $s_I \le s \le s_F$ with *fixed* values of the q_i at the endpoints and we look for that set of q_i for which the variation of J is zero. It follows from (30) that we must have

$$\int_{s_I}^{s_F} \sum_{j=1}^{n} \left[\frac{\partial L}{\partial q_j} - \frac{d}{ds}\left(\frac{\partial L}{\partial \dot{q}_j} \right) \right] \delta q_j \, ds = 0$$

This integral will vanish for arbitrary δq_j only if the coefficient of each δq_j in the integrand vanishes—that is,

$$\frac{\partial L}{\partial q_j} - \frac{d}{ds}\left(\frac{\partial L}{\partial \dot{q}_j} \right) = 0 \tag{32}$$

This system of n second-order equations, attributed to Euler and Lagrange, gives a *necessary condition* for the solution of the variational problem (31).

We see from (32) that whenever a particular q_j is absent from the Lagrangian, the assoicated momentum is a constant along the motion. In dynamics, $L \equiv T - V$, where T is the kinetic energy and V is the potential energy; the variational principle (31) is called Hamilton's principle, and (32) are the Lagrange equations governing the evolution of the coordinates q_i. For example, see Chapter 2 of [1].

Fermat's principle and the equations for the light rays in two dimensions discussed in Section 6.1.3 provide another special case. In the general three-dimensional problem for a medium with speed of light $c(x_1, x_2, x_3)$, the time elapsed for light to travel along a ray, defined parametrically in the form $x_1(s)$, $x_2(s)$, $x_3(s)$, is given by

$$J = \int_{s_I}^{s_F} \frac{\left(\sum_{j=1}^{3} \dot{x}_j^2 \right)^{1/2}}{c(x_1, x_2, x_3)} ds \tag{33a}$$

where s is any parameter that varies monotonically along the ray, such as the arc length or the time. Fermat's principle is just (31) and leads to the system (32) with $L = (\sum_{j=1}^{3} \dot{x}_j^2)^{1/2}/c$. As expected, the Euler-Lagrange equations associated with (33a) correspond to the equations for light rays that we obtain from (10) (see Problem 1).

A related more general problem that arises in geometry has

$$L = \left[\sum_{j=1}^{n} \sum_{k=1}^{n} g_{jk}(q_i) \dot{q}_j \dot{q}_k \right]^{1/2} \tag{33b}$$

Here, $L\,ds$ is the infinitesimal displacement in the n-dimensional space spanned by the curvilinear coordinates q_i, and the $g_{jk} = g_{kj}$ define the fundamental metric tensor (see Review Problem 3 of Chapter 2). To fix ideas, let q_1, q_2 denote the spherical polar coordinates on $\mathscr{S}$, the unit sphere centered at origin. We express the Cartesian x, y, z coordinates in terms of q_1, q_2 by

$$x = \sin q_1 \cos q_2; \qquad y = \sin q_1 \sin q_2, \qquad z = \cos q_1 \tag{34}$$

Thus, q_1 is the co-latitude and q_2 is the longitude on $\mathscr{S}$ measured from the x-axis. Now the infinitesimal displacement between two neighboring points on $\mathscr{S}$ is

$$|d\mathbf{x}| = \sqrt{dx^2 + dy^2 + dz^2} = \sqrt{dq_1^2 + \sin^2 q_1 \, dq_2^2} \tag{35}$$

Therefore, for this example $n = 2$, $g_{11} = 1$, $g_{12} = g_{21} = 0$, and $g_{22} = \sin^2 q_1$. If we regard s as an arbitrary parameter along a curve on $\mathscr{S}$, then J is the distance between s_I and s_F. The solution of (32) for a given Lagrangian (33b) is called a *geodesic* on the corresponding surface, and we refer to J as the *geodetic distance*. Thus, for the special case (35), the geodesics are great circles. In general, the solution of (32) associated with the functional (19) is called an *extremal*, and we next derive an alternate representation of the equations governing extremals.

6.2.3 Hamiltonian Form of the Variational Problem

We seek an alternate representation of the equations (32) as a system of $2n$ first-order equations. We accomplish this by eliminating the $\dot{q}_i$ in favor of a new set of n variables. One choice is to introduce a *Legendre transformation* defined by the *Hamiltonian*

$$H = \sum_{j=1}^{n} p_j \dot{q}_j - L \tag{36a}$$

where the p_i are the momenta defined in (29)

$$p_j \equiv \frac{\partial L}{\partial \dot{q}_j} \tag{36b}$$

The transformation (36) is implemented as follows. We solve the n equations (36b) for the $\dot{q}_i$ in terms of the q_i and p_i in the form

$$\dot{q}_j = f_j(s, q_i, p_i) \tag{37}$$

and then use this result in (36a) to express the Hamiltonian as a function of s, q_i, p_i. Note that a necessary and sufficient condition for being able to solve the system (36b) in the form (37) is that the determinant of the Jacobian matrix for the system (36b) be nonzero; that is,

$$\det \left\{ \frac{\partial^2 L}{\partial \dot{q}_j \partial \dot{q}_k} \right\} \neq 0 \tag{38}$$

Note in particular that if any one of the $\dot{q}_j$ is absent from L, then a Legendre transformation does not exist. See also the discussion following (46) for another example where the determinant in (38) vanishes.

If (38) is satisfied, a Legendre transformation exists and we can express the Hamiltonian in the form $H(s, q_i, p_i)$. Now, if we calculate the differential of H, we have

$$dH = \frac{\partial H}{\partial s} ds + \sum_{j=1}^{n} \left(\frac{\partial H}{\partial q_j} dq_j + \frac{\partial H}{\partial p_j} dp_j \right) \tag{39a}$$

But according to (36a), we must also have

$$dH = \sum_{j=1}^{n} \left(p_j d\dot{q}_j + \dot{q}_j dp_j - \frac{\partial L}{\partial q_j} dq_j - \frac{\partial L}{\partial \dot{q}_i} d\dot{q}_j \right) - \frac{\partial L}{\partial s} ds$$

or

$$dH = -\frac{\partial L}{\partial s} ds + \sum_{j=1}^{n} \left(-\frac{\partial L}{\partial q_j} dq_j + \dot{q}_j dp_j \right) \tag{39b}$$

when we use (36b). Therefore, equating the coefficients of ds, dq_j, and dp_j in the two expressions for dH gives

$$\frac{\partial H}{\partial s} = -\frac{\partial L}{\partial s} \tag{40a}$$

$$\frac{\partial H}{\partial q_j} = -\frac{\partial L}{\partial q_j} \tag{40b}$$

$$\frac{\partial H}{\partial p_j} = \dot{q}_j \tag{40c}$$

The formulas (36) and (40) define a Legendre transformation from $L(s, q_i, \dot{q}_i)$ and the $(q_i, \dot{q}_i)$ variables to $H(s, q_i, p_i)$ and the (q_i, p_i) variables. Repeating the Legendre transformation—that is, eliminating the p_i in favor of new variables, say u_i—we arrive back at the $L, q_i, \dot{q}_i$ set. Therefore, a Legendre transformation is its own inverse.

Now, if the q_i satisfy the Euler-Lagrange equations (32), then (36b) and (40b) lead to the system of n equations

$$\dot{p}_j = -\frac{\partial H}{\partial q_j} \tag{41a}$$

Equations (40c) give n equations for the $\dot{q}_i$,

$$\dot{q}_j = \frac{\partial H}{\partial p_j} \tag{41b}$$

and (40a) relates $\partial H/\partial s$ to $\partial L/\partial s$. Equations (41) are *Hamilton's differential equations* associated with the Hamiltonian $H(s, q_i, p_i)$.

Again, we note that if a particular coordinate is absent from H, the corresponding momentum is a constant according to (41a). This is consistent with the observation made earlier based on the Euler-Lagrange equations: If q_k is absent from L, then it is also absent from H according to (36a). Similarly, if a particular p_k is absent from H, (41b) implies that q_k is constant. Another important property of the Hamiltonian function is that along a solution of (41),

$$\dot{H} = \frac{\partial H}{\partial s} \tag{42}$$

To show this result, we compute the general expression for the derivative with respect to s of $H(s, q_i(s), p_i(s))$:

$$\dot{H} = \frac{\partial H}{\partial s} + \sum_{j=1}^{n} \left(\frac{\partial H}{\partial q_j} \dot{q}_j + \frac{\partial H}{\partial p_j} \dot{p}_j \right) \tag{43}$$

Along a solution of (41), the terms under the summation sign in (43) cancel identically, and we obtain (42). An important special case has L, and hence H, independent of s; that is, $\partial H/\partial s = 0$ [see (40a)]. In this case $H(q_i, p_i)$-constant is an integral of the system (41).

We shall now show that the Hamiltonian system of differential equations (41) also results directly from a variational principle. We introduce the functional

$$I = \int_{s_I}^{s_F} \left\{ \sum_{j=1}^{n} p_j \dot{q}_j - H(s, q_i, p_i) \right\} ds \tag{44}$$

which equals J if the Legendre transformation (36) exists. Consider the variational principle

$$\delta I = 0 \tag{45a}$$

subject to

$$\delta \kappa_j = \delta Q_j = 0 \tag{45b}$$

and

$$\delta s_I = \delta s_F = 0 \tag{45c}$$

In view of (45b) and (45c), δI is simply

$$\delta I = \int_{s_I}^{s_F} \left\{ \sum_{j=1}^{n} p_j^* \dot{q}_j^* - H(s, q_i^*, p_i^*) \right\} ds - \int_{s_I}^{s_F} \left\{ \sum_{j=1}^{n} p_j \dot{q}_j - H(s, q_i, p_i) \right\} ds \quad (46)$$

where the $q_j^*, \dot{q}_j^*$ are defined in (21) and $p_j^* \equiv p_j + \delta p_j$. If we ignore quadratic terms in perturbation quantities, (46) reduces to

$$\delta I = \int_{s_I}^{s_F} \left\{ \sum_{j=1}^{n} p_j \delta \dot{q}_j + \dot{q}_j \delta p_j - \frac{\partial H}{\partial q_j} \delta q_j - \frac{\partial H}{\partial p_j} \delta p_j \right\} ds$$

Integrating the first term by parts and then using (45b) gives

$$\delta I = \int_{s_I}^{s_F} \left\{ -\sum_{j=1}^{n} \left(\dot{p}_j + \frac{\partial H}{\partial q_j} \right) \delta q_j + \sum_{j=1}^{n} \left(\dot{q}_j - \frac{\partial H}{\partial p_j} \right) \delta p_j \right\} ds$$

Therefore, the variational principle (45) gives the Hamiltonian system (41). In dynamics (45) is called *Hamilton's extended principle*.

A Legendre transformation need not exist. For example, consider the case where the Lagrangian is a homogeneous function of degree one in the $\dot{q}_i$; that is, L has the property $L(s, q_i, \alpha \dot{q}_i) = \alpha L(s, q_i, \dot{q}_i)$ for any $\alpha > 0$. The Lagrangian (33) has this property. According to Euler's theorem for homogeneous functions, we have the identity

$$\sum_{j=1}^{n} \frac{\partial L}{\partial \dot{q}_j} \dot{q}_j = L$$

It then follows from the definition (36) for H that $H \equiv 0$, so a Legendre transformation is not possible.

At the risk of digressing somewhat from the main goal of this section, which is to derive the partial differential equations governing a field of extremals, it is instructive to look at some further specific examples to illustrate ideas.

The Newtonian Approximation for a Minimum-Drag Body

Sir Isaac Newton proposed that the force acting on a body in a flow is due to the body's suppression of the component of fluid momentum in the direction of the local normal to the surface. In effect, the normal component of velocity is locally annihilated at the surface, and the tangential component is left unchanged. This model is incorrect at all but hypersonic speeds, where a bow shock wave lies very close to the body and produces a flow that essentially agrees with Newton's description if one also ignores the centrifugal force on the fluid particles flowing along the surface. For more details see [2].

In spite of these objections, the Newtonian description is appealing as we can derive the formula for the drag using simple arguments. Also, the calculation of the minimum-drag body of revolution, which is attributed to Newton, is the earliest example of the use of calculus of variations.

We consider a body of revolution of length ℓ, zero nose radius, and maximum radius $R(\ell) = A$. The density ρ is assumed to be constant, and the flow speed U is taken in the axial direction (see Figure 6.4).

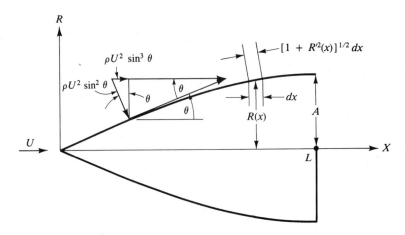

Figure 6.4

The area of the annular element of width dx is $dS = 2\pi R(x)[1 + R'^2(x)]^{1/2}\,dx$, where $' \equiv d/dx$. The normal component of the momentum flux per unit area at the surface is $M = \rho U^2 \sin^2 \theta$, where θ is the angle between the local tangent to the body and the horizontal. Therefore, the force (drag) acting in the horizontal direction for the annular element is $dD = M \sin \theta \, dS$. Note that in the Newtonian approximation, $dD = 0$ if $\sin \theta < 0$. Since $\sin \theta = R'/(1 + R'^2)^{1/2}$, we obtain the following formula for the drag:

$$D = 2\pi\rho U^2 \int_0^\ell \frac{R(x)R'^3(x)}{1 + R'^2(x)}\,dx \tag{47a}$$

Using dimensionless variables $r \equiv R/\ell$, $x \equiv X/\ell$, and $a \equiv A/\ell$ and expressing the drag in the usual form

$$D \equiv \tfrac{1}{2}\rho U^2 S C_D$$

where $S \equiv \pi A^2$ and C_D is a dimensionless drag coefficient, gives

$$C_D = \frac{4}{a^2} \int_0^1 \frac{r(x)r'^3(x)}{1 + r'^2(x)}\,dx \tag{47b}$$

Suppose that we fix a and ask for the shape $r(x)$ that results in minimum drag. This is the variational problem (31) with $J = a^2 C_D/4$, $n = 1$, $s = x$, $r = q$, $s_I = 0$, $s_F = 1$, $\kappa = 0$, $Q = a$, and $L(r, r') = rr'^3/(1 + r'^2)$; that is, L does not depend explicitly on x. The Euler-Lagrange equation (32) becomes

$$\frac{r'^3}{1 + r'^2} - \frac{d}{dx}\left[\frac{rr'^2(3 + r'^2)}{(1 + r'^2)^2}\right] = 0 \tag{48a}$$

which after some algebra simplifies to

$$r'^2(1 + r'^2) + rr''(3 - r'^2) = 0 \tag{48b}$$

This nasty nonlinear equation is to be solved subject to the boundary conditions

$r(0) = 0, r(1) = a$. Actually, (48b) has an integral that is not directly obvious from this expression but is immediately apparent from the fact that the Hamiltonian is independent of x.

To show this, we first compute [see (36b)]

$$p \equiv \frac{\partial L}{\partial r'} = \frac{rr'^2(3 + r'^2)}{(1 + r'^2)^2} \tag{49}$$

Then we solve for r'^2 in terms of p and r:

$$r'^2 = \frac{3r - 2p - (9r^2 - 8rp)^{1/2}}{2(p - r)} \tag{50}$$

Therefore, $H = pr' - L$ has the following form in terms of r and r':

$$H = \frac{rr'^3(3 + r'^2)}{(1 + r'^2)^2} - \frac{rr'^3}{1 + r'^2} = \frac{2rr'^3}{(1 + r'^2)^2} \tag{51a}$$

Using (50) to express r' in terms of r and p gives

$$H(r, p) = \left[\frac{3r - 2p - (9r^2 - 8rp)^{1/2}}{2(p - r)} \right]^{3/2} [5r - 4p + (9r^2 - 8rp)^{1/2}] \tag{51b}$$

The Hamiltonian form (41) of the equations governing the extremal is not particularly useful in this example. However, the result (51a) with $H = \alpha = $ constant does provide an integral for the Euler-Lagrange equation (48b), as can be verified by showing that $dH/dx = 0$ reduces to (48b). Thus, we need only to integrate the first-order equation

$$\frac{2rr'^3}{(1 + r'^2)^2} = \alpha = \text{constant} \tag{52}$$

to calculate the extremal. The constant α and the second constant of integration resulting from integrating (52) are to be evaluated using the two boundary conditions $r(0) = 0, r(1) = a$. The integration of (52) for $r(x)$ cannot be carried out explicitly but is easily worked out numerically.

A Problem in Dynamics

Consider a dynamical system having two degrees of freedom characterized by the coordinates q_1, q_2, which evolve as functions of the time t. For the time being, we describe this system somewhat generally by assuming that its Lagrangian has the form

$$L = T - V \tag{53a}$$

where T is the kinetic energy, with the form

$$T \equiv \tfrac{1}{2}[\phi_1(q_1) + \phi_2(q_2)][\dot{q}_1^2 + \dot{q}_2^2] \tag{53b}$$

and V is the potential energy, with the form

$$V(q_1, q_2) \equiv -\frac{V_1(q_1) + V_2(q_2)}{\phi_1(q_1) + \phi_2(q_2)} \tag{53c}$$

for prescribed functions $\phi_1(q_1)$, $\phi_2(q_2)$, $V_1(q_1)$, and $V_2(q_2)$. Since L does not depend on t explicitly, H is constant along a solution, so it is useful to calculate H first. We have

$$p_j \equiv \frac{\partial L}{\partial \dot{q}_j} = (\phi_1 + \phi_2)\dot{q}_j \tag{54}$$

Therefore, in this case

$$H \equiv \sum_{j=1}^{2} p_j\dot{q}_j - L = T + V = E = \text{total energy} = \text{constant} \tag{55}$$

To derive the Lagrange equations, we compute

$$\frac{\partial L}{\partial q_j} = \frac{1}{2}\frac{d\phi_j}{dq_j}(\dot{q}_1^2 + \dot{q}_2^2) + \frac{1}{\phi_1 + \phi_2}\frac{dV_j}{dq_j} - \frac{V_1 + V_2}{(\phi_1 + \phi_2)^2}\frac{d\phi_j}{dq_j}$$

and in view of (55), this is just

$$\frac{\partial L}{\partial q_j} = \frac{E}{\phi_1 + \phi_2}\frac{d\phi_j}{dq_j} + \frac{1}{\phi_1 + \phi_2}\frac{dV_j}{dq_j} \tag{56}$$

Using (54) and (56) in Lagrange's equations (32) gives

$$\frac{d}{dt}[(\phi_1 + \phi_2)\dot{q}_j] - \frac{E}{Q_1 + Q_2}\frac{d\phi_j}{dq_j} - \frac{1}{\phi_1 + \phi_2}\frac{dV_j}{dq_j} = 0, \qquad j = 1, 2 \tag{57}$$

These equations admit a second integral, as can be seen by multiplying (57) by $2(\phi_1 + \phi_2)\dot{q}_j$, to obtain

$$\frac{d}{dt}[(\phi_1 + \phi_2)^2\dot{q}_j^2 - 2E\phi_j - 2V_j] = 0, \qquad j = 1, 2 \tag{58a}$$

or

$$(\phi_1 + \phi_2)^2\dot{q}_j^2 - 2E\phi_j - 2V_j = \delta_j = \text{constant}, \qquad j = 1, 2 \tag{58b}$$

The two integrals in (58b) are not independent, as can be seen by adding the two expressions for $j = 1, 2$ to obtain $\delta_1 + \delta_2 = 0$ when the energy integral (55) is used. The significance of the integrals (58b) will become clear later on when we discuss the complete integral in Section 6.4.3.

The Hamiltonian form of the evolution equations are derived from the expression for H in (55) written as a function of the q_i, p_i; that is,

$$H(q_i, p_i) = \frac{1}{2}\frac{p_1^2 + p_2^2}{\phi_1 + \phi_2} - \frac{V_1 + V_2}{\phi_1 + \phi_2} \tag{59}$$

We find

$$\dot{q}_j = p_j/(\phi_1 + \phi_2) \tag{60a}$$

$$\dot{p}_j = \frac{H(q_i, p_i)}{\phi_1 + \phi_2}\frac{d\phi_j}{dq_j} + \frac{1}{\phi_1 + \phi_2}\frac{dV_j}{dq_j}, \qquad j = 1, 2 \tag{60b}$$

which are equivalent to (57).

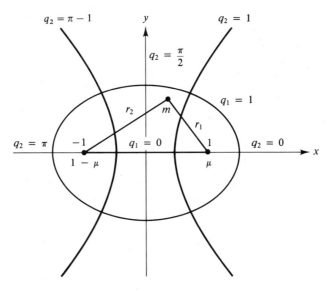

Figure 6.5

A specific example of a Lagrangian of the form (53) is given by the *Euler problem* when appropriate q_1, q_2 variables are chosen. Euler's problem consists of the motion of a particle of mass m in the field of two *fixed* Newtonian centers of gravitation having masses M_1 and M_2 and located at $X = \pm D$, $Y = 0$, $Z = 0$. It is clear from symmetry that the particle will remain in the $Z = 0$ plane if Z and dZ/dT both vanish simultaneously, and we shall consider only this planar case. We normalize the X- and Y-coordinates using D and the time T using $[D^3/\gamma(M_1 + M_2)]^{1/2}$, where γ is the universal gravitational constant [see (74) of Chapter 2]. The equations of motion (mass $\times$ acceleration = force) then take the dimensionless form:

$$\ddot{x} = -\mu \frac{(x-1)}{r_1^3} - (1-\mu)\frac{(x+1)}{r_2^3} \tag{61a}$$

$$\ddot{y} = -\mu \frac{y}{r_1^3} - (1-\mu)\frac{y}{r_2^3} \tag{61b}$$

where

$$r_1^2 \equiv (x-1)^2 + y^2 \tag{61c}$$

$$r_2^2 \equiv (x+1)^2 + y^2 \tag{61d}$$

and $\mu \equiv M_1/(M_1 + M_2)$ (see Figure 6.5).

We must note at the outset that having the two gravitational centers M_1 and M_2 fixed in an inertial frame is dynamically inconsistent, so Euler's problem does not model an actual gravitational three-body problem. A more appropriate model that does correspond to a limiting case of the gravitational three-body problem is the so-called restricted three-body problem (see Problem 2). Here, M_1

and M_2 move in circular orbits under their mutual gravitation, unaffected by m, which is assumed to be very small. Thus, m moves in the gravitational field that results from the motion of M_1 and M_2.

A limiting case of (61), which does have some physical significance, corresponds to letting $\mu \to 0$ with $\bar{x}$, $\bar{y}$, $\bar{t}$ fixed, where $\bar{x} \equiv (x - 1)/\mu$, $\bar{y} \equiv y/\mu$, $\bar{t} \equiv t/\mu$. In this limit, (61) tends to

$$\frac{d^2\bar{x}}{d\bar{t}^2} = -\frac{\bar{x}}{\bar{r}^3} - 1 \tag{62a}$$

$$\frac{d^2\bar{y}}{d\bar{t}^2} = -\frac{\bar{y}}{\bar{r}^3} \tag{62b}$$

where $\bar{r}^2 \equiv \bar{x}^2 + \bar{y}^2$. In this limit, the particle of mass m moves in the field of a Newtonian gravitational center at the origin plus a uniform horizontal field represented by the term (-1) in (62a). Such a uniform field would result, for example, if we were to account for the light pressure due to the sun on a near-earth satellite. A second interpretation of (62) is the motion of an electron under the added effect of a uniform external electrostatic field.

In writing (61), we appealed to Newton's second law of motion. An equivalent starting point would be to construct the Lagrangian for Hamilton's principle. Equations (61) would then correspond to Lagrange's equations (32). The advantage of having the Lagrangian in some given coordinate system, such as the Cartesian xy-system, is that we can then transform the equations of motions to any other coordinate system by first transforming the Lagrangian and then evaluating (32) for the new coordinates. We illustrate this feature next.

The kinetic energy T and potential energy V of the particle in terms of the dimensionless (x, y) variables are

$$T = \tfrac{1}{2}(\dot{x}^2 + \dot{y}^2) \tag{63a}$$

$$V = -\frac{\mu}{r_1} - \frac{(1 - \mu)}{r_2} \tag{63b}$$

Therefore, the Lagrangian is $L \equiv T - V$, and it is easily seen that Lagrange's equations with the $(x, y, \dot{x}, \dot{y})$ variables are just (61). It is also clear that V, as given in (63b), is not in the form (53c), so the integrals (58b) are not immediately available; the only obvious integral is the Hamiltonian.

Suppose that instead of (x, y), we introduce the curvilinear coordinates (q_1, q_2) defined by (see Figure 6.5)

$$x = \cosh q_1 \cos q_2 \tag{64a}$$

$$y = \sinh q_1 \sin q_2 \tag{64b}$$

Since it follows from (64) that

$$\frac{x^2}{\cosh^2 q_1} + \frac{y^2}{\sinh^2 q_1} = 1 \tag{64c}$$

$$\frac{x^2}{\cos^2 q_2} - \frac{y^2}{\sin^2 q_2} = 1 \tag{64d}$$

we see that curves of q_1 = constant are confocal ellipses in the xy-plane with foci at $x = \pm 1$, $y = 0$, semimajor axes $\cosh q_1$, and semiminor axes $\sinh q_1$. Similarly, curves of q_2 = constant are confocal hyperbolas orthogonal to the family of ellipses.

If we now compute $(\dot{x}, \dot{y})$ and use these expressions, together with (64), in (63), we find

$$T(q_1, q_2, \dot{q}_1, \dot{q}_2) = \tfrac{1}{2}(\cosh^2 q_1 - \cos^2 q_2)(\dot{q}_1^2 + \dot{q}_2^2)$$

$$V(q_1, q_2) = -\frac{\cosh q_1 + (2\mu - 1)\cos q_2}{\cos^2 q_1 - \cos^2 q_2}$$

Therefore, the Lagrangian in terms of the $(q_i, \dot{q}_i)$ variables of (64) is in the form (53) with $\phi_1 = \cosh^2 q_1$, $\phi_2 = -\cos^2 q_1$, $V_1 = \cosh q_1$, and $V_2 = (2\mu - 1)\cos q_2$.

The Hamiltonian (59) is given by

$$H(q_1, q_2, p_1, p_2) = \frac{1}{\cosh^2 q_1 - \cos^2 q_2}\left[\frac{p_1^2 + p_2^2}{2} - \cosh q_1 - (2\mu - 1)\cos q_2\right]$$

$$= E = \text{constant} \tag{65a}$$

where (p_1, p_2) are related to $(q_1, q_2, \dot{q}_1, \dot{q}_2)$ according to

$$p_1 \equiv \frac{\partial L}{\partial \dot{q}_1} = \frac{\partial T}{\partial \dot{q}_1} = (\cosh^2 q_1 - \cos^2 q_2)\dot{q}_1 \tag{65b}$$

$$p_2 \equiv \frac{\partial L}{\partial \dot{q}_2} = \frac{\partial T}{\partial \dot{q}_2} = (\cosh^2 q_1 - \cos^2 q_2)\dot{q}_2 \tag{65c}$$

The integrals (58b) have the form:

$$(\cosh^2 q_1 - \cos^2 q_2)^2 \dot{q}_1^2 - 2E \cosh^2 q_1 - 2\cosh q_1 = \delta_1 = \text{constant}$$

$$(\cosh^2 q_1 - \cos^2 q_2)^2 \dot{q}_2^2 + 2E \cos^2 q_1 - -2(2\mu - 1)\cos q_2 = \delta_2 = \text{constant}$$

which can also be written in the form

$$p_1^2 - 2E \cosh^2 q_1 - 2\cosh q_1 = \delta_1 = \text{constant} \tag{65d}$$

$$p_2^2 + 2E \cos^2 q_1 - 2(2\mu - 1)\cos q_2 = \delta_2 = \text{constant} \tag{65e}$$

In Section 6.2.6 we shall give an alternate derivation of these integrals and show that they allow us to reduce to quadrature the solution for the coordinates as functions of time and four constants of integration.

6.2.4 Field of Extremals from a Point; the Hamilton-Jacobi Equation

In a sense all our developments so far in this section have been preliminaries in preparation for deriving a partial differential equation that governs certain families of extremals.

Let $s = s_I$, $q_i = \kappa_i$ define a fixed point in the $(n + 1)$-dimensional space of (s, q_i); that is, we specify the $(n + 1)$ constants $(s_I, \kappa_1, \ldots, \kappa_n)$. For each choice of endpoint $(s_F, Q_1, \ldots, Q_n)$, if a solution of the two-point boundary-value problem for (32) exists, we obtain an extremal. Along such an extremal, J, as defined by the integral (19), can be expressed as a function of the endpoint; that is, $J = J(s_F, Q_i)$.

To see this, note that the extremals emerging from the point (s_I, κ_i) area an n-parameter family of functions $q_i = f_i(s, c_i)$ involving n constants of integration c_i; the other n constants of integration have been determined by the requirement $f_i(s_I, c_i) = \kappa_i$. Thus, along any one such extremal, the Lagrangian is $L(s, q_i, \dot{q}_i) = L(s, f_i(s, c_i), (\partial f_i / \partial s)(s, c_i)) \equiv \mathscr{L}(s, c_i)$, a function of s and n constants. Therefore, upon evaluating the integral in (39) along an extremal, we compute J as a function of s and n constants. But since the c_i are independent, we can invert the expressions $q_i = f_i(s, c_i)$ to compute $c_i = g_i(s, q_i)$. When these expressions are substituted into the result for J, one obtains J as a function of s and the q_i. The details for a particular example are worked out later on.

Now consider all possible values of (s_F, Q_i); that is, construct the *field of extremals* through the fixed point (s_I, κ_i). Clearly, J is a scalar function of the $(n + 1)$ variables (s, q_i), and according to (30), we have

$$\delta J(s, q_i) = \left[L(s, q_i, \dot{q}_i) - \sum_{j=1}^{n} p_j \dot{q}_j \right] \delta s + \sum_{j=1}^{n} p_j \delta q_j \tag{66a}$$

The terms in (30) multiplied by δs_I and $\delta \kappa_i$ vanish because the point (s_I, κ_i) is fixed. The integral in (30) also vanishes because we consider only the extremals emerging from (s_I, κ_i).

If a Legendre transformation exists for the given Lagrangian, we use (36a) to write (66a) in the form

$$\delta J(s, q_i) = -H(s, q_i, p_i) \, \delta s + \sum_{j=1}^{n} p_j \delta q_j \tag{66b}$$

It then follows from (66b) that

$$\frac{\partial J}{\partial s} = -H(s, q_i, p_i) \tag{67a}$$

$$\frac{\partial J}{\partial q_j} = p_j \tag{67b}$$

if a Legendre transformation exists. Combining the two equations, (67) gives the *Hamilton-Jacobi equation*

$$\frac{\partial J}{\partial s} + H\left(s, q_i, \frac{\partial J}{\partial q_i}\right) = 0 \tag{68}$$

for $J(s, q_i)$. This is a nonlinear first-order equation for the scalar J over the field of extremals. It is a fundamental equation in a number of applications, and we shall study its significance later on when we derive solutions.

Homogeneous Lagrangian

Now, suppose that a Legendre transformation does not exist—say, for example, L is a homogeneous function of degree 1 in the $\dot{q}_i$, as in (33b). In this case, equations (36) hold with $H \equiv 0$, and the p_i have the form

$$p_j \equiv \frac{\partial L}{\partial \dot{q}_j} = \sum_{\ell=1}^{n} \frac{g_{j\ell} \dot{q}_\ell}{L} \tag{69a}$$

Also, if we divide (36a) by L and use the definition (36b) for the p_i, we have

$$1 = \sum_{k=1}^{n} \frac{p_k \dot{q}_k}{L} \tag{69b}$$

Let the matrix $\{b_{jk}\}$ be the inverse of the $\{g_{jk}\}$ matrix. Multiplying (69a) by b_{kj} and summing over j gives (for each $k = 1, \ldots, n$)

$$\sum_{j=1}^{n} b_{kj} p_j = \sum_{j=1}^{n} \sum_{\ell=1}^{n} \frac{b_{kj} g_{j\ell} \dot{q}_\ell}{L} = \frac{\dot{q}_k}{L}$$

Multiplying this by p_k, summing over k, and using (69b) gives

$$\sum_{j=1}^{n} \sum_{k=1}^{n} b_{kj} p_j p_k = 1 \tag{70}$$

Finally, using (67b) for the p_i gives

$$\sum_{j=1}^{n} \sum_{k=1}^{n} b_{kj} \frac{\partial J}{\partial q_j} \frac{\partial J}{\partial q_k} = 1 \tag{71a}$$

This is a generalized eikonal equation; see (4b). For the case of light rays, J is the time and L is specified by (33a). Therefore, comparing (33a) and (33b) we have $\{g_{jk}\} = \delta_{jk}/c^2$, so $\{b_{jk}\} = c^2 \delta_{jk}$, where δ_{jk} is the Kronecker delta. Equation (71a) now reduces to

$$\left(\frac{\partial J}{\partial x_1}\right)^2 + \left(\frac{\partial J}{\partial x_2}\right)^2 + \left(\frac{\partial J}{\partial x_3}\right)^2 = \frac{1}{c^2(x_1, x_2, x_3)} \tag{71b}$$

The Analogy between Dynamics and Geometrical Optics

There is a mathematical analogy between the family of trajectories emerging from a fixed point in certain dynamical systems and the family of light rays emerging from a fixed point in a medium with a given speed of light c.

For simplicity, let us restrict attention to the two-dimensional problem and consider a particle moving in the $x_1 x_2$-plane under the influence of a conservative force field; that is, in appropriate dimensionless variables the equations of motion are

$$\ddot{x}_1 = -\frac{\partial V}{\partial x_1} \tag{72a}$$

$$\ddot{x}_2 = -\frac{\partial V}{\partial x_2} \tag{72b}$$

for a given potential energy $V(x_1, x_2)$. It then follows that the Lagrangian $L \equiv T - V$ has the form

$$L(x_1, x_2, \dot{x}_1, \dot{x}_2) = \tfrac{1}{2}(\dot{x}_1^2 + \dot{x}_2^2) - V(x_1, x_2) \tag{73}$$

and the Hamiltonian is

$$H(x_1, x_2, p_1, p_2) = \tfrac{1}{2}(p_1^2 + p_2^2) + V(x_1, x_2) = E = \text{constant} \tag{74}$$

The Hamilton-Jacobi equation associated with motion originating from some fixed point $t^{(0)}$, $x_1^{(0)}$, $x_2^{(0)}$ is

$$\frac{\partial S}{\partial t} + \frac{1}{2}\left(\frac{\partial S}{\partial x_1}\right)^2 + \frac{1}{2}\left(\frac{\partial S}{\partial x_2}\right)^2 + V(x_1, x_2) = 0 \tag{75}$$

where [see (68)] we are using $S(t, x_1, x_2)$ instead of J to indicate the integral of the Lagrangian from the fixed point to a variable point t, x_1, x_2 along solutions of (72).

Now, if we denote

$$S(t, x_1, x_2) \equiv -Et + W(x_1, x_2) \tag{76}$$

and consider only the family of solutions with the same energy E, it is easily seen that W obeys the two-dimensional eikonal equation [see (4b) with $\partial/\partial z = 0$],

$$\left(\frac{\partial W}{\partial x_1}\right)^2 + \left(\frac{\partial W}{\partial x_2}\right)^2 = 2[E - V(x_1, x_2)] \tag{77}$$

Thus, we identify $W \leftrightarrow u$ and $c^{-2} \leftrightarrow 2[E - V(x_1, x_2)]$, and we note that the equations for the light rays (6) are then identical to Hamilton's equations corresponding to (72); that is,

$$\frac{dx_j}{dt} = p_j, \qquad j = 1, 2 \tag{78a}$$

$$\frac{dp_j}{dt} = -\frac{\partial V}{\partial x_j}, \qquad j = 1, 2 \tag{78b}$$

if we identify $d\sigma$ in (6) with dt/c. This apparent discrepancy in units is due to the fact that u has units of time whereas W has units of action. One can avoid this discrepancy by either considering the eikonal equation in geometrical optics for the variable $U = uc_0$, where c_0 is some constant reference value of the speed of light, or by using the variable W/E (which has units of time) in (77).

An Example

A simple example that illustrates the preceding is motion starting from the origin under the influence of uniform gravity. We have $\ddot{x}_1 = 0$, $\ddot{x}_2 = -1$. Therefore, $x_1 = ut$ and $x_2 = vt - t^2/2$, where u and v are the arbitrary constant initial values of $\dot{x}_1$ and $\dot{x}_2$ at $t = 0$. The Lagrangian is $L(x_1, x_2, \dot{x}_1, \dot{x}_2) = (\dot{x}_1^2 + \dot{x}_2^2)/2 - x_2$. Therefore, along any motion emerging from the origin, we can express L as a function of the time and the two constants u, v in the form

$$L = \frac{u^2 + (v - t)^2}{2} - \left(vt - \frac{t^2}{2}\right)$$

$$= \frac{u^2 + v^2}{2} - 2vt + t^2$$

Integrating this expression with respect to time over the interval $(0, t)$ gives

$$S = \left(\frac{u^2 + v^2}{2}\right)t - vt^2 + \frac{t^3}{3}$$

Now, to express S as a function of the endpoint t, x, y, we use $u = x_1/t$,

$v = x_2/t - t/2$ to eliminate u and v and we find

$$S(t, x_1, x_2) = \frac{x_1^2}{2t} + \frac{x_2^2}{2t} - \frac{x_2 t}{2} - \frac{t^3}{24} \tag{79}$$

The Hamilton-Jacobi equation is

$$\frac{\partial S}{\partial t} + \frac{1}{2}\left(\frac{\partial S}{\partial x_1}\right)^2 + \frac{1}{2}\left(\frac{\partial S}{\partial x_2}\right)^2 + x_2 = 0 \tag{80}$$

and it is easily seen that (79) satisfies (80). Also, using (76), we find the following eikonal equation for W:

$$\left(\frac{\partial W}{\partial x_1}\right)^2 + \left(\frac{\partial W}{\partial x_2}\right)^2 = 2(E - x_2) \tag{81}$$

In classical dynamics, the wave fronts $W = $ constant are somewhat artificial mathematically defined surfaces with no particular physical significance. For further details, the reader is referred to [1].

6.2.5 Extremals from a Manifold; Transversality

We can generalize the idea of a field of extremals from a fixed point $s = s_I, q_i = \kappa_i$ to consider extremals originating from the *n-dimensional manifold* defined by the relation

$$\Gamma(s_I, \kappa_i) = 0 \tag{82}$$

in the $(n + 1)$-dimensional space of s and the q_i. We also assume that Γ is a continuously differentiable function of its arguments, so the surface is smooth. Thus, for example, if $n = 1$, (82) defines a smooth curve of possible initial points in s, q space. If $n = 2$, (82) defines a smooth surface in the three-dimensional space of s, q_1, q_2 and, in general, (82) defines an n-dimensional manifold in the $(n + 1)$-dimensional space of s, q_i.

Let B denote the endpoint s_F, Q_i, and consider an extremal for the functional (19) subject to B being fixed and the initial point A being allowed to lie anywhere on $\Gamma = 0$. Thus, we want $\delta J = 0$ with $\delta s_F = \delta Q_i = 0$. It then follows from (30) that we must satisfy (32). In addition, we must have

$$H(s_i, \kappa_i, \phi_i)\,\delta s_I - \sum_{j=1}^{n} \phi_j \delta \kappa_j = 0 \tag{83a}$$

where we have assumed that the Legendre transformation (36) exists and have used the definition (36a) for H evaluated on $\Gamma = 0$. But the variations δs_I and $\delta \kappa_i$ are not independent; they must be consistent with the requirement that the initial point s_I, κ_i lie on $\Gamma = 0$. Therefore, we must have

$$\delta\Gamma = 0 = \frac{\partial \Gamma}{\partial s_I}\delta s_I + \sum_{j=1}^{n}\frac{\partial \Gamma}{\partial \kappa_j}\delta \kappa_j \tag{83b}$$

Since $H \neq 0$, we can solve for δs_I from (83a) and substitute this into (83b). The resulting linear homogeneous expression for the $\delta \kappa_i$ is satisfied for arbitrary variations $\delta \kappa_j$, $j = 1, \ldots, n$, if

$$-\frac{H(s_I, \kappa_i, \phi_i)}{\phi_j} = \frac{\partial \Gamma / \partial s_I}{\partial \Gamma / \partial \kappa_j} \tag{84}$$

Condition (84), called the *transversality condition*, fixes the values of the ϕ_i (or, equivalently, the $\dot{\kappa}_i$) for any point on the manifold $\Gamma = 0$. An extremal emerging from $\Gamma = 0$ that satisfies (84) is said to be *transverse* to this manifold. If a Legendre transformation does not exist, we can still use the two requirements (83) to derive the transversality conditions.

To illustrate these ideas, consider the Lagrangian (33b) for the case $n = 2$. Let the function $\Gamma(q_1, q_2) = 0$, independent of s, be given. Thus $\Gamma = 0$ defines the same curve $\mathscr{C}$ in the $q_1 q_2$-plane for all s. We wish to calculate the necessary conditions governing the geodetic distance from the fixed point $B = (q_1^{(0)}, q_2^{(0)})$ to the curve $\mathscr{C}$. The geodesics from B must satisfy the Euler-Lagrange equations, and these follow immediately once we have calculated the expressions:

$$\frac{\partial L}{\partial q_1} = \frac{1}{2L} \left[\frac{\partial g_{11}}{\partial q_1} \dot{q}_1^2 + \frac{2 \partial g_{12}}{\partial q_1} \dot{q}_1 \dot{q}_2 + \frac{\partial g_{12}}{\partial q_1} \dot{q}_2^2 \right] \tag{85a}$$

$$\frac{\partial L}{\partial q_2} = \frac{1}{2L} \left[\frac{\partial g_{11}}{\partial q_2} \dot{q}_1^2 + \frac{2 \partial g_{12}}{\partial q_2} \dot{q}_1 \dot{q}_2 + \frac{\partial g_{22}}{\partial q_2} \dot{q}_2^2 \right] \tag{85b}$$

$$\frac{\partial L}{\partial \dot{q}_1} = \frac{1}{L} [g_{11} \dot{q}_1 + g_{12} \dot{q}_2] \tag{85c}$$

$$\frac{\partial L}{\partial \dot{q}_2} = \frac{1}{L} [g_{12} \dot{q}_1 + g_{22} \dot{q}_2] \tag{85d}$$

As shown earlier, $H = 0$ in this case, and a Legendre transformation does not exist. However, (83a) with $H = 0$ and (83b) with $\partial \Gamma / \partial s_I = 0$ must still hold on $\mathscr{C}$, and combining these two expressions, we obtain the transversality condition

$$\frac{g_{11} \dot{q}_1 + g_{22} \dot{q}_2}{g_{12} \dot{q}_1 + g_{22} \dot{q}_2} = \frac{\partial \Gamma / \partial \kappa_1}{\partial \Gamma / \partial \kappa_2} \tag{86}$$

For the special case of Cartesian coordinates $q_1 = x_1$, $q_2 = x_2$, $g_{11} = g_{22} = 1$, $g_{12} = 0$, (86) reduces to

$$\frac{dx_1}{dx_2} = \frac{\partial \Gamma / \partial \kappa_1}{\partial \Gamma / \partial \kappa_2}$$

—that is, the geometrically obvious statement that geodesics are normal to $\mathscr{C}$. For the case of spherical polar coordinates q_1, q_2 on the surface of the unit sphere [see (35)], the geodesics are the one-parameter family of great circles passing through B. The transversality condition isolates the great circles that intersect a given curve $\mathscr{C}$ at right angles (see Problem 3).

It is clear from the preceding simple examples that the transversality condition introduces a *local* requirement for an allowable extremal emerging from $\Gamma = 0$; there may be a number of points on $\Gamma = 0$ for which the transversality condition is satisfied. In fact, for the case of the unit sphere, we may regard the fixed point B as the north pole: $q_1 = 0$, with no loss of generality, and we see that any simple closed curve $\mathscr{C}$ on the sphere will contain at least two points (defining the closest and farthest points from B) where the transversality condition is satisfied. In the degenerate case, where $\mathscr{C}$ is the equator, this condition is satisfied at every point on $\mathscr{C}$.

Consider now how we might go about calculating an extremal from the manifold $\Gamma = 0$ to a fixed point B. One approach consists of "forward-shooting" in the following sense. At $s = s_I$ we guess the n initial values κ_i and use (84) to compute the ϕ_i (or the $\dot{\kappa}_i$). Using these values of κ_i, ϕ_i (or $\dot{Q}_i$) at s_I, we integrate Hamilton's equations (41) (or the Euler-Lagrange equations (32)) forward up to $s = s_F$ and check whether our solution satisfies the n end conditions $q_i(s_F) = Q_i$. If not, we revise our initial guess and repeat this process until the end conditions are satisfied (if possible). An alternate approach is to guess the n unknown end values P_i (or $\dot{Q}_i$) at B and integrate backward to $s = s_I$, where we check whether the n transversality conditions are satisfied.

Of course, in a particular problem, one or the other approach may be preferable depending on the structure of the solution. At any rate, a solution need not exist nor be unique. More details would require digressing into an area that is beyond the scope of this text. Here our main goal is the derivation of the partial differential equation associated with the field of extremals emerging from the manifold $\Gamma = 0$.

We proceed as in the last section except that now the initial point, instead of being fixed, is allowed to range over the manifold $\Gamma = 0$. For every such point, we construct an extremal that is transverse to $\Gamma = 0$. The set of all transverse extremals emerging from $\Gamma = 0$ defines a field (over which J is a scalar function of s and the q_i, with $J = 0$ on $\Gamma = 0$) as long as the extremals *do not lie entirely on $\Gamma = 0$*. We exclude this possibility and argue, as in the previous section, that $J(s, q_i)$ obeys the Hamilton-Jacobi equation (68) if the Legendre transformation (36) exists. We shall discuss the situation where the extremals lie entirely on $\Gamma = 0$ (characteristic initial manifold) or the case where the extremals are all tangent to $\Gamma = 0$ without lying entirely on it (caustic manifold) when we study the solution details of (68).

6.2.6 Canonical Transformations

Consider the Hamiltonian system of $2n$ first-order equations (41) associated with a given Hamiltonian $H(s, q_i, p_i)$. We have shown that equations (41) define the extremals associated with the variational principle (45).

A *canonical* transformation is a special transformation of the $2n$ variables $\{q_i, q_i\}$ to a new set $\{\bar{q}_i, \bar{p}_i\}$ such that the Hamiltonian form of the governing equations *is preserved*. More precisely, let

$$\bar{q}_j = F_j(s, q_i, p_i) \tag{87a}$$

$$\bar{p}_j = G_j(s, q_i, p_i) \tag{87b}$$

represent a general transformation of the $2n$ variables $\{q_i, p_i\}$. Note that we do not transform s itself, but instead we allow the relations linking the two sets of coordinates and momenta $\{q_i, p_i\}$ and $\{\bar{q}_i, \bar{p}_i\}$ to depend on s. The only restriction on the functions F_i and G_i for a general transformation is that the associated Jacobian be nonvanishing for all s in the interval of interest. For such a transformation, the equations governing the $\bar{q}_i$, $\bar{p}_i$ have the form

$$\dot{\bar{q}}_j = \frac{\partial F_j}{\partial s} + \sum_{k=1}^{n} \left(\frac{\partial F_j}{\partial q_k} \dot{q}_k + \frac{\partial F_j}{\partial p_k} \dot{p}_k \right)$$

$$\dot{\bar{p}}_j = \frac{\partial G_j}{\partial s} + \sum_{k=1}^{n} \left(\frac{\partial G_j}{\partial q_k} \dot{q}_k + \frac{\partial G_j}{\partial p_k} \dot{p}_k \right)$$

and if the $\{q_i, p_i\}$ satisfy (41), then the $\{\bar{q}_i, \bar{p}_i\}$ satisfy

$$\dot{\bar{q}}_j = \frac{\partial F_j}{\partial s} + \sum_{k=1}^{n} \left(\frac{\partial F_j}{\partial q_k} \frac{\partial H}{\partial p_k} - \frac{\partial F_j}{\partial p_k} \frac{\partial H}{\partial q_k} \right) \tag{88a}$$

$$\dot{\bar{p}}_j = \frac{\partial G_j}{\partial s} + \sum_{k=1}^{n} \left(\frac{\partial G_j}{\partial q_k} \frac{\partial H}{\partial p_k} - \frac{\partial G_j}{\partial p_k} \frac{\partial H}{\partial q_k} \right) \tag{88b}$$

The right-hand sides of (87) are known functions of the s, q_i, p_i for a given H and a given transformation (87). Therefore, inverting (87) and substituting the result into (88) gives $2n$ first-order equations of the form

$$\dot{\bar{q}}_j = \Lambda_j(s, \bar{q}_i, \bar{p}_i) \tag{89a}$$

$$\dot{\bar{p}}_j = \Delta_j(s, \bar{q}_i, \bar{p}_i) \tag{89b}$$

for functions Λ_i and Δ_i, which can be calculated in principle.

The transformation (87) is said to be canonical if there exists a new Hamiltonian $\bar{H}(s, \bar{q}_i, \bar{p}_i)$, depending on s and the new coordinates $\bar{q}_i$ and momenta $\bar{p}_i$ such that (89) is in Hamiltonian form—that is,

$$\Lambda_j = \frac{\partial \bar{H}}{\partial \bar{p}_j}; \qquad \Delta_j = -\frac{\partial \bar{H}}{\partial \bar{q}_j} \tag{90}$$

In general, a transformation (87) is not canonical. To illustrate ideas, consider the Hamiltonian form of the equations describing simple harmonic oscillations. We have $H(q, p) = \frac{1}{2}(p^2 + q^2)$ and $\dot{q} = p$, $\dot{p} = -q$—that is, $\ddot{q} + q = 0$. Now suppose we introduce the transformation $\bar{q} = q^3$, $\bar{p} = p$, which assigns a unique $(\bar{q}, \bar{p})$ to every (q, p) and vice versa according to $q = \bar{q}^{1/3}$, $p = \bar{p}$. The transformed differential equations are calculated as follows:

$$\dot{\bar{q}} = 3q^2 \dot{q} = 3q^2 p = 3\bar{q}^{2/3} \bar{p}$$

$$\dot{\bar{p}} = \dot{p} = -q = -\bar{q}^{1/3}$$

Now, in order for this system to be Hamiltonian, we must be able to find a function $\bar{H}(\bar{q}, \bar{p})$ such that $\partial \bar{H}/\partial \bar{p} = 3\bar{q}^{2/3} \bar{p}$ and $\partial \bar{H}/\partial \bar{q} = \bar{q}^{1/3}$. A necessary condition for the existence of such a function is the consistency condition

$$\frac{\partial}{\partial \bar{q}} (3\bar{q}^{2/3} \bar{p}) = \frac{\partial}{\partial \bar{p}} (\bar{q}^{1/3})$$

which is clearly violated. Admittedly, this is a somewhat contrived example proposed only to show that a transformation of dependent variables need not be canonical. Actually, a large class of transformations for the example in question are indeed canonical. In particular, the reader can verify that the linear transformation $\bar{q} = A_{11}q + A_{12}p$, $\bar{p} = A_{21}q + A_{22}p$ is canonical for any non-singular constant matrix $\{A_{ij}\}$.

The laborious calculations that we have outlined to check whether or not a given transformation is canonical are inefficient and unnecessary, and we next discuss an approach based on the coordinate-invariance of the variational principle that governs the system (41).

The variational principle (45) is independent of the variables that we choose to represent a given Hamiltonian system. Let $\{s, q_i, p_i\}$ and $\{s, \bar{q}_i, \bar{p}_i\}$ be two such sets of variables. Then we must have

$$\delta \int_{s_I}^{s_F} \left\{ \sum_{j=1}^{n} p_i \dot{q}_i - H(s, q_i, p_i) \right\} ds = 0 \tag{91a}$$

subject to $\delta\kappa_i = \delta Q_i = 0$ and $\delta s_I = \delta s_F = 0$. Similarly, we must also have

$$\delta \int_{s_I}^{s_F} \left\{ \sum_{j=1}^{n} \bar{p}_i \dot{\bar{q}}_i - \bar{H}(s, \bar{q}_i, \bar{p}_i) \right\} ds = 0 \tag{91b}$$

subject to $\delta\bar{\kappa}_i = \delta\bar{q}_i = 0$ and $\delta s_I = \delta s_F = 0$. Since (91b) is true for any canonical transformation, we conclude by subtracting (91b) from (91a) that the difference in integrands can at most be the total derivative of an arbitrary function of the two sets of coordinates, momenta and s. In this case, the integral of the difference will depend only on the endpoints, and since these are fixed, the variation in question will vanish. In summary, if the two sets of coordinates and momenta are related by a canonical transformation, we must have

$$\sum_{j=1}^{n} p_j \dot{q}_j - H(s, q_i, p_i) - \sum_{j=1}^{n} \bar{p}_j \dot{\bar{q}}_j + \bar{H}(s, \bar{q}_i, \bar{p}_i) = \dot{K} \tag{92}$$

where K, called the *generating function*, is an arbitrary function of the two sets of coordinates momenta and s. But, since a canonical transformation must also satisfy the $2n$ conditions (87), K can depend only on s and $2n$ of the $4n$ variables. In order that K define a transformation, it must involve n of the old variables and n of the new ones, which means it must have one of the following four possible forms:

$$K_1(s, q_i, \bar{q}_i), \qquad K_2(s, q_i, \bar{p}_i), \qquad K_3(s, \bar{q}_i, p_i), \quad \text{or} \quad K_4(s, p_i, \bar{p}_i)$$

We now show how an arbitrary function $K_1(s, q_i, \bar{q}_i)$ generates a canonical transformation. The total derivative of K_1 is given by

$$\dot{K}_1 = \frac{\partial K_1}{\partial s} + \sum_{j=1}^{n} \left(\frac{\partial K_1}{\partial q_j} \dot{q}_j + \frac{\partial K_1}{\partial \bar{q}_j} \dot{\bar{q}}_j \right)$$

and when we use this for the right-hand side of (92), multiply the result by ds, and collect terms, we find

$$\sum_{j=1}^{n}\left(p_j - \frac{\partial K_1}{\partial q_j}\right)dq_j - \sum_{j=1}^{n}\left(\bar{p}_j + \frac{\partial K_1}{\partial \bar{q}_j}\right)d\bar{q}_j + \left(\bar{H} - H - \frac{\partial K_1}{\partial s}\right)ds = 0$$

Since the q_i, $\bar{q}_i$ and s are $(2n + 1)$ independent quantities that can be varied arbitrarily, we conclude that we must have

$$p_j = \frac{\partial K_1}{\partial q_j} \tag{93a}$$

$$\bar{p}_j = -\frac{\partial K_1}{\partial \bar{q}_j} \tag{93b}$$

$$\bar{H} = H + \frac{\partial K_1}{\partial s} \tag{93c}$$

Equations (93) define a canonical transformation implicitly. To obtain the explicit form (87), we solve the n equations (93a) for the $\bar{q}_i$ as functions of s, q_i, p_i; this gives (87a). We then use the result just computed in the right-hand sides of the n equations (93b) to obtain (87b). Finally, the new Hamiltonian $\bar{H}$ is obtained from (93c), in which H and $\partial K_1/\partial s$ are expressed as functions of the new coordinates, new momenta and s.

To derive the transformation formulas for the case $K_2(s, q_i, \bar{p}_i)$, we appeal to our knowledge of Legendre transformations. Notice that in going from K_1 to K_2, we are eliminating the $\bar{q}_i$ variables in favor of the $\bar{p}_i$. Moreover, the relationship defining the $\bar{p}_i$ is just (93b). Thus, aside from the minus sign in this equation, we are dealing with precisely the same kind of transformation as we used in (36). In particular, we define K_2 by [see (36a)]

$$K_2(s, q_i, \bar{p}_i) = \sum_{j=1}^{n} \bar{p}_j\bar{q}_j + K_1(s, q_i, \bar{q}_i) \tag{94}$$

Now, if we let K in (92) be the expression given by (94) for K_1, we find

$$\sum_{j=1}^{n} p_j\dot{q}_j - H(s, q_i, p_i) - \sum_{j=1}^{n} \bar{p}_j\dot{\bar{q}}_j + \bar{H}(s, \bar{q}_i, \bar{p}_i) = \frac{d}{ds}\left[K_2(s, q_i, \bar{p}_i) - \sum_{j=1}^{n} \bar{p}_j\bar{q}_j\right]$$

Upon evaluating the right-hand side and collecting terms, we find

$$\sum_{j=1}^{n}\left(p_j - \frac{\partial K_2}{\partial q_j}\right)dq_j + \sum_{j=1}^{n}\left(\bar{q}_j - \frac{\partial K_2}{\partial \bar{p}_j}\right)d\bar{p}_j + \left(\bar{H} - H - \frac{\partial K_2}{\partial s}\right)ds = 0$$

Therefore, the implicit definition of the canonical transformation for a given K_2 is

$$p_j = \frac{\partial K_2}{\partial q_j} \tag{95a}$$

$$\bar{q}_j = \frac{\partial K_2}{\partial \bar{p}_j} \tag{95b}$$

$$\bar{H} = H + \frac{\partial K_2}{\partial s} \tag{95c}$$

In this case, we calculate the explicit form (87b) by solving the system (95a); then we use this result in (95b) to obtain (87a).

Similar results can be derived for the generating functions K_3 and K_4, but we do not list these. The interested reader can find these formulas in [1].

Let us concentrate on the canonical transformation defined by (95). We have argued that these equations define, in principle, an explicit canonical transformation of the form (87). Its inverse would be in the form

$$q_j = \bar{F}_j(s, \bar{q}_i, \bar{p}_i) \tag{96a}$$

$$p_j = \bar{G}_j(s, \bar{q}_i, \bar{p}_i) \tag{96b}$$

Faced with the task of solving the Hamiltonian system of $2n$ equations (41), we might ask if it is possible to transform (41) canonically to a simpler form that can be solved more readily. The simplest new Hamiltonian is $\bar{H} \equiv 0$, in which case the $\bar{q}_i$ and $\bar{p}_i$ are $2n$ constants. More importantly, the solution of the original system (41) is just (96), which gives the q_i and p_i as functions of s and $2n$ constants of integration!

In essense then, we can solve (41) if we can find the canonical transformation that renders $\bar{H} \equiv 0$. But, this transformation must obey (95), which gives

$$0 = H\left(s, q_i, \frac{\partial K_2}{\partial q_i}\right) + \frac{\partial K_2}{\partial s} \tag{97a}$$

the Hamilton-Jacobi equation for K_2. In Section 6.4, we demonstrate in detail how we can solve the system (41) if a *complete integral*, which is a solution $K_2(s, q_i, \alpha_i)$ of (97a) involving n independent constants α_i, is found.

An important special case has $\partial H/\partial s = 0$, so that $H(q_i, p_i) = \alpha_1 = $ constant. Suppose that we now seek a canonical transformation generated by a function of the type K_2—that is, one that depends on the q_i and $\bar{p}_i$, so that $\bar{H}$ is independent of the $\bar{q}_i$. In this case, the $\bar{p}_i = $ constant $= \gamma_i$. If we denote the generating function for such a canonical transformation by $W(q_i, \bar{p}_i)$, we see from (95a) and (95c) that W obeys

$$H\left(q_i, \frac{\partial W}{\partial q_i}\right) = \alpha_1 \tag{97b}$$

the time-independent Hamilton-Jacobi equation.

A complete integral of (97b) is a solution $W(q_i, \alpha_i)$ involving α_1 and $(n-1)$ additional independent constants $\alpha_2, \ldots, \alpha_n$. Again, we shall show that having such a solution is equivalent to being able to solve the system (41) associated with H.

At this point, we note that if H is independent of s, then the substitution $K_2 \equiv -\alpha_1 s + W$ reduces (97a) to (97b). We now illustrate ideas with two examples.

The Linear Oscillator

Consider the linear oscillator with variable frequency $\omega(s)$:

$$\ddot{q} + \omega^2(s)q = 0 \tag{98}$$

This equation also follows from Hamilton's differential equations for

$$H(s, q, p) \equiv \frac{p^2 + \omega^2(s)q^2}{2} \tag{99}$$

We find

$$\dot{q} = \frac{\partial H}{\partial p} = p \tag{100a}$$

$$\dot{p} = -\frac{\partial H}{\partial q} = -\omega^2(s)q \tag{100b}$$

and eliminating p gives (98). Note that H is not constant if ω depends on s. In fact,

$$\dot{H} = \frac{\partial H}{\partial s} = \omega \dot{\omega} q^2 \tag{100c}$$

and once $q(s)$ is calculated, (100c) defines $H(s)$ by quadrature along a solution.

Let us now study the properties of the canonical transformation generated by the function

$$K_1(s, q, \bar{q}) \equiv \frac{\omega(s)}{2} q^2 \cot \bar{q} \tag{101}$$

Equations (93a) and (93b) give the following two relations linking q and p to $\bar{q}$ and $\bar{p}$:

$$p \equiv \frac{\partial K_1}{\partial q} = \omega q \cot \bar{q} \tag{102a}$$

$$\bar{p} \equiv -\frac{\partial K_1}{\partial \bar{q}} = \frac{\omega q^2}{2} \csc^2 \bar{q} \tag{102b}$$

If we want to express $\bar{q}$ and $\bar{p}$ as functions of q and p, we first solve (102a) for $\bar{q}$ to obtain (87a) for this case:

$$\bar{q} = \cot^{-1} \frac{p}{\omega q} \tag{103a}$$

We then use this in (102b) to derive (87b):

$$\bar{p} = \frac{p^2 + \omega^2 q^2}{2\omega} \tag{103b}$$

The inverse transformations (96) are given by

$$q = \left(\frac{2\bar{p}}{\omega} \right)^{1/2} \sin \bar{q} \tag{104a}$$

$$p = (2\omega \bar{p})^{1/2} \cos \bar{q} \tag{104b}$$

The new Hamiltonian is

$$\bar{H} = H + \frac{\partial K_1}{\partial s} = \frac{1}{2}p^2 + \frac{\omega^2}{2}q^2 + \frac{\dot{\omega}}{2}q^2 \cot \bar{q}$$

$$= \frac{1}{2}[2\omega\bar{p}\cos^2\bar{q}] + \frac{\omega^2}{2}\left[\frac{2\bar{p}}{\omega}\sin^2\bar{q}\right] + \frac{\dot{\omega}}{2}\left[\frac{2\bar{p}}{\omega}\sin^2\bar{q}\right]\cot\bar{q}$$

$$= \omega(s)\bar{p} + \frac{\dot{\omega}(s)\bar{p}}{2\omega(s)}\sin 2\bar{q} \tag{105}$$

Therefore, Hamiltion's differential equations for the new variables are

$$\dot{\bar{q}} = \omega(s) + \frac{\dot{\omega}(s)}{2\omega(s)}\sin 2\bar{q} \tag{106a}$$

$$\dot{\bar{p}} = -\frac{\dot{\omega}(s)\bar{p}}{\omega(s)}\cos 2\bar{q} \tag{106b}$$

We note that for $\omega = $ constant, $\bar{q} = \omega t + \bar{q}_0$. Thus, $\bar{q}$ is the phase and $\bar{q}_0 = $ constant is the phase shift. Also, $\bar{p} = $ constant $= E/\omega$, where E is the constant energy. There is no particular advantage associated with the $\bar{p}, \bar{q}$ variables if ω depends on s, except if $\dot{\omega}$ is small. In this case, one may construct a perturbation solution having the form

$$\bar{q} = \bar{q}_0 + \int_0^s \omega(\sigma)\,d\sigma + \cdots \tag{107a}$$

$$\bar{p} = \bar{p}_0 + \cdots \tag{107b}$$

In fact, a Hamiltonian is said to be in *standard form* if it is independent of the $\bar{q}_i$ to $O(1)$ and is a 2π-periodic function of the $\bar{q}_i$ to higher order. For a large class of problems, one can transform the Hamiltonian to such a standard form as a starting point for a perturbation solution (see [3] and Section 8.4.3 for more details).

In the preceding discussion, the function K_1 in (101) was just "pulled out of the hat." Any function K_1 having continuous first partial derivates with respect to its arguments generates a canonical transformation. Suppose that instead of K_1, we were given the explicit transformation (103) or (104). How would we test whether such a transformation is canonical without having to find the associated generating function or having to transform the system (100) to (106) and show that this latter is derivable from a Hamiltonian? Again, we can appeal to the invariance of the expression $p\,dq - H\,ds$. We have

$$p\,dq - H\,ds = (2\omega\bar{p})^{1/2}\cos\bar{q}\left[\left(\frac{2\bar{p}}{\omega}\right)^{1/2}\cos\bar{q}\,d\bar{q} + \frac{1}{(2\omega\bar{p})^{1/2}}\sin\bar{q}\,d\bar{p}\right.$$

$$\left. - (2\bar{p})^{1/2}\sin\bar{q}\frac{d\omega}{2\omega^{3/2}}\right] - \frac{1}{2}(2\omega\bar{p})\cos^2\bar{q}\,ds - \frac{\omega^2}{2}\frac{2\bar{p}}{\omega}\sin^2\bar{q}\,ds$$

$$= 2\bar{p}\cos^2\bar{q}\,d\bar{q} + \sin\bar{q}\cos\bar{q}\,d\bar{q} - \bar{p}\sin\bar{q}\cos\bar{q}\frac{\dot{\omega}}{\omega}\,ds - \omega\bar{p}\,ds$$

Anticipating that the right-hand side must be in the form $\bar{p}\,d\bar{q} - \bar{H}\,ds + dK$, we use trigonometric identities to write $2\bar{p}\cos^2\bar{q}\,d\bar{q} = \bar{p}d\bar{q} + \bar{p}\cos 2\bar{q}\,d\bar{q}$, and $\sin\bar{q}\cos\bar{q} = \frac{1}{2}\sin 2\bar{q}$. Therefore,

$$p\,dq - H\,ds = \bar{p}\,d\bar{q} - \left(\omega\bar{p} + \frac{\dot{\omega}}{2\omega}\bar{p}\sin 2\bar{q}\right)ds + \left(\bar{p}\cos 2\bar{q}\,d\bar{q} + \frac{1}{2}\sin 2\bar{q}\,d\bar{p}\right) \tag{108}$$

We identify the second term on the right-hand side with $\bar{H}$ and note that the third term in parentheses is indeed the differential $d[(\bar{p}/2)\sin 2\bar{q}]$. This demonstrates that (103)–(104) is a canonical transformation without exhibiting K_1 or calculating the differential equations satisfied by the $\bar{q}$, $\bar{p}$ variables.

Now suppose that we want to find the canonical transformation that results in $\bar{H} = 0$ for new variables $\bar{q}$, $\bar{p}$. The Hamilton-Jacobi equation (97a) for this case is

$$\frac{1}{2}\left(\frac{\partial K_2}{\partial q}\right)^2 + \frac{1}{2}\omega^2(s)q^2 + \frac{\partial K_2}{\partial s} = 0 \tag{109}$$

Consider the case $\omega = $ constant, where setting $K_2 \equiv -\alpha s + W(q,\alpha)$ with $\alpha = $ constant gives

$$\frac{1}{2}\left(\frac{\partial W}{\partial q}\right)^2 + \frac{\omega^2}{2}q^2 = \alpha \tag{110}$$

We can integrate this equation and find

$$W = \int^q (2\alpha - \omega^2\xi^2)^{1/2}\,d\xi = \frac{\alpha}{\omega}\left[\sin^{-1}\frac{\omega q}{(2\alpha)^{1/2}} + \frac{\omega q}{(2\alpha)^{1/2}}\left(1 - \frac{\omega^2 q^2}{2\alpha}\right)^{1/2}\right] \tag{111}$$

which gives W as a function of s, q, and α, where α is the energy. In order to define the generating function K_2, we must decide what the constant $\bar{p}$ is; this constant can be any function of α that we wish to prescribe. One obvious choice is to let $\bar{p} = \alpha$, the energy, in which case

$$K_2(s,q,\bar{p}) \equiv -\bar{p}s + W(q,\bar{p}) \tag{112a}$$

where

$$W(q,\bar{p}) = \frac{\bar{p}}{\omega}\left[\sin^{-1}\frac{\omega q}{(2\bar{p})^{1/2}} + \frac{\omega q}{(2\bar{p})^{1/2}}\left(1 - \frac{\omega^2 q^2}{2\bar{p}}\right)^{1/2}\right] \tag{112b}$$

The canonical transformation generated by K_2 obeys [see (95)]

$$p = \frac{\partial K_2}{\partial q} = \frac{\partial W}{\partial q} = (2\bar{p} - \omega^2 q^2)^{1/2} \tag{113a}$$

$$\bar{q} = \frac{\partial K_2}{\partial \bar{p}} = -s + \frac{\partial W}{\partial \bar{p}} = -s + \frac{1}{\omega}\sin^{-1}\frac{\omega q}{(2\bar{p})^{1/2}} \tag{113b}$$

$$\bar{H} = 0 = H + \frac{\partial K_2}{\partial s} = H - \bar{p} \tag{113c}$$

The explicit form of this transformation is obtained by solving (113a)–(113b) for q and p. We find

$$q = \frac{(2\bar{p})^{1/2}}{\omega} \sin \omega(s + \bar{q}) \tag{114a}$$

$$p = (2\bar{p})^{1/2} \cos \omega(s + \bar{q}) \tag{114b}$$

Since $\bar{H} = 0$, $\bar{p} = $ constant $=$ energy and $\bar{q} = $ constant $=$ phase shift. Thus, the transformation (114) is in fact the solution of (100) with $\omega = $ constant.

A second choice is to set $\bar{p} = \alpha/\omega$, and it is easily seen that now

$$K_2(s, q, \bar{p}) \equiv -\omega \bar{p} s + W(q, \bar{p}) \tag{115a}$$

where

$$W(q, \bar{p}) = \bar{p} \left[\sin^{-1} \left(\frac{\omega}{2\bar{p}} \right)^{1/2} q + \left(\frac{\omega}{2\bar{p}} \right)^{1/2} \left(1 - \frac{\omega}{2\bar{p}} \bar{q}^2 \right)^{1/2} \right] \tag{115b}$$

and that W in (115b) generates the same canonical transformation as K_1 in (101).

Although we have evaluated the integral defining W in (112b) and (115b) explicitly, this calculation is not needed to define the canonical transformation; we need only to evaluate the integral resulting for $\partial W/\partial \bar{p}$.

The variables $\bar{q}$, $\bar{p}$ defined by (103) are normalized *angle and action* variables, which are important for the asymptotic solution of (106) for the case where $\dot{\omega}$ is small. In fact, it is in this context and in other perturbation problems (rather than the solution of linear constant coefficient equations) that canonical transformations play a crucial role (see [3] and Section 8.4.3).

If $\dot{\omega}$ is not small, the solution of the Hamilton-Jacobi equation (109) is no easier and certainly much less direct than the solution of (98). We pointed out earlier that there exists a connection between the solvability of a given Hamiltonian system of differential equations (41) on the one hand and the solvability (through the availability of a complete integral) of the associated Hamilton-Jacobi equation (97). The simple example of the linear oscillator gives a hint of this connection. Euler's problem discussed next provides a less trivial illustration. The detailed discussion of this question will be given in Section 6.4.

Euler's Problem

In Section 6.2.3 (page 338) we showed by direct calculation that Euler's problem has two independent integrals. It is interesting to see how this property emerges from the solvability of the Hamilton-Jacobi equation in the curvilinear coordinates (64).

We identify the energy E in (65a) with α_1 and write the Hamiltonian in the form

$$H(q_1, q_2, p_1, p_2) = \frac{1}{2} \frac{p_1^2 + p_2^2}{\cosh^2 q_1 - \cos^2 q_2} - \frac{\cosh q_1 + (2\mu - 1)\cos q_2}{\cosh^2 q_1 - \cos^2 q_2}$$

$$= \alpha_1 = \text{constant} \tag{116}$$

Therefore, the time-independent Hamilton-Jacobi equation for W is [after multiplying by $2(\cosh^2 q_1 - \cos^2 q_2)$]

$$\left(\frac{\partial W}{\partial q_1}\right)^2 + \left(\frac{\partial W}{\partial q_2}\right)^2 - 2[\cosh q_1 + (2\mu - 1)\cos q_2] - 2\alpha_1(\cosh^2 q_1 - \cos^2 q_2) = 0$$

(117)

We have already observed that use of the elliptic-hyperbolic coordinates leads in a rather straightforward way to two independent integrals for the solution. This result is also a consequence of the remarkable simplification of the structure of the solution of (117) for W in terms of the q_1, q_2 variables. In particular, we see that assuming a solution for W in the *separated form*

$$W = W_1(q_1, \alpha_1, \alpha_2) + W_2(q_1, \alpha_1, \alpha_2)$$

(118a)

is consistent with (117) because, upon substitution of (118a) into (117) and rearrangement of terms, we find

$$\left(\frac{\partial W_1}{\partial q_1}\right)^2 - 2\cosh q_1 - 2\alpha_1 \cosh^2 q_1$$

$$= -\left(\frac{\partial W_2}{\partial q_2}\right)^2 + 2(2\mu - 1)\cos q_2 - 2\alpha_1 \cos^2 q_2$$

(118b)

The right-hand side of (118b) depends only on the variable q_2, whereas the left-hand side depends only on q_1. Therefore, each side is equal to a constant, say α_2. We can then calculate W_1 and W_2 by quadrature in the form assumed in (118a). At this point, we may express each of the α_i as any desired function of the new momenta $\bar{p}_i$, which are also constants, since the new Hamiltonian is independent of the $\bar{q}_i$. A simple choice has $\alpha_1 = \bar{p}_1$ and $\alpha_2 = \bar{p}_2$, and we find the generating function in the separated form.

$$W(q_1, q_2, \bar{p}_1, \bar{p}_2) = W_1(q_1, \bar{p}_1, \bar{p}_2) + W_2(q_2, \bar{p}_1, \bar{p}_2)$$

(119a)

where

$$W_1(q_1, \bar{p}_1, \bar{p}_2) \equiv \int^{q_1} (\bar{p}_2 + 2\cosh \xi + 2\bar{p}_1 \cosh^2 \xi)^{1/2} \, d\xi$$

(119b)

$$W_2(q_1, \bar{p}_1, \bar{p}_2) \equiv \int^{q_2} [-\bar{p}_2 + 2(2\mu - 1)\cos \eta - 2\bar{p}_1 \cos^2 \eta]^{1/2} \, d\eta$$

(119c)

The canonical transformation generated by W satisfies [see (95)]

$$p_1 \equiv \frac{\partial W}{\partial q_1} = \frac{\partial W_1}{\partial q_1} = (\bar{p}_2 + 2\cosh q_1 + 2\bar{p}_1 \cosh^2 q_1)^{1/2}$$

(120a)

$$p_2 \equiv \frac{\partial W}{\partial q_2} = \frac{\partial W_2}{\partial q_2} = [-\bar{p}_2 + 2(2\mu - 1)\cos q_2 - 2\bar{p}_1 \cos^2 q_2]^{1/2}$$

(120b)

$$\bar{q}_1 \equiv \frac{\partial W}{\partial \bar{p}_1} = \int^{q_1} \frac{\cosh^2 \xi}{(\bar{p}_2 + 2\cosh \xi + 2\bar{p}_1 \cosh^2 \xi)^{1/2}} \, d\xi$$

$$- \int^{q_2} \frac{\cos^2 \eta}{[-\bar{p}_2 + 2(2\mu - 1)\cos \eta - 2\bar{p}_1 \cos^2 \eta]^{1/2}} \, d\eta$$

(121a)

$$\bar{q}_2 \equiv \frac{\partial W}{\partial \bar{p}_2} = \frac{1}{2} \int^{q_1} \frac{d\xi}{(\bar{p}_2 + 2\cosh\xi + 2\bar{p}_1\cosh^2\xi)^{1/2}}$$

$$-\frac{1}{2} \int^{q_2} \frac{d\eta}{[-\bar{p}_2 + 2(2\mu - 1)\cos\eta - 2\bar{p}_1\cos^2\eta]^{1/2}} \qquad (121b)$$

The new Hamiltonian equals the old one, since W does not involve s explicitly, and we have

$$\bar{H} = \alpha_1 = \bar{p}_1 \qquad (122)$$

Therefore, Hamilton's equations associated with (122) are

$$\dot{\bar{q}}_1 = \frac{\partial \bar{H}}{\partial \bar{p}_1} = 1 \qquad (123a)$$

$$\dot{\bar{q}}_2 = \frac{\partial \bar{H}}{\partial \bar{p}_2} = 0 \qquad (123b)$$

$$\dot{\bar{p}}_1 = -\frac{\partial \bar{H}}{\partial \bar{q}_1} = 0 \qquad (123c)$$

$$\dot{\bar{p}}_2 = -\frac{\partial \bar{H}}{\partial \bar{q}_2} = 0 \qquad (123d)$$

These have the solutions

$$\bar{q}_1 = s + \bar{q}_1^{(0)}, \bar{q}_1^{(0)} = \text{constant} \qquad (124a)$$

$$\bar{q}_2 = \bar{q}_2^{(0)} = \text{constant} \qquad (124b)$$

$$\bar{p}_1 = \text{constant} \qquad (124c)$$

$$\bar{p}_2 = \text{constant} \qquad (124d)$$

involving the four arbitrary constants $\bar{q}_1^{(0)}, \bar{q}_2^{(0)}, \bar{p}_1, \bar{p}_2$.

Now we show that the solution in terms of the original q_1, q_2, p_1, p_2 variables can be calculated in principle. First note that squaring (120a)–(120b), adding, and solving for $\bar{p}_1$ gives (116)—that is, that energy is conserved. Squaring (120a)–(120b) and subtracting the result gives

$$\bar{p}_2 = \tfrac{1}{2}(p_1^2 - p_2^2) - \cosh q_1 + (2\mu - 1)\cos q_2 - \bar{p}_1(\cosh^2 q_1 + \cos^2 q_2) \quad (125)$$

and we identify the constant $\bar{p}_2$ with $(\delta_1 - \delta_2)/2$ of (65d)–(65e). Thus, the two independent integrals we derived in Section 6.2.3 arise almost automatically according to this formulation.

To define the explicit canonical transformation for q_1, q_2, p_1, and p_2 as functions of $\bar{q}_1, \bar{q}_2, \bar{p}_1$, and $\bar{p}_2$, we proceed as follows. First we evaluate the integrals (121), which can be expressed in terms of elliptic functions, to give $\bar{q}_1$ and $\bar{q}_2$ in the form

$$\bar{q}_1 = \psi_1(q_1, q_2, \bar{p}_1, \bar{p}_2) \qquad (126a)$$

$$\bar{q}_2 = \psi_2(q_1, q_2, \bar{p}_1, \bar{p}_2) \qquad (126b)$$

Solving these for q_1 and q_2 gives the explicit form [see (96a)]

$$q_1 = \bar{F}_1(\bar{q}_1, \bar{q}_2, \bar{p}_1, \bar{p}_2) \tag{127a}$$

$$q_2 = \bar{F}_2(\bar{q}_1, \bar{q}_2, \bar{p}_1, \bar{p}_2) \tag{127b}$$

Although this is a straightforward calculation in principle, the details are rather messy and are omitted.

The expressions for p_1 and p_2 [see (96b)] now follow in the form

$$p_1 = [\bar{p}_2 + 2\cosh \bar{F}_1(\bar{q}_1, \bar{q}_2, \bar{p}_1, \bar{p}_2) + 2\bar{p}_1 \cosh^2 \bar{F}_1(\bar{q}_1, \bar{q}_2, \bar{p}_1, \bar{p}_2)]^{1/2}$$
$$\equiv \bar{G}_1(\bar{q}_1, \bar{q}_2, \bar{p}_1, \bar{p}_2) \tag{128a}$$

$$p_2 = [-\bar{p}_2 + 2(2\mu - 1)\cos \bar{F}_2(\bar{q}_1, \bar{q}_2, \bar{p}_1, \bar{p}_2) - 2\bar{p}_1 \cos^2 \bar{F}_2(\bar{q}_1, \bar{q}_2, \bar{p}_1, \bar{p}_2)]^{1/2}$$
$$\equiv \bar{G}_2(\bar{q}_1, \bar{q}_2, \bar{p}_1, \bar{p}_2) \tag{128b}$$

when we use (127) in (120).

Equations (127) and (128) define the explicit canonical transformation generated by (119). The solution for the q_i and p_i as functions of s and four integration constants now follows immediately when we substitute (124) into (127) and (128).

6.3   The Nonlinear Equation

The essential difference between the quasilinear and nonlinear problems is that in the first instance the partial differential equation specifies a unique characteristic direction at each point, whereas for the nonlinear case, we have a "cone" of possible characteristic directions. This feature necessitates that we keep track of certain characteristic strips (which are characteristic curves imbedded in an infinitesimal surface strip) in order to construct a solution. We begin our discussion with the case of two independent variables for which the geometry is easily visualized in the three-dimensional space of x, y, u.

6.3.1 The Geometry of Solutions

We consider the general nonlinear equation

$$F(x, y, u, p, q) = 0 \tag{129}$$

for the independent variables x and y and the dependent variable u, and we let $p \equiv u_x$ and $q \equiv u_y$. It will be useful to refer to the special case of the eikonal equation for which F does not depend on u and has the form [see (4b)]

$$F \equiv p^2 + q^2 - \frac{1}{c^2(x, y)} = 0 \tag{130}$$

As in Section 5.2.1, let us examine the geometrical constraints imposed by (129) on possible solution surfaces through a given point $P = (x_0, y_0, u_0)$. A normal vector $\mathbf{n}$ to a possible solution surface at P again has components $\mathbf{n} = (p, q, -1)$, but now the relation between p and q that is dictated by (129) is nonlinear. In particular, the family of possible normals will, in general, not lie in a plane; instead this family generates a curved surface centered at P. This surface

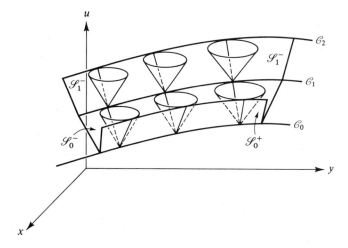

Figure 6.6

is actually a right circular cone for the eikonal equation [because the equation linking p and q defines a circle of radius $1/c(x_0, y_0)$], but in general the relation $F(x_0, y_0, u_0, p, q) = 0$ describes a more complicated surface. The tangent vectors to a possible solution surface through P, each of which is perpendicular to the cone of normals, will therefore also lie on a curved surface, called the *Monge cone* (see Figure 6.6).

Again, the Monge cone is only an inverted right circular cone with apex at P for the case (130). In general, it is a more complicated local surface dictated by the dependence of F on p and q. Notice that for the quasilinear problem, the Monge cone degenerates to a single characteristic ray. To see how we can isolate a solution surface in this space "filled" with Monge cones, let us assume, as we did in Section 5.2.1, that we want to consider only those solutions that pass through a given curve $\mathscr{C}_0$. We see immediately that even for the simple case of (130) depicted in Figure 6.6, *specifying the curve $\mathscr{C}_0$ does not isolate a unique solution surface near $\mathscr{C}_0$*. In fact, for the case of (130), there are two possible infinitesimal tangent surfaces $\mathscr{S}_0^+$ and $\mathscr{S}_0^-$, each of which contains $\mathscr{C}_0$ and is tangent to all the Monge cones on $\mathscr{C}_0$. The reason we have two surfaces is because (130) is quadratic in p and q; in the general case (129), we may have more possible infinitesimal strips through $\mathscr{C}_0$.

Based on the preceding observation, we conclude that, in addition to specifying $\mathscr{C}_0$, we must also specify a *strip condition* that isolates the particular strip we wish to follow. The analytical details of this strip condition are given in Section 6.3.3.

Next, we proceed a distance Δs along the generators of each of the Monge cones that are imbedded in the strip we have chosen, say $\mathscr{S}_0^-$, as indicated in Figure 6.6. This takes us to the new curve $\mathscr{C}_1$, where we repeat our construction of the Monge cones. But now, there is no longer any ambiguity as to which strip we must choose; *only one strip joins smoothly with $\mathscr{S}_0^-$*. In our case this is the strip $\mathscr{S}_1^-$, because if we were to choose the new strip $\mathscr{S}_1^+$ (which is omitted from

Figure 6.6 for clarity), the rays would have a finite discontinuity in the first derivative along $\mathscr{C}_1$ (the apex angle of each of the Monge cones on $\mathscr{C}_1$ is finite for a nonlinear equation), and this is inconsistent with a continuously differentiable solution surface.

6.3.2 Focal Strips and Characteristic Strips

According to our geometrical description, a solution surface u is everywhere tangent to a Monge cone along one of its generators. This particular generator, together with the associated infinitesimal tangent plane, forms a strip that is used to construct the solution surface.

We shall define the generator of a Monge cone as the intersection of two planes tangent to the cone in the limit as the lines of tangency approach one another.

In Figure 6.7, we show a portion of the Monge cone with apex at the point $P = (x_0, y_0, u_0)$. We identify the one-parameter family of generators by the parameter λ and consider two planes A and B tangent to the Monge cone; the plane A is tangent along the generator λ and the plane B is tangent along the neighboring generator $\lambda + \Delta\lambda$.

The points x, y, u lying on A satisfy

$$u - u_0 = (x - x_0)p(\lambda) + (y - y_0)q(\lambda) \tag{131a}$$

where p and q in (131a) are values consistent with (129) at P. Similarly, points lying on B satisfy

$$\begin{aligned} u - u_0 &= (x - x_0)p(\lambda + \Delta\lambda) + (y - y_0)q(\lambda + \Delta\lambda) \\ &= (x - x_0)p(\lambda) + (y - y_0)q(\lambda) \\ &\quad + \{(x - x_0)p'(\lambda) + (y - y_0)q'(\lambda)\}\,\Delta\lambda + O((\Delta\lambda)^2) \end{aligned} \tag{131b}$$

Therefore, points x, y, u that lie on the intersection PQ (see Figure 6.7) satisfy

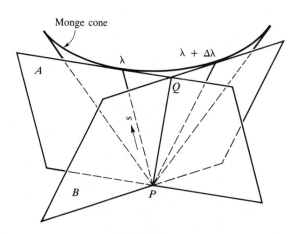

Figure 6.7

both equations. Subtracting these and taking the limit $\Delta\lambda \rightarrow 0$ gives the condition

$$(x - x_0)p'(\lambda) + (y - y_0)q'(\lambda) = 0 \tag{132}$$

which must hold on the generator of a Monge cone. We must also satisfy (129) on the generator; that is

$$F(x_0, y_0, u_0, p(\lambda), q(\lambda)) = 0 \tag{133a}$$

from which it follows by differentiation with respect to λ that

$$F_p p'(\lambda) + F_q q'(\lambda) = 0 \tag{133b}$$

Let s be a parameter that varies along a generator. We see that for x, y, u close to P, (132) and (133b) give

$$\frac{dx}{ds} = F_p \tag{134a}$$

$$\frac{dy}{ds} = F_q \tag{134b}$$

when we let $x - x_0 = dx$, $y - y_0 = dy$ and identify the coefficients of p' and q' in these two equations. Along a fixed generator, we must also satisfy the limiting form of (131a)—that is,

$$\frac{du}{ds} = p\frac{dx}{ds} + q\frac{dy}{ds}$$

—which, in view of (134a)–(134b) gives

$$\frac{du}{ds} = pF_p + qF_q \tag{134c}$$

The condition (134c) is called a *strip condition* because [see (131)] it assigns an infinitesimal plane to the curve $x(s)$, $y(s)$, $u(s)$.

A curve $x(s)$, $y(s)$, $u(s)$, which *for some given* $p(s)$, $q(s)$ satisfies the three equations (134), is called a *focal curve*. We see that for the quasilinear problem, the system (134) together with (21) of Chapter 5 corresponds to the characteristic system (23) of Chapter 5, and this system defines a unique curve passing through a given point P. In contrast, for the nonlinear problem, the four equations (134), (129) do not define $x(s)$ $y(s)$, $u(s)$, $p(s)$, and $q(s)$ uniquely; we need one more condition to ensure that the *focal strip* defined by (134) and (129) is tangent to a solution surface.

To illustrate this point, consider the counterexample that results if we choose an *arbitrary surface* $u = \phi(x, y)$ that is not a consistent solution of (129). If we ignore the requirement $p = \phi_x$, $q = \phi_y$, we can satisfy (129), (134) as follows. Equation (129) gives one algebraic relation linking p, q, x, and y in the form

$$F(x, y, \phi(x, y), p, q) = 0 \tag{135a}$$

Equation (134c) gives

$$\frac{du}{ds} = pF_p(x, y, \phi(x, y), p, q) + qF_q(x, y, \phi(x, y), p, q)$$

But since $u = \phi(x, y)$, we also have $du/ds = \phi_x(dx/ds) + \phi_y(dy/ds)$, and using (134a)–(134b) gives $du/ds = \phi_x F_p + \phi_y F_q$. Subtracting the last two expressions for du/ds, we obtain the second algebraic equation linking p, q:

$$[p - \phi_x(x, y)]F_p(x, y, \phi(x, y), p, q) + [q - \phi_y(x, y)]F_q(x, y, \phi(x, y), p, q) = 0$$
(135b)

Solving (135a)–(135b) gives $p(x, y)$, $q(x, y)$; using these in (134a)–(134b) leads to $x(s)$, $y(s)$. The strip $x(s)$, $y(s)$, $u(s)$, $p(s)$, $q(s)$ that results is *not necessarily tangent* to the surface $u = \phi(x, y)$. In summary, each of the many possible solutions of (129), (134) defines a focal strip that is not necessarily tangent to a solution surface. Our next task is to isolate from the preceding family of focal strips the one strip along which $p = \phi_x$, $q = \phi_y$ for a solution surface: $u = \phi(x, y)$.

On a given solution surface $u = \phi(x, y)$, (129) must be satisfied identically; that is,

$$F(x, y, \phi(x, y), \phi_x(x, y), \phi_y(x, y)) \equiv 0$$
(136a)

Moreover, the partial derivatives of (136a) with respect to x and y must also vanish; that is, denoting $\phi_x = p$, $\phi_y = q$, we must have

$$F_x + F_u p + F_p p_x + F_q q_x = 0$$
(136b)

$$F_y + F_u q + F_p p_y + F_q q_y = 0$$
(136c)

The solution surface $u = \phi(x, y)$ must contain the focal curves; hence $F_p = dx/ds$, $F_q = dy/ds$. In addition, we must have $p_y = q_x$ for consistency. Therefore, (136b) may also be written in the form

$$F_x + F_u p + p_x \frac{dx}{ds} + p_y \frac{dy}{ds} = 0$$

or

$$\frac{dp}{ds} = -(F_x + pF_u)$$
(137a)

Similarly, (136c) gives

$$\frac{dq}{ds} = -(F_y + qF_u)$$
(137b)

The crux of the derivation of (137) is the fact that we have identified p and q with ϕ_x and ϕ_y on a given solution surface $u = \phi(x, y)$.

In Section 6.1.2 we used physical arguments to derive the system (6) for the light rays associated with the two-dimensional eikonal equation

$$F \equiv p^2 + q^2 - \frac{1}{c^2(x, y)} = 0$$
(138)

Using our general theory for (138), we compute the following special case of (134), (137) with $F_u = 0$:

$$\frac{dx}{ds} = F_p = 2p \tag{139a}$$

$$\frac{dy}{ds} = F_q = 2q \tag{139b}$$

$$\frac{du}{ds} = pF_p + qF_q = 2p^2 + 2q^2 = \frac{2}{c^2(x, y)} \tag{139c}$$

$$\frac{dp}{ds} = -F_x = -\frac{2c_x}{c^3} \tag{139d}$$

$$\frac{dq}{ds} = -F_y = -\frac{2c_y}{c^3} \tag{139e}$$

If we identify $(2/c)\,ds$ with $d\sigma$, the infinitesimal distance along a light ray, we see that (139) is identical with (6). In particular, we note that the light rays are mathematically the projections of the characteristic curves on the xy-plane.

In view of the mathematical analogy between (138) and (77), we also conclude that (139a)–(139b) correspond to Hamilton's equations (78a) for the coordinates, and (139d)–(139e) correspond to (78b) for the momenta. The role of the Hamilton-Jacobi equation in dynamics is explored in more detail in Section 6.4.3.

The system (134) and (137) of five equations for the five variables x, y, u, p, q defines a four-parameter family of strips; one of the five integration constants is s_0, which appears only in the additive form $(s - s_0)$ in the solution because this system is autonomous.

We show next that (129) is an integral of the system (134), (137)—that is, that $F(x, y, u, p, q)$ is a constant along any solution of (134), (137). To prove this, we differentiate the expression for F with respect to s and obtain

$$\frac{dF}{ds} = F_x \frac{dx}{ds} + F_y \frac{dy}{ds} + F_u \frac{du}{ds} + F_p \frac{dp}{ds} + F_q \frac{dq}{ds}$$

Along a solution of (134), (137), this vanishes identically because

$$\frac{dF}{ds} = F_x F_p + F_y F_q + F_u(pF_p + qF_q) - F_p(F_x + pF_u) - F_q(F_y + qF_u) = 0$$

Therefore, $F = \text{constant}$, and when we require this constant to be zero [in order to conform with (129)], we reduce the solutions of (134), (137) to a three-parameter family. We shall refer to a solution of (134), (137) along which $F = 0$ as a *characteristic strip*. In the next section, we show how an integral surface that passes through a prescribed initial strip can be isolated from the three-parameter family of characteristic strips associated with (129).

6.3.3 The Initial-Value Problem

We are given a noncharacteristic initial strip $\mathscr{S}_0$ defined parametrically in the form

$$x = x_0(\tau), \qquad y = y_0(\tau), \qquad u = u_0(\tau), \qquad p = p_0(\tau), \qquad q = q_0(\tau) \qquad (140)$$

for functions x_0, y_0, u_0, p_0, q_0 that are continuous and have a continuous first derivative.

The five functions in (140) are not entirely arbitrary. To begin with, we must again exclude the situation where the ground curve $x_0(\tau), y_0(\tau)$ has intersections. More importantly, we must require the strip (140) to be (1) self-consistent—that is, to satisfy the strip condition

$$\frac{du_0}{d\tau} = p_0(\tau)\frac{dx_0}{d\tau} + q_0(\tau)\frac{dy_0}{d\tau} \qquad (141a)$$

—and (2) consistent with (129)—that is,

$$F(x_0(\tau), y_0(\tau), u_0(\tau), p_0(\tau), q_0(\tau)) = 0 \qquad (141b)$$

Equations (140), (141) impose three independent conditions to be satisfied by the three-parameter family of characteristic strips obtained by solving (134), (137) subject to $F = 0$. We shall demonstrate next that these three conditions specify a unique solution surface as long as a certain Jacobian does not vanish.

We interpret the construction of the solution surface u, in the sense depicted in Figure 6.8, of smoothly joining all the characteristic strips $\mathscr{C}_i$ that emerge from the initial strip $\mathscr{S}_0$. Thus, for any fixed value τ_j of the parameter along $\mathscr{S}_0$, we generate a characteristic strip $\mathscr{C}_j$, which for $s = 0$ coincides with $\mathscr{S}_0$ at τ_j. The converse of this construction, which will be useful in later discussion, is that whenever two integral surfaces join smoothly, this juncture occurs along a characteristic strip.

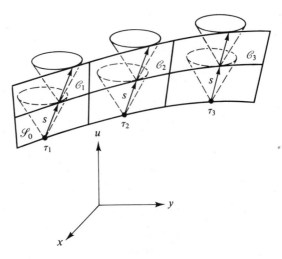

Figure 6.8

To implement the construction of a solution surface that passes through $\mathscr{S}_0$, we need to express the solution of (134), (137) in the form of a one-parameter family of characteristic strips. Let the general solution of (134), (137), subject to (129), be given in the form

$$x = \bar{X}(s - s_0, c_1, c_2, c_3) \tag{142a}$$

$$y = \bar{Y}(s - s_0, c_1, c_2, c_3) \tag{142b}$$

$$u = \bar{U}(s - s_0, c_1, c_2, c_3) \tag{142c}$$

$$p = \bar{P}(s - s_0, c_1, c_2, c_3) \tag{142d}$$

$$q = \bar{Q}(s - s_0, c_1, c_2, c_3) \tag{142e}$$

involving the three arbitrary constants c_1, c_2, c_3 and the additive constant s_0. A general one-parameter subfamily of (142) is obtained by regarding s_0, c_1, c_2, and c_3 to be arbitrary functions of the parameter τ; that is,

$$x = X(s, \tau) \equiv \bar{X}(s - s_0(\tau), c_1(\tau), c_2(\tau), c_3(\tau)) \tag{143a}$$

$$y = Y(s, \tau) \equiv \bar{Y}(s - s_0(\tau), c_1(\tau), c_2(\tau), c_3(\tau)) \tag{143b}$$

$$u = U(s, \tau) \equiv \bar{U}(s - s_0(\tau), c_1(\tau), c_2(\tau), c_3(\tau)) \tag{143c}$$

$$p = P(s, \tau) \equiv \bar{P}(s - s_0(\tau), c_1(\tau), c_2(\tau), c_3(\tau)) \tag{143d}$$

$$q = Q(s, \tau) \equiv \bar{Q}(s - s_0(\tau), c_1(\tau), c_2(\tau), c_3(\tau)) \tag{143e}$$

The dependence of the functions X, Y, U, P, Q on τ is specified by requiring the one-parameter family (143) to contain the initial strip (140); that is,

$$X(0, \tau) = x_0(\tau), \qquad Y(0, \tau) = y_0(\tau), \qquad U(0, \tau) = u_0(\tau)$$

$$P(0, \tau) = p_0(\tau), \qquad Q(0, \tau) = q_0(\tau)$$

Again, in practice, we calculate the solution directly in the form $x = X(s, \tau)$, and so on.

If the Jacobian

$$\Delta(s, \tau) \equiv X_s Y_\tau - Y_s X_\tau \tag{144}$$

does not vanish, we can invert (143a)–(143b) to express s and τ as functions of x, y. Using this result in (143c)–(143e) gives $u(x, y), p(x, y), q(x, y)$. It is easy to prove that this result satisfies (129) with $p = u_x$ and $q = u_y$ (for example, see Section 3.2 of [4].

If $\Delta(0, \tau) \equiv 0$ and $\mathscr{S}_0$ is a characteristic strip, the solution of (129) is non-unique, as in the quasilinear problem. In the exceptional case, where $\Delta(0, \tau) = 0$ but $\mathscr{S}_0$ is not a characteristic strip, $\mathscr{S}_0$ must be a focal strip because $\Delta(0, \tau) = 0$ implies that (134a)–(134b) are satisfied on $\mathscr{S}_0$, and the strip condition (134c) is always required on $\mathscr{S}_0$ [see (141a)]. Thus, in this case we are unable to satisfy (137) on $\mathscr{S}_0$, and the projections of the characteristic curves must therefore have an envelope (a "caustic") along the projection of the initial curve on the xy-plane. In the context of geometrical optics, a "focal curve" is indeed a curve along which light rays envelope (the degenerate case corresponds to a focal point), and this is the origin of the terminology used in describing the curves obeying the general system (134).

The extension of the above results to n independent variables is straightforward, and we list the results with no further discussion. The partial differential equation in n variables has the form

$$F(x_i, u, p_i) = 0 \tag{145}$$

where x_i indicates $x_1, \ldots, x_n$ and $p_j = \partial u / \partial x_j$. The characteristic strips are governed by the system of $(2n + 1)$ equations

$$\frac{dx_j}{ds} = \frac{\partial F}{\partial x_j} \tag{146a}$$

$$\frac{du}{ds} = \sum_{j=1}^{n} p_j \frac{\partial F}{\partial p_j} \tag{146b}$$

$$\frac{dp_j}{ds} = -\left(\frac{\partial F}{\partial x_j} + p_j \frac{\partial F}{\partial u}\right) \tag{146c}$$

6.3.4 Example Problems for the Eikonal Equation

The simplest initial-value problem for the eikonal equation consists of a *point* disturbance initially. The resulting wave front is called an *integral conoid*, and we calculate this solution next for the case $n = 2$ and $c = c_0 = $ constant.

Integral Conoid

The initial strip degenerates in the case of the integral conoid to just the Monge cone, which we take at $x = y = u = 0$. Thus, $x_0 = 0$, $y_0 = 0$, $u_0 = 0$. We see that the strip condition (141a) is then identically satisfied so the only restriction on $p_0(\tau)$ and $q_0(\tau)$ is (141b); that is, $p_0^2(\tau) + q_0^2(\tau) = 1/c_0^2 = $ constant. The characteristic strips obey

$$\frac{dx}{ds} = 2p; \quad \frac{dy}{ds} = 2q; \quad \frac{du}{ds} = 2(p^2 + q^2) = \frac{2}{c_0^2}; \quad \frac{dp}{ds} = 0; \quad \frac{dq}{ds} = 0 \tag{147}$$

Therefore, $p = p_0(\tau)$; $q = q_0(\tau)$; $x = 2p_0(\tau)s$; $y = 2q_0(\tau)s$; $u = 2s/c_0^2$. Squaring the expressions for x and y and adding gives $x^2 + y^2 = 4s^2(p_0^2 + q_0^2) = 4s^2/c_0^2$. Therefore, $u = \sqrt{x^2 + y^2}/c_0$, as is obvious from Huyghens' construction [see (5)].

Moving Disturbance

Suppose we have a point disturbance that moves with constant speed v along a straight line in the xy-plane. With no loss of generality, we may assume the motion to occur along the x-axis and to pass through the origin when $u = 0$. The initial values for x_0, y_0 and u_0 are then $x_0 = \tau$, $y_0 = 0$, $u_0 = \tau/v$ because $dx_0/du_0 = (dx_0/d\tau)/(du_0/d\tau) = v$, and $dy_0/du_0 = 0$.

The strip condition (141a) requires that $1/v = p_0(\tau)$, whereas (141b) requires $p_0^2(\tau) + q_0^2(\tau) = 1/c_0^2$. This latter condition is expressed more conveniently if we denote $p_0 \equiv (\cos\theta)/c_0$ and $q_0 \equiv (\sin\theta)/c_0$ in terms of a new parameter θ, in which case the strip condition defines the parameter θ as $\theta \equiv \cos^{-1}(c_0/v)$. This has two real values in $(0, 2\pi)$ as long as $v > c_0$. Let us restrict our attention to the value where $0 \le \cos^{-1}(c_0/v) \le \pi/2$. The requirement $v > c_0$ is geometrically obvious because the wave fronts have no envelope for finite x, y if $v < c_0$. The requirement $v > c_0$ violates physical law for light, so in this particular example we may wish to regard the physical problem as one in acoustics, where $v > c_0$ corresponds to a supersonic disturbance speed.

The solution of (147) now takes the form

$$p = \frac{1}{v}; \qquad q = \frac{(v^2 - c_0^2)^{1/2}}{c_0 v_0}; \qquad x = \frac{2s}{v} + \tau$$

$$y = \frac{2(v^2 - c_0^2)^{1/2}s}{c_0 v_0}; \qquad u = \frac{2s}{c_0^2} + \frac{\tau}{v}$$

and we see that

$$\Delta(s, \tau) \equiv \frac{2(v^2 - c_0^2)^{1/2}}{c_0 v}$$

does not vanish as long as $v > c_0$.

The solution for $u(x, y)$ is the plane

$$u = \frac{1}{v}[x + (M^2 - 1)^{1/2}y] \tag{148a}$$

where $M \equiv v/c_0$ is the Mach number. The other choice of θ results in a negative q_0, and

$$u = \frac{1}{v}[x - (M^2 - 1)^{1/2}y] \tag{148b}$$

Variable c

As a final illustration, consider the case where $c = |x|$ and let the initial front be the straight line $y = ax$, where $a =$ constant. The two unit normals to the front are $\mathbf{n}_1 \equiv (a/\sqrt{1 + a^2}, -1/\sqrt{1 + a^2})$ and $\mathbf{n}_2 \equiv (-a/\sqrt{1 + a^2}, 1/\sqrt{1 + a^2})$. If we choose $\mathbf{n}_1$, our initial strip will be defined parametrically in the form

$$x_0 = \tau; \qquad y_0 = a\tau; \qquad u_0 = 0$$

$$p = \frac{a}{|\tau|(1 + a^2)^{1/2}}; \qquad q = \frac{1}{|\tau|(1 + a^2)^{1/2}}$$

The forms for p_0 and q_0 ensure that the strip condition associated with $\mathbf{n}_1$ and the eikonal equation are initially satisfied. Let us temporarily consider only the characteristics with $\tau > 0$ and omit the absolute value signs.

The characteristic strips obey

$$\frac{dx}{ds} = 2p; \qquad \frac{dy}{ds} = 2q; \qquad \frac{du}{ds} = \frac{2}{x^2}; \qquad \frac{dp}{ds} = -\frac{2}{x^3}; \qquad \frac{dq}{ds} = 0 \tag{149}$$

Solving for q gives $q = q_0 = -1/\tau(1 + a)^{1/2}$. Therefore, we find p directly from the eikonal equation,

$$p = \pm\left[\frac{1}{x^2} - \frac{1}{\tau^2(1 + a^2)}\right]^{1/2} = \frac{[\tau^2(1 + a^2) - x^2]^{1/2}}{|x\tau|(1 + a^2)^{1/2}} \tag{150}$$

or less directly by integrating the equation $dp/dx = -1/px^3$, which results from dividing dp/ds by dx/ds. Initially we must use the plus sign in (150) in front of the radical in conformity with our choice of initial strip, where p is positive. The sign for p along a given ray $\tau = $ constant changes whenever $x = \pm\tau(1 + a^2)^{1/2}$.

We now express dx/ds in the form

$$\frac{dx}{ds} = \pm 2\frac{[\tau^2(1 + a^2) - x^2]^{1/2}}{x\tau(1 + a^2)^{1/2}}$$

and integrate this subject to $x = \tau$ at $s = 0$ to find

$$s = \mp 2\{\tau[\tau^2(1 + a^2) - x^2]^{1/2} + \tau^2 a\}(1 + a^2)^{1/2} \tag{151}$$

The solution for y is just

$$y = 2q_0 s + a\tau = -\frac{2s}{\tau(1 + a^2)^{1/2}} + a\tau$$

or

$$s = -\frac{\tau}{2}(y - a\tau)(1 + a^2)^{1/2} \tag{152}$$

If we now eliminate s from (151) and (152), we find that the projections of the characteristics on the xy-plane are the one-parameter family of concentric circles centered at the origin with radius $\tau(1 + a^2)^{1/2}$; that is,

$$\tau = \pm\left[\frac{x^2 + y^2}{1 + a^2}\right]^{1/2} \tag{153}$$

The remaining equation for u may be expressed in the form

$$\frac{du}{dx} = \frac{du/ds}{dx/ds} = \frac{\tau(1 + a^2)^{1/2}}{x[\tau^2(1 + a^2) - x^2]^{1/2}} \tag{154}$$

The integral of (154), subject to $u = 0$, $x = \tau$, gives

$$u = -\log\frac{\tau(1 + a^2)^{1/2} + [\tau^2(1 + a^2) - x^2]^{1/2}}{x} + \log[(1 + a^2)^{1/2} + a]$$

and when the expression (153) for τ is used, we find the following result valid for $x > 0$ or $x < 0$:

$$u(x, y) = -\log\frac{(x^2 + y^2)^{1/2} + y}{x[(1 + a^2)^{1/2} + a]}, \qquad x \neq 0 \tag{155}$$

The solution becomes infinite as $x \to 0$, and this is expected because the disturbance speed $c \to 0$. Therefore, it takes an infinite time for the initial front to approach the y-axis from either side. We note the similarity behavior $u = $ constant on rays $(y/x) = $ constant, and it is easily verified that in this case the eikonal equation can be solved directly using similarity arguments.

6.4   The Complete Integral; Solutions by Envelope Formation

In this section, we shall study an alternate approach that bypasses the necessity of integrating the system (134), (137) wherever a *complete integral* of (129) is available.

A complete integral of (129) is simply a solution involving two arbitrary constants a, b:

$$u = \phi(x, y, a, b) \tag{156}$$

We ensure that these constants are independent [that is, do not occur in a particular combination $f(a, b)$ in (156)] by requiring the determinant

$$D \equiv \phi_{xa}\phi_{yb} - \phi_{ya}\phi_{xb} \tag{157}$$

to be nonzero.

In the special case where F is independent of u (for example, the eikonal equation or the Hamilton-Jacobi equation), one of the constants in the complete integral is additive and (156) has the form

$$u = \phi(x, y, a) + b \tag{158}$$

This follows immediately from the fact that if $u = \phi(x, y, a)$ solves (129) for arbitrary a, then $u = \phi(x, y, a) + b$ is also a solution for arbitrary b if u does not occur in F.

6.4.1 Envelope Surfaces Associated with the Complete Integral

Suppose we are given a complete integral of (129) in the form (156). We shall show now that we can construct another solution by a process of envelope formation.

Let us specify an arbitrary relation

$$b = w(a) \tag{159}$$

linking b to a, so that (156) now reads

$$u = \phi(x, y, a, w(a)) \tag{160a}$$

If this one-parameter family of surfaces has an envelope, we must be able to solve

$$\phi_a(x, y, a, w(a)) + \phi_b(x, y, a, w(a))w'(a) = 0 \tag{160b}$$

for $a(x, y)$. Substituting this expression in (160a) gives

$$u = \Psi(x, y) \equiv \phi(x, y, a(x, y), w(a(x, y))) \tag{161}$$

which is a surface involving the arbitrary function w. We now show that $u = \Psi(x, y)$, as given by (161), also solves (129). We compute

$$\Psi_x = \phi_x + (\phi_a + \phi_b w'(a))a_x \tag{162}$$

$$\Psi_y = \phi_y + (\phi_a + \phi_b w'(a))a_y \tag{163}$$

But, $a(x, y)$ was determined from the requirement (160b). Therefore, the coefficients of a_x and a_y vanish identically in (162) and (163), and we have $\Psi_x = \phi_x$, $\Psi_y = \phi_y$. We know that $F(x, y, \phi, \phi_x, \phi_y) = 0$, since ϕ is a complete integral. We also know that $\phi = \Psi$ and have shown that $\phi_x = \Psi_x$, $\phi_y = \Psi_y$. Therefore, $\Psi(x, y)$, as defined by (161), solves (129).

The *singular integral* of (129) is the envelope of the two-parameter family of solutions resulting from varying a and b independently in the complete integral; that is, we eliminate a and b from the three expressions

$$u = \phi(x, y, a, b); \qquad \phi_a(x, y, a, b) = 0; \qquad \phi_b(y, y, a, b) = 0 \tag{164}$$

If a singular integral exists, we can also derive it directly from (129) by eliminating p and q from the three relations

$$F(x, y, u, p, q) = 0; \qquad F_p(x, y, u, p, q) = 0; \qquad F_q(x, y, u, p, q) = 0 \tag{165}$$

To show that (164) and (165) define the same function, note that the function of x, y, a, b defined by substituting the complete integral into (129), vanishes identically—that is,

$$F(x, y, \phi, \phi_x, \phi_y) \equiv 0 \tag{166}$$

In particular, the total derivatives dF/da and dF/db of (166) also vanish, where

$$\frac{dF}{da} = F_u\phi_a + F_p\phi_{xa} + F_q\phi_{ya} = 0 \tag{167a}$$

$$\frac{dF}{db} = F_u\phi_b + F_p\phi_{xb} + F_q\phi_{yb} = 0 \tag{167b}$$

On the singular integral, $\phi_a = 0$ and $\phi_b = 0$; the homogeneous system, which results from (167), can hold with $D \neq 0$ only if $F_p = F_q = 0$. Thus, we have shown that the two conditions (164), (165) are equivalent.

For purposes of illustration, consider the paraboloid of revolution defined by

$$u = 1 - (x - a)^2 - (y - b)^2 \equiv \phi(x, y, a, b) \tag{168}$$

where a and b are arbitrary constants. The surface $u = \phi(x, y, a, b)$ is generated by rotating the parabola $u = 1 - x^2$ around the vertical axis $x = a$, $y = b$ in x, y, u space. Working backward, we compute $\phi_x = -2(x - a)$, $\phi_y = -2(y - a)$, from which it is easily seen that (168) is a complete integral of the nonlinear equation

$$F \equiv 1 - u - \tfrac{1}{4}(p^2 + q^2) = 0 \tag{169}$$

It is geometrically obvious that the singular integral of (169) is the horizontal plane $u = 1$, and this result follows immediately either from the conditions (164) applied to (168) or the conditions (165) applied to (169).

We can also construct the envelope of the one-parameter family of surfaces

$$u = 1 - (x - a)^2 - [y - w(a)]^2 \tag{170}$$

which corresponds to requiring the axis of the paraboloid to lie on the curve $y = w(x)$. This envelope is defined by (170) and

$$2(x - a) + 2[y - w(a)]w'(a) = 0 \tag{171}$$

For example, if w is the straight line $y = \alpha x$, $\alpha = $ constant, (171) gives $(x - a) + (y - \alpha a)\alpha = 0$, or $a(x, y) = (x + \alpha y)/(1 + \alpha^2)$, for which (170) defines the surface

$$u = 1 - \left[x - \frac{x + \alpha y}{1 + \alpha^2} \right]^2 - \left[y - \frac{\alpha(x + \alpha y)}{1 + \alpha^2} \right]^2$$

$$= 1 - \frac{(y - \alpha x)^2}{1 + \alpha^2} \tag{172}$$

We verify that this surface is a solution of (169), since (172) implies that $1 - u = (y - \alpha x)^2/(1 + \alpha^2)$ and $p = 2\alpha(y - \alpha x)/(1 + \alpha^2)$, $q = -2(y - \alpha x)/(1 + \alpha^2)$. Therefore, $1 - u = (p^2 + q^2)/4$.

6.4.2 Relationship between Characteristic Strips and the Complete Integral

In the previous section, we demonstrated that the complete integral can be used to generate *certain* solutions by envelope formation. Actually, the complete integral is more far-reaching; it can be used to generate the *three-parameter* family of characteristic strips that gives the general solution of the system (134), (137) discussed in Section 6.3.2.

As a first step in demonstrating this property, we recall that if two solution surfaces join smoothly along a curve, this curve, along with the attached strip, is characteristic. Next, suppose that we generate a solution surface from the envelope of the family of complete integrals

$$u = \phi(x, y, a, w_1(a))$$

for a given $w_1(a)$. A second solution surface can be generated from the family

$$u = \phi(x, y, a, w_2(a))$$

If, for some $a = a_0$, we have $w_1(a_0) = w_2(a_0)$ and $w_1'(a_0) = w_2'(a_0)$, then the two envelope surfaces join smoothly along the intersection strip associated with the *three* constants a_0, $b_0 \equiv w(a_0)$, $c_0 \equiv w'(a_0)$. Since w and w' can be chosen arbitrarily, a_0, b_0, c_0 are arbitrary, and we conclude that the strip that smoothly joins any two envelope surfaces of a complete integral is a characteristic strip.

We now prove this result formally for the equation

$$F(x, y, u, p, q) = 0 \tag{173a}$$

for which we assume to have found the complete integral

$$u = \phi(x, y, a, b) \tag{173b}$$

Let us define a three-parameter family of strips, $x(\sigma; a, b, c), y(\sigma; a, b, c), u(\sigma; a, b, c),$

$p(\sigma; a, b, c)$, and $q(\sigma; a, b, c)$, using the following five conditions:

$$\phi_a(x, y, a, b) = \alpha\sigma; \qquad \alpha = \text{constant} \tag{174a}$$

$$\phi_b(x, y, a, b) = \beta\sigma; \qquad \beta = \text{constant} \tag{174b}$$

$$p = \phi_x(x, y, a, b) \tag{174c}$$

$$q = \phi_y(x, y, a, b) \tag{174d}$$

$$0 = \alpha\sigma + \beta w'\sigma \quad \text{or} \quad \frac{\alpha}{\beta} = -w' \equiv -c \tag{174e}$$

To see that (174) indeed defines a three-parameter family of strips in parametric form [(a, b, c) are constants that identify each member of the family and σ varies along each strip], note that we can solve (174a)–(174b) for x and y, since (157) does not vanish. Moreover, using (174e) we can express this result in the form $x(\sigma; a, b, c)$, $y(\sigma; a, b, c)$. Substituting these expressions for x and y into (173b) gives $u(\sigma; a, b, c)$, and (174c)–(174d) give $p(\sigma; a, b, c)$, $q(\sigma; a, b, c)$.

Having shown that (174) does indeed define a three-parameter family of strips, our next task is to show these strips are characteristic. We differentiate (174a)–(174b) with respect to σ to obtain

$$\phi_{ax} \frac{dx}{d\sigma} + \phi_{ay} \frac{dy}{d\sigma} = \alpha \tag{175a}$$

$$\phi_{bx} \frac{dx}{d\sigma} + \phi_{by} \frac{dy}{d\sigma} = \beta \tag{175b}$$

Next, we note that (173b) satisfies (173a) identically for any a and b; therefore,

$$\frac{dF}{da} = F_u \phi_a + F_p \phi_{xa} + F_q \phi_{ya} = 0 \tag{176a}$$

$$\frac{dF}{db} = F_u \phi_b + F_p \phi_{xb} + F_q \phi_{yb} = 0 \tag{176b}$$

Solving (175) for $(dx/d\sigma)$ and $(dy/d\sigma)$ gives

$$\frac{dx}{d\sigma} = \frac{1}{D}(\alpha\phi_{by} - \beta\phi_{ay}) \tag{177a}$$

$$\frac{dy}{d\sigma} = \frac{1}{D}(\beta\phi_{ax} - \alpha\phi_{bx}) \tag{177b}$$

where D is the determinant (157). Similarly, solving (176) for F_p and F_q gives

$$F_p = -\frac{F_u}{D}(\phi_a \phi_{yb} - \phi_b \phi_{ya}) = -F_u \sigma \frac{dx}{d\sigma} \tag{178a}$$

$$F_q = -\frac{F_u}{D}(\phi_b \phi_{xa} - \phi_a \phi_{xb}) = -F_u \sigma \frac{dy}{d\sigma} \tag{178b}$$

when (174a)–(174b) and (177) are used.

So far, we have shown that the three-parameter family of strips defined by (174) satisfy

$$\frac{dx}{d\sigma} = -\frac{1}{\sigma F_u} F_p \tag{179a}$$

$$\frac{dy}{d\sigma} = -\frac{1}{\sigma F_u} F_q \tag{179b}$$

which, after the change of variable $\sigma \to s$ defined by $(ds/d\sigma) = -1/\sigma F_u$, give (134a), (134b). The proof that the remaining three equations, (134c), (137a), and (137b), are also satisfied parallels the steps used in Section 6.3.2 and is not repeated.

We conclude this section by demonstrating that each member of the three-parameter family of characteristic strips associated with the example problem (169) is simply the branch strip that smoothly joins any two envelope surfaces of (170). The characteristic strips of (169) satisfy the system

$$\frac{dx}{ds} = -\frac{p}{2}; \quad \frac{dy}{ds} = -\frac{q}{2}; \quad \frac{du}{ds} = -\frac{1}{2}(p^2 + q^2) = 2(u - 1)$$

$$\frac{dp}{ds} = p; \quad \frac{dq}{ds} = q$$

We can solve these equations in the form

$$x = \frac{p_0}{2}(1 - e^s) + x_0 \tag{180a}$$

$$y = \frac{q_0}{2}(1 - e^s) + y_0 \tag{180b}$$

$$u = 1 + (u_0 - 1)e^{2s} \tag{180c}$$

$$p = p_0 e^s \tag{180d}$$

$$q = q_0 e^s \tag{180e}$$

where x_0, y_0, u_0, p_0, q_0 must satisfy

$$\dot{u}_0 = p_0 \dot{x}_0 + q_0 \dot{y}_0 \tag{181a}$$

$$1 - u_0 - \tfrac{1}{4}(p_0^2 + q_0^2) = 0 \tag{181b}$$

If we let $p_0/2 + x_0 = a$ and $q_0/2 + y_0 = b$, (180a)–(180b) give $(x - a)^2 + (y - b)^2 = (e^{2s}/4)(p_0^2 + q_0^2)$. Using (181b) to eliminate p_0, q_0 from the right-hand side of the last equation gives

$$(x - a)^2 + (y - b)^2 = e^{2s}(1 - u_0)$$

and using (180c), we find the complete integral

$$(x - a)^2 + (y - b)^2 = 1 - u \tag{182}$$

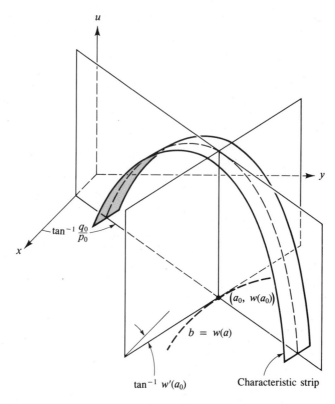

Figure 6.9

Now, (180a)–(180b) also imply that we must have

$$\frac{y - b}{x - a} = \frac{q_0}{p_0} = \text{constant} \tag{183}$$

This defines a vertical plane in xyu-space; it intersects the $u = 0$ plane along the straight line (183) with slope $(dy/dx) = q_0/p_0$. Thus, the characteristic strip is the infinitesimal surface tangent to (182) along the curve where the vertical plane defined by (183) and the surface defined by (182) intersect (see Figure 6.9, which shows the $u \geq 0$ half-space).

We have argued earlier that by displacing the complete integral an infinitesimal amount in some arbitrary direction associated with the tangent $w'(a)$ to a curve $w(a)$, we generate a characteristic strip. It then follows that the line with slope $w'(a)$ passing through $x = a_0$, $y = b_0 = w(a_0)$ in the xy-plane *must be normal* to the plane (183). We now demonstrate this for our example, where

$$w'(a) = \frac{db}{da} = \frac{\dot{b}}{\dot{a}} = \frac{\dot{q}_0/2 + \dot{y}_0}{\dot{p}_0/2 + \dot{x}_0}$$

But, subtracting the derivative of (181b) with respect to τ from (181a) gives

$$0 = p_0 \left(\dot{x}_0 + \frac{\dot{p}_0}{2} \right) + q_0 \left(\dot{y}_0 + \frac{\dot{q}_0}{2} \right)$$

Therefore, we find that

$$w'(a) = -\frac{p_0}{q_0} \tag{184}$$

which is the negative of the reciprocal of the slope of the line (183). This proves that the vertical plane tangent to the curve that generates the envelope is indeed normal to the plane in which the characteristic curve lies (see Figure 6.9).

6.4.3 The Complete Integral of the Hamilton-Jacobi Equation

In Section 6.2.4 we showed that the Hamilton-Jacobi equation [see (68) with $s = t$]

$$\frac{\partial J}{\partial t} + H\left(t, q_i, \frac{\partial J}{\partial q_i}\right) = 0 \tag{185}$$

for the scalar $J(t, q_i)$ governs the field of extremals from a fixed point $t = t_1$, $q_i = \kappa_i$ for the variational principle (31). We also saw [see (97a) with $J = K_2$ and $s = t$] that (185) defines the generating function, K_2, of a canonical transformation to a new Hamiltonian, which vanishes identically. As a result, the new set of coordinates and momenta are constants, and the transformation relations (96) define the solution of Hamilton's differential equations (41).

Let us now study the partial differential equation (185) written in the form (145). We use the notation

$$u \equiv J \tag{186a}$$

$$x_j \equiv q_j; \qquad j = 1, \ldots, n \tag{186b}$$

$$p_j \equiv \frac{\partial J}{\partial q_j}; \qquad j = 1, \ldots, n \tag{186c}$$

$$x_{n+1} \equiv t \tag{186d}$$

$$p_{n+1} \equiv \frac{\partial J}{\partial t} \tag{186e}$$

and observe that the definition (186c) for the p_j is notationally consistent with (67b) or (95a) and that regarding $t = x_{n+1}$ implies that (186e) is the extension of (186c) to $j = n + 1$. In effect, we have the $(n + 1)$-dimensional equation

$$F(x_1, \ldots, x_{n+1}, p_1, \ldots, p_{n+1}) \equiv p_{n+1} + H(x_1, \ldots, x_{n+1}, p_1, \ldots, p_n) = 0 \tag{187}$$

which does not involve u explicitly.

According to (146), the characteristics of (187) obey

$$\frac{dx_j}{ds} = \frac{\partial H}{\partial p_j}; \qquad j = 1, \ldots, n \tag{188a}$$

$$\frac{dx_{n+1}}{ds} = 1 \tag{188b}$$

$$\frac{du}{ds} = \sum_{j=1}^{n} p_j \frac{\partial H}{\partial p_j} + p_{n+1} \tag{188c}$$

$$\frac{dp_j}{ds} = -\frac{\partial H}{\partial x_j}; \qquad j = 1, \dots, n \tag{188d}$$

$$\frac{dp_{n+1}}{ds} = -\frac{\partial H}{\partial x_{n+1}} \tag{188e}$$

Thus, (188a) and (188d) give Hamilton's differential equations (41), in which we use (188b) to set $x_{n+1} = s - s_0$. This system of $2n$ equations does not involve u or p_{n+1}. Once the x_j and p_j have been computed, (188e) gives p_{n+1} by quadrature. In fact, (188e) is just (42). We can then compute u by quadrature from (188c).

We see that the basic system of $2n$ equations (188a) and (188d) defining a given dynamical system recurs as the essential part of characteristic system for (187). It would therefore appear to be of no particular advantage to recast a given Hamiltonian system of $2n$ ordinary differential equations in terms of the associated Hamilton-Jacobi partial differential equation. This is generally true if a complete integral of the Hamilton-Jacobi equation is not available. However, if a complete integral of (187) can be derived directly—that is, without relying on a solution of (188)—we expect [based on the examples that we studied in Sections 6.2.6 (page 355) and 6.4 (page 373)] to be able to *bypass having to solve* (188a) and (188d). The proof of this statement is discussed next.

In direct analogy with the two-dimensional case (158), since u does not occur explicitly in (187), a complete integral is the $(n+1)$-dimensional manifold

$$u = \phi(x_1, \dots, x_{n+1}, a_1, \dots, a_n) + a_{n+1} \tag{189}$$

which satisfies (187) identically. Here, $a_1, \dots, a_n$ are n independent constants—that is,

$$\det \left\{ \frac{\partial^2 \phi}{\partial x_j \partial a_k} \right\} \neq 0 \tag{190}$$

—and a_{n+1} is an arbitrary additive constant. We shall now prove that given a complete integral, the solution of Hamilton's equations (188a) and (188d) are defined implicitly by the $2n$ algebraic relations

$$\frac{\partial \phi}{\partial a_j} = b_j = \text{constant} \tag{191a}$$

$$\frac{\partial \phi}{\partial x_j} = p_j \tag{191b}$$

A solution of the system (188a) and (188d) consists of $2n$ functions x_i, p_i of s and $2n$ arbitrary constants. Let us first show that (191) defines such a set of functions and then show that the result satisfies (188a) and (188d). The system (191a) can be solved for the x_i because (190) holds. This results in n functions x_i of s and the $2n$ constants a_i, b_i. Substituting this result into the left-hand side of (191b) directly gives a set of n functions p_i of s and the a_i, b_i. Here we use (188b) to identify $x_{n+1} = s$.

To prove that the $2n$ functions q_i, p_i of s defined by (191) satisfy (188a) and (188d), we begin by taking the total derivative of (191a) with respect to s:

$$\frac{\partial^2 \phi}{\partial a_j \partial s} + \sum_{k=1}^{n} \frac{\partial^2 \phi}{\partial a_j \partial x_k} \frac{dx_k}{ds} = 0 \tag{192}$$

Now, since (189) is a solution of (187), we have the identity

$$\frac{\partial \phi}{\partial s} + H\left(s, x_i, \frac{\partial \phi}{\partial x_i}\right) = 0 \tag{193}$$

Therefore, the partial derivative of this with respect to a_j gives

$$\frac{\partial^2 \phi}{\partial a_j \partial s} + \sum_{k=1}^{n} \frac{\partial H}{\partial p_k} \frac{\partial^2 \phi}{\partial a_j \partial x_k} = 0 \tag{194}$$

Subtracting (194) from (192) gives the identity

$$\sum_{k=1}^{n} \frac{\partial^2 \phi}{\partial a_j \partial x_k} \left(\frac{dx_k}{ds} - \frac{\partial H}{\partial p_k}\right) = 0 \tag{195}$$

which is a homogeneous system of n algebraic equations for the $z_i \equiv (dx_i/ds - \partial H/\partial p_i)$. Since the determinant (190) of coefficients does not vanish, we conclude that each of the z_i must vanish, and this gives (188a).

To prove that (188d) holds, we take the total derivative of (191b) with respect to s:

$$\frac{dp_j}{ds} = \frac{\partial^2 \phi}{\partial x_j \partial s} + \sum_{k=1}^{n} \frac{\partial^2 \phi}{\partial x_j \partial x_k} \frac{dx_k}{ds} \tag{196}$$

Next, we take the partial derivative of (193) with respect to x_j:

$$0 = \frac{\partial^2 \phi}{\partial x_j \partial s} + \sum_{k=1}^{n} \frac{\partial H}{\partial p_k} \frac{\partial^2 \phi}{\partial x_k \partial x_j} + \frac{\partial H}{\partial x_j} \tag{197}$$

Subtracting (197) from (196) and noting (195) gives (188d).

This completes the proof that knowing the complete integral leads (after some algebra) to the general solution of the Hamiltonian system of equations (188a) and (188d). This result means that the search for a canonical transformation generated by K_2 to a new Hamiltonian that vanishes identically (see (97a)] is exactly equivalent to finding the complete integral for the Hamiltonian in a given set of variables.

It is instructive, although somewhat repetitive, to rederive the solution of Euler's problem [see (116) and the discussion in Section 6.2.6 (page 357)] from the point of view of calculating the complete integral for the Hamilton-Jacobi equation

$$\frac{\partial u}{\partial x_3} + \frac{1}{2(\cosh^2 x_1 - \cos^2 x_2)} [p_1^2 + p_2^2 - 2\cosh x_1 - 2(2\mu - 1)\cos x_2] = 0 \tag{198}$$

We assume that the complete integral of (198)

$$u = \phi(x_1, x_2, x_3, a_1, a_2) + a_3 \tag{199}$$

has the separated form

$$\phi = \phi_1(x_1, a_1, a_2) + \phi_2(x_2, a_1, a_2) + \phi_3(x_3, a_1, a_2) \tag{200}$$

Substituting (200) into (198) gives [see the calculations that lead to (119)]

$$\phi_1 = \int^{x_1} (a_2 + 2\cosh\xi + 2a_1\cosh^2\xi)^{1/2} \, d\xi \tag{201a}$$

$$\phi_2 = \int^{x_2} [-a_2 + 2(2\mu - 1)\cos\eta - 2a_1\cos^2\eta]^{1/2} \, d\eta \tag{201b}$$

$$\phi_3 = -a_1 x_3 \tag{201c}$$

and this defines the complete integral.

Equations (191a) give

$$\frac{\partial\phi}{\partial a_1} = -x_3 + \int^{x_1} \frac{\cosh^2\xi}{(a_2 + 2\cosh\xi + 2a_1\cosh^2\xi)^{1/2}} \, d\xi$$
$$- \int^{x_2} \frac{\cos^2\eta}{[-a_2 + 2(2\mu - 1)\cos\eta - 2a_1\cos^2\eta]^{1/2}} \, d\eta = b_1 \tag{202a}$$

$$\frac{\partial\phi}{\partial a_2} = \frac{1}{2} \int^{x_1} \frac{d\xi}{(a_2 + 2\cosh\xi + 2a_1\cosh^2\xi)^{1/2}}$$
$$- \frac{1}{2} \int^{x_2} \frac{d\eta}{[-a_2 + 2(2\mu - 1)\cos\eta - 2a_1\cos^2\eta]^{1/2}} = b_2 \tag{202b}$$

and equations (191b) give

$$\frac{\partial\phi}{\partial x_1} = (a_2 + 2\cosh x_1 + 2a_1\cosh^2 x_1)^{1/2} = p_1 \tag{203a}$$

$$\frac{\partial\phi}{\partial x_2} = [-a_2 + 2(2\mu - 1)\cos x_2 - 2a_1\cos^2 x_2]^{1/2} = p_2 \tag{203b}$$

Comparing (202)–(203) with (120)–(124), we see that we have derived identical results when we identify $p_1 \to p_1, p_2 \to p_2, x_1 \to q_1, x_2 \to q_2, x_3 \to s, a_1 \to \bar{p}_1$, $a_2 \to \bar{p}_2$, $b_1 \to \bar{q}_1^{(0)}$, and $b_2 \to \bar{q}_2^{(0)}$. Thus, the calculation of the solution via canonical transformation to a zero Hamiltonian is exactly equivalent to the calculation using (191) for the complete integral.

We have seen that solvability of the Hamilton-Jacobi equation is intimately connected with the integrability of the Hamiltonian system (188a) and (188d) of $2n$ differential equations. We have used the idea of separation of variables to solve the Hamilton-Jacobi equation directly in the form of a complete integral. Whether a given Hamiltonian is separable or not [in the sense discussed in Section 6.2.6 (page 356) and above] is easy to establish by trial substitution. As we have observed, separability is a property of the particular choice of variables. Therefore, the question of whether or not we can compute the complete integral directly also depends on this choice of variables. The more fundamental question of whether for a given Hamiltonian there exists a set of variables in terms of which the Hamilton-Jacobi equation becomes separable (hence solvable) is not known in general.

Problems

1. Consider light rays in two dimensions, with the speed of light $c(x, y)$ given.
   a. Show that the Euler-Lagrange equations (32) associated with the Lagrangian

   $$L(x, y, \dot{x}, \dot{y}) = \sqrt{\dot{x}^2 + \dot{y}^2}/c(x, y) \tag{204}$$

   reduce to

   $$c\dot{y}^2\ddot{x} - c\dot{x}\dot{y}\ddot{y} = -\dot{y}^2(\dot{x}^2 + \dot{y}^2)c_x + \dot{x}\dot{y}(\dot{x}^2 + \dot{y}^2)c_y \tag{205a}$$

   $$-c\dot{x}\dot{y}\ddot{x} - c\dot{x}^2\ddot{y} = \dot{x}\dot{y}(\dot{x}^2 + \dot{y}^2)c_x - \dot{x}^2(\dot{x}^2 + \dot{y}^2)c_y \tag{205b}$$

   and that these two equations are not independent in the sense that we cannot solve for $\ddot{x}$ and $\ddot{y}$ as functions of $x, y, \dot{x}$ and $\dot{y}$. This is to be expected, since we must have $dx^2 + dy^2 \equiv d\sigma^2$ along a light ray, where σ is the arc length, and this implies that $\dot{x}$ and $\dot{y}$ are related by $\dot{x}^2 + \dot{y}^2 = \dot{\sigma}^2$, where $\dot{} \equiv d/ds$.
   b. Let $s = \sigma$, and show that (205) reduces to

   $$c\frac{d^2x}{d\sigma^2} + c_x\left[1 - \left(\frac{dx}{d\sigma}\right)^2\right] - c_y\frac{dx}{d\sigma}\frac{dy}{d\sigma} = 0 \tag{206a}$$

   $$c\frac{d^2y}{d\sigma^2} + c_y\left[1 - \left(\frac{dy}{d\sigma}\right)^2\right] - c_x\frac{dx}{d\sigma}\frac{dy}{d\sigma} = 0 \tag{206b}$$

   and that (206) also follows from (6) when p and q are eliminated.
   c. Now let $s = t$, the time, and show that (205) reduces to

   $$c\frac{d^2x}{dt^2} + \left[c^2 - 2\left(\frac{dx}{dt}\right)^2\right]c_x - 2c_y\frac{dx}{dt}\frac{dy}{dt} = 0 \tag{207a}$$

   $$c\frac{d^2y}{dt^2} + \left[c^2 - 2\left(\frac{dy}{dt}\right)^2\right]c_y - 2c_x\frac{dx}{dt}\frac{dy}{dt} = 0 \tag{207b}$$

   which also follows from (6) when p and q are eliminated and $u = t$ is chosen as the independent variable.
   d. Finally, let $s = x$, and write $(dy/dx) \equiv y' = \dot{y}/\dot{x}$. Therefore, $y'' = \ddot{y}/\dot{x}^2 - \dot{y}\ddot{x}/\dot{x}^3$. Show that (205a) reduces to (13) and, using corresponding expressions with x and y interchanged in (205a), gives a formula of the form (13) with x and y interchanged.
2. The *circular restricted three-body problem* is a dynamically consistent generalization of Euler's problem, where a point of mass μ and a point of mass $(1 - \mu)$ describe circular orbits about their common mass center. In the resulting gravitational field we introduce a particle that does not disturb the

motion of the two circling masses (the primaries); this particle merely moves under the influence of the Newtonian gravitational forces exerted by the primaries.

If we choose dimensionless variables such that lengths are normalized by the constant distance between the primaries and the time is normalized by the reciprocal angular velocity of the circular motion, we obtain the following equations for the special case where the particle moves in the plane of the circular motion:

$$\ddot{\bar{x}} = -\frac{(1-\mu)(\bar{x}-\bar{\xi}_1)}{\bar{r}_1^3} - \frac{\mu(\bar{x}-\bar{\xi}_2)}{\bar{r}_2^3} \tag{208a}$$

$$\ddot{\bar{y}} = -\frac{(1-\mu)(\bar{y}-\bar{\eta}_1)}{\bar{r}_1^3} - \frac{\mu(\bar{y}-\bar{\eta}_2)}{\bar{r}_2^3} \tag{208b}$$

Here, $\bar{x}, \bar{y}$ are Cartesian coordinates in the inertial frame with origin at the center of mass of the primaries. Hence, we have

$$\bar{r}_1^2 \equiv (\bar{x}-\bar{\xi}_1)^2 + (\bar{y}-\bar{\eta}_1)^2 \tag{209a}$$

$$\bar{r}_2^2 \equiv (\bar{x}-\bar{\xi}_2)^2 + (\bar{y}-\bar{\eta}_2)^2 \tag{209b}$$

where

$$\bar{\xi}_1 \equiv -\mu\cos t \tag{210a}$$

$$\bar{\eta}_1 \equiv -\mu\sin t \tag{210b}$$

$$\bar{\xi}_2 \equiv (1-\mu)\cos t \tag{210c}$$

$$\bar{\eta}_2 \equiv (1-\mu)\sin t \tag{210d}$$

(See Figure 6.10).

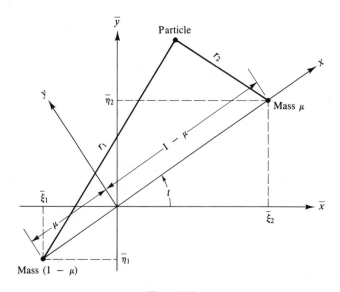

Figure 6.10

a. Show that equations (208) follow from the time-dependent Lagrangian

$$\bar{L}(t,\bar{x},\bar{y},\dot{\bar{x}},\dot{\bar{y}}) = \frac{1}{2}(\dot{\bar{x}}^2 + \dot{\bar{y}}^2) + \frac{(1-\mu)}{\bar{r}_1} + \frac{\mu}{\bar{r}_2} \tag{211}$$

b. Introduce the coordinate system x, y defined by

$$\bar{x} \equiv x\cos t - y\sin t \tag{212a}$$

$$\bar{y} \equiv x\sin t + y\cos t \tag{212b}$$

in which the two primaries lie at $x = 1 - \mu$, $y = 0$, and $x = -\mu$, $y = 0$; then show that the Lagrangian that results is time-independent:

$$L(x,y,\dot{x},\dot{y}) = \frac{1}{2}(\dot{x}^2 + \dot{y}^2) + \frac{1}{2}(x^2 + y^2) + (x\dot{y} - y\dot{x}) + \frac{(1-\mu)}{r_1} + \frac{\mu}{r_2} \tag{213}$$

where

$$r_1^2 \equiv (x+\mu)^2 + y^2 \tag{214a}$$

$$r_2^2 \equiv (x-1+\mu)^2 + y^2 \tag{214b}$$

Derive (213) directly in the rotating x, y frame by introducing appropriate centrifugal and Coriolis forces.

c. Show that the Hamiltonian corresponding to (213) is

$$H(q_i,p_i) = \frac{1}{2}(p_1^2 + p_2^2) - p_1 q_2 + p_2 q_1 - \frac{(1-\mu)}{r_1} - \frac{\mu}{r_2} \tag{215}$$

where

$$p_1 = \dot{x} - y \tag{216a}$$

$$p_2 = \dot{y} + x \tag{216b}$$

$$q_1 = x \tag{216c}$$

$$q_2 = y \tag{216d}$$

Since (215) is time-independent, it is a constant of the motion called the *Jacobi* integral. Verify that the Hamilton-Jacobi equation associated with (215) for W *cannot be solved* by separation of variables. The existence of a coordinate system in which the associated Hamilton-Jacobi equation is solvable is a famous problem in celestial mechanics (see [5] for more details).

3. The Lagrangian for geodesics on the unit sphere is [see (35)]

$$L = (\dot{q}_1^2 + \dot{q}_2^2\sin^2 q_1)^{1/2} \tag{217}$$

a. Derive the Euler-Lagrange equations and show that the equation for q_2 integrates to

$$\dot{q}_2\sin^2 q_1 = \lambda L \tag{218}$$

where λ is an arbitrary constant.

b. Show that the Euler-Lagrange equation for q_1 is identically satisfied by (218). Therefore, (218) suffices to define the geodesics. Guided by the special cases $\lambda = 0$ and $\lambda = 1$, interpret λ geometrically in the three-dimensional space containing the unit sphere. In particular, show that $\lambda \equiv \cos i$, where i is the inclination angle between the polar axis and the normal to the local geodesic plane. This plane is defined by $\mathbf{r}$, the unit displacement vector from the origin to the geodesic and $\dot{\mathbf{r}}$, the vector tangent to the geodesic. Thus, since $i = $ constant, the geodesic is given by the intersection with the unit sphere of a plane inclined at the angle i to the equatorial plane.

c. Let $\Gamma(q_1, q_2) = 0$ be a given curve on the surface of the unit sphere. Show that the transversality condition on $\Gamma = 0$ reduces to

$$\frac{\partial\Gamma/\partial q_1}{\partial\Gamma/\partial q_2} = \frac{\dot{q}_1}{\dot{q}_2 \sin^2 q_1} \tag{219}$$

Show that (219) merely states that a transversal geodesic must be normal to $\Gamma = 0$.

4. In (81), fix the constant E, so you restrict attention to planar motions of a given energy under the influence of gravity.

   a. Find the integral conoid centered at the origin in parametric form, and indicate how we could, in principle, calculate $W(x, y)$.

   b. Let x and y be small and show that the approximate expression for the conoid is

   $$W(x, y) \approx \frac{r(2E - y)}{\sqrt{2E}} \tag{220}$$

   where $r^2 = x^2 + y^2$. Sketch curves of $W = $ constant in the xy-plane and argue that the conoid is not a cone.

   c. Give an optical as well as a dynamical interpretation for your results. In particular, identify the trajectories emerging from the origin in terms of the solutions of the characteristic equations.

5. The eikonal equation in cylindrical polar coordinates (r, θ) is

$$\left(\frac{\partial u}{\partial r}\right)^2 + \frac{1}{r^2}\left(\frac{\partial u}{\partial\theta}\right)^2 = \frac{1}{c^2(r, \theta)} \tag{221}$$

   a. Without taking advantage of cylindrical symmetry, show that the integral conoid from the origin for the case where $c = (1 + r^2)^{1/2}$ is

   $$u(r, \theta) = \log[r + (1 + r^2)^{1/2}] \tag{222}$$

   b. Rederive (222) by noting that if c depends only on r, then u does not depend on θ for the integral conoid. Therefore, (221) reduces to

   $$\frac{du}{dr} = \frac{1}{c(r)} \tag{223}$$

   and integrating this for $c = (1 + r^2)^{1/2}$ gives (222).

6. The three-dimensional eikonal equation for a medium with constant signal speed (normalized to equal unity) is

$$u_x^2 + u_y^2 + u_z^2 = 1 \qquad (224)$$

a. Solve the initial-value problem $u = k = $ constant on the plane $\alpha x + \beta y + \gamma z = 0$.

b. Construct the integral conoid by a process of envelope formation using your solution in part (a). What is the projection of the integral conoid on the manifold $u = $ constant?

7. Given a two-parameter family of surfaces

$$u = \Phi(x, y, a, b) \qquad (225)$$

show that there exists a unique partial differential equation of the first order for which (225) is a complete integral. Derive this partial differential equation. Specialize your results to the case

$$u = abxy + ax^2 \qquad (226)$$

8. Consider the equation

$$u^2(1 + p^2 - q^2) - 1 = 0 \qquad (227)$$

a. Calculate the complete integral and describe it geometrically. Calculate the singular integral.

b. Construct a solution involving an arbitrary function by envelope formation from the complete integral.

c. Calculate the characteristics strips of (227), and discuss how they are formed from the complete integral.

9. Clairaut's equation is defined as

$$F(x, y, u, p, q) \equiv xp + yq + f(p, q) - u = 0 \qquad (228)$$

for a given function f.

a. Use separation of variables to show that the two-parameter family of planes

$$u = ax + by + f(a, b) \qquad (229)$$

is a complete integral of (228).

b. For the case $f(a, b) \equiv -\frac{1}{2}(a^2 + b^2)$, calculate the singular integral of (228) and interpret this result geometrically.

c. Calculate the characteristic strips of (228) and discuss how they are formed from the complete integral.

10. For a given function $H(x, y, p)$, we want to study the nonlinear equation

$$F \equiv H(x, y, p) + q = 0 \qquad (230)$$

where $p = u_x$, $q = u_y$.

a. Show that the equations for the characteristic strips for (230) can be separated into two equations involving x and p as dependent variables and y as independent variable.

b. Let

$$u = \phi(x, y, a) + b \tag{231}$$

be a complete integral of (230).

Consider now the two equations

$$\phi_a = \alpha = \text{constant} \tag{232a}$$

$$\phi_x = p \tag{232b}$$

Assume that $x = X(y, a, \alpha)$ is the solution of (232a) for x and also assume that if the preceding $X(y, a, \alpha)$ is substituted into (232b) for x, this equation takes the form $p = P(y, a, \alpha)$. Prove that the two functions $X(y, a, \alpha)$ and $P(y, a, \alpha)$ so calculated define the general solution of the two ordinary differential equations you derived in part (a).

c. In this part, interpret y as the time, and consider the nonlinear oscillator defined by

$$\frac{d^2x}{dy^2} + f'(x) = 0 \tag{233}$$

with $p = x'$. The energy equation for (233) is

$$H = \frac{p^2}{2} + f(x) = \text{constant} \tag{234}$$

What are the two differential equations for x and p corresponding to those in part (a)? Use (234) for H in (230) and derive an expression for the complete integral for this case. Show by explicit calculation that use of the complete integral, as indicated in part (b), leads to a solution of either (233) or, equivalently, of the two differential equations for x and p.

11. In a two-dimensional isotropic medium, the light rays emanating from some initial wave front are given by the one-parameter family

$$y - ke^x = 0 \tag{235}$$

where k is an arbitrary constant. Show that this information specifies the speed of light $c(x, y)$ in the form

$$c(x, y) = \frac{g(y^2 + 2x)}{(1 + y^2)^{1/2}} \tag{236}$$

where g is an arbitrary function of its argument. Derive (236) in two ways:

a. Use the eikonal equation

$$u_x^2 + u_y^2 = \frac{1}{c^2(x, y)} \tag{237}$$

and the fact that light rays are the orthogonal trajectories of the $u = $ constant curves. In this case, verify that (235) is a solution of the differential equation (13) for the light rays with c given by (236).

b. Regard (13) as a quasilinear first-order partial differential equation for $c(x, y)$ and solve it for an unknown initial curve.

c. What additional information is needed in order to specify g?

References

1. H. Goldstein, *Classical Mechanics*, 2d ed., Addison-Wesley, Reading, Mass., 1980.
2. J. D. Cole, "Newtonian Flow Theory for Slender Bodies," *J. Aeronaut. Sci.* 24 (1957): 448–55.
3. J. Kevorkian, "Perturbation Techniques for Oscillatory Systems with Slowly Varying Coefficients," *SIAM Rev.* 29 (1987): 391–461.
4. R. Courant, and D. Hilbert, *Methods of Mathematical Physics*, Vol. II: *Partial Differential Equations*, by R. Courant, Wiley-Interscience, New York, 1962.
5. V. Szebehely, *Theory of Orbits, the Restricted Problem of Three Bodies*, Academic Press, New York, 1967.

Quasilinear Hyperbolic Systems

Much of Chapter 4 concerned the linear hyperbolic equation of second order, or the system of two first-order equations. Typically, such linear equations govern the perturbation to a known solution of a quasilinear problem. In Chapter 5 we discussed some aspects of weak solutions for quasilinear equations associated with systems of integral conservation laws. In particular, we derived the shock conditions and used these results to study simple solutions consisting of uniformly propagating shocks.

In this chapter we consider more general initial- and boundary-value problems for quasilinear hyperbolic equations, but we restrict the discussion to the case of two independent variables. Our approach parallels that used in Chapter 4 in many respects. The fundamental difference is that for the quasilinear problem, the characteristics depend on the solution and are therefore defined only locally by the governing equations. The geometrical concepts remain valid locally as long as the characteristics are well defined; the essential added complication is that the calculations for the characteristic curves are coupled with those for the solution. As in the scalar quasilinear problem, characteristics of a given family may cross, in which event we look for weak solutions containing discontinuities.

7.1 The Quasilinear Second-Order Hyperbolic Equation

The general quasilinear second-order equation in two independent variables has the form [see (1) of Chapter 4]

$$au_{xx} + 2bu_{xy} + cu_{yy} + d = 0 \tag{1}$$

where a, b, c, d are now functions of x, y, u, u_x and u_y, and we shall consider only the hyperbolic problem, where $\Delta \equiv b^2 - ac > 0$ in some solution domain. Note that since Δ now also depends on u, u_x, and u_y, the type of a given equation is generally not defined by the functional form of the coefficients a, b, c, as in the linear case; one must also account for the solution in evaluating Δ. Thus, Δ may well be positive in some domain for certain Cauchy data and negative for other choices.

The geometrical interpretation of the characteristic curves is locally identical to the situation for the linear problem. More precisely, we again define a characteristic curve $x(s)$, $y(s)$ as one for which knowledge of u, u_x, and u_y does not define the second derivative leading out of the curve (see Section 4.3.1).

7.1.1 Transformation to Characteristic Variables

Let $\mathscr{C}$ be a smooth nonintersecting curve defined in the plane in parametric form by $x = X(s)$, $y = Y(s)$. Denote $u_x = p$, $u_y = q$, and assume that $u = U(s)$, $p = P(s)$, $q = Q(s)$ are also specified on $\mathscr{C}$ in a consistent manner—that is, that we require

$$\dot{U}(s) = P(s)\dot{X}(s) + Q(s)\dot{Y}(s) \tag{2}$$

We wish to use the preceding Cauchy data together with (1) to calculate all three second derivatives u_{xx}, u_{xy}, and u_{yy} on $\mathscr{C}$. In addition to (1), we have at our disposal the two equations that result from differentiating $p = u_x(x, y)$ and $q = u_y(x, y)$ with respect to s on $\mathscr{C}$. Thus, we must have

$$A(s)u_{xx} + 2B(s)u_{xy} + C(s)u_{yy} = -D(s) \tag{3a}$$

$$\dot{X}(s)u_{xx} + \dot{Y}(s)u_{xy} = \dot{P}(s) \tag{3b}$$

$$\dot{X}(s)u_{xy} + \dot{Y}(s)u_{yy} = \dot{Q}(s) \tag{3c}$$

where we use capital letters to denote the coefficients calculated along $\mathscr{C}$. For instance, $A(s) \equiv a(X(s), Y(s), U(s), P(s), Q(s))$, and so on.

If for a given $\mathscr{C}$ and given Cauchy data U, P, Q, the determinant of coefficients in (3) does not vanish, we can calculate u_{xx}, u_{xy}, and u_{yy} on $\mathscr{C}$. A characteristic curve is defined as a curve along which Cauchy data does not determine these three unknowns. This occurs only if the following determinant vanishes:

$$\begin{vmatrix} A & 2B & C \\ \dot{X} & \dot{Y} & 0 \\ 0 & \dot{X} & \dot{Y} \end{vmatrix} = A\dot{Y}^2 - 2B\dot{X}\dot{Y} + C\dot{X}^2 = 0 \tag{4}$$

Let us divide (4) by $\dot{X}$ and denote $\dot{Y}/\dot{X} = dy/dx = y'$. We see that the characteristic slope y' obeys [see (13b) of Chapter 4]

$$Ay'^2 - 2By' + C = 0 \tag{5}$$

which has real solutions if $B^2 - AC > 0$. In this case, we denote the two characteristic slopes by λ^+ and λ^-, where

$$\lambda^+ = \frac{B + \sqrt{B^2 - AC}}{A} \tag{6a}$$

$$\lambda^- = \frac{B - \sqrt{B^2 - AC}}{A} \tag{6b}$$

An alternate way of expressing this result is to denote by ξ the parameter s that varies along the characteristic curve having slope λ^-, and let η be the parameter that varies along the curve with slope λ^+. We may then regard x and y as functions of ξ and η. Using the notation $x = X(\xi, \eta)$, $y = Y(\xi, \eta)$ as before, we have the following pair of partial differential equations governing the characteristic curves:

$$Y_\xi(\xi, \eta) - \lambda^-(\xi, \eta)X_\xi = 0 \tag{7a}$$

$$Y_\eta(\xi, \eta) - \lambda^+(\xi, \eta)X_\eta = 0 \tag{7b}$$

Henceforth, for brevity, we shall also refer to the curves on which $\eta =$ constant as the λ^- *characteristics* and the curves on which $\xi =$ constant as the λ^+ *characteristics*. At this stage λ^+ and λ^- are unknown because they involve u, p, q in addition to x, y. Thus (7) defines the characteristic curves only locally in a small neighborhood of a point where u, p, q are given.

Now suppose we reexamine the system (3) along a particular characteristic strip (that is, an infinitesimal solution surface attached to a characteristic curve). If a solution $u(x, y)$ that contains this characteristic strip exists, the three algebraic relations that result from solving (3) for u_{xx}, u_{xy}, u_{yy} must be compatible. Each of these relations is in the form of a numerator determinant divided by the same denominator determinant (4). In particular, since the denominator determinant (4) vanishes along a characteristic curve, each of the three numerator determinants associated with the solutions for u_{xx}, u_{xy}, and u_{yy} must also vanish in order for a solution to exist. For example, in the solution for u_{xy} from (3), we have the numerator determinant

$$N \equiv \begin{vmatrix} A & -D & C \\ \dot{X} & \dot{P} & 0 \\ 0 & \dot{Q} & \dot{Y} \end{vmatrix} \tag{8a}$$

which must vanish. This gives

$$A\dot{P}\dot{Y} + D\dot{X}\dot{Y} + C\dot{X}\dot{Q} = 0 \tag{8b}$$

Dividing (8b) by $A\dot{Y}$ gives

$$\dot{P} + \frac{C}{A}\frac{\dot{X}}{\dot{Y}}\dot{Q} + \frac{D}{A}\dot{X} = 0 \tag{8c}$$

which must hold along either the $\eta =$ constant or $\xi =$ constant characteristic. Along the $\eta =$ constant characteristic, we regard $d/ds \to \partial/\partial\xi$, and (8c) gives

$$P_\xi + \frac{C}{A}\frac{1}{\lambda^-}Q_\xi + \frac{D}{A}X_\xi = 0 \tag{8d}$$

Now, according to (6b),

$$\frac{C}{A\lambda^-} = \frac{CA}{A[B-(B^2-AC)^{1/2}]} = \frac{C[B+(B^2-AC)^{1/2}]}{B^2-(B^2-AC)} = \frac{B+(B^2-AC)^{1/2}}{A} = \lambda^+$$

Therefore, (8d) may also be written as

$$P_\xi + \lambda^+ Q_\xi + \frac{D}{A} X_\xi = 0 \qquad (9a)$$

Similarly, along the $\xi = $ constant characteristic, we have

$$P_\eta + \lambda^- Q_\eta + \frac{D}{A} X_\eta = 0 \qquad (9b)$$

We can also verify that setting the numerator determinants for either u_{xx} or u_{yy} equal to zero gives result (9), as expected.

 To complete the system of characteristic equations, we need to use the consistency condition (2) written either in terms of ξ,

$$U_\xi = P X_\xi + Q Y_\xi \qquad (10a)$$

or η,

$$U_\eta = P X_\eta + Q Y_\eta \qquad (10b)$$

 In summary, we have derived the system of six partial differential equations (7), (9), and (10) for the five variables X, Y, U, P, Q. It is easily seen that this system is not overdetermined and that, in fact, any five of these equations imply the sixth. For example, let us show that (10b) is a consequence of (7), (9), and (10a). We denote

$$F(\xi, \eta) \equiv U_\eta - P X_\eta - Q Y_\eta \qquad (11a)$$

and wish to prove that $F(\xi, \eta) \equiv 0$ if (7), (9), and (10a) hold. We calculate

$$F_\xi = U_{\xi\eta} - P_\xi X_\eta - P X_{\xi\eta} - Q_\xi Y_\eta - Q Y_{\xi\eta} \qquad (11b)$$

Taking the partial derivative of (10a) with respect to η gives

$$0 = U_{\xi\eta} - P_\eta X_\xi - P X_{\xi\eta} - Q_\eta Y_\xi - Q Y_{\xi\eta} \qquad (11c)$$

and when (11c) is subtracted from (11b), we find

$$F_\xi = P_\eta X_\xi - P_\xi X_\eta + Q_\eta Y_\xi - Q_\xi Y_\eta \qquad (12)$$

Now we use (7) to express Y_ξ and Y_η in terms of X_ξ and X_η in (12):

$$F_\xi = P_\eta X_\xi - P_\xi X_\eta + Q_\eta \lambda^- X_\xi - Q_\xi \lambda^+ X_\eta \qquad (13)$$

Using (9a) and (9b) to eliminate $\lambda^+ Q_\xi$ and $\lambda^- Q_\eta$, we find $F_\xi = 0$. Therefore, $F = f(\eta)$, a constant along each characteristic curve $\eta = $ constant. But, since F vanishes on some noncharacteristic initial curve, we conclude that $f \equiv 0$; hence, $F(\xi, \eta) \equiv 0$, which is just (10b).

 For the special case where the coefficients of (1) do not depend on u, (7) and (9) decouple from (10), and we can solve these first for X, Y, P, Q. Either of equations (10) then gives u by quadrature.

7.1.2   The Cauchy Problem; the Numerical Method of Characteristics

Let $\mathscr{C}_0$ be a noncharacteristic initial curve on which u, p, q are specified consistently with (2). We discretize this curve by selecting along it a spacing of points that is appropriate to the rate of change of the given data (see Figure 7.1a).

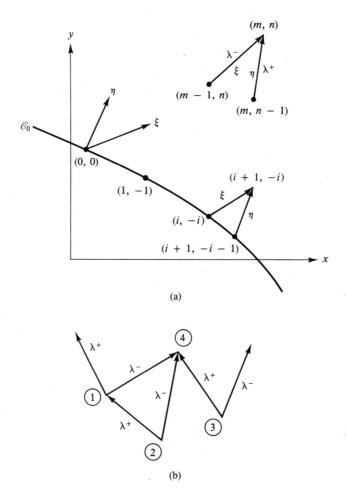

(a)

(b)

Figure 7.1

Let us use the same subscript notation as in Section 4.4.2 and indicate different values of ξ and η by the first and second subscripts, respectively. With no loss of generality, we may choose the point $\xi = 0$ and $\eta = 0$ to lie on $\mathscr{C}_0$. Points on $\mathscr{C}_0$ will then be denoted as follows:

$$x = X(\xi_i, \eta_{-i}) \equiv X_{i,-i}, \qquad y = Y(\xi_i, \eta_{-i}) \equiv Y_{i,-i}, \ldots$$

which assumes that the directions of increasing ξ and η on $\mathscr{C}_0$ are both above it, as indicated in Figure 7.1a.

Consider now two adjacent points $(m - 1, n)$ and $(m, n - 1)$ at which the values of x, y, u, p, q are known. Assuming that $\lambda^+_{m,n-1} > \lambda^-_{m-1,n} > 0$, the point (m, n) is located by the intersection of the $\eta = $ constant characteristic curve from $(m - 1, n)$ with the $\xi = $ constant characteristic from $(m, n - 1)$, as shown in Fig. 7.1a.

Therefore, the x- and y-coordinates of the point (m, n) are approximately defined by the following forward difference formulas associated with (7):

$$Y_{m,n} - Y_{m-1,n} - \lambda_{m-1,n}^-(X_{m,n} - X_{m-1,n}) = 0 \tag{14a}$$

$$Y_{m,n} - Y_{m,n-1} - \lambda_{m,n-1}^+(X_{m,n} - X_{m,n-1}) = 0 \tag{14b}$$

Since $\lambda_{m-1,n}^-$ and $\lambda_{m,n-1}^+$ are known [they depend on the known values of x, y, u, p, q at $(m-1, n)$ and $(m, n-1)$, respectively], we can solve the two linear equations (14) for $X_{m,n}$ and $Y_{m,n}$ to obtain

$$X_{m,n} = \frac{1}{\lambda_{m,n-1}^+ - \lambda_{m-1,n}^-} [Y_{m-1,n} - Y_{m,n-1} + \lambda_{m,n-1}^+ X_{m,n-1} - \lambda_{m-1,n}^- X_{m-1,n}] \tag{15a}$$

$$\begin{aligned} Y_{m,n} = \frac{1}{\lambda_{m,n-1}^+ - \lambda_{m-1,n}^-} [&\lambda_{m,n-1}^+ Y_{m-1,n} - \lambda_{m-1,n}^- Y_{m,n-1} \\ &+ \lambda_{m,n-1}^+ \lambda_{m-1,n}^-(X_{m,n-1} - X_{m-1,n})] \end{aligned} \tag{15b}$$

This result breaks down when $\lambda_{m,n-1}^+ \approx \lambda_{m-1,n}^-$, but this can occur only if $B^2 - AC = 0$ at some point between $(m, n-1)$ and $(m-1, n)$, a situation which we have ruled out for a solution that remains hyperbolic in some domain.

The finite difference form of (9) is

$$P_{m,n} - P_{m-1,n} + \lambda_{m-1,n}^+(Q_{m,n} - Q_{m-1,n}) = -\tilde{D}_{m-1,n}(X_{m,n} - X_{m-1,n}) \tag{16a}$$

$$P_{m,n} - P_{m,n-1} + \lambda_{m,n-1}^-(Q_{m,n} - Q_{m,n-1}) = -\tilde{D}_{m,n-1}(X_{m,n} - X_{m,n-1}) \tag{16b}$$

where $\tilde{D} \equiv D/A$ and the right-hand sides of (16) are known. These two linear equations for $P_{m,n}$, $Q_{m,n}$ can also be easily solved, and we find

$$P_{m,n} = \frac{1}{\lambda_{m-1,n}^+ - \lambda_{m,n-1}^-} [\beta_{m,n}\lambda_{m-1,n}^+ - \alpha_{m,n}\lambda_{m,n-1}^-] \tag{17a}$$

$$Q_{m,n} = \frac{1}{\lambda_{m-1,n}^+ - \lambda_{m,n-1}^-} [\alpha_{m,n} - \beta_{m,n}] \tag{17b}$$

where

$$\begin{aligned} \alpha_{m,n} \equiv P_{m-1,n} + \lambda_{m-1,n}^+ Q_{m-1,n} - &\frac{\tilde{D}_{m-1,n}}{\lambda_{m,n-1}^+ - \lambda_{m-1,n}^-}[Y_{m-1,n} - Y_{m,n-1} \\ &+ \lambda_{m,n-1}^+(X_{m,n-1} - X_{m-1,n})] \end{aligned} \tag{18a}$$

$$\begin{aligned} \beta_{m,n} \equiv P_{m,n-1} + \lambda_{m,n-1}^- Q_{m,n-1} - &\frac{\tilde{D}_{m,n-1}}{\lambda_{m,n-1}^+ - \lambda_{m-1,n}^-}[Y_{m-1,n} - Y_{m,n-1} \\ &+ \lambda_{m-1,n}^-(X_{m,n-1} - X_{m-1,n})] \end{aligned} \tag{18b}$$

Equations (17) and (18) define P and Q at (m, n) in terms of known quantities at $(m-1, n)$ and $(m, n-1)$. Again, we rule out $\lambda_{m-1,n}^+ = \lambda_{m,n-1}^-$ and $\lambda_{m,n-1}^+ = \lambda_{m-1,n}^-$ for hyperbolic problems.

To complete the solution at (m, n), we need u there, and this can be calculated from either (10a) or (10b); that is,

$$U_{m,n} = U_{m-1,n} + P_{m-1,n}(X_{m,n} - X_{m-1,n}) + Q_{m-1,n}(Y_{m,n} - Y_{m-1,n}) \qquad (19a)$$

or

$$U_{m,n} = U_{m,n-1} + P_{m,n-1}(X_{m,n} - X_{m,n-1}) + Q_{m,n-1}(Y_{m,n} - Y_{m,n-1}) \qquad (19b)$$

All the terms on the right-hand sides of (19) have been calculated at this stage. As in (59) of Chapter 4, we may use the average of the two expressions in (19) to define $U_{m,n}$.

This procedure defines x, y, u, p, q uniquely as long as characteristics of the same family from adjacent gridpoints do not intersect. For example, consider the situation depicted in Figure 7.1b, where the values of x, y, u, p, q are given (from previous calculations) at the three adjacent points ①, ②, ③. We have the λ^- characteristics from ① and ② intersecting at the same point, ④, as the λ^+ characteristic from ③. The values of p and q (and u) at ④ are ambiguous because they depend on whether we use the pair of points ①, ③ or ②, ③ to compute them.

A similar situation was encountered in Chapter 5, where one-parameter families of characteristic curves had intersections beyond a certain envelope curve. We then approached the problem from the vantage of weak solutions and prevented the crossing of characteristics by inserting appropriate shocks. We shall postpone discussion of weak solutions here until we have considered hyperbolic systems in the next section.

7.2   Systems of *n* First-Order Equations

The discussion in this section parallels that in Section 4.5. As in the case of the quasilinear second-order equation, the essential difference from the corresponding linear problem is that characteristic slopes depend on the solution, and therefore the geometry of characteristics can be derived only locally.

The physical problem is modeled by a system of n first-order equations of the form

$$\frac{\partial u_j}{\partial t} + \sum_{k=1}^{n} A_{jk} \frac{\partial u_k}{\partial x} = f_j, \qquad j = 1, \ldots, n \qquad (20)$$

for the n dependent variables $u_1(x, t), \ldots, u_n(x, t)$. The matrix components A_{jk} and the components f_j depend on x, t as well as $u_1, \ldots, u_n$, indicated by the subscript i in $A_{jk}(x, t, u_i)$, $f_j(x, t, u_i)$.

As pointed out in Section 4.5.2, we may regard (20) as the component form (with respect to some unspecified *constant* basis $\mathbf{b}_1, \ldots, \mathbf{b}_n$) of the vector equation

$$\mathbf{u}_t + A\mathbf{u}_x = \mathbf{f} \qquad (21a)$$

Thus,

$$\mathbf{u} = \sum_{k=1}^{n} u_k \mathbf{b}_k \qquad (21b)$$

$$\mathbf{f} = \sum_{k=1}^{n} f_k \mathbf{b}_k \qquad (21c)$$

and the components of the linear operator A with respect to the $\{\mathbf{b}_i\}$ basis are defined in the usual way by

$$A\mathbf{b}_j = \sum_{k=1}^{n} A_{kj}\mathbf{b}_k \tag{21d}$$

The fact that the basis $\{\mathbf{b}_i\}$ does not depend on x, t is reflected by the statements

$$\mathbf{u}_t = \sum_{k=1}^{n} \frac{\partial u_k}{\partial t} \mathbf{b}_k \tag{21e}$$

$$\mathbf{u}_x = \sum_{k=1}^{n} \frac{\partial u_k}{\partial x} \mathbf{b}_k \tag{21f}$$

that follow from (21b) and in which the derivatives of the $\mathbf{b}_i$ are all equal to zero.

The system (20) is said to be hyperbolic in some solution domain if the eigenvalues of the $\{A_{ij}\}$ matrix are *real and distinct*. Since this matrix depends on the u_i, this statement makes sense only if we specify a solution $u_i(x, t)$ of (20) in some domain of the xt-plane. In this case, each of the n^2 components A_{ij} are defined at each point in this domain, and we can evaluate the eigenvalues there.

7.2.1 Characteristic Curves and the Normal Form

We define a characteristic curve $\mathscr{C}$, as in Section 4.5.2, by the property that along $\mathscr{C}$ Cauchy data together with (20) do not define the outward derivative of each of the u_i. The calculation of the characteristic condition for $\mathscr{C}$ is identical to that for the linear problem, since the A_{ij} are known along $\mathscr{C}$ once the u_i are specified there. Therefore, the characteristic slopes at a given point x, t where the u_i are known are defined by the n roots for the following nth degree polynomial in λ:

$$\det\{A_{ij} - \delta_{ij}\lambda\} = 0 \tag{22}$$

where δ_{ij} is the Kronecker delta.

The determinant in (22) is the one that defines the eigenvalues of the $\{A_{ij}\}$ matrix. We assume that these eigenvalues are real and distinct, and this property implies the following consequences for the operator A and its eigenvalues:

A is a symmetric operator; that is,

$$A\mathbf{u} \cdot \mathbf{v} \equiv \mathbf{u} \cdot A\mathbf{v} \tag{23}$$

for any pair of vectors $\mathbf{u}$ and $\mathbf{v}$. Note that the matrix components of A defined by the right-hand side of (21d) need not be symmetric ($A_{ij} \neq A_{ji}$) for the $\{\mathbf{b}_i\}$ basis.

Let $\mathbf{w}_j$ denote an eigenvector of A associated with the eigenvalue λ_j—that is,

$$A\mathbf{w}_j = \lambda_j\mathbf{w}_j \tag{24}$$

Then we have n mutually orthogonal eigenvectors $\mathbf{w}_1, \ldots, \mathbf{w}_n$; that is,

$$\mathbf{w}_i \cdot \mathbf{w}_j = 0 \quad \text{if } i \neq j \tag{25}$$

In our discussion of the linear problem with $n = 2$ in Section 4.5.2, we proceeded from the component form of the governing equations and verified explicitly that using a basis of eigenvectors resulted in a diagonal normal form for the new components of A. Here, it will be more convenient first to calculate the corresponding normal form without specifying the components of A, $\mathbf{u}$, and $\mathbf{f}$.

We take the dot product of (21a) with respect to $\mathbf{w}_j$ for a fixed j to obtain

$$\mathbf{w}_j \cdot \mathbf{u}_t + \mathbf{w}_j \cdot A\mathbf{u}_x = \mathbf{w}_j \cdot \mathbf{f} \tag{26}$$

But since A is symmetric, (23) and (24) imply that

$$\mathbf{w}_j \cdot A\mathbf{u}_x = A\mathbf{w}_j \cdot \mathbf{u}_x = \lambda_j \mathbf{w}_j \cdot \mathbf{u}_x = \mathbf{w}_j \cdot \lambda_j \mathbf{u}_x \tag{27}$$

Therefore, using (27) in (26) gives the following *normal form*, expressed without the use of components:

$$\mathbf{w}_j \cdot (\mathbf{u}_t + \lambda_j \mathbf{u}_x) = \mathbf{w}_j \cdot \mathbf{f}; \qquad j = 1, \ldots, n \tag{28}$$

Henceforth, we shall omit the reminder $j = 1, \ldots, n$ [as in (28)] in our results.

The deceptively simple looking and easily derived result (28) masks the fact that more work needs to be done in order to relate this expression to the given components u_i, f_i, A_{ij}. This further calculation is crucial if one is to use (28) for a given application. In particular, we need to determine the components of $\mathbf{w}_j$ and the various dot products indicated in (28).

Recall that the starting point of this discussion is the component system (20); the basis vectors $\{\mathbf{b}_i\}$ do not have an intrinsic meaning. More precisely, we are not given the fundamental metric tensor

$$g_{ij} = \mathbf{b}_i \cdot \mathbf{b}_j \tag{29}$$

for this basis. (See Review Problem 3 of Chapter 2.) We do know however that the eigenvectors $\mathbf{w}_j$ are mutually orthogonal and can be derived in terms of their components with respect to the $\{\mathbf{b}_i\}$ basis once the eigenvalues have been computed. Equation (109) of Chapter 4 gives this result for the case $n = 2$, and we shall work out some examples for $n = 3$ later on. Thus, we have n expressions of the form

$$\mathbf{w}_j = \sum_{k=1}^{n} W_{kj}\mathbf{b}_k \tag{30}$$

which can be inverted to give

$$\mathbf{b}_j = \sum_{k=1}^{n} V_{kj}\mathbf{w}_k \tag{31}$$

Here $\{V_{kj}\}$ is the inverse matrix of $\{W_{kj}\}$. Note again that since for any scalar $\alpha(x, t, u_i)$, the vector $\tilde{\mathbf{w}}_j \equiv \alpha_j \mathbf{w}_j$ is also an eigenvector belonging to the eigenvalue λ_j, each column vector in the matrix $\{W_{ij}\}$ may be multiplied by an arbitrary scalar without altering the property that the resulting n vectors, as given by (30), define a basis of eigenvectors. This arbitrariness in the choice of eigenvectors cannot be removed by normalizing these to have unit magnitude because $\mathbf{w}_j \cdot \mathbf{w}_j$ depends on the g_{ij}, which are undefined. However, once we express $\mathbf{u}_t$, $\mathbf{u}_x$, and $\mathbf{f}$ in (28) in terms of components with respect to the basis $\{\mathbf{w}_i\}$, orthogonality implies that, for each j, only the scalar $(\mathbf{w}_j \cdot \mathbf{w}_j)$ will remain as a common factor in the equation for the *j*th component and will then cancel out of the final result.

We compute

$$\mathbf{w}_j \cdot \mathbf{u}_t = \mathbf{w}_j \cdot \sum_{k=1}^{n} \frac{\partial u_k}{\partial t} \mathbf{b}_k = \mathbf{w}_j \cdot \sum_{k=1}^{n} \sum_{\ell=1}^{n} \frac{\partial u_k}{\partial t} V_{\ell k} \mathbf{w}_\ell$$

$$= (\mathbf{w}_j \cdot \mathbf{w}_j) \sum_{k=1}^{n} V_{jk} \frac{\partial u_k}{\partial t}$$

Similarly,

$$\mathbf{w}_j \cdot \lambda_j \mathbf{u}_x = (\mathbf{w}_j \cdot \mathbf{w}_j) \lambda_j \sum_{k=1}^{n} V_{jk} \frac{\partial u_k}{\partial x}$$

and

$$\mathbf{w}_j \cdot \mathbf{f} = (\mathbf{w}_j \cdot \mathbf{w}_j) \sum_{k=1}^{n} V_{jk} f_k$$

Therefore, the scalars $(\mathbf{w}_j \cdot \mathbf{w}_j)$ indeed cancel out of each of the *n* equations (28), leaving

$$\sum_{k=1}^{n} V_{jk} \left(\frac{\partial u_k}{\partial t} + \lambda_j \frac{\partial u_k}{\partial x} \right) = \sum_{k=1}^{n} V_{jk} f_k \tag{32}$$

Notice that the directional derivative

$$\partial_j \equiv \frac{\partial}{\partial t} + \lambda_j \frac{\partial}{\partial x} \tag{33}$$

along the *j*th characteristic is the only one that occurs in the *j*th equation.

Unfortunately, if $n > 2$, we have more characteristic curves passing through a given point than we have coordinates (x, t), so we cannot use the *n* families of characteristic curves as a set of curvilinear coordinates. We can, however, exploit the fact that the *j*th equation involves only the derivative along the *j*th characteristic curve in the following way: We select for each *j* the pair of coordinate families consisting of the family $\phi_j(x, t) = \xi_j = $ constant and the family $t = \sigma_j = $ constant. Here, the curves $\phi_j(x, t) = \xi_j = $ constant make up the family of characteristic curves having slope $dx/dt = \lambda_j$—that is,

$$\frac{dx}{dt} = \lambda_j = -\frac{\partial \phi_j / \partial t}{\partial \phi_j / \partial x} \tag{34}$$

—and the lines $t = \sigma_j = $ constant are not tangent to any of the characteristic curves as long as λ_j is finite.

If we now regard all the u_k in the *j*th equation to be functions of the two variables ξ_j and σ_j, we can write (32) in the following form, which is convenient for solution using finite differences along characteristic directions:

$$\sum_{k=1}^{n} V_{jk} \frac{\partial u_k}{\partial \sigma_j} = \sum_{k=1}^{n} V_{jk} f_j \tag{35}$$

The notation $\partial u_k/\partial \sigma_j$ in (35) is somewhat misleading. We reiterate that it means $\partial u_k/\partial t$, holding ξ_j fixed in the jth equation. Of course, instead of t we could have chosen some other monotonically increasing parameter for σ_j. The choice $\sigma_j = t$ gives the simplest form for (35).

For the quasilinear problem with $n > 2$, this is as far as we can simplify the system (20); an illustrative example is discussed later on. In Section 7.3.1 we show that if $n = 2$, we can use the ξ_j as coordinates instead of x, t and simplify the results further.

It is also interesting to show that if (20) is semilinear (that is, A does not depend on the u_i, but f_j may depend on the u_i in an arbitrary way), one can introduce the components of $\mathbf{u}$ with respect to the basis of eigenvectors and derive a *diagonal form* from (27). This diagonal form generalizes the result in (113) of Chapter 4 to the case $n > 2$.

We express $\mathbf{u}$ in the form

$$\mathbf{u} = \sum_{k=1}^{n} U_k \mathbf{w}_k \tag{36a}$$

and note that according to (216) and (31) the components U_j are defined by

$$U_j = \sum_{k=1}^{n} V_{jk} u_k \tag{36b}$$

Equation (28) can be written in the form

$$\mathbf{w}_j \cdot \partial_j \mathbf{u} = \mathbf{w}_j \cdot \mathbf{f} \tag{37}$$

using the notation (33). The crucial calculation concerns the component form of the left-hand side of (37).

We first evaluate $\partial_j \mathbf{u}$ using (36a):

$$\partial_j \mathbf{u} = \partial_j \sum_{k=1}^{n} U_k \mathbf{w}_k = \sum_{k=1}^{n} (\partial_j U_k) \mathbf{w}_k + \sum_{k=1}^{n} U_k (\partial_j \mathbf{w}_k)$$

But according to (30),

$$\partial_j \mathbf{w}_k = \partial_j \sum_{\ell=1}^{n} W_{\ell k} \mathbf{b}_\ell = \sum_{\ell=1}^{n} (\partial_j W_{\ell k}) \mathbf{b}_\ell \tag{38a}$$

and converting back to $\{\mathbf{w}_i\}$ components, this gives

$$\partial_j \mathbf{w}_k = \sum_{\ell=1}^{n} \sum_{m=1}^{n} (\partial_j W_{\ell k}) V_{m\ell} \mathbf{w}_m \tag{38b}$$

Therefore,

$$\mathbf{w}_j \cdot \partial_j \mathbf{u} = (\mathbf{w}_j \cdot \mathbf{w}_j)(\partial_j U_j) + \mathbf{w}_j \cdot \sum_{k=1}^{n} U_k \sum_{\ell=1}^{n} \sum_{m=1}^{n} (\partial_j W_{\ell k}) V_{m\ell} \mathbf{w}_m$$

$$= (\mathbf{w}_j \cdot \mathbf{w}_j) \left[\partial_j U_j + \sum_{k=1}^{n} \sum_{\ell=1}^{n} (\partial_j W_{\ell k}) V_{j\ell} U_k \right] \tag{39a}$$

Similarly,

$$\mathbf{w}_j \cdot \mathbf{f} = (\mathbf{w}_j \cdot \mathbf{w}_j) \sum_{\ell=1}^{n} V_{j\ell} f_\ell \tag{39b}$$

We see that the common factor $(\mathbf{w}_j \cdot \mathbf{w}_j)$ cancels out of (37), and this expression reduces to the diagonal component form

$$\partial_j U_j + \sum_{k=1}^{n} C_{jk}(x, t) U_k = F_j(x, t, U_i) \tag{40}$$

where

$$C_{jk} = \sum_{\ell=1}^{n} V_{j\ell}(\partial_j W_{\ell k}) \tag{41a}$$

$$F_j = \sum_{\ell=1}^{n} V_{j\ell} f_\ell \tag{41b}$$

It is easy to show that the expressions in (41) for C_{jk} and F_j are in fact the n-dimensional versions of (113) of Chapter 4. This result can also be obtained by direct matrix manipulations starting from the component form (20).

Finally, note that if A depends on the $\mathbf{u}_i$, so do the $\mathbf{w}_i$. Therefore, in deriving (38a), we also obtain derivatives of the U_i with respect to x and t, which precludes the diagonal form (40) in this case. We now illustrate the calculations leading to (32) and (40) for two examples with $n = 3$. See also Problem 2, where shallow-water flow for a two-layer model is outlined. This model leads to a hyperbolic system with $n = 4$.

7.2.2 Unsteady Nonisentropic Flow

In Chapter 5, we derived the three equations governing unsteady flow where the entropy is allowed to be a different constant along each particle path [see (77) of Chapter 5]. In order to illustrate the role of the f_j in (20), let us assume an arbitrary distribution of mass and energy sources whose strengths depend on x, t and the local flow speed. Then the dimensional equations describing mass, momentum, and energy conservation become

$$\rho_t + (\rho u)_x = Q(x, t, u) \tag{42a}$$

$$u_t + u u_x + \frac{px}{\rho} = 0 \tag{42b}$$

$$(p/\rho^\gamma)_t + u(p/\rho^\gamma)_x = E(x, t, u) \tag{42c}$$

It is convenient to introduce the entropy s instead of ρ as a dependent variable. The relation between S, p, and ρ is

$$s \equiv \log\left(\frac{p}{\rho^\gamma}\right) \tag{43}$$

(see (202) of Chapter 5). Equation (42c) becomes

$$s_t + u s_x = e^{-s} E(x, t, u) \equiv h(x, t, u, s) \tag{44}$$

Definition (43) implies that ρ is a function of p and s:

$$\rho(p, s) \equiv (pe^{-s})^{1/\gamma} \tag{45a}$$

and so is the speed of sound c [see (61) of Chapter 3].

$$c(p, s) = \sqrt{\gamma p^{(\gamma - 1)/\gamma} e^{s/\gamma}} \tag{45b}$$

The flow is defined in terms of the three variables u, p, s, and we need to transform (42a). Equation (43) implies that $\rho^{\gamma} = p e^{-s}$. Therefore,

$$\rho_t = \frac{e^{-s}}{\gamma \rho^{\gamma - 1}} (p_t - ps_t) = \frac{1}{c^2} (p_t - ps_t)$$

and

$$\rho_x = \frac{1}{c^2} (p_x - ps_x)$$

Using these expressions in (42a) gives

$$\frac{1}{c^2} (p_t - ps_t + up_x - ups_x) + \rho u_x = Q$$

and using (44), this simplifies to

$$p_t + up_x + \gamma pu_x = c^2 Q + ph \equiv q(x, t, u, p, s) \tag{46}$$

The system of equations corresponding to (20) is given by (42b), (46), and (44) for the dependent variables u, p, and s, respectively. In particular,

$$\mathbf{u} = u\mathbf{b}_1 + p\mathbf{b}_2 + s\mathbf{b}_3 \tag{47a}$$

$$\{A_{ij}\} = \begin{pmatrix} u & \rho^{-1} & 0 \\ \gamma p & u & 0 \\ 0 & 0 & u \end{pmatrix} \tag{47b}$$

$$\mathbf{f} = q\mathbf{b}_2 + h\mathbf{b}_3 \tag{47c}$$

The eigenvalues of A are defined by the vanishing of the determinant (22), which reduces to

$$(u - \lambda)^3 - (u - \lambda)c^2 = 0$$

That is,

$$\lambda_1 = u + c \tag{48a}$$

$$\lambda_2 = u - c \tag{48b}$$

$$\lambda_3 = u \tag{48c}$$

Therefore, the eigenvalues of A are real and distinct.

If we use the component form (30) for the eigenvectors, the defining statement (24) results in the following homogeneous linear algebraic equations governing the $\{W_{ij}\}$.

$$(u - \lambda_j)W_{1j} + \frac{1}{\rho} W_{2j} = 0 \tag{49a}$$

$$\rho c^2 W_{1j} + (u - \lambda_j) W_{2j} = 0 \tag{49b}$$

$$(u - \lambda_j) W_{3j} = 0 \tag{49c}$$

For $j = 1$, (49a) or (49b) gives $W_{21}/W_{11} = \rho c$, whereas (49c) gives $W_{31} = 0$, since $u - \lambda_1 = -c \neq 0$. We *choose*

$$\mathbf{w}_1 = \mathbf{b}_1 + \rho c \mathbf{b}_2 \tag{50a}$$

Similarly, we find $W_{22}/W_{12} = -\rho c$, $W_{32} = 0$ for $j = 2$, and $W_{13} = W_{23} = 0$, $W_{33} = $ arbitrary for $j = 3$. We then choose

$$\mathbf{w}_2 = \mathbf{b}_1 - \rho c \mathbf{b}_2 \tag{50b}$$

$$\mathbf{w}_3 = \mathbf{b}_3 \tag{50c}$$

—that is,

$$\{W_{ij}\} = \begin{pmatrix} 1 & 1 & 0 \\ \rho c & -\rho c & 0 \\ 0 & 0 & 1 \end{pmatrix} \tag{50d}$$

Note again that the condition (24) does not specify all three components of the $\mathbf{w}_j$; it gives only two independent conditions for these three quantities.

The inverse of (50d) is

$$\{V_{ij}\} = \begin{pmatrix} \dfrac{1}{2} & \dfrac{1}{2\rho c} & 0 \\ \dfrac{1}{2} & -\dfrac{1}{2\rho c} & 0 \\ 0 & 0 & 1 \end{pmatrix} \tag{51a}$$

—that is,

$$\mathbf{b}_1 = \frac{\mathbf{w}_1}{2} + \frac{\mathbf{w}_2}{2} \tag{51b}$$

$$\mathbf{b}_2 = \frac{\mathbf{w}_1}{2\rho c} - \frac{\mathbf{w}_2}{2\rho c} \tag{51c}$$

$$\mathbf{b}_3 = \mathbf{w}_3 \tag{51d}$$

Using (51a) in (32) gives the three equations

$$\frac{1}{2}[u_t + (u + c)u_x] + \frac{1}{2\rho c}[p_t + (u + c)p_x] = \frac{q}{2\rho c} \tag{52a}$$

$$\frac{1}{2}[u_t + (u - c)u_x] - \frac{1}{2\rho c}[p_t + (u - c)p_x] = -\frac{q}{2\rho c} \tag{52b}$$

$$s_t + u s_x = h \tag{52c}$$

This can also be written in the form (35):

$$\rho c \frac{\partial u}{\partial \sigma_1} + \frac{\partial p}{\partial \sigma_1} = q \tag{53a}$$

$$\rho c \frac{\partial u}{\partial \sigma_2} - \frac{\partial p}{\partial \sigma_2} = -q \tag{53b}$$

$$\frac{\partial s}{\partial \sigma_3} = h \tag{53c}$$

where ρ and c are the given functions of p, s defined in (45).

Equations (53) are a convenient starting point for a numerical solution using differencing along characteristics, as sketched in Figure 7.2.

Consider the two adjacent points ① and ② in the xt-plane and assume u, p, and s are known there (for instance, these may be two adjacent points on an initial curve). Assume also that $0 < u < c$ at each of these points so that the characteristic directions that emerge have $\lambda_2 < 0$, $0 < \lambda_3 < \lambda_1$. We wish to calculate the values of u, p, q at ③, which is located by the intersection of the λ_1 characteristic from ① with the λ_2 characteristic from ②.

First we locate ③ using the finite difference approximation of (34) with $j = 1$ and $j = 2$:

$$x^{(3)} - x^{(1)} = \lambda_1^{(1)}[t^{(3)} - t^{(1)}] \tag{54a}$$

$$x^{(3)} - x^{(2)} = \lambda_2^{(2)}[t^{(3)} - t^{(2)}] \tag{54b}$$

where we are using superscripts to indicate the values at a given point. In (54), the unknowns are $x^{(3)}$ and $t^{(3)}$, since all the other terms are evaluated at ① or ②, where u, p, s as well as x, t are specified. Solving these linear equations

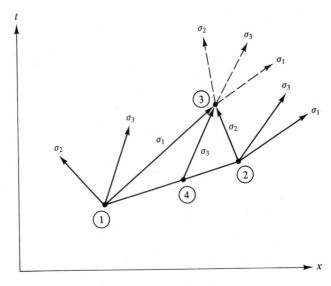

Figure 7.2

gives

$$x^{(3)} = \frac{1}{\lambda_1^{(1)} - \lambda_2^{(2)}} \left[\lambda_1^{(1)}(x^{(2)} - \lambda_2^{(2)}t^{(2)}) - \lambda_2^{(2)}(x^{(1)} - \lambda_1^{(1)}t^{(1)}) \right] \tag{55a}$$

$$t^{(3)} = \frac{1}{\lambda_1^{(1)} - \lambda_2^{(2)}} \left[x^{(2)} - \lambda_2^{(2)}t^{(2)} - x^{(1)} + \lambda_1^{(1)}t^{(1)} \right] \tag{55b}$$

Next, we consider the following difference form of (53a) and (53b) to compute u and p at ③:

$$\rho^{(1)}c^{(1)}(u^{(3)} - u^{(1)}) + p^{(3)} - p^{(1)} = q^{(1)}(t^{(3)} - t^{(1)}) \tag{56a}$$

$$\rho^{(2)}c^{(2)}(u^{(3)} - u^{(2)}) - p^{(3)} + p^{(2)} = -q^{(2)}(t^{(3)} - t^{(2)}) \tag{56b}$$

where $t^{(3)}$ has been determined in (55b). Solving (56a)–(56b) for $u^{(3)}$ and $p^{(3)}$ gives

$$u^{(3)} = \frac{1}{\rho^{(1)}c^{(1)} + \rho^{(2)}c^{(2)}} \left[\rho^{(1)}u^{(1)}c^{(1)} + p^{(1)} + \rho^{(2)}u^{(2)}c^{(2)} - p^{(2)} \right.$$
$$\left. + q^{(1)}(t^{(3)} - t^{(1)}) - q^{(2)}(t^{(3)} - t^{(2)}) \right] \tag{57a}$$

$$p^{(3)} = \frac{1}{\rho^{(1)}c^{(1)} + \rho^{(2)}c^{(2)}} \left\{ \rho^{(1)}c^{(1)} \left[-\rho^{(2)}c^{(2)}u^{(2)} + p^{(2)} + q^{(2)}(t^{(3)} - t^{(2)}) \right] \right.$$
$$\left. + \rho^{(2)}c^{(2)} \left[\rho^{(1)}c^{(1)}u^{(1)} + p^{(1)} + q^{(1)}(t^{(3)} - t^{(1)}) \right] \right\} \tag{57b}$$

To complete the solution at ③, we need to calculate s there. Since s is propagated along the λ_3 characteristic, we need to define starting values at an intermediate point ④. One possible approach is to use linear interpolation to determine ④ and the values of u, p, s there, based on data at ① and ②.

The straight-line approximation for the λ_3 characteristic joining ④ to ③ follows from (53f):

$$x^{(3)} - x^{(4)} = \lambda_3^{(4)} \left[t^{(3)} - t^{(4)} \right] \tag{58a}$$

but now $\lambda_3^{(4)}$, $x^{(4)}$, and $t^{(4)}$ are not known directly. We use linear interpolation between ① and ② to write

$$\frac{\lambda_3^{(4)} - \lambda_3^{(1)}}{\sigma_{14}} = \frac{\lambda_3^{(2)} - \lambda_3^{(1)}}{\sigma_{12}} \tag{58b}$$

$$\frac{t^{(4)} - t^{(1)}}{x^{(4)} - x^{(1)}} = \frac{t^{(2)} - t^{(1)}}{x^{(2)} - x^{(1)}} \tag{58c}$$

where σ_{ij} denotes

$$\sigma_{ij} \equiv \left\{ [x^{(i)} - x^{(j)}]^2 + [t^{(i)} - t^{(j)}]^2 \right\}^{1/2} \tag{58d}$$

Equations (58a)–(58c) define $x^{(4)}$, $t^{(4)}$, and $\lambda_3^{(4)}$ in principle, but an explicit solution is not practical. An efficient iterative approach is to guess a value of $\lambda_3^{(4)}$ (for example, the average of the two known values $\lambda_3^{(1)}$ and $\lambda_3^{(2)}$) and use this initial guess in the two linear equations (58a) and (58c) for $x^{(4)}$ and $t^{(4)}$. Solving these gives

$$x^{(4)} = \frac{[x^{(2)} - x^{(1)}][\lambda_3^{(4)}t^{(3)} - x^{(3)}] + \lambda_3^{(4)}[x^{(1)}t^{(2)} - x^{(2)}t^{(1)}]}{\lambda_3^{(4)}(t^{(2)} - t^{(1)}) + x^{(1)} - x^{(2)}} \tag{59a}$$

$$t^{(4)} = \frac{x^{(1)}t^{(2)} - x^{(2)}t^{(1)} + [t^{(2)} - t^{(1)}][\lambda_3^{(4)}t^{(3)} - x^{(3)}]}{\lambda_3^{(4)}(t^{(2)} - t^{(1)}) + x^{(1)} - x^{(2)}} \tag{59b}$$

Next, we use these values of $x^{(4)}$, $t^{(4)}$ to compute σ_{14} from (58d). We then use the expression

$$\lambda_3^{(4)} = \lambda_3^{(1)} + \frac{\sigma_{14}}{\sigma_{12}}[\lambda_3^{(2)} - \lambda_3^{(1)}] \tag{60}$$

obtained from (58d), to calculate an improved value of $\lambda_3^{(4)}$, and so on.

Once the values of $x^{(4)}$, $t^{(4)}$, and $\lambda_3^{(4)}$ have converged, the final calculation for $s^{(3)}$ follows from the difference form of (53c)—that is

$$s^{(3)} = s^{(4)} + h^{(4)}(t^{(2)} - t^{(4)}) \tag{61a}$$

where $s^{(4)}$ is obtained by linear interpolation between the points ① and ②:

$$s^{(4)} = s^{(1)} + \frac{\sigma_{14}}{\sigma_{12}}[s^{(2)} - s^{(1)}] \tag{61b}$$

This procedure defines u, p, s at ③ uniquely as long as characteristics of the same family emerging from points adjacent to ① or ② do not intersect on or inside the triangle ①–②–③.

7.2.3 A Semilinear Example

The model system

$$\frac{\partial u_1}{\partial t} + x\frac{\partial u_1}{\partial x} + x^2\frac{\partial u_2}{\partial x} = f_1(x, t, u_1, u_2, u_3) \tag{62a}$$

$$\frac{\partial u_2}{\partial t} + (1 + t)^2\frac{\partial u_1}{\partial x} + x\frac{\partial u_2}{\partial x} = f_2(x, t, u_1, u_2, u_3) \tag{62b}$$

$$\frac{\partial u_3}{\partial t} + x\frac{\partial u_3}{\partial x} = f_3(x, t, u_1, u_2, u_3) \tag{62c}$$

is similar in structure to the one just discussed except the u_i occur only in the right-hand sides. The matrix $\{A_{ij}\}$ is now a function of x, t only:

$$\{A_{ij}(x, t)\} = \begin{pmatrix} x & x^2 & 0 \\ (1 + t)^2 & x & 0 \\ 0 & 0 & x \end{pmatrix} \tag{63}$$

and its eigenvalues are

$$\lambda_1 = x(2 + t) \tag{64a}$$

$$\lambda_2 = -xt \tag{64b}$$

$$\lambda_3 = x \tag{64c}$$

The equations corresponding to (49) are now

$$(x - \lambda_j)W_{1j} + x^2 W_{2j} = 0 \tag{65a}$$

$$(1 + t)^2 W_{1j} + (x - \lambda_j)W_{2j} = 0 \tag{65b}$$

$$(x - \lambda_j)W_{3j} = 0 \tag{65c}$$

and we choose the eigenvectors having the following components:

$$\mathbf{w}_1 = x\mathbf{b}_1 + (1 + t)\mathbf{b}_2 \tag{66a}$$

$$\mathbf{w}_2 = x\mathbf{b}_1 - (1 + t)\mathbf{b}_2 \tag{66b}$$

$$\mathbf{w}_3 = \mathbf{b}_3 \tag{66c}$$

consistent with (65). Therefore, the $\{W\}_{ij}$ matrix in (30) is given by

$$\{W_{ij}\} = \begin{pmatrix} x & x & 0 \\ 1 + t & -1 - t & 0 \\ 0 & 0 & 1 \end{pmatrix} \tag{67a}$$

and its inverse is

$$\{V_{ij}\} = \begin{pmatrix} \dfrac{1}{2x} & \dfrac{1}{2(1 + t)} & 0 \\ \dfrac{1}{2x} & \dfrac{-1}{2(1 + t)} & 0 \\ 0 & 0 & 1 \end{pmatrix} \tag{67b}$$

It follows from (67a) that

$$\left\{ \frac{\partial W_{ij}}{\partial t} \right\} = \begin{pmatrix} 0 & 0 & 0 \\ 1 & -1 & 0 \\ 0 & 0 & 0 \end{pmatrix}; \qquad \left\{ \frac{\partial W}{\partial x} \right\}_{ij} = \begin{pmatrix} 1 & 1 & 0 \\ 0 & 0 & 0 \\ 0 & 0 & 0 \end{pmatrix}$$

and (41a) gives

$$\{C_{ij}\} = \frac{1}{2(1 + t)} \begin{pmatrix} t^2 + 3t + 3 & t^2 + 3t + 1 & 0 \\ -t^2 - t - 1 & -t^2 - t + 1 & 0 \\ 0 & 0 & 0 \end{pmatrix} \tag{68}$$

The components F_i are given by

$$F_1 = \frac{1}{2x}f_1 + \frac{1}{2(1 + t)}f_2 \tag{69a}$$

$$F_2 = \frac{1}{2x}f_1 - \frac{1}{2(1 + t)}f_2 \tag{69b}$$

$$F_3 = f_3 \tag{69c}$$

and the u_1, u_2, u_3 arguments of the f_j in (69) are as follows:

$$f_j = f_j(x, t, xU_1 + xU_2, (1 + t)U_1 - (1 + t)U_2, U_3) \tag{70}$$

7.3   Systems of Two First-Order Equations

If the system under consideration has only two dependent variables, there are two distinct families of characteristic curves for a given solution of (20). These two families of curves may be regarded as a curvilinear coordinate system, in terms of which the normal form (28) will simplify further.

7.3.1 Characteristic Coordinates

The eigenvalues of the 2×2 matrix $\{A_{ij}\}$ are [see (107b) of Chapter 4]

$$\lambda_1 = \tfrac{1}{2}\{A_{11} + A_{22} + [(A_{11} - A_{22})^2 + 4A_{12}A_{21}]^{1/2}\} \tag{71a}$$

$$\lambda_2 = \tfrac{1}{2}\{A_{11} + A_{22} - [(A_{11} - A_{22})^2 + 4A_{12}A_{21}]^{1/2}\} \tag{71b}$$

These are real and distinct if $(A_{11} - A_{12})^2 + 4A_{12}A_{21} > 0$, and they define the slopes of the two families of characteristic curves

$$\phi(x, t) = \xi = \text{constant} \tag{71c}$$

$$\psi(x, t) = \eta = \text{constant} \tag{71d}$$

according to $\lambda_1 = -(\phi_t/\phi_x) = (dx/dt)$ and $\lambda_2 = -(\psi_t/\psi_x) = (dx/dt)$. Now, the λ_i are given functions of x, t, u_1, and u_2, so the curves (71c)–(71d) depend on the solution. Let us regard u_1, u_2, x, and t as functions of ξ and η for a given solution. We shall use the notation

$$u_1 = U(\xi, \eta); \qquad u_2 = V(\xi, \eta); \qquad x = X(\xi, \eta); \qquad t = T(\xi, \eta)$$

The characteristic curves thus satisfy the pair of partial differential equations [see (7)]

$$X_\eta = \lambda_1(X, T, U, V)T_\eta \tag{72a}$$

$$X_\xi = \lambda_2(X, T, U, V)T_\xi \tag{72b}$$

To solve these, we need to derive the rule for the variation of U and V along the characteristic curves from (28). In preparation for this, we first compute the expressions for $\mathbf{u}_x$ and $\mathbf{u}_t$:

$$\mathbf{u}_x = \mathbf{u}_\xi \phi_x + \mathbf{u}_\eta \psi_x \tag{73a}$$

$$\mathbf{u}_t = \mathbf{u}_\xi \phi_t + \mathbf{u}_\eta \psi_t = -\lambda_1 \phi_x \mathbf{u}_\xi - \lambda_2 \psi_x u_\eta \tag{73b}$$

Equation (28) for $j = 1$ and $j = 2$ gives

$$\mathbf{w}_1 \cdot (\lambda_1 \mathbf{u}_x + \mathbf{u}_t) = \mathbf{w}_1 \cdot \mathbf{f} \tag{74a}$$

$$\mathbf{w}_2 \cdot (\lambda_2 \mathbf{u}_x + \mathbf{u}_t) = \mathbf{w}_2 \cdot \mathbf{f} \tag{74b}$$

As in Chapter 4, the idea is to express all vectors in terms of components with respect to a basis of eigenvectors.

We know that

$$\mathbf{w}_1 = -A_{12}\mathbf{b}_1 + (A_{11} - \lambda_1)\mathbf{b}_2 \tag{75a}$$

$$\mathbf{w}_2 = -A_{12}\mathbf{b}_1 + (A_{11} - \lambda_2)\mathbf{b}_2 \tag{75b}$$

are orthogonal eigenvectors [see (109) of Chapter 4], and solving these for $\mathbf{b}_1$ and $\mathbf{b}_2$ gives

$$\mathbf{b}_1 = V_{11}\mathbf{w}_1 + V_{21}\mathbf{w}_2 \tag{76a}$$

$$\mathbf{b}_2 = V_{12}\mathbf{w}_1 + V_{22}\mathbf{w}_2 \tag{76b}$$

where [see (112) of Chapter 4 and (31)]

$$\{V_{ij}\} = \frac{1}{A_{12}(\lambda_2 - \lambda_1)}\begin{pmatrix} A_{11} - \lambda_2 & A_{12} \\ \lambda_1 - A_{11} & -A_{12} \end{pmatrix} \tag{76c}$$

We are given the components of $\mathbf{u}_\xi$, $\mathbf{u}_\eta$, and $\mathbf{f}$ in terms of the $\mathbf{b}_1$, $\mathbf{b}_2$ basis—that is,

$$\mathbf{u}_\xi = U_\xi \mathbf{b}_1 + V_\xi \mathbf{b}_2 \tag{77a}$$

$$\mathbf{u}_\eta = U_\eta \mathbf{b}_1 + V_\eta \mathbf{b}_2 \tag{77b}$$

$$\mathbf{f} = f_1 \mathbf{b}_1 + f_2 \mathbf{b}_2 \tag{77c}$$

Therefore, $\mathbf{u}_\xi$, $\mathbf{u}_\eta$, and $\mathbf{f}$ have the following components relative to the $\mathbf{w}_1$, $\mathbf{w}_2$ basis:

$$\mathbf{u}_\xi = (U_\xi V_{11} + V_\xi V_{12})\mathbf{w}_1 + (U_\xi V_{21} + V_\xi V_{22})\mathbf{w}_2 \tag{78a}$$

$$\mathbf{u}_\eta = (U_\eta V_{11} + V_\eta V_{12})\mathbf{w}_1 + (U_\eta V_{21} + V_\eta V_{22})\mathbf{w}_2 \tag{78b}$$

$$\mathbf{f} = (f_1 V_{11} + f_2 V_{12})\mathbf{w}_1 + (f_1 V_{21} + f_2 V_{22})\mathbf{w}_2 \tag{78c}$$

Upon substituting the expressions in (78) into (74), noting the orthogonality of $\mathbf{w}_1$ and $\mathbf{w}_2$, and canceling out the common factors $\mathbf{w}_1 \cdot \mathbf{w}_1$ and $\mathbf{w}_2 \cdot \mathbf{w}_2$ that arise on both sides, we find

$$U_\eta V_{11} + V_\eta V_{12} = (f_1 V_{11} + f_2 V_{12})T_\eta \tag{79a}$$

$$U_\xi V_{21} + V_\xi V_{22} = (f_1 V_{21} + f_2 V_{22})T_\xi \tag{79b}$$

The system of four equations (72) and (79) (together with the subsidiary equations (71) and (76c) that define λ_1, λ_2 and the $\{V_{ij}\}$ as functions of X, T, U and V) govern the solution.

A numerical solution using finite differences along the local characteristic directions can be easily implemented for given Cauchy data (U, V) along a noncharacteristic spacelike arc. The procedure is analogous to that discussed in Section 7.1.2 and is left as an exercise (see Problem 3).

Further progress is possible for the special case where $f_1 = f_2 = 0$ and the $\{A_{ij}\}$ [and hence the $\{V_{ij}\}$] do not depend on x, t. We simplify the notation for this case and set $u_1 = u(x, t)$, $u_2 = v(x, t)$, $A_{11} = a(u, v)$, $A_{12} = b(u, v)$, $A_{21} = c(u, v)$, and $A_{22} = d(u, v)$, so that (20) becomes

$$u_t + a(u, v)u_x + b(u, v)v_x = 0 \tag{80a}$$

$$v_t + c(u, v)u_x + d(u, v)v_x = 0 \tag{80b}$$

Similarly, equations (79) in terms of the characteristic coordinates become

$$A(U, V)U_\eta + B(U, V)V_\eta = 0 \tag{81a}$$

$$C(U, V)U_\xi + D(U, V)V_\xi = 0 \tag{81b}$$

where we have set $V_{11} = A(U, V)$, $V_{12} = B(U, V)$, $V_{21} = C(U, V)$, and $V_{22} = D(U, V)$.

7.3.2 The Hodograph Transformation

The hodograph transformation is a transformation that reverses the roles of the dependent and independent variables in (80). Thus, we regard x and t as functions of u and v. Let us assume for the time being that this is possible and proceed with the formal transformation; later on we shall examine the conditions under which our procedure is feasible.

If we regard x and t as functions of u, v denoted by

$$x = \tilde{X}(u, v) \tag{82a}$$

$$t = \tilde{T}(u, v) \tag{82b}$$

we compute

$$\begin{pmatrix} dx \\ dt \end{pmatrix} = \begin{pmatrix} \tilde{X}_u & \tilde{X}_v \\ \tilde{T}_u & \tilde{T}_v \end{pmatrix} \begin{pmatrix} du \\ dv \end{pmatrix} \tag{83}$$

In (80), we have $u(x, t)$, $v(x, t)$, which implies that

$$\begin{pmatrix} du \\ dv \end{pmatrix} = \begin{pmatrix} u_x & u_t \\ v_x & v_t \end{pmatrix} \begin{pmatrix} dx \\ dt \end{pmatrix} \tag{84}$$

Combining (83) and (84) gives the identity

$$\begin{pmatrix} \tilde{X}_u & \tilde{X}_v \\ \tilde{T}_u & \tilde{T}_v \end{pmatrix} \begin{pmatrix} u_x & u_t \\ v_x & v_t \end{pmatrix} = \begin{pmatrix} 1 & 0 \\ 0 & 1 \end{pmatrix} \tag{85}$$

Therefore, the two matrices on the left-hand side of (85) are each other's inverses. Let us denote the Jacobians of the two transformations by

$$\tilde{J} \equiv \tilde{X}_u \tilde{T}_v - \tilde{T}_u \tilde{X}_v \tag{86a}$$

$$J \equiv u_x v_t - v_x u_t \tag{86b}$$

The product rule for determinants gives $J = 1/\tilde{J}$, and using the expression for the inverse of a 2×2 matrix in (85) gives

$$\begin{pmatrix} u_x & u_t \\ v_x & v_t \end{pmatrix} = J \begin{pmatrix} \tilde{T}_v & -\tilde{X}_v \\ -\tilde{T}_u & \tilde{X}_u \end{pmatrix} \tag{87}$$

The component form of (87) is

$$u_x = J\tilde{T}_v \tag{88a}$$

$$u_t = -J\tilde{X}_v \tag{88b}$$

$$v_x = -J\tilde{T}_u \tag{88c}$$

$$v_t = J\tilde{X}_u \tag{88d}$$

When we use these expressions in (80) and cancel out J, we obtain the *linear system* in the hodograph plane:

$$-\tilde{X}_v + a(u, v)\tilde{T}_v - b(u, v)\tilde{T}_u = 0 \tag{89a}$$

$$\tilde{X}_u + c(u, v)\tilde{T}_v - d(u, v)\tilde{T}_u = 0 \tag{89b}$$

Note that if (80) contains inhomogeneous terms (that is, f_1 and f_2 are not both identically equal to zero), then $J(x,t)$ does not cancel out of (89), and the transformed system remains quasilinear.

The linear system (89) has the same type (hyperbolic, elliptic, or parabolic) as (80). To show this, we write (89) in matrix form as

$$\begin{pmatrix} 1 & -d \\ 0 & -b \end{pmatrix}\begin{pmatrix} \tilde{X}_u \\ \tilde{T}_u \end{pmatrix} + \begin{pmatrix} 0 & c \\ -1 & a \end{pmatrix}\begin{pmatrix} \tilde{X}_v \\ \tilde{T}_v \end{pmatrix} = 0 \tag{90}$$

Then, assuming $b \neq 0$, we multiply (90) by the inverse of the first matrix to obtain

$$\begin{pmatrix} \tilde{X}_u \\ \tilde{T}_u \end{pmatrix} + \begin{pmatrix} \dfrac{d}{b} & \dfrac{bc - ad}{b} \\ \dfrac{1}{b} & -\dfrac{a}{b} \end{pmatrix}\begin{pmatrix} \tilde{X}_v \\ \tilde{T}_v \end{pmatrix} = 0 \tag{91}$$

The type of the system in (91) depends on the sign of $(d - a)^2/b^2 + 4c/b$, which is the same as the sign of $(d - a)^2 + 4cb$, which characterizes the type of the system (80). If $b \equiv 0$ but $c \neq 0$, we multiply (90) by the inverse of the second matrix to arrive at the same conclusion. If $b \equiv c \equiv 0$, both systems (80) and (91) are in diagonal form and are hyperbolic. Thus, a hodograph transformation is possible as long as $J \neq 0$, regardless of the type of the system (80), and it preserves this type in the hodograph plane.

We show next that if (80) is hyperbolic, its characteristics map onto characteristics of the hodograph system (91). To prove this result, recall first that the two families of characteristic curves for (80) have slopes $(dx/dt) = \lambda_1$ and $(dx/dt) = \lambda_2$, where λ_1 and λ_2 are the two roots of

$$(a - \lambda)(d - \lambda) - bc = 0 \tag{92a}$$

—that is,

$$\lambda_1 = \frac{d + a + \sqrt{\Delta}}{2} \tag{92b}$$

$$\lambda_2 = \frac{d + a - \sqrt{\Delta}}{2} \tag{92c}$$

$$\Delta \equiv (d - a)^2 + 4bc \tag{92d}$$

The two families of characteristic curves associated with (91) have slopes $(dv/du) = \mu_1$ and $(dv/du) = \mu_2$ in the hodograph plane, where μ_1 and μ_2 are the two roots of

$$\left(\frac{d}{b} - \mu\right)\left(-\frac{a}{b} - \mu\right) - \frac{bc - ad}{b^2} = 0 \tag{93a}$$

or

$$\mu_1 = \frac{d - a + \sqrt{\Delta}}{2b} \tag{93b}$$

$$\mu_2 = \frac{d - a - \sqrt{\Delta}}{2b} \tag{93c}$$

Let $\mathscr{C}$ be a characteristic curve with slope $(dx/dt) = \lambda$ in the physical plane, and let its image in the hodograph plane be the curve $\mathscr{C}^*$. Now, along $\mathscr{C}^*$ we have the slope

$$\frac{dv}{du} = \frac{v_x dx + v_t dt}{u_x dx + u_t dt} = \frac{v_x \lambda + v_t}{u_x \lambda + u_t}$$

and using (80), this takes the form

$$\frac{dv}{du} = \frac{-cu_x + (\lambda - d)v_x}{(\lambda - a)u_x - bv_x} \tag{94a}$$

But, it follows from (92a) that

$$(\lambda - a) = \frac{bc}{\lambda - d}$$

Substituting this in (94a) and canceling out the common factors from the numerator and denominator gives

$$\frac{dv}{du} = \frac{d - \lambda}{b} \tag{94b}$$

In particular, according to (92b)–(92c) and (93b)–(93c), the right-hand side of (94b) becomes

$$\frac{d - \lambda_1}{b} = \frac{d - a - \sqrt{\Delta}}{2b} = \mu_2; \qquad \frac{d - \lambda_2}{b} = \frac{d - a + \sqrt{\Delta}}{2b} = \mu_1$$

Therefore, we have shown that the characteristics of (80) with slope λ_1 map to characteristics of (91) with slope μ_2 (and $\lambda_2 \to \mu_1$). The implications of this result are explored further in the next section.

The requirement $J \neq 0$ excludes use of a hodograph transformation in regions where u or v are constant or when u can be expressed in terms of v. This is the case for a simple wave region, as discussed in Section 7.3.4. However, as we shall see, one can solve the system (80) exactly in such regions. A more serious drawback of the solution in the hodograph plane is that, as boundaries are unknown in general, it is difficult to satisfy given boundary condition, Also, the interpretation of discontinuities in U and V becomes troublesome.

Steady two-dimensional transonic flow is an example where the hodograph transformation is quite useful, and the reader is referred to Sections 3.5–3.8 of [1] for a detailed discussion of various problems. Here, we demonstrate only how the Tricomi equation [see (21) of Chapter 4] arises in the hodograph plane for this problem.

As shown in [1], the governing equations for steady two-dimensional transonic flow with small disturbances are (see (3.5.1) of [1])

$$u_y - v_x = 0; \qquad \text{irrotational flow} \tag{95a}$$

$$v_y - uu_x = 0; \qquad \text{mass conservation} \tag{95b}$$

where u, v are rescaled velocity components and x, y are Cartesian coordinates in the plane. This quasilinear system can also be expressed as a quasilinear second-order equation for the velocity potential.

Comparing (80) with (95), we identify $x \to x$, $t \to y$, $u \to u$, $v \to v$, $a \to 0$, $b \to -1$, $c \to -u$, and $d \to 0$. We also denote $\tilde{T}$ in (89) by $\tilde{Y}$ and obtain the following pair of equations in the hodograph plane:

$$-\tilde{X}_v + \tilde{Y}_u = 0 \tag{96a}$$

$$\tilde{X}_u - u\tilde{Y}_v = 0 \tag{96b}$$

Eliminating $\tilde{X}$ from this pair results in the Tricomi equation for $\tilde{Y}$:

$$u\tilde{Y}_{vv} - \tilde{Y}_{uu} = 0$$

We also verify immediately from (92) and (93) that

$$\lambda_1 = u^{1/2} = \mu_2; \qquad \lambda_2 = -u^{1/2} = \mu_1$$

7.3.3   The Riemann Invariants

In this section, we study the implications of the property that if (80) is hyperbolic, its characteristics map onto the characteristics of (89). Let us denote the characteristic curves of the hodograph system (89) having slopes $dv/du = \mu_1$ and $dv/du = \mu_2$ by $S(u, v) = s = $ constant and $R(u, v) = r = $ constant, respectively. Recall that the characteristic curves of (80) with slopes $dx/dt = \lambda_1$ and $dx/dt = \lambda_2$ are denoted by $\phi(x, t) = \xi = $ constant and $\psi(x, t) = \eta = $ constant, respectively. We have shown that curves $\phi(x, t) = $ constant map onto curves $R(u, v) = $ constant and that curves of $\psi(x, t) = $ constant map onto curves $S(u, v) = $ constant.

The preceding implies that (81) *must reduce to*

$$\frac{\partial R(U, V)}{\partial \eta} = 0 \tag{97a}$$

$$\frac{\partial S(U, V)}{\partial \xi} = 0 \tag{97b}$$

—that is,

$$R(U, V) = r = \text{constant on } \xi = \text{constant} \tag{97c}$$

$$S(U, V) = s = \text{constant on } \eta = \text{constant} \tag{97d}$$

The functions $R(U, V) = $ constant and $S(U, V) = $ constant are the *Riemann invariants*, which play roles analogous to integrals for systems of ordinary differential equations. One approach for deriving the Riemann invariants is to simply calculate the characteristic curves in the hodograph plane. A second equivalent approach, which provides insight into their role as integrals of the system (81), is discussed next.

The idea is to identify (81a) with (97a) and (89b) with (97b) for appropriate functions R and S. We have

$$\frac{\partial R}{\partial \eta} = R_U U_\eta + R_V V_\eta$$

and if we identify $(\partial R/\partial\eta) = 0$ in the preceding expression with (81a), we conclude that we must set

$$R_U = G(U,V)A(U,V) \tag{98a}$$

$$R_V = G(U,V)B(U,V) \tag{98b}$$

for some function $G(U,V)$ analogous to an integrating factor. To specify G, we require (98) to be consistent—that is, $(R_U)_V = (R_V)_U$—which gives the following quasilinear first-order equation for G:

$$BG_U - AG_V = G(A_V - B_U) \tag{99}$$

Any nontrivial solution of (99) will define $G(U,V)$, which can then be used in (98) to obtain R by quadrature.

Similarly, in order for (81b) to reduce to (97b), we must have

$$S_U = K(U,V)C(U,V) \tag{100a}$$

$$S_V = K(U,V)D(U,V) \tag{100b}$$

which implies that K must satisfy

$$DK_U - CK_V = K(C_V - D_U) \tag{101}$$

It is easy to show that being able to solve (99) and (101) is exactly equivalent to being able to calculate the characteristic curves of (89) by quadrature. So there is no particular advantage in using one approach in favor of the other for calculating the Riemann invariants.

It is also important to note that Riemann invariants need not exist if $n > 2$. To illustrate this point, let $n = 3$ and consider (32) with $f_1 = f_2 = f_3 = 0$, where the V_{jk} do not depend on x, t. If we attempt to identify the jth equation (32) with the directional derivative ∂_j of a function $R_j(u_1, u_2, u_3)$, we obtain the following three conditions analogous to (98) involving the integrating factor $G_j(u_1, u_2, u_3)$:

$$\frac{\partial R_j}{\partial u_1} = G_j V_{j1}; \qquad \frac{\partial R_j}{\partial u_2} = G_j V_{j2}; \qquad \frac{\partial R_j}{\partial u_3} = G_j V_{j3}$$

Consistency of these conditions introduces the *three* requirements

$$\frac{\partial}{\partial u_2}(G_j V_{j1}) = \frac{\partial}{\partial u_1}(G_j V_{j2}); \qquad \frac{\partial}{\partial u_2}(G_j V_{j3}) = \frac{\partial}{\partial u_3}(G_j V_{j2});$$

$$\frac{\partial}{\partial u_1}(G_j V_{j3}) = \frac{\partial}{\partial u_3}(G_j V_{j1})$$

This implies that each G_j must satisfy *three independent* first-order equations analogous to (99), and this is not possible in general.

The availability of the Riemann invariants (97c)–(97d) simplifies the solution of the system (72), (81) considerably; we shall explore this further in Section 7.3.4. Here, we note that we no longer need consider (81); these two equations are already integrated in the form (97c)–(97d), and the solution is actually governed by the four relations

$$X_\eta - \lambda^+(U, V)T_\eta = 0 \qquad \text{on} \quad \xi = \text{constant} \tag{102a}$$

$$R(U, V) = \text{constant} \quad \text{on} \quad \xi = \text{constant} \tag{102b}$$

$$X_\xi - \lambda^-(U, V)T_\xi = 0 \qquad \text{on} \quad \eta = \text{constant} \tag{102c}$$

$$S(U, V) = \text{constant} \quad \text{on} \quad \eta = \text{constant} \tag{102d}$$

where we have set $\lambda_1 = \lambda^+$ and $\lambda_2 = \lambda^-$.

Now, suppose that we use

$$r = R(U, V) \tag{103a}$$

$$s = S(U, V) \tag{103b}$$

as independent variables instead of ξ, η. This is a generalized hodograph trans-formation in the sense that the independent variables are certain *functions* of the u, v. The preceding is permissible as long as the functions R or S are not identically constant in a given solution domain [see the discussion of simple waves in Section 7.3.4)]. Let us denote

$$X(\xi, \eta) \equiv \bar{X}(r, s) \tag{104a}$$

$$T(\xi, \eta) \equiv \bar{T}(r, s) \tag{104b}$$

and we compute

$$X_\eta = \bar{X}_r R_\eta + \bar{X}_s S_\eta = \bar{X}_s S_\eta$$

$$T_\eta = \bar{T}_r R_\eta + \bar{T}_s S_\eta = \bar{T}_s S_\eta$$

because $R_\eta = 0$. Similarly, we have

$$X_\xi = \bar{X}_r R_\xi; \qquad T_\xi = \bar{T}_r R_\xi$$

Therefore, (102a) and (102c) reduce to the following pair of linear equations, which are simpler than (89):

$$\bar{X}_s - \bar{\lambda}^+(r, s)\bar{T}_s = 0 \tag{105a}$$

$$\bar{X}_r - \bar{\lambda}^-(r, s)\bar{T}_s = 0 \tag{105b}$$

Here we have solved (104) for U, V in terms of r, s in the form

$$U = \bar{U}(r, s); \qquad V = \bar{V}(r, s)$$

and have substituted these expressions into the definitions (71) for $\lambda_1(U, V)$ and $\lambda_2(U, V)$, where we have denoted

$$\bar{\lambda}^+(r, s) \equiv \lambda_1(\bar{U}(r, s), \bar{V}(r, s)) \tag{106a}$$

$$\bar{\lambda}^-(r, s) \equiv \lambda_2(\bar{U}(r, s), \bar{V}(r, s)) \tag{106b}$$

We can combine the two equations (105) into a single linear second-order equation for $\bar{T}$ by differentiating (105a) with respect to r and (105b) with respect to s and subtracting the result, to obtain

$$\bar{T}_{rs} + \frac{1}{\bar{\lambda}^- - \bar{\lambda}^+}(\bar{\lambda}_s^- \bar{T}_r - \bar{\lambda}_r^+ \bar{T}_s) = 0 \tag{107}$$

To illustrate these ideas, let us reconsider the transonic flow equations (95) for the hyperbolic case—that is, $u > 0$—and where we identify y with t. The $\{A_{ij}\}$ matrix elements are $A_{11} = 0$, $A_{12} = -1$, $A_{21} = -u$, and $A_{22} = 0$, and we compute $\lambda_1 = u^{1/2}$, $\lambda_2 = -u^{1/2}$ from (71a)–(71b). The inverse matrix $\{V_{ij}\}$ has the elements $A = C = \frac{1}{2}$, $B = -D = -1/2u^{1/2}$. Therefore, the homogeneous system (81) for this example is

$$\frac{1}{2} U_\eta - \frac{1}{2U^{1/2}} V_\eta = 0 \tag{108a}$$

$$\frac{1}{2} U_\xi + \frac{1}{2U^{1/2}} V_\xi = 0 \tag{108b}$$

and (99) becomes

$$-\frac{1}{2U^{1/2}} G_U - \frac{1}{2} G_V = -\frac{1}{4U^{3/2}} G \tag{109}$$

The solution of (109) has the form (chosen for convenience)

$$G(U, V) = 2U^{1/2}\Gamma'(\tfrac{2}{3} U^{3/2} - V) \tag{110}$$

where Γ' is an arbitrary function. Using (110) in (98) gives

$$R_U = U^{1/2}\Gamma'(\tfrac{2}{3} U^{3/2} - V) \tag{111a}$$

$$R_V = -\Gamma'(\tfrac{2}{3} U^{3/2} - V) \tag{111b}$$

which implies that $R = \Gamma$. Similarly, we find that S is an arbitrary function of $(\tfrac{2}{3} U^{3/2} + V)$. Thus, as expected, each Riemann invariant is constant along the corresponding characteristic of the hodograph equations. For simplicity, let us choose $\Gamma(z) = z$ and write (103) as

$$r = \tfrac{2}{3} U^{3/2} - V \tag{112a}$$

$$s = \tfrac{2}{3} U^{3/2} + V \tag{112b}$$

We then find

$$V = \frac{s - r}{2} \tag{113a}$$

$$U = \left(\frac{3s + 3r}{4}\right)^{2/3} \tag{113b}$$

and (105) becomes

$$\bar{X}_s - \left(\frac{3s + 3r}{4}\right)^{1/3} \bar{Y}_s = 0 \tag{114a}$$

$$\bar{X}_r + \left(\frac{3s + 3r}{4}\right)^{1/3} \bar{Y}_r = 0 \tag{114b}$$

We can also derive this result by transforming the hodograph equations (96) to characteristic variables using the procedure described in Section 4.5.2.

The characteristics of the system (96) are the curves in the hodograph plane with slope $dv/du = \pm u^{1/2}$—that is, the curves $\frac{2}{3}u^{3/2} - v = r = $ constant and $\frac{2}{3}u^{3/2} + v = s = $ constant. If in (96) we express $\tilde{X}(u,v) \equiv \bar{X}(r,s)$, $\tilde{Y}(u,v) \equiv \bar{Y}(r,s)$, we compute

$$\tilde{X}_u = \bar{X}_r r_u + \bar{X}_s s_u = \bar{X}_r u^{1/2} + \bar{X}_s u^{1/2} \tag{115a}$$

$$\tilde{X}_v = \bar{X}_r r_v + \bar{X}_s s_v = -\bar{X}_r + \bar{X}_s \tag{115b}$$

$$\tilde{Y}_u = \bar{Y}_r r_u + \bar{Y}_s s_u = \bar{Y}_r u^{1/2} + \bar{Y}_s u^{1/2} \tag{115c}$$

$$\tilde{Y}_v = \bar{Y}_r r_v + \bar{Y}_s s_v = -\bar{Y}_r + \bar{Y}_s \tag{115d}$$

Substituting (115b) and (115c) into (96a) gives

$$\bar{X}_r - \bar{X}_s + u^{1/2}\bar{Y}_r + u^{1/2}\bar{Y}_s = 0 \tag{116a}$$

Substituting (115a) and (115d) into (96b) and dividing by $u^{1/2}$ gives

$$\bar{X}_r + \bar{X}_s + u^{1/2}\bar{Y}_r - u^{1/2}\bar{Y}_s = 0 \tag{116b}$$

If we now subtract (116a) from (116b) and note the definition (113b) for u, we obtain (114a). Similarly, if we add (116a) to (116b), we find (114b).

This demonstrates that use of the Riemann invariants as independent variables is equivalent to use of characteristic coordinates in the hodograph plane. The second-order equation (107) for $\bar{Y}$ (which is just the canonical form of the Tricomi equation) is

$$\bar{Y}_{rs} + \frac{\bar{Y}_r + \bar{Y}_s}{6(r+s)} = 0 \tag{117}$$

7.3.4 Applications of the Riemann Invariants

In most applications, it is preferable to work with the physical variables u, v, x, t as functions of ξ and η, and the governing equations are (102). We now discuss some specific consequences of the existence of the Riemann invariants.

Cauchy Problem on a Spacelike Arc

Consider the solution of the Cauchy problem on the spacelike arc $\mathscr{C}_0$, where U and V are prescribed, as shown in Figure 7.3. If we use the subscript notation of Section 7.1.2, we prescribe $\mathscr{C}_0$ in the discrete form:

$$x = X(\xi_i, \eta_{-i}) \equiv X_{i,-i}; \qquad t = T(\xi_i, \eta_{-i}) \equiv T_{i,-i}; \qquad -N \le i \le M$$

The values of U, V are also prescribed on $\mathscr{C}_0$: $U_{i,-i} \equiv U(\xi_i, \eta_{-i})$, $V_{i,-i} \equiv V(\xi_i, \eta_{-i})$, which means that we know the values of

$$R_{i,-i} \equiv R(U_{i,-i}, V_{i,-i}), \qquad S_{i,-i} \equiv S(U_{i,-i}, V_{i,-i}) \qquad \text{for } -N \le i \le M$$

The constancy of the Riemann invariants implies that for any (m,n) such that $-N \le m \le M$, $-m \le n \le N$, we have

$$R_{m,n} = R_{m,-m} \tag{118a}$$

$$S_{m,n} = S_{-n,n} \tag{118b}$$

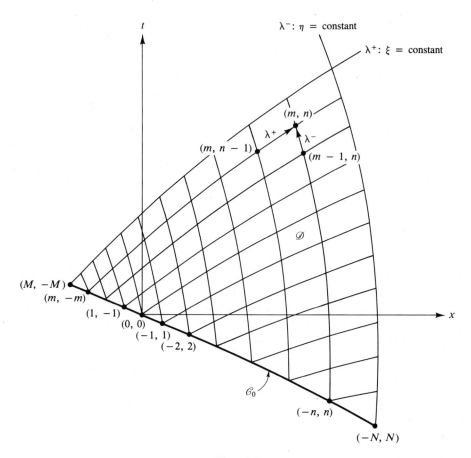

Figure 7.3

Thus, $R_{m,n}$ and $S_{m,n}$ are known at all the gridpoints in the zone of influence $\mathcal{D}$ of the initial arc. This means that solving the two *algebraic* equations (102b) and (102d) gives U and V at each gridpoint in $\mathcal{D}$.

It remains to establish the location of the various gridpoints in the xt-plane, and this is easily accomplished using the finite difference form of (102a) and (102c).

If we know X, T, U, V at the two adjacent gridpoints $(m - 1, n)$ and $(m, n-1)$, then we also know λ^- and λ^+ there. Solving the two linear algebraic equations that result from the finite difference form of (102a) and (102c) then gives [see (15) with $X \to T$, $Y \to X$]:

$$T_{m,n} = \frac{1}{\lambda^+_{m,n-1} - \lambda^-_{m-1,n}} [X_{m-1,n} - X_{m,n-1} + \lambda^+_{m,n-1} T_{m,n-1} - \lambda^-_{m-1,n} T_{m-1,n}]$$

(119a)

$$X_{m,n} = \frac{1}{\lambda^+_{m,n-1} - \lambda^-_{m-1,n}} [\lambda^+_{m,n-1} X_{m-1,n} - \lambda^-_{m-1,n} X_{m,n-1}$$
$$+ \lambda^+_{m,n-1} \lambda^-_{m-1,n} (T_{m,n-1} - T_{m-1,n})]$$

(119b)

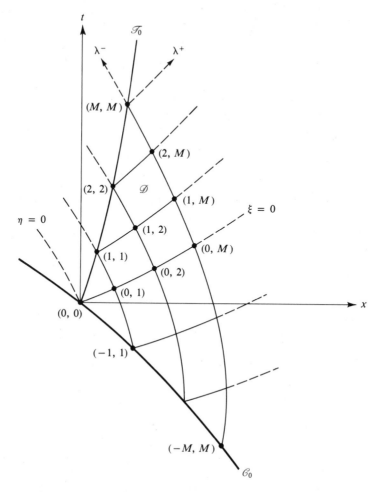

Figure 7.4

Boundary Condition on a Timelike Arc

Consider now the situation illustrated in Figure 7.4, where we prescribe a relationship between U and V on the timelike arc $\mathcal{T}_0$, which is the left boundary of our domain $\mathcal{D}$. Such an arc is characterized by the property that the characteristics on it with slope λ^- all point to the left as t increases. We must keep in mind that since λ^- depends in general on both U and V, we *cannot* establish whether a given arc in the xt-plane is timelike until we know both U *and* V along it. But, our boundary condition consists of only one condition between U and V, so we actually need to solve the problem in order to ascertain whether it is well-posed. Therefore, the validity of the remainder of our arguments is contingent on meeting this criterion for λ^- all along $\mathcal{T}_0$ for the solution that we calculate.

The Cauchy data are now given as the set of discrete values $X_{0,0} = 0$, $T_{0,0} = 0$, $U_{0,0}$, $V_{0,0}$; $X_{-1,1}$, $T_{-1,1}$, $U_{-1,1}$, $V_{-1,1}$; $\dots$; $X_{-M,M}$, $T_{-M,M}$, $U_{-M,M}$, $V_{-M,M}$. In addition, the boundary data on $\mathcal{T}_0$ consist of the following set of discrete values: $X_{0,0} = 0$, $T_{0,0} = 0$, $\alpha_{0,0}U_{0,0} + \beta_{0,0}V_{0,0} = \gamma_{0,0}$; $X_{1,1}$, $T_{1,1}$, $\alpha_{1,1}U_{1,1} + \beta_{1,1}V_{1,1} = \gamma_{1,1}$; ...; $X_{M,M}$, $T_{M,M}$, $\alpha_{M,M}U_{M,M} + \beta_{M,M}V_{M,M} = \gamma_{M,M}$, for given constants $X_{m,m}$, $T_{m,m}$, $\alpha_{m,m}$, $\beta_{m,m}$, $\gamma_{m,m}$. We have assumed a linear relation between U and V along $\mathcal{T}_0$ for simplicity; we could equally easily handle a nonlinear boundary condition of the form $G(U, V, T) = $ constant for a prescribed function G on $\mathcal{T}_0$.

The solution procedure for all gridpoints not adjacent to $\mathcal{T}_0$ are as in the Cauchy problem, and we need to discuss only how we specify both U and V on $\mathcal{T}_0$ as well as on the adjacent gridpoints $(0, 1), (1, 2), \ldots, (m, m + 1), \ldots, (M-1, M)$.

So far in this chapter we have restricted our attention to smooth solutions. In the present context, the solution in the neighborhood of the origin is smooth if the Cauchy data at the origin are consistent with the boundary data in the limit as a point approaches the origin along $\mathcal{T}_0$. More precisely, we assume that using the values of $U_{0,0}$, $V_{0,0}$ specified at the origin on $\mathcal{C}_0$ to compute $(\alpha_{0,0}U_{0,0} + \beta_{0,0}V_{0,0})$ gives the prescribed value $\gamma_{0,0}$. We shall illustrate the case where a shock is introduced at the origin later on for specific examples.

It is easy to see that the solution procedure for U and V at the $(m, m + 1)$ gridpoints is no different than the procedure we outlined in discussing the Cauchy problem. In fact, since we know U and V at $(0, 0)$ and $(-1, 1)$, we can compute U and V at $(0, 1)$ using the Riemann invariants $R_{0,1} = R_{0,0}$ and $S_{0,1} = S_{-1,1}$. Thus, we need discuss only how to compute U and V at each of the (m, m) gridpoints on $\mathcal{T}_0$, assuming we know their values at the $(m-1, m)$ points.

We have the Riemann invariant $S_{m,m} = S_{-m,m}$, which gives one relation linking $U_{m,m}$ to $V_{m,m}$. The second relation is provided by the boundary condition $\alpha_{m,m}U_{m,m} + \beta_{m,m}V_{m,m} = \gamma_{m,m}$. These two conditions define $U_{m,m}$ and $V_{m,m}$. Before proceeding to larger values of t, one must check that $\lambda_{m,m}^-$ is indeed less than $(X_{m+1,m+1} - X_{m,m})/(T_{m+1,m+1} - T_{m,m})$ to verify that $\mathcal{T}_0$ is timelike beyond (m, m).

Simple Waves

Consider the special case of Figure 7.4, illustrated in Figure 7.5, where U and V are constant on $\mathcal{C}_0$.

The solution in the domain $\mathcal{D}_0$ above $\mathcal{C}_0$ and to the right of the λ^+ characteristic emerging from the origin is constant, since the values of the two Riemann invariants remain unchanged throughout this region. Furthermore, since this implies that λ^+ and λ^- are both also constant, the ξ and η characteristics are straight lines in $\mathcal{D}_0$.

Let us assume that $\mathcal{T}_0$ is a timelike arc—that is, that all the $\eta = $ constant characteristics emerging from $\mathcal{T}_0$ lie to its left, as indicated in Figure 7.5. Thus, the domain $\mathcal{D}_1$ is entirely covered by the $\eta = $ constant characteristics that originate from $\mathcal{C}_0$. Therefore, the relation $S(U, V) = $ constant holds throughout $\mathcal{D}_1$, and we can, in principle, solve for V as a function of U in $\mathcal{D}_1$, for example, $V = F(U)$. Moreover, all the $\xi = $ constant characteristics that emerge from $\mathcal{T}_0$ into $\mathcal{D}_1$ are straight lines. To see this, note that (102b) gives $R(U, F(U)) = $ constant on $\xi = $ constant; that is, U *does not vary on a given* $\xi = $ *constant characteristic*. This, in turn, implies that $\lambda^+(U, F(U)) = $ *constant on a given* $\xi = $ *constant characteristic*; that is, the $\xi = $ constant characteristics are straight lines,

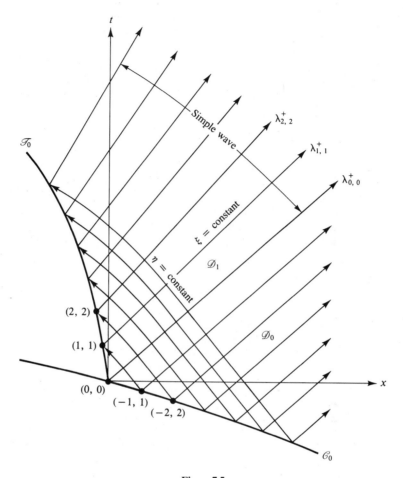

Figure 7.5

as claimed. The solution is defined by the given boundary values for U that propagate unchanged along the $\xi = $ constant characteristics and the values $V = F(U)$, which also propagate unchanged along each $\xi = $ constant ray.

In effect, we can use the expression $v = f(u)$ that results from the constancy of $S(u, v)$ throughout $\mathscr{D}_1$ to eliminate v in favor of u from the governing system

$$u_t + A_{11}(u, v)u_x + A_{12}(u, v)v_x = 0 \tag{120a}$$

$$v_t + A_{21}(u, v)u_x + A_{22}(u, v)v_x = 0 \tag{120b}$$

to obtain the single first-order equation for u:

$$u_t + \{A_{11}(u, f(u)) + A_{12}(u, f(u))f'(u)\}u_x = 0 \tag{121}$$

Note incidentally that we cannot use the Riemann invariants as independent variables in a simple wave region (as either R or S is constant throughout) or in a region of constant u and v (where both R and S are constant).

A simple wave solution fails to exist whenever the λ^- that we calculate on $\mathcal{T}_0$ imply that the $\eta = $ constant characteristics emerge to the right of $\mathcal{T}_0$ for increasing t. We have also tacitly assumed that the values of λ^- decrease as ξ increases; that is, the $\xi = $ constant characteristics "fan out." Both of these features depend on the boundary values prescribed on $\mathcal{T}_0$ for U. In the first instance, if the $\eta = $ constant characteristics emerge to the right of $\mathcal{T}_0$, then this arc is spacelike, and we need to specify a second condition there. In the second instance, if the λ^+ characteristics converge and intersect in $\mathcal{D}_1$, a solution with continuous values of U and V is not possible, and we need to introduce shocks.

A limiting case for which a shock is needed has the boundary value of U prescribed to be a constant U_1 on $\mathcal{T}_0$, in addition to having $U = U_0 = $ constant $\neq U_1$, $V = V_0 = $ constant on $\mathcal{C}_0$.

If we assume that the solution is continuous across the λ^+ characteristics emerging from the origin, we would conclude that the Riemann invariant S holds in $\mathcal{D}_0$ and $\mathcal{D}_1$. This means that in $\mathcal{D}_1$, V is given by $V = F(U)$ obtained by solving

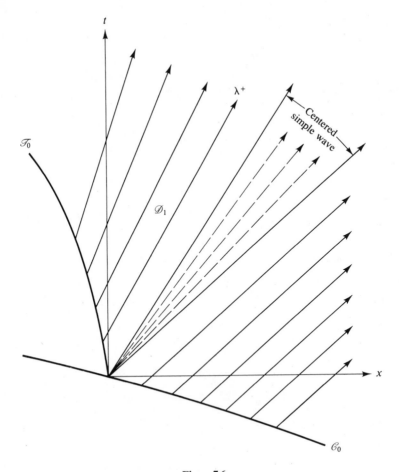

Figure 7.6

the expression $S(U, V) = $ constant $= S_0(U_0, V_0)$ for V in terms of U. In particular, since $U = U_1$ in $\mathscr{D}_1$, we must set $V_1 = F(U_1)$. We then compute the slopes $\lambda^+(U_0, V_0)$ and $\lambda^+(U_1, F(U_1))$ for the $\xi = $ constant characteristics emerging from $\mathscr{C}_0$ and $\mathscr{T}_0$, respectively. These characteristics *intersect* if

$$\lambda^+(U_0, V_0) < \lambda^+(U_1, F(U_1)) \tag{122}$$

and we need to introduce a shock starting at $x = t = 0$. In general, we must introduce a shock whenever characteristics of the same family cross. Special cases were discussed in Chapter 5 (see Figures 5.10 and 5.12) and will be reconsidered later on.

The reverse situation of (122) has

$$\lambda^+(U_0, V_0) > \lambda^+(U_1, F(U_1)) \tag{123}$$

with $U = U_0 = $ constant $\neq U_1$, $V = V_0 = $ constant on $\mathscr{C}_0$; this results in a *centered simple wave* that corresponds to the following limiting case of the smooth problem in Figure 7.5. Assume that U is prescribed as a monotone function of the arc length σ over an ε interval above the origin on $\mathscr{T}_0$, with $U(0) = U_0$, $U(\varepsilon) = U_1$, and $U = U_1$ for $\sigma \geq \varepsilon$. If we solve this problem and then let $\varepsilon \to 0$, we obtain the *centered simple wave* shown in Figure 7.6.

7.4   Shallow-Water Waves

In this section we illustrate the results derived in Section 7.3 for the one-dimensional flow of shallow water. The integral conservation laws of mass and momentum were derived using dimensionless variables in Chapter 3 in the following form [see (16) and (20) of Chapter 3]:

$$\frac{d}{dt} \int_{x_1}^{x_2} h(x, t)\, dx + u(x, t)h(x, t) \bigg|_{x=x_1}^{x=x_2} = 0 \tag{124a}$$

$$\frac{d}{dt} \int_{x_1}^{x_2} u(x, t)h(x, t)\, dx + \{u^2(x, t)h(x, t) + \tfrac{1}{2}h^2(x, t)\} \bigg|_{x=x_1}^{x=x_2} = 0 \tag{124b}$$

where x_1 and x_2 are two fixed points.

The bore conditions associated with (124) are [see (69) of Chapter 5]

$$V[h] = [uh] \tag{125a}$$

$$V[uh] = \left[u^2 h + \frac{h^2}{2} \right] \tag{125b}$$

where [ ] denotes the jump in a quantity across a bore that propagates with speed $(dx/dt) = V$ [see (49) and (51) of Chapter 5].

For smooth solutions equations (124) imply [see (22) of Chapter 3 in reverse order]

$$u_t + uu_x + h_x = 0 \tag{126a}$$

$$h_t + hu_x + uh_x = 0 \tag{126b}$$

Therefore, the vector $\mathbf{u}$ and operator A in (21) have the following components:

$$\mathbf{u} = u\mathbf{b}_1 + h\mathbf{b}_2 \tag{127a}$$

$$\{A_{ij}\} = \begin{pmatrix} u & 1 \\ h & u \end{pmatrix} \tag{127b}$$

7.4.1 Characteristic Coordinates; Riemann Invariants

The eigenvalues of A are defined by the vanishing of the determinant

$$\det\{A_{ij} - \delta_{ij}\lambda\} \equiv (u - \lambda)^2 - h = 0 \tag{128}$$

which gives

$$\lambda_1 = u + \sqrt{h} \tag{129a}$$

$$\lambda_2 = u - \sqrt{h} \tag{129b}$$

Using the values of the A_{ij} and λ_i in (76c) results in

$$\{V_{ij}\} = \begin{pmatrix} -\dfrac{1}{2} & -\dfrac{1}{2\sqrt{h}} \\[2ex] -\dfrac{1}{2} & \dfrac{1}{2\sqrt{h}} \end{pmatrix} \tag{130}$$

Therefore, the characteristic system (72), (79) takes the form:

$$X_\eta - (U + H^{1/2})T_\eta = 0$$

$$X_\xi - (U - H^{1/2})T_\xi = 0$$

$$-\frac{1}{2}U_\eta - \frac{1}{2H^{1/2}}H_\eta = 0$$

$$-\frac{1}{2}U_\xi + \frac{1}{2H^{1/2}}H_\xi = 0$$

where we are denoting $x = X(\xi, \eta)$, $t = T(\xi, \eta)$, $u = U(\xi, \eta)$, and $h = H(\xi, \eta)$.

The hodograph form (89) of the system (126) is

$$-\tilde{X}_h + u\tilde{T}_h - \tilde{T}_u = 0 \tag{131a}$$

$$\tilde{X}_u + h\tilde{T}_h - u\tilde{T}_u = 0 \tag{131b}$$

We have argued that the Riemann invariants for this problem can be interpreted as the characteristics of (131)—that is, curves with slope $(dh/du) = \mu_1$ and $(dh/du) = \mu_2$. Using (93b)–(93c) we find $\mu_1 = \sqrt{h}$ and $\mu_2 = -\sqrt{h}$. Therefore, integrating gives

$$R(u, h) = u + 2\sqrt{h} \tag{132a}$$

$$S(u, h) = u - 2\sqrt{h} \tag{132b}$$

This result also follows from the solutions of (99) and (101) for G and K, respectively. In particular, (99) reduces to

$$-\frac{1}{2H^{1/2}}G_U + \frac{1}{2}G_H = 0$$

The solution of this has G equal to an arbitrary function of $(U + 2H^{1/2})$, and for convenience, we choose

$$G(U, H) = -2\Gamma'(U + 2H^{1/2})$$

which, when used in (98), gives $R = U + 2\sqrt{H}$. A similar calculation gives (132b).

7.4.2 Simple Wave Solutions

In this section we study two examples where the solution is made up of simple wave regions bounded by constant states, so that we can calculate an exact solution explicitly.

The Dam-Breaking Problem

The exact solution for this problem was used in Chapter 4 to study linear small disturbance equations [see Figure 4.12 and (135) of Chapter 4]. We shall now derive this exact solution.

The initial conditions are:

$$u(x, 0) = 0 \tag{133a}$$

$$h(x, 0) = \begin{cases} 1, & \text{if } x < 0 \\ 0, & \text{if } x > 0 \end{cases} \tag{133b}$$

For the time being, let us consider only the characteristics that emerge from $x < 0$. These characteristics have constant slopes $\lambda^+ = \sqrt{h} = 1$ and $\lambda^- = -\sqrt{h} = -1$ in the domain $\mathcal{D}_0$ to the left of $t = -x$, where the solution is the uniform state $u = 0$, $h = 1$ [see Figure 7.7]. The Riemann invariant on the λ^+ characteristics is

$$R = u + 2\sqrt{h} = u(x, 0) + 2\sqrt{h(x, 0)} = 2 \tag{134}$$

Thus, the relation (134) links u and h in the domain $\mathcal{D}_1$ to the right of $t = -x$ that is covered by the λ^+ characteristics emerging from $x < 0$, $t = 0$. We do not yet know the extent of this domain.

At this point, we anticipate the occurrence of a centered simple wave at the origin. This simple wave is defined by the λ^- characteristics emerging from the origin and forming the family of rays

$$\frac{x}{t} = u - \sqrt{h} \tag{135}$$

The solution of the two algebraic relations (134) and (135) for u and h gives the result [see (135) of Chapter 4]

$$u = \frac{2}{3}\left(\frac{x}{t} + 1\right) \tag{136a}$$

$$h = \frac{1}{9}\left(2 - \frac{x}{t}\right)^2 \tag{136b}$$

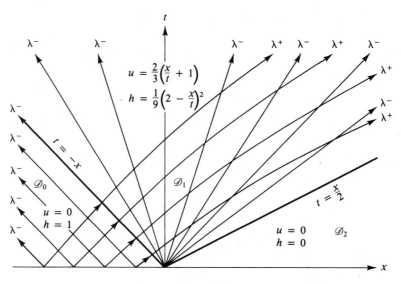

Figure 7.7

Using these values of u and h in the expression $dx/dt = u + \sqrt{h}$ and integrating, gives the curved λ^+ characteristics sketched in $\mathscr{D}_1$ [see also (144a) and Figure 4.14 of Chapter 4].

The solution (136) remains valid as long as $h \geq 0$ and (136b) shows that $h \to 0$ as $x/t \to 2$. Therefore, $\mathscr{D}_1$ consists of the triangular domain $-1 \leq x/t \leq 2$. Since $u = h = 0$ in $\mathscr{D}_2$ and $u = 2$, $h = 0$ on $x = 2t$, we see that R, S, and u are discontinuous on $x = 2t$, but h remains continuous.

The fact that h must vanish at the right boundary of $\mathscr{D}_1$ is also confirmed by the bore conditions (125). Let us assume that the right boundary of $\mathscr{D}_1$ is a bore traveling with the constant speed V and let $u_2 = h_2 = 0$ to its right, whereas $u_1 > 0$, $h_1 > 0$ to its left. The bore conditions (125) then give

$$Vh_1 = u_1 h_1 \tag{137a}$$

$$Vu_1 h_1 = u_1^2 h_1 + \frac{h_1^2}{2} \tag{137b}$$

and these imply that $h_1 = 0$.

Once we note that h must be continuous, it is also possible to derive the solution (136) by using (134) to eliminate u from either of the governing equations (126a)–(126b) to obtain

$$h_t + (2 - 3\sqrt{h})h_x = 0 \tag{138}$$

This equation is to be solved subject to the initial condition (133b).

The change of variable $v = 2 - 3\sqrt{h}$ reduces (138) and (133b) to

$$v_t + vv_x = 0 \tag{139a}$$

$$v(x,0) = \begin{cases} -1, & \text{if } x < 0 \\ 2, & \text{if } x > 0 \end{cases} \tag{139b}$$

and we recognize this as (61c) of Chapter 5 for the special case of the initial condition (82). The solution for v is the centered fan

$$v(x,t) = \begin{cases} -1, & \text{if } x \le -t \\ \dfrac{x}{t}, & \text{if } -t \le x \le 2t \\ 2, & 2t \le x \end{cases} \tag{140a}$$

—that is,

$$h = \frac{1}{9}(v-2)^2 = \begin{cases} 1, & \text{if } x \le -t \\ \dfrac{1}{9}\left(2 - \dfrac{x}{t}\right)^2, & \text{if } -t \le x \le 2t \\ 0, & \text{if } 2t \le x \end{cases} \tag{140b}$$

We use (140b) in the Riemann invariant (134), which is valid in $\mathscr{D}_0$ and $\mathscr{D}_1$, to obtain u there:

$$u(x,t) = \begin{cases} 0, & \text{if } x \le t \\ \dfrac{2}{3}\left(\dfrac{x}{t} + 1\right), & \text{if } -t \le x < 2t \end{cases} \tag{141}$$

and we know that $u = 0$ if $2t < x$.

Retracting Piston

Consider the flow generated in a semi-infinite body of shallow water at rest when a wavemaker (piston) begins to move to the left (that is, away from the fluid) along the path $x_p = -\varepsilon g(t) < 0$ (see Figure 3.7). In order to ensure that the flow be free of bores (see Section 7.3.4, page 418), we require εg to be a monotone nonincreasing function of t for $t > 0$.

Thus, we wish to solve (126) subject to the initial conditions

$$u(x,0) = 0 \tag{142a}$$

$$h(x,0) = 1 \tag{142b}$$

and the boundary condition

$$u(-\varepsilon g(t), t) = -\varepsilon \dot{g}(t), \qquad t > 0 \tag{143}$$

Here ε is the dimensionless ratio of the wavemaker characteristic speed to the characteristic flow speed (see Section 3.2.8), and $-\varepsilon g(t)$ is the piston trajectory with $g(0) = 0$. We do not assume that ε is small in this analysis. The geometry is sketched in Figure 7.8a.

In the domain $\mathscr{D}_0$ to the right of the λ^+ characteristic $x = t$, the solution is the constant state $u = 0$, $h = 1$. Therefore, the Riemann invariant

$$S = u - 2\sqrt{h} = -2 \tag{144}$$

holds in the domain covered by the λ^- characteristics

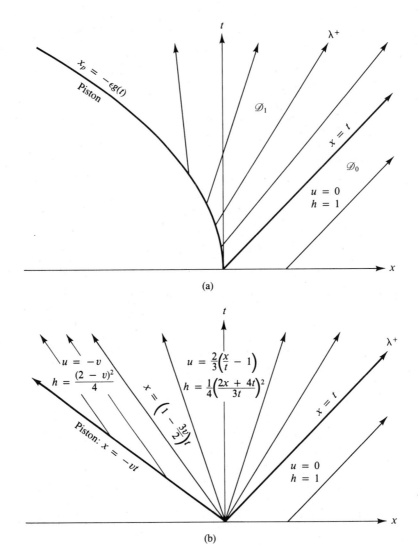

Figure 7.8

$$\frac{dx}{dt} = u - \sqrt{h} \tag{145}$$

that emerge from $t = 0$, $x > 0$.

The λ^+ characteristics that originate from the piston curve $x_p = -\varepsilon g(t)$ have slope

$$\frac{dx}{dt} = u + \sqrt{h} \tag{146a}$$

and this slope equals

$$\frac{dx}{dt} = u + \frac{(u+2)}{2} = \frac{3u+2}{2} = \frac{2-3\varepsilon\dot{g}}{2} \tag{146b}$$

when we use (144) to express h in terms of u and impose the boundary condition (143). If $\varepsilon\dot{g}(0^+) = 0$—that is, the piston starts out with zero speed—then the λ^+ characteristic emerging from $x = -\varepsilon g(0^+)$, $t = 0^+$ coincides with λ^+ characteristic emerging from $x = 0^+$, $t = 0$. However, if $\dot{g}(0^+) > 0$, we must insert a centered simple wave in the triangular domain $[(2 - 3\varepsilon\dot{g}(0^+))/2]t \le x \le t$ bounded by these two characteristics [see (150)].

The solution for u in $\mathscr{D}_1$ follows from the fact that $R = u + 2\sqrt{h}$ is a constant on each λ^+ characteristic in $\mathscr{D}_1$. Since (144) holds throughout $\mathscr{D}_1$, we have $\sqrt{h} = (u+2)/2$ there. Therefore, the statement $R = $ constant means that $u + (u+2) = 2u + 2 = $ constant; that is, u is a constant equal to its boundary value on each λ^+ characteristic (146b). This result may be expressed more precisely in the parametric form

$$u = -\varepsilon\dot{g}(\tau) \tag{147a}$$

on

$$x = \frac{2 - 3\varepsilon\dot{g}(\tau)}{2}(t - \tau) - \varepsilon g(\tau) \tag{147b}$$

For a given $g(\tau)$, we can solve (147b) for τ as a function of x and t. Substituting this into (147a) gives u as a function of x and t (see Problem 4). Once u is known, we compute h from (144):

$$h = \tfrac{1}{4}(u + 2)^2 = \tfrac{1}{4}[2 - \varepsilon\dot{g}(\tau)]^2 \tag{147c}$$

on the lines defined by (147b).

The preceding assumes that $\varepsilon\dot{g}(0^+) = 0$ and that the piston trajectory is a timelike arc. This is the requirement that λ^- be less than dx_p/dt on $x = x_p$—that is,

$$u - \sqrt{h} = u - \left(\frac{u+2}{2}\right) = \frac{u}{2} - 1 = \frac{-\varepsilon\dot{g}}{2} - 1 < -\varepsilon\dot{g} \tag{148}$$

This inequality is satisfied as long as $\varepsilon\dot{g} < 2$, but if the piston is pulled away with a speed that exceeds 2, the λ^- characteristics will propagate to the right and render $x = -\varepsilon g(t)$ a spacelike arc. Note also from (147a) and (147c) that as $\varepsilon\dot{g} \to 2$, $h \to 0, u \to -2$, so that in this limit, we have the mirror image of the dam-breaking problem. Thus, if the piston is pulled with a speed faster than the critical speed 2, the flow is the same as if the piston were suddenly removed at $t = 0$.

The result (147a)–(147b) also follows from the single equation for u that results when (144) is used to eliminate h from either (126a) or (126b) to obtain

$$u_t + \frac{(3u+2)}{2}u_x = 0 \tag{149}$$

This is to be solved subject to the boundary condition (143), which we write in parametric form as

$$t = \tau; \qquad x = -\varepsilon g(\tau); \qquad u = -\varepsilon \dot{g}(\tau)$$

The characteristic equations of (149) are [see (23) of Chapter 5]

$$\frac{dt}{ds} = 1; \qquad \frac{dx}{ds} = \frac{3u + 2}{2}; \qquad \frac{du}{ds} = 0$$

which can be immediately solved in the form (147a)–(147b).

If $\dot{g}(0^+) > 0$, we insert the centered simple wave solution obtained from $x/t = (3u + 2)/2$; that is,

$$u = \frac{2}{3}\left(\frac{x}{t} - 1\right) \tag{150a}$$

$$h = \frac{1}{4}\left(\frac{2x + 4t}{3t}\right)^2 \tag{150b}$$

for $[(2 - 3\varepsilon \dot{g}(0^+))/2]t \le x \le t$.

For the special case where the piston moves with constant speed v, the solution (150) terminates at the ray $x = (1 - 3v/2)t$. To the left of this ray we have the constant state $u = -v$, $h = (2 - v)^2/4$, as shown in Figure 7.8b.

Interacting Simple Waves

Consider the problem of two pistons initially at rest a unit distance apart and retaining quiescent water of unit height. At $t = 0$, the pistons are impulsively retracted with equal and opposite speeds v. By symmetry, this also corresponds to the problem for a single piston with a vertical wall at $x = 0$ (see Figure 7.9).

Each piston motion initially generates a centered simple wave with origin at P: $x = -1/2$, $t = 0$, and Q: $x = 1/2$, $t = 0$. These simple waves are unaffected by each other until the fastest λ^+ characteristic from P intersects the fastest λ^- characteristic from Q at the point A. Let us first define the solution before this interaction starts.

Consider the equilibrium domain ⓪ in which $u = 0$ and $h = 1$. The boundary characteristic between ⓪ and ② has the slope $\lambda^+ = \sqrt{h} = 1$; that is, it is the straight line $x = -\frac{1}{2} + t$, and by symmetry, the bounding characteristic between ⓪ and ④ is $x = \frac{1}{2} - t$. These two characteristics intersect at the point A: $x = 0$, $t = \frac{1}{2}$.

The Reimann invariant that holds in ② and ① is

$$S(u, h) = u - 2\sqrt{h} = -2 \tag{151a}$$

which gives h in terms of u in the form

$$h = \frac{(u + 2)^2}{4} \tag{151b}$$

in ② and ①. In particular, $u = -v$ in ①, so that $h = (2 - v)^2/4$ there.

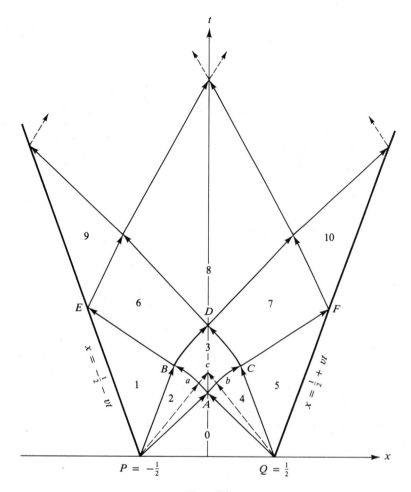

Figure 7.9

The characteristic rays emerging from P are given by [see (135)]:

$$\frac{x + 1/2}{t} = u + \sqrt{h} \tag{152}$$

Upon substituting (151b) for h into (152) and solving the result for u, we obtain

$$u(x, t) = -\frac{2}{3} + \frac{2x + 1}{3t} \qquad \text{in } ② \tag{153a}$$

When this is substituted in (151b), we find

$$h(x, t) = \left(\frac{4t + 2x + 1}{6t}\right)^2 \qquad \text{in } ② \tag{153b}$$

The ray PB, which is the left boundary of ②, is given by (152), in which we set

$u = -v$ and $\sqrt{h} = (2 - v)/2$—that is,

$$x = -\frac{1}{2} + \frac{(2 - 3v)}{2}t \tag{154}$$

Thus, in the limit $v \to 2$, ② extends all the way to the piston curve, and if $v > 2$, the flow is the same as if the piston were suddenly removed at $t = 0$ (dambreaking problem).

Similarly, in region ④, we have

$$u(x, t) = \frac{2}{3} + \frac{2x - 1}{3t} \tag{155a}$$

$$h(x, t) = \left(\frac{4t - 2x + 1}{6t}\right)^2 \tag{155b}$$

with the boundary ray QC defined by

$$x = \frac{1}{2} + \left(\frac{3v - 2}{2}\right)t \tag{156}$$

Now consider the solution in ③ that is bounded by the arcs AB, BD, DC, CA. We can define the arc AB in terms of its slope $dx/dt = u - \sqrt{h}$, as given in ②:

$$\frac{dx}{dt} = -\frac{2}{3} + \frac{2x + 1}{3t} - \frac{4t + 2x + 1}{6t} = \frac{2x - 8t + 1}{6t}$$

—that is,

$$\frac{dx}{dt} - \frac{x}{3t} = \frac{1 - 8t}{6t}$$

Solving this linear equation subject to the initial condition $t = 1/2$, $x = 0$, gives

$$x = -\frac{1}{2} - 2t + \frac{3t^{1/3}}{2^{2/3}} \tag{157}$$

The intersection of this curve with the ray (154) defines the coordinates of B to be

$$x = -\frac{1}{2} + \frac{(2 - 3v)}{2^{1/2}(2 - v)^{3/2}} \tag{158a}$$

$$t = \frac{2^{1/2}}{(2 - v)^{3/2}} \tag{158b}$$

One can define the arc AC from the preceding by symmetry.

The solution in ③ can be computed numerically in a straightforward way for the given values of u and h on AB and AC using essentially the approach outlined in Section 7.3.4. The fact that AB and AC are intersecting characteristic arcs does not alter the solution procedure. One starts with the values of u and h at the points (a) and (b) to compute the location of (c) and the values of u and h there, and so on. The known data on AB and AC then define the solution in all of ③.

It is also instructive to examine the solution in ③ from the point of view of the general result (107), which defines t as a function of r and s, regarded as independent variables. For our special case, u and $\sqrt{h}$ are given in terms of r and s by

$$u = \frac{r + s}{2} \tag{159a}$$

$$\sqrt{h} = \frac{r - s}{4} \tag{159b}$$

Therefore [see (106)],

$$\bar{\lambda}^+ = \frac{r + s}{2} + \frac{r - s}{4} = \frac{3r + s}{4} \tag{160a}$$

$$\bar{\lambda}^- = \frac{r + s}{2} - \frac{r - s}{4} = \frac{3s + r}{4} \tag{160b}$$

and (107) becomes [see (117) for the corresponding result for the Tricomi equation]

$$\bar{T}_{rs} + \frac{3}{2(s - r)}(\bar{T}_r - \bar{T}_s) = 0 \tag{161}$$

This linear second-order hyperbolic equation is to be solved subject to prescribed values for $\bar{T}$ on the two arcs AB and AC. The arc AB is a λ^- characteristic on which $s = -2$. Similarly, AC is a λ^+ characteristic with $r = 2$. But $r = $ constant and $s = $ constant are also characteristics of (161), so we need to solve a *characteristic boundary-value problem* for this linear equation (see Section 4.4.4).

We can actually compute $\bar{T}$ as a function of r on AB and $\bar{T}$ as a function of s on AC from the two known solutions in ② and ④. For example, AB is defined by (105b), which for our case gives

$$(u - \sqrt{h})\bar{T}_r = \bar{X}_r \tag{162a}$$

and using (159) this becomes

$$\frac{r + 3s}{4}\bar{T}_r = \bar{X}_r \tag{162b}$$

In region ②, x is given by (152)—that is,

$$x = -\tfrac{1}{2} + (u + \sqrt{h})t$$

which, using (159), becomes

$$\bar{X}(r, s) = -\frac{1}{2} + \left(\frac{r + s}{2} + \frac{r - s}{4}\right)\bar{T}$$

$$= -\frac{1}{2} + \frac{3r + s}{4}\bar{T}$$

Therefore,

$$\bar{X}_r = \frac{3}{4}\bar{T} + \frac{3r}{4}\bar{T}_r$$

and (162b) reduces to the linear equation for $\bar{T}$;

$$\bar{T}_r + \frac{3}{2(r-s)}\bar{T} = 0 \tag{163}$$

Solving this on $s = -2$, subject to $\bar{T} = \frac{1}{2}$ at $r = 2$, gives

$$\bar{T}(r, -2) = \frac{4}{(r+2)^{3/2}} \tag{164a}$$

Similarly, on AC, we find

$$\bar{T}(2, s) = \frac{4}{(2-s)^{3/2}} \tag{164b}$$

The Riemann function (see Section 4.4.6) for (161) can be calculated in terms of the hypergeometric function, so that we can, in principle, solve for t as a function of r and s in ③. The calculation of the Riemann function for a general form of (161) is outlined in Problem 9 of Section 5.1 of [2]. We do not go into the details here because from a practical point of view, such a solution, although elegant, is not very useful; we have to next calculate $x(r, s)$, invert the result numerically to find $r(x, t)$, $s(x, t)$, and then use this to compute $u(x, t)$ and $h(x, t)$. A direct numerical solution by the method outlined in Section 7.3.4 gives u and h directly and efficiently.

Once the solution in ③ has been calculated, we can compute the simple wave solutions in ⑥ and ⑦ explicitly. Region ⑥ is covered by the λ^+ characteristics emerging from BE, on which u and h have the constant values $u = -v$, $h = (2-v)^2/4$. Therefore, the Riemann invariant

$$u + 2\sqrt{h} = -v + (2-v) = 2(1-v) \tag{165}$$

holds throughout ⑥, and this allows us to express h in terms of u there. Now, each of the λ^- characteristics emerging from BD has the constant slope

$$\frac{dx}{dt} = u_0(\sigma) - \sqrt{h_0(\sigma)} \tag{166}$$

where σ is a parameter that varies along the arc BD and $u_0(\sigma)$, $h_0(\sigma)$ are the values of u and h on BD, as computed in ③. On each of the straight lines defined by (166) for a fixed σ, we have $u = u_0(\sigma)$, $h = h_0(\sigma)$. Similar remarks apply to region ⑦.

In region ⑧, we have the uniform flow $u = u_0^*$, $h = h_0^*$ corresponding to the values of u and h at the point D as computed in ③. One can continue the solution above the λ^+ characteristics emerging from E and the λ^- characteristics emerging from F, and so on. The details of the calculation procedure are left as an exercise (see Problem 5).

7.4.3 Solutions with Bores

In this section we study some examples where bores are needed in order to prevent the crossing of characteristics of the same family.

The Uniformly Propagating Bore

In Section 5.3.4 (page 293), we showed that the bore conditions (125) admit physically realistic solutions consisting of constant-speed bores propagating into a uniform flow of lower height. Let us focus on the problem (see Problem 9 of Chapter 5) of a piston set impulsively into motion with constant speed $v > 0$ into water of unit height at rest. We argued that this left boundary condition produces a bore that propagates to the right with constant speed, say, V, and that the speed of the water u_1 and the height h_1 between the piston and the bore are constants with $u_1 = v$ and h_1 given by the larger positive root $h_1^{(2)}$ of the cubic

$$h_1^3 - h_1^2 - (1 + 2v^2)h_1 + 1 = 0 \tag{167}$$

and V given in terms of $h_1^{(2)}$ by

$$V = \left[h_1^{(2)} \frac{h_1^{(2)} + 1}{2} \right]^{1/2} \tag{168}$$

To confirm the need for a bore, let us assume a continuous solution across the characteristic $x = t$ separating regions $\mathscr{D}_0$ and $\mathscr{D}_1$ as on page 425. If we now used the Riemann invariant (144) to express h in terms of u in $\mathscr{D}_1$, we would compute the following expression for λ_p^+ all along the piston curve:

$$\lambda_p^+ = u + \sqrt{h} = u + \frac{u + 2}{2} = \frac{3u + 2}{2} = 1 + \frac{3v}{2} > 1$$

But, the value of λ^+ in $\mathscr{D}_0$ is $\lambda_0^+ = 1$. Therefore, since $\lambda_p^+ > \lambda_0^+$, the limiting characteristics immediately cross [see (122)], and we need to introduce a bore. The fact that the values of u and h on either side of this bore remain the same implies that the bore speed must be constant (see Figure 7.10).

It is instructive to study the solution (167), (168) for the case of a weak disturbance. Assume that the piston velocity v is small. Since $h_1 \to 1$ as $v \to 0$, we expect the root $h_1^{(2)}$ of (167) to be a function of v, which may be expanded in powers of v in the form

$$h_1^{(2)} = 1 + c_1 v + c_2 v^2 + c_3 v^3 + O(v^4) \tag{169}$$

Substituting this series into (167) and collecting like powers of v shows that we must have

$$v^2(2c_1^2 - 2) + v^3(c_1^3 + 4c_1 c_2 - 2c_1) = O(v^4) \tag{170}$$

This power series must vanish identically. Therefore, each coefficient must vanish, and we find $c_1 = 1$, $c_2 = \frac{1}{4}$ for the larger of the two positive roots. The result $c_1 = 1$ agrees with what we calculated in Chapter 3 for the small disturbance theory [see (137b) of Chapter 3].

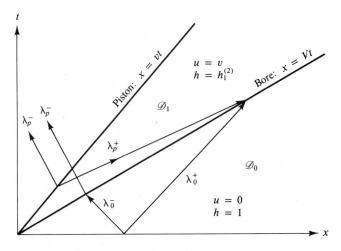

Figure 7.10

Using this result, we derive the following expansions for the bore speed, the characteristic speeds λ^+, λ^-, and the Riemann invariants S and R behind the bore:

$$V = 1 + \frac{3}{4}v + \frac{5}{32}v^2 + O(v^3) \tag{171a}$$

$$\lambda^+ = 1 + \frac{3}{2}v + O(v^3) \tag{171b}$$

$$\lambda^- = -1 + \frac{v}{2} + O(v^3) \tag{171c}$$

$$S = -2 + O(v^3) \tag{171d}$$

$$R = 2 + 2v + O(v^3) \tag{171e}$$

where $\lambda^+ = 1$, $\lambda^- = -1$, $S = -2$, and $R = 2$ in front of the bore.

Equations (171a)–(171b) show that to $O(v)$, the bore speed is the average of the λ^+ speeds on each side. The other noteworthy observation is that although S is indeed discontinuous across the bore, the jump in its value is only $O(v^3)$; the terms proportional to v and v^2 cancel out identically in the expression (171d).

Variable-Speed Piston

The variable-speed piston is the counterpart of the problem discussed in Section 7.4.2 (page 423). Now, the piston moves to the right, into the fluid at rest according to

$$x_p(\varepsilon g(t), t) = \varepsilon \dot{g}(t), \qquad t > 0 \tag{172}$$

where $g(0) = 0$, $\varepsilon g(t) > 0$; again, we do not assume that ε is small. We do assume, however, that $\varepsilon \dot{g}$ is a nondecreasing function of t with $\varepsilon \dot{g}(0^+) > 0$ to ensure

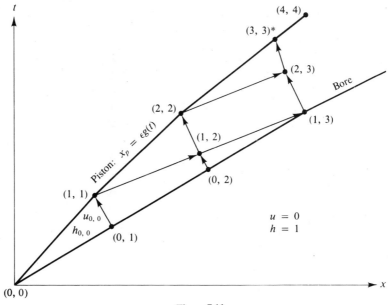

Figure 7.11

that we have only one bore starting at the origin. The geometry is sketched in Figure 7.11.

An analytic solution is not possible (unless g is constant) and we outline a numerical procedure based on the Riemann invariants.

First we need to establish how the solution is started from the origin. We assume that in some small neighborhood of $(0, 0)$, the solution corresponds to *a constant-speed* bore associated with the initial value of the piston speed $\varepsilon \dot{g}(0^{+})$. Thus, in Figure 7.11, we regard $u = \text{constant} = \varepsilon \dot{g}(0)$ along the straight-line segment joining the origin to the known point $(1, 1)$ on the piston curve, and we update the value of u at $(1, 1)$.

The successive points on the piston curve at which the value of u is updated are indicated in Figure 7.11 by $(1, 1), (2, 2), \ldots, (m, m)$ without asterisks, to indicate that these are a priori fixed points. The piston curve is approximated by straight-line segments joining these fixed gridpoints, and we assume for the sake of simplicity that u varies linearly on these straight-line segments. Points on the piston curve marked by an asterisk [for example, $(3, 3)^*$ in Figure 7.11] represent the intersections of reflected λ^{-} characteristics that originated at the bore. As mentioned earlier, we assume that the value of u at these boundary gridpoints is derived by linear interpolation from the given values at the two adjacent gridpoints.

We also approximate the bore trajectory by a piecewise linear curve and update the bore speed (and the values of u and h behind the bore) whenever a λ^{+} characteristic originating from the piston arrives. Since the bore propagates into a region of uniform flow, the values of u and h behind the bore *remain the same on each straight segment*. Thus, in Figure 7.11, the gridpoints $(0, 0), (0, 1), (0, 2),$ and $(1, 3)$ all lie on the same straight segment associated with the starting uniform flow solution inside the triangle $(0,0)$, $(1,3)$, $(1,1)$. We first update u, h and the bore speed at $(1,3)$, where the λ^+ characteristic from $(1,2)$ intersects with the bore trajectory.

To derive the starting solution, we use the given value of $u_{0,0} = \varepsilon \dot{g}(0^+)$ for v in (167) to compute $h_{0,0}$, and then use this in (168) to compute $V_{0,0}$. The next calculation concerns $h_{1,1}$ and the location of the point $(0,1)$ on the bore from which the λ^- characteristic reaches $(1,1)$. Thus, we have three unknowns: $x_{0,1}$, $t_{0,1}$, and $h_{1,1}$. We know $u_{0,0}$, $h_{0,0}$, $u_{1,1}$, $x_{1,1}$, $t_{1,1}$, $V_{0,0}$ and have the following three conditions:

$$u_{1,1} - 2\sqrt{h_{1,1}} = S_{0,0} \tag{173a}$$

$$x_{1,1} - x_{0,1} = \lambda^-_{0,0}(t_{1,1} - t_{0,1}) \tag{173b}$$

$$x_{0,1} = V_{0,0} t_{0,1} \tag{173c}$$

where

$$S_{0,0} \equiv u_{0,0} - 2\sqrt{h_{0,0}} \tag{174a}$$

$$\lambda^-_{0,0} \equiv u_{0,0} - \sqrt{h_{0,0}} \tag{174b}$$

Equation (173a) states that S is constant along the λ^- characteristic emerging from $(0,1)$ (where the values of u and h are the same as those at $(0,0)$). Equations (173b) and (173c) are the straight-line approximations of the characteristic, $(0,1)$–$(1,1)$, and bore, $(0,0)$–$(0,1)$, respectively.

Solving (173b)–(173c) gives

$$t_{0,1} = \frac{x_{1,1} - \lambda^-_{0,0} t_{1,1}}{V_{0,0} - \lambda^-_{0,0}} \tag{175a}$$

$$x_{0,1} = \frac{V_{0,0}(x_{1,1} - \lambda^-_{0,0} t_{1,1})}{V_{0,0} - \lambda^-_{0,0}} \tag{175b}$$

and (173a) defines $h_{1,1}$ as

$$h_{1,1} = \tfrac{1}{4}(u_{1,1} - S_{0,0})^2 \tag{175c}$$

We can now compute $\lambda^-_{1,1}$, $\lambda^+_{1,1}$, $R_{1,1}$, and $S_{1,1}$ to proceed with the calculation to adjacent points. Note, incidentally, that for this problem the condition $\lambda^-_{m,m} < u_{m,m}$, needed to ensure that the boundary curve is timelike, is automatically satisfied, since $\lambda^-_{m,m} = u_{m,m} - \sqrt{h_{m,m}} < u_{m,m}$.

We observe from Figure 7.11 that the domain of interest can be covered entirely by λ^- characteristics. These either originate from a corner in the bore trajectory (and hence end up, in general, somewhere between two fixed boundary points), or we require them to end up at a fixed boundary point, and therefore they originate from somewhere on a straight segment of bore. We shall next illustrate the calculation details for both these types of points using the sequence $(0,2)$, $(1,2)$, and $(2,2)$, followed by the sequence $(1,3)$, $(2,3)$, and $(3,3)^*$. The calculations for the remainder of the solution domain will belong to one or the other of these categories.

Along the sequence of points $(0,2)$, $(1,2)$, and $(2,2)$, the unknowns are $x_{0,2}$, $t_{0,2}$, $x_{1,2}$, $t_{1,2}$, $u_{1,2}$, $h_{1,2}$, and $h_{2,2}$. We know $u_{0,2} = u_{0,0}$, $h_{0,2} = h_{0,0}$, $V_{0,2} = V_{0,0}$, $x_{1,1}$, $t_{1,1}$, $u_{1,1}$, $h_{1,1}$, $x_{2,2}$, $t_{2,2}$, and $u_{2,2}$. Thus, we need seven independent conditions to define our seven unknowns.

Two of these are the two Riemann invariants reaching $(1,2)$,

$$R_{1,2} = R_{1,1} \tag{176a}$$

$$S_{1,2} = S_{0,2} = S_{0,0} \tag{176b}$$

They give $u_{1,2} = u_{1,1}$ and $h_{1,2} = h_{1,1}$ which imply that $\lambda_{1,2}^- = \lambda_{1,1}^-$ and $\lambda_{1,2}^+ = \lambda_{1,1}^+$. Two other conditions locate $(1,2)$ from $(1,1)$ via a λ^+ characteristic, and $(2,2)$ from $(1,2)$ via a λ^- characteristic; that is,

$$x_{1,2} - x_{1,1} = \lambda_{1,1}^+(t_{1,2} - t_{1,1}); \qquad x_{2,2} - x_{1,2} = \lambda_{1,1}^-(t_{2,2} - t_{1,2})$$

Solving these defines $x_{1,2}$ and $t_{1,2}$ in terms of known quantities in the form

$$x_{1,2} = \frac{1}{\lambda_{1,1}^+ - \lambda_{1,1}^-}[\lambda_{1,1}^+(x_{2,2} - \lambda_{1,1}^- t_{2,2}) - \lambda_{1,1}^-(x_{1,1} - \lambda_{1,1}^+ t_{1,1})] \tag{177a}$$

$$t_{1,2} = \frac{1}{\lambda_{1,1}^+ - \lambda_{1,1}^-}[(x_{2,2} - \lambda_{1,1}^- t_{2,2}) - (x_{1,1} - \lambda_{1,1}^+ t_{1,1})] \tag{177b}$$

The fifth condition is the invariances of S along $(1,2)$–$(2,2)$, which gives:

$$h_{2,2} = \tfrac{1}{4}(u_{2,2} - S_{0,0})^2 \tag{178}$$

The sixth and seventh conditions define the slopes of the segments $(0,2)$–$(1,2)$ and $(0,1)$–$(0,2)$:

$$x_{1,2} - x_{0,2} = \lambda_{0,2}^-(t_{1,2} - t_{0,2}) = \lambda_{0,0}^-(t_{1,2} - t_{0,2})$$

$$x_{0,2} = V_{0,0} t_{0,2}$$

Solving these gives

$$x_{0,2} = \frac{V_{0,0}(x_{1,2} - \lambda_{0,0}^- t_{1,2})}{V_{0,0} - \lambda_{0,0}^-} \tag{179a}$$

$$t_{0,2} = \frac{x_{1,2} - \lambda_{0,0}^- t_{1,2}}{V_{0,0} - \lambda_{0,0}^-} \tag{179b}$$

Such a sequence of calculations can always be implemented along a λ^- characteristic that terminates at a fixed piston gridpoint. Notice that this sequence of calculations always breaks up into a subsequence involving at most the solution of two linear algebraic equations. The converse approach, which will generally not terminate on a fixed piston gridpoint, is to start the sequence of calculations from the bore. We illustrate this for the points $(1,3)$, $(2,3)$, and $(3,3)$*.

We begin by fixing the point $(1,3)$ according to

$$x_{1,3} = V_{0,0} t_{1,3}$$

since this point is the endpoint of the first bore segment. The second condition is that $(1, 3)$ lies on the λ^+ characteristic form $(1, 2)$—that is,

$$x_{1,3} - x_{1,2} = \lambda_{1,2}^+(t_{1,3} - t_{1,2})$$

Having already computed $x_{1,2}$, $t_{1,2}$ and $\lambda_{1,2}^+ = u_{1,2} + \sqrt{h_{1,2}}$, we solve for $x_{1,3}$ and $t_{1,3}$ in the form [see (179)]

$$x_{1,3} = \frac{V_{0,0}(x_{1,2} - \lambda_{1,2}^+ t_{1,2})}{V_{0,0} - \lambda_{1,2}^+} \tag{180a}$$

$$t_{1,3} = \frac{x_{1,2} - \lambda_{1,2}^+ t_{1,2}}{V_{0,0} - \lambda_{1,2}^+} \tag{180b}$$

Next we solve the pair of nonlinear algebraic equations

$$u_{1,3} + 2\sqrt{h_{1,3}} = R_{1,2} = R_{1,1} \tag{181a}$$

$$u_{1,3} = (h_{1,3} - 1)\left(\frac{h_{1,3} + 1}{2h_{1,3}}\right)^{1/2} \tag{181b}$$

for $u_{1,3}$ and $h_{1,3}$. Equation (181b) is just the expression we obtain by solving (167) for the flow speed behind the bore in terms of the height [see (90a) of Chapter 5]. One approach for solving this pair is to guess a value of $u_{1,3}^{(1)}$ and calculate $h_{1,3}^{(1)}$ from (181a)

$$h_{1,3}^{(1)} = \frac{[R_{1,1} - u_{1,3}^{(1)}]^2}{4} \tag{182}$$

Then substitute this into (181b) to derive

$$\bar{u}_{1,3}^{(1)} = (h_{1,3}^{(1)} - 1)\left(\frac{h_{1,3}^{(1)} + 1}{2h_{1,3}^{(1)}}\right)^{1/2} \tag{183}$$

The iteration converges if we let $u_{1,3}^{(2)}$ be the average:

$$u_{1,3}^{(2)} = \frac{u_{1,3}^{(1)} + \bar{u}_{1,3}^{(1)}}{2} \tag{184}$$

Then we repeat the process. Once the values of $u_{1,3}$ and $h_{1,3}$ have converged, we compute $\lambda_{1,3}^-$, $\lambda_{1,3}^+$, $S_{1,3}$, $R_{1,3}$; the updated bore speed is given by (168). We calculate $u_{2,3}$ and $h_{2,3}$ from

$$u_{2,3} = \tfrac{1}{2}(S_{1,3} + R_{2,2}) \tag{185a}$$

$$h_{2,3} = \tfrac{1}{4}(R_{2,2} - S_{1,3}) \tag{185b}$$

The values of $x_{2,3}$ and $t_{2,3}$ follow from the solution of

$$x_{2,3} - x_{1,3} = \lambda_{1,3}^-(t_{2,3} - t_{1,3}) \tag{186a}$$

$$x_{2,3} - x_{2,2} = \lambda_{2,2}^+(t_{2,3} - t_{2,2}) \tag{186b}$$

Next, we fix the location of the terminal point $(3, 3)^*$ using the two conditions

$$x_{3,3} - x_{2,3} = \lambda_{2,3}^-(t_{3,3} - t_{2,3}) \qquad (187a)$$

$$\frac{x_{3,3} - x_{2,2}}{t_{3,3} - t_{2,2}} = \frac{x_{4,4} - x_{2,2}}{t_{4,4} - t_{2,2}} \qquad (187b)$$

where (187b) is just the equation for a point $(x_{3,3}, t_{3,3})$ lying on the straight line joining the two fixed points $(2,2)$, $(4,4)$.

Finally, we compute $u_{3,3}$ by linear interpolation:

$$\frac{u_{3,3} - u_{2,2}}{t_{3,3} - t_{2,2}} = \frac{u_{4,4} - u_{2,2}}{t_{4,4} - t_{2,2}} \qquad (188a)$$

We use this in

$$h_{3,3} = (S_{1,3} - u_{3,3})^2/4 \qquad (188b)$$

to obtain $h_{3,3}$.

The following table gives the numerical results that we obtain for the case

$$x_p = \frac{\sqrt{3}}{2}(t + t^2) \qquad (189)$$

for fixed gridpoints at $t = 0, 0.1, 0.2, 0.3, \dots$ along the piston curve.

Point	V	R	S	u	h	x	t	λ^+	λ^-
(0,0)	1.732	3.694	−1.962	0.866	2.000	0	0	2.280	−0.548
(0,1)	1.732	3.694	−1.962	0.866	2.000	0.114	0.066	2.280	−0.548
(1,1)	—	4.040	−1.962	1.039	2.252	0.095	0.100	2.540	−0.462
(0,2)	1.732	3.694	−1.962	0.866	2.000	0.239	0.138	2.280	−0.548
(1,2)	—	4.040	−1.962	1.039	2.252	0.230	0.153	2.540	−0.462
(2,2)	—	4.368	−1.962	1.212	2.519	0.208	0.200	2.799	−0.375
(1,3)	1.901	4.040	−1.940	1.050	2.235	0.341	0.197	2.545	−0.445
(2,3)	—	4.386	−1.940	1.223	2.501	0.321	0.241	2.804	−0.358
(3,3)*	—	4.386	−1.940	1.346	2.699	0.307	0.277	2.988	−0.297

The Dam Breaking over Water Downstream

The dam-breaking problem discussed in Section 7.4.2 (page 421) must be modified when there is a body of quiescent water of height $a < 1$ downstream $(x > 0)$. The initial conditions that replace (133) are now

$$u(x,0) = 0 \qquad (190a)$$

$$h(x,0) = \begin{cases} 1, & \text{if } x < 0 \\ a < 1, & \text{if } x > 0 \end{cases} \qquad (190b)$$

The gas-dynamic counterpart of this problem is the shock-tube problem where

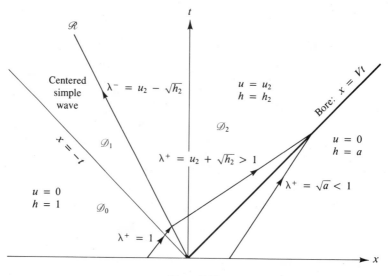

Figure 7.12

a diaphragm separating gases of different densities is suddenly removed at $t = 0$. This problem is discussed in Section 7.5.

We recall that for $a \to 0$, we had a discontinuity in u propagating along the bounding λ^+ characteristic, but h was continuous across this characteristic. Now, for $t = 0$, $\lambda^+ = \sqrt{a} < 1$ for $x > 0$, and $\lambda^+ = 1$ for $x = 0^-$. Actually, we shall see from our results that $\lambda^+ = u_2 + \sqrt{h_2} > 1$ in $\mathscr{D}_2$. Therefore, the λ^+ characteristics emerging from either side of the origin immediately cross, and a bore must start out from there (see Figure 7.12).

We also know that the solution in $\mathscr{D}_0$ to the left of $x = -t$ is the quiescent state $u = 0$, $h = 1$. Therefore, as in the case $a = 0$, the Reimann invariant

$$u + 2\sqrt{h} = 2 \tag{191}$$

must hold in the entire domain $\mathscr{D}_0$, $\mathscr{D}_1$, $\mathscr{D}_2$ to the left of the bore, covered by the λ^+ characteristic emerging from $t = 0$, $x < 0$. This means that the characteristic $x = -t$ is the left boundary of the centered simple wave domain $\mathscr{D}_1$, which terminates along some ray $\mathscr{R}$ from the origin. In $\mathscr{D}_1$ the solution is exactly the one given by (136). The ray $\mathscr{R}$ is the left boundary of a uniform flow domain $\mathscr{D}_2$, which extends up to the bore and in which $u = u_2 = $ constant and $h = h_2 = $ constant. Once $\mathscr{R}$ is identified, we will know u_2 and h_2, since these values must be the same as those predicted by the simple wave solution (136) on its right boundary $\mathscr{R}$.

The crucial question is to identify $\mathscr{R}$, and we do so by combining the information provided by the two bore conditions (125) with the requirement (191), which must persist into $\mathscr{D}_2$. For our case, the bore conditions are:

$$V = \frac{u_2 h_2}{h_2 - a} \tag{192a}$$

$$V = \frac{u_2^2 h_2 + h_2^2/2 - a^2/2}{u_2 h_2} \tag{192b}$$

Eliminating V gives

$$h_2^3 - ah_2^2 - (a^2 + 2au_2^2)h_2 + a^3 = 0 \tag{193}$$

Now, using (191) to express u_2 in terms of h_2 in (193) gives

$$h_2^3 - 9ah_2^2 + 16ah_2^{3/2} - (a^2 + 8a)h_2 + a^3 = 0 \tag{194}$$

We note that as $a \to 1$, (194) has a root $h_2 = 1$, as expected. Also, as $a \to 0, h_2 \to 0$, in agreement with the result in Section 7.4.2 (page 422). The appropriate root of (194) must be larger than a for $a > 0$, and we compute this using Newton's method for a range of values of a. Having h_2, we obtain u_2 from (191) and the slope of the ray $\mathscr{R}$ from

$$\lambda^- = u_2 - \sqrt{h_2} \tag{195}$$

The bore speed is defined by either (192a) or (192b). These results are tabulated next for a range of values of a.

a	h_2	u_2	V	λ^-	h_ℓ	u_ℓ
0.9	0.94933	0.05132	0.98763	−0.92302	0.95	0.05
0.8	0.89715	0.10564	0.97555	−0.84154	0.90	0.10
0.7	0.84309	0.16360	0.96394	−0.67949	0.85	0.15
0.6	0.78661	0.22618	0.95340	−0.56043	0.80	0.20
0.5	0.72692	0.29480	0.94437	−0.43212	0.75	0.25
0.4	0.66268	0.37190	0.93822	−0.29078	0.70	0.30
0.3	0.59143	0.46192	0.93742	−0.12951	0.65	0.35
0.2	0.50787	0.57470	0.94804	0.06683	0.60	0.40
0.1	0.39617	0.74116	0.99141	0.34499	0.55	0.45
0.06	0.33080	0.84970	1.03796	0.51890		
0.03	0.25811	0.98390	1.11330	0.72579		
0.01	0.17118	1.17252	1.24527	1.0013		
0	0	2	2	2		

In the last two columns of this table, we compare our nonlinear results with those obtained in Chapter 3 using the linear theory [see Figures 3.14–3.16]. We set $\tilde{h}_2 = \tilde{u}_2 = \tilde{u}_1 = 0$ in the results given for u and h and consider region ⑤ of Figure 3.15 to obtain

$$u = \frac{\varepsilon \tilde{h}_1}{2} \tag{196a}$$

$$h = a + \frac{\varepsilon \tilde{h}_1}{2} \tag{196b}$$

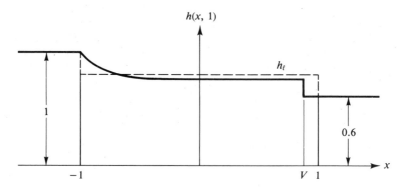

Figure 7.13

corresponding to the renormalized initial conditions

$$u(x,0) = 0 \tag{197a}$$

$$h(x,0) = \begin{cases} a, & \text{if } x > 0 \\ a + \varepsilon\tilde{h}_1, & \text{if } x < 0 \end{cases} \tag{197b}$$

The normalization used here has $a + \varepsilon\tilde{h} = 1$. Therefore, we must set $\varepsilon\tilde{h} = 1 - a$ in the linear results (196) for comparison. This gives

$$u_\ell = \frac{1 - a}{2} \tag{198a}$$

$$h_\ell = \frac{a + 1}{2} \tag{198b}$$

where we use the subscript ℓ to indicate the results of the linear theory.

As seen in the preceding table, the linear results for u and h deteriorate rapidly for $a < \frac{1}{2}$. This is not surprising because $(1 - a)/a$ measures the perturbations and $(1 - a)/a \to 1$ as $a \to \frac{1}{2}$. In the linear theory, the domain $\mathcal{D}_1$ gets squeezed into the characteristic $x = -t$, across which u and h jump from the values given in (198) to the equilibrium values $u = 0$, $h = 1$. Although this result is correct in the limit $a \to 1$, the linear theory does not give any indication of the simple wave behavior. In Figure 7.13, we compare the "exact" variation of h as a function of x with that predicted by the linear theory (dashed profile) for $t = 1$ and the extreme case $a = 0.6$ in order to highlight the features pointed out earlier.

7.5   Compressible Flow Problems

7.5.1 One-Dimensional Unsteady Flow

We return to the problem of one-dimensional unsteady compressible flow for an ideal, inviscid non-heat-conducting gas considered in Chapters 3 and 5 to illustrate a second application area for the general results derived in Section 7.3. For nonisentropic flows we must keep track of *three* dependent variables, as in our calculations in Section 7.2.2. Here our first goal is to establish the conditions under which one can describe the flow in terms of two variables; in this case the results are analogous to those derived for shallow-water waves in Section 7.4.

Problem Formulation

We begin as in Section 7.4 with a summary of the governing equations.

The dimensionless divergence forms for the laws of mass, momentum, and energy conservation are [see (93) of Chapter 5]

$$\rho_t + (u\rho)_x = 0; \qquad \text{mass} \tag{199a}$$

$$(\rho u)_t + (\rho u^2 + p/\gamma)_x = 0; \qquad \text{momentum} \tag{199b}$$

$$\left(\frac{\rho u^2}{2} + \frac{p}{\gamma(\gamma - 1)}\right)_t + \left(\frac{\rho u^3}{2} + \frac{pu}{(\gamma - 1)}\right)_x = 0; \qquad \text{energy} \tag{199c}$$

The temperature θ for an ideal gas is defined in terms of the pressure and density by the equation of state (54a) of Chapter 3.

The shock conditions associated with (199) are

$$V[\rho] = [\rho u] \tag{200a}$$

$$V[\rho u] = \left[\rho u^2 + \frac{p}{\gamma}\right] \tag{200b}$$

$$V\left[\frac{\rho u^2}{2} + \frac{p}{\gamma(\gamma - 1)}\right] = \left[\frac{\rho u^3}{2} + \frac{pu}{\gamma - 1}\right] \tag{200c}$$

where $V \equiv (dx/dt)$ is the shock speed.

For smooth solutions, equations (199) simplify to

$$\rho_t + (\rho u)_x = 0 \tag{201a}$$

$$u_t + uu_x + \frac{p_x}{\gamma\rho} = 0 \tag{201b}$$

$$\left(\frac{p}{\rho^\gamma}\right)_t + u\left(\frac{p}{\rho^\gamma}\right)_x = 0 \tag{201c}$$

Again, we point out that the factor $1/\gamma$ multiplying the dimensionless pressure in (199b)–(199c), (200b)–(200c), and (201b) is due to our choice of the ambient speed of sound $a_0 \equiv \sqrt{\gamma p_0/\rho_0}$ as the velocity scale. This factor divides out of (201c).

Equation (201c) states that (p/ρ^γ) remains constant on particle paths [These are curves in the xt-plane along which $(dx/dt) = u$]. Since the entropy is a function of p/ρ^γ, (201c) also implies that the entropy remains constant on particle paths. But (201c) is valid only if the flow is smooth. Therefore, the statement that the entropy remains constant on particle paths is strictly correct only as long as these paths do not cross a shock. For the special case where p/ρ^γ = constant over some initial arc, a smooth flow has the *same* entropy in the domain swept out by all the particle paths emerging from this arc. This is called *isentropic* flow, and if the initial arc is the ambient state $u = 0$, we have $p/\rho^{\gamma} = 1$ for our choice of dimensionless variables. Therefore, we can replace p with ρ^{γ} in (201b) to obtain [see (73) of Chapter 3]

$$u_t + uu_x + \rho^{\gamma-2}\rho_x = 0 \tag{202}$$

The pair of equations (201a) and (202) govern u and ρ for isentropic flow. An alternate description involves the dimensionless speed of sound

$$a \equiv (p/\rho)^{1/2} = \rho^{(\gamma-1)/2} \tag{203}$$

instead of ρ as the second variable, and the two equations become [see (75) of Chapter 3]

$$u_t + uu_x + \frac{2}{\gamma - 1}aa_x = 0 \tag{204a}$$

$$a_t + \frac{\gamma - 1}{2}au_x + ua_x = 0 \tag{204b}$$

Flows with Shocks

Under what conditions is it correct to use the pair of equations (201a), (202) or (204a)–(204b) for describing flows which contain shocks?

To begin with, consider a flow in which we have one constant speed shock; that is, the flow is uniform on either side of the shock, so the entropy is a *different* constant on either side. We can still use the isentropic equation to describe the flow on either side of the shock as long as we relate the two states through the shock conditions (200). This was discussed in Chapter 5, where we viewed the flow in a reference frame moving with the speed of the flow in front of the shock. Thus, our dimensionless variables have $u = 0$, $p = 1$, $\rho = 1$ in front of the shock. The shock speed V, density ρ, and pressure p behind the shock can then be expressed in terms of the speed $u > 0$ behind the shock in the form [see (96) of Chapter 5]:

$$V = \frac{\gamma + 1}{4}u + \frac{1}{4}[(\gamma + 1)^2u^2 + 16]^{1/2} \tag{205a}$$

$$\rho = \frac{4 + u[(\gamma + 1)^2u^2 + 16]^{1/2} + (\gamma + 1)u^2}{4 + 2(\gamma - 1)u^2} \tag{205b}$$

$$p = 1 + \frac{\gamma(\gamma + 1)}{4}u^2 + \frac{\gamma u}{4}[(\gamma + 1)^2u^2 + 16]^{1/2} \tag{205c}$$

Equations (205) also define the *local* conditions just behind a *variable speed* shock propagating into a uniform flow region. Such a flow could be generated by impulsively setting a piston into motion with a prescribed variable speed into a gas at rest. In this case $V(t)$ is the local shock speed and $\rho(t), p(t)$ are the local values just behind the shock. In Chapter 5 we noted the remarkable fact that whereas $p - 1$, $\rho - 1$, and $V - 1$ are all of order u for small u, the entropy change across a shock is of order u^3. This means that if the piston speed is of order ε [see

(67) of Chapter 3], u will also be of order ε and we can still use (204) to describe the flow behind the shock correctly up to terms of order ε^2. A more accurate statement is that the $O(\varepsilon)$ and $O(\varepsilon^2)$ perturbation equations that we derive from (204) remain correct; we need to account only for entropy changes in the equations governing the $O(\varepsilon^3)$ terms. In Chapter 8 we shall show for the specific example of shallow-water flow that having the equations to $O(\varepsilon^2)$ allows us to compute the solution to $O(\varepsilon)$ in the far field (that is, for $x = O(\varepsilon^{-1})$ and $t = O(\varepsilon^{-1})$). This condition also holds for the compressible flow problems discussed in this section.

Now suppose that the piston speed is not small, but the piston acceleration (or deceleration) is small. The entropy jump *across* the shock is not negligible, but the difference in the entropy between different particle paths behind the shock is of third order. We can therefore still use (204) to describe the flow behind a strong shock as long as the curvature of this shock is small.

In view of this wide range of applicability for (204), and the qualitative analogy between (204) and the shallow-water equations that we have discussed thoroughly, we shall briefly review the results and outline some sample problems with little further discussion.

Characteristic Coordinates, the Hodograph Transformation, the Riemann Invariants

The vector $\mathbf{u}$ and operator A in (21a) now have the components

$$\mathbf{u} = u\mathbf{b}_1 + a\mathbf{b}_2 \tag{206a}$$

$$\{A_{ij}\} = \begin{pmatrix} u & \dfrac{2}{\gamma - 1}a \\ \dfrac{\gamma - 1}{2}a & u \end{pmatrix} \tag{206b}$$

so the eigenvalues of (206b) are

$$\lambda_1 = u + a \tag{207a}$$

$$\lambda_2 = u - a \tag{207b}$$

Equation (76c) gives

$$\{V_{ij}\} = -\frac{1}{2a}\begin{pmatrix} \dfrac{\gamma - 1}{2} & 1 \\ \dfrac{\gamma - 1}{2} & -1 \end{pmatrix} \tag{208}$$

and the characteristic system (72), (79) becomes

$$x_\eta - (u + a)t_\eta = 0 \tag{209a}$$

$$x_\xi - (u - a)t_\xi = 0 \tag{209b}$$

$$\frac{\gamma - 1}{2}u_\eta + a_\eta = 0 \tag{210a}$$

$$\frac{\gamma - 1}{2} u_\xi - a_\xi = 0 \tag{210b}$$

In (209), (210) and for the remainder of this chapter, we shall use the same lowercase symbols for functions of (x, t), (ξ, η), (u, v), or (r, s) to simplify the notation. The particular choice of independent variables will be clear from the context.

The hodograph form of (204) is

$$-x_a + ut_a - \frac{2}{\gamma - 1} at_u = 0 \tag{211a}$$

$$x_u + \frac{\gamma - 1}{2} at_a - ut_u = 0 \tag{211b}$$

The characteristics of (211) have slopes $\mu_1 \equiv da/du = (\gamma - 1)/2$ and $\mu_2 \equiv da/du = -(\gamma - 1)/2$. Therefore, integrating these simple expressions gives the Riemann invariants in the form:

$$r = a + \frac{\gamma - 1}{2} u \tag{212a}$$

$$s = a - \frac{\gamma - 1}{2} u \tag{212b}$$

The system (209), (210) thus reduces to the statements

$$a + \frac{\gamma - 1}{2} u = \text{constant} \quad \text{on} \quad \xi = \text{constant} \tag{213a}$$

$$a - \frac{\gamma - 1}{2} u = \text{constant} \quad \text{on} \quad \eta = \text{constant} \tag{213b}$$

Centered Simple Wave

Consider the analogue for the problem discussed in Section 7.4.2 (page 426) and illustrated in Figure 7.8b. Now we have gas at rest ($u = 0$, $a = 1$) over $x \geq 0$, and we consider the special case of a piston being impulsively retracted (to the left) with constant speed v. The solution domains in the xt-plane are qualitatively the same as those in Figure 7.8b. The two conditions that determine the flow in the centered simple wave region are

$$a - \frac{\gamma - 1}{2} u = 1 \tag{214a}$$

$$\frac{x}{t} = u + a \tag{214b}$$

Equation (214a) is the Riemann invariant along the λ_2 characteristics, and (214b) relates the λ_1 characteristic slopes to the rays from the origin. Solving these expressions for u and a gives

$$u = \frac{2}{\gamma + 1}\left(\frac{x}{t} - 1\right) \tag{215a}$$

$$a = \frac{x}{t}\left(\frac{\gamma - 1}{\gamma + 1}\right) + \frac{2}{\gamma + 1} \tag{215b}$$

for $1 - vt(\gamma + 1)/2 < x < t$, where the boundary ray $x/t = 1 - v(\gamma + 1)/2$ is obtained by setting $u = -v$ in (215a). To the left of this ray we have the uniform flow $u = -v$, $a = 1 - (\gamma - 1)v/2$. For $x > t$ we have the ambient state $u = 0$, $a = 1$.

The Shock-Tube Problem

In this problem we have a stationary gas at a given pressure p_1, and density ρ_1 (hence temperature $\theta_1 = p_1/R\rho_1$) in the domain $x < 0$, which is separated by a thin diaphragm from $x > 0$, where there is a stationary gas with properties $p_0 < p_1$, $\rho_0 < \rho_1$. The flow is initiated by suddenly removing the diaphragm at time $t = 0$. For simplicity we shall discuss the case where the two gases are the same, and we nondimensionalize pressures, densities, and speeds using p_0, ρ_0, and $a_0 \equiv (\gamma p_0/\rho_0)^{1/2}$, respectively. Thus, we have the dimensionless quantities $u_0 = 0$, $p_0 = 1$, $\rho_0 = 1$, $a_0 = 1$, and $p_1 > 1$, $\rho_1 > 1$. The case where the two gases are different is easy to work out but it is more convenient to use dimensional quantities in the calculations (for example, see Section 6.13 of [2]).

Although the shock-tube problem is qualitatively analogous to the problem of a dam breaking over water downstream [Section 7.4.3, page 437), there is one important distinction due to the fact that we now have to keep track of three variables (u, p, ρ) here instead of two (u, h). When the diaphragm is removed the interface separating the two gases can no longer support a pressure difference as it moves to the right at the constant speed of the gases on either side. The flow therefore consists of the five regions shown in Figure 7.14.

We have a shock propagating with constant speed V to the right into the gas at rest, and we have a centered simple wave propagating to the left into the high-pressure state. In addition, and unlike the dam-breaking problem, we also have the interface that is the boundary between the two initial states propagating to the right with speed $u = u_3 = u_4$. Across the interface $p_3 = p_4$, but $\rho_3 \neq \rho_4$ (hence, $\theta_3 \neq \theta_4$). The interface is kinematically equivalent to a piston moving with speed u_3 in the sense that it produces a shock to its right and a centered simple wave to its left.

The solution in the centered simple wave region $\mathcal{D}_2$ is obtained from the Riemann invariant

$$a + \frac{\gamma - 1}{2}u = \text{constant} = a_1 \equiv \left(\frac{p_1}{\rho_1}\right)^{1/2} \tag{216}$$

and the equation for the rays through the origin

$$\frac{x}{t} = u - a \tag{217}$$

Solving (216)–(217) for u and a gives

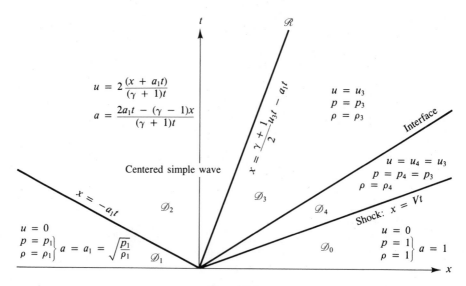

Figure 7.14

$$u = \frac{2(x + a_1 t)}{(\gamma + 1)t} \tag{218a}$$

$$a = \frac{2a_1 t - (\gamma - 1)x}{(\gamma + 1)t} \tag{218b}$$

Again (218) is valid to the right of the known boundary ray $x = -a_1 t$, but at this stage we do not know the right boundary ray $\mathcal{R}$ for $\mathcal{D}_2$. To determine $\mathcal{R}$, we must establish the uniform flow solution $u = u_3$, $p = p_3$, $\rho = \rho_3$ in $\mathcal{D}_3$. To do so we first list all the information for flow quantities in the various domains.

First, the Riemann invariant (217a) holds on $\mathcal{R}$; therefore,

$$a_3 + \frac{\gamma - 1}{2} u_3 = a_1 \tag{219a}$$

Secondly, p/ρ^γ is constant throughout $\mathcal{D}_1, \mathcal{D}_2, \mathcal{D}_3$; in particular,

$$\frac{p_1}{\rho_1^\gamma} = \frac{p_3}{\rho_3^\gamma} \tag{219b}$$

The speed of sound in $\mathcal{D}_3$ and $\mathcal{D}_4$ is given by

$$a_3^2 \equiv \frac{p_3}{\rho_3} \tag{220a}$$

$$a_4^2 \equiv \frac{p_4}{\rho_4} \tag{220b}$$

The interface conditions are

$$u_3 = u_4 \tag{221a}$$

$$p_3 = p_4 \tag{221b}$$

The density and pressure in $\mathcal{D}_4$ behind the shock are given by (205b)–(205c):

$$\rho_4 = \frac{4 + u_4[(\gamma + 1)^2 u_4^2 + 16]^{1/2} + (\gamma + 1)u_4^2}{4 + 2(\gamma - 1)u_4^2} \tag{222a}$$

$$p_4 = 1 + \frac{\gamma(\gamma + 1)}{4} u_4^2 + \frac{\gamma u_4}{4}[(\gamma + 1)^2 u_4^2 + 16]^{1/2} \tag{222b}$$

Using (221b) and (220a) in (220b) gives

$$a_4^2 = \frac{a_3^2 \rho_3}{\rho_4} \tag{223}$$

But, according to (219a) and (221a) we have

$$a_3^2 = \left[a_1 - \frac{\gamma - 1}{2} u_4 \right]^2 \tag{224}$$

Also, according to (219b) and (221b) we have

$$\rho_3 = \rho_1 \left(\frac{p_4}{p_1} \right)^{1/\gamma} \tag{225}$$

Therefore, (223) becomes

$$a_4^2 = \frac{\rho_1}{\rho_4} \left(\frac{p_4}{p_1} \right)^{1/\gamma} \left[a_1 - \frac{\gamma - 1}{2} u_4 \right]^2 \tag{226a}$$

We also have the expression (222b) for p_4, and (220b) becomes

$$a_4^2 = \frac{1}{\rho_4} \left\{ 1 + \frac{\gamma(\gamma + 1)}{4} u_4^2 + \frac{\gamma u_4}{4}[(\gamma + 1)^2 u_4^2 + 16]^{1/2} \right\} \tag{226b}$$

Equating the two expressions (226) for a_4 and simplifying gives the following implicit relation defining u_4 in terms of known quantities:

$$\frac{\rho_1^{(\gamma-1)/2\gamma}}{a_1^{1/\gamma}} \left(a_1 - \frac{\gamma - 1}{2} u_4 \right) = \left\{ 1 + \frac{\gamma(\gamma + 1)}{4} u_4^2 + \frac{\gamma u_4}{4}[(\gamma + 1)^2 u_4^2 + 16]^{1/2} \right\}^{(\gamma-1)/2\gamma} \tag{227}$$

Thus, given ρ_1 and a_1 (or p_1) we can calculate u_4 from (227). Here we choose to express the final result (227) for u_4 in terms of given values of ρ_1 and a_1. The corresponding result can also be derived for p_4 in terms of p_1 and a_1 in a somewhat simpler form (see [2]). Once u_4 is known, we compute the shock speed from (205a). The speed of the interface is u_4, and the ray $\mathcal{R}$ is obtained from (218a) with $u = u_4$; that is,

$$\frac{x}{t} = \frac{\gamma + 1}{2} u_4 - a_1 \tag{228}$$

The density ρ_4 behind the shock is given by (222a) and a_3 by (224), and so on. This completes the solution.

The following numerical example illustrates our results. We choose $p_1 = 40$ and $\rho_1 = 10$—that is, $a_1 = 2$—and find that (227) gives $u_4 = u_3 = 1.9756$. With this value of u_4 we compute $\rho_4 = 3.5974$ from (222a), $p_4 = p_3 = 8.5679$ from (222b), and $V = 2.7362$ from (205a). The ray $\mathscr{R}$ has speed $x/t = 0.3707$ according to (228). Equation (225) gives $\rho_3 = 3.3267$. It is also interesting to note that $p_1/\rho_1^\gamma = p_3/\rho_3^\gamma = 1.5924$, whereas $p_4/\rho_4^\gamma = 1.4272$. Thus, the entropy rises in going from $\mathscr{D}_0$ to $\mathscr{D}_4$ across the shock, and it rises again in going from $\mathscr{D}_4$ to $\mathscr{D}_3$ across the interface. Note that the $p_4/\rho_4^\gamma - p_0/\rho_0^\gamma = 0.4272$, a number that is less than $1/2$ even though u_4 is nearly equal to 2.

In an actual shock tube we have end walls at some positive x_r and negative x_ℓ, so that our results are valid only for $t < x_r/V$ when $x < 0$ and $t < x_\ell/a_1$ when $x < 0$. The problem of a shock reflecting from an end wall is outlined in Problem 7. The reflection process from an end wall for a centered simple wave is analogous to the case for water waves discussed in Section 7.4.2 (page 426). See also Section 6.12 of [2].

Spherically Symmetric Isentropic Flow

We use the radial distance r and the time t as independent variables and find that (87) and (88) of Chapter 3 reduce to

$$\rho_t + u\rho_r + \rho u_r + \frac{2\rho u}{r} = 0; \qquad \text{mass} \tag{229a}$$

$$u_t + uu_r + \frac{p_r}{\gamma\rho} = 0; \qquad \text{momentum} \tag{229b}$$

where ρ, u, p are made dimensionless as in (201). For isentropic flow, we can dispense with the energy equation, which for smooth solutions implies that $p = \rho^\gamma$. Using this in (229b) and introducing the dimensionless speed of sound $a = (p/\rho)^{1/2}$, we obtain the system

$$u_t + uu_r + \frac{2}{\gamma - 1} aa_r = 0 \tag{230a}$$

$$a_t + \frac{\gamma - 1}{2} au_r + ua_r = -(\gamma - 1)\frac{ua}{r} \tag{230b}$$

which generalizes (204) for smooth spherically symmetric solutions. We do not discuss solutions with shocks here. The interested reader can find an account of certain aspects of spherically symmetric flow with shocks in Section 6.16 of [2].

The extra term on the right-hand side of (230b), which arises from the expression of the divergence with spherical symmetry, now complicates the solution considerably because Riemann invariants do not exist. We can transform (230) only to the form (79) in terms of the characteristic independent variables.

Equations (206)–(208) still hold, since they involve only the coefficients of the left-hand sides of (230), which are the same as those in (204). In addition, the two components of **f** in (21c) are given by

$$f_1 = 0 \tag{231a}$$

$$f_2 = -(\gamma - 1)\frac{ua}{r} \tag{231b}$$

Therefore, the characteristic system (72), (79) becomes

$$r_\eta = (u + a)t_\eta \tag{232a}$$

$$r_\xi = (u - a)t_\xi \tag{232b}$$

$$\frac{\gamma - 1}{2}u_\eta + a_\eta = -(\gamma - 1)\frac{ua}{r}t_\eta \tag{232c}$$

$$\frac{\gamma - 1}{2}u_\xi - a_\xi = (\gamma - 1)\frac{ua}{r}t_\xi \tag{232d}$$

These equations are in a form convenient for solution by the method of characteristics.

7.5.2 Steady Irrotational Two-Dimensional Flow

The formulation of this problem was outlined in Problem 6 of Chapter 4, and the reader is again referred to Section 2.4 of [1] for more details. Here we shall restrict our discussion to shock-free solutions for brevity. The treatment of flows with shocks is entirely analogous to the cases discussed in Section 7.4 and 7.5.1. In fact, the shock conditions follow directly from the Rankine-Hugoniot relations [(75) of Chapter 5] in an appropriate frame (see Section 6.17 of [2]).

Characteristics

The flow is governed by the equation for mass conservation

$$u_x + 2\frac{uv}{u^2 - a^2}u_y + \frac{v^2 - a^2}{u^2 - a^2}v_y = 0 \tag{233a}$$

and the irrotationality condition

$$v_x - u_y = 0 \tag{233b}$$

Here u and v are the Cartesian components of the velocity vector and a is the local speed of sound, governed by

$$a^2 \equiv 1 - \frac{\gamma - 1}{2}(q^2 - M^2) \tag{234}$$

We have used (233b) to simplify (233a), which we have divided by the coefficient of u_x. We are using dimensionless variables so u, v, and a are normalized by the speed of sound at some reference point, say upstream at $x = -\infty$, where the dimensionless flow velocity (Mach number) in the x direction is M. The magnitude of the dimensionless local velocity is $q \equiv (u^2 + v^2)^{1/2}$.

The system (233) is in the standard form (80) if we identify t, x in (80) with x, y respectively in (233). Therefore,

$$\{A_{ij}\} = \begin{pmatrix} \dfrac{2uv}{u^2 - a^2} & \dfrac{v^2 - a^2}{u^2 - a^2} \\ -1 & 0 \end{pmatrix} \tag{235}$$

The eigenvalues of (235) are the roots of

$$\lambda^2 - \frac{2uv}{u^2 - a^2}\lambda + \frac{v^2 - a^2}{u_2 - a^2} = 0$$

—that is,

$$\lambda_1 = \frac{uv + a(u^2 + v^2 - a^2)^{1/2}}{u^2 - a^2} \tag{236a}$$

$$\lambda_2 = \frac{uv - a(u^2 + v^2 - a^2)^{1/2}}{u^2 - a^2} \tag{236b}$$

These are real and distinct if the local flow is supersonic—that is, $u^2 + v^2 > a^2$—and we assume that this is the case for the remainder of this discussion.

To understand the geometrical meaning of the characteristic curves defined by the two slopes $(dy/dx) = \lambda_1$, $(dy/dx) = \lambda_2$, we examine the following simple example. Suppose that a two-dimensional point disturbance is moving with constant speed M in the negative x direction in a gas at rest. Thus, $y = 0$, $x = -Mt$ locates this point disturbance, which may be thought of as a distribution of mass sources of constant strength along the infinite straight line $x = -Mt$, $y = 0$ in xyz-space. The disturbance that was generated by this point source at time t_0 will be located on the circle of radius $(t - t_0)$ at the time t (recall that the ambient sound speed equals unity in our dimensionless variables). The envelope of disturbances then consists of the two straight lines at the angles $\pm\alpha$ relative to the x-axis, where

$$\alpha \equiv \sin^{-1}\frac{1}{M}; \qquad 0 < \alpha < \frac{\pi}{2} \tag{237}$$

as shown in Figure 7.15a. The angle α is called the Mach angle. This result was also derived in Chapter 6 using the eikonal equation [see (148) of Chapter 6].

An observer moving with the point source experiences a steady flow with speed M to the right and sees the *fixed* straight lines, called Mach lines, with the constant slopes $dy/dx = \pm\tan\alpha$. More generally, if the disturbance is moving along a straight line inclined at the angle θ relative to the x-axis, then the Mach lines have slopes $(dy/dx) = \tan(\theta \pm \alpha)$.

We shall now show that the characteristic curves have a local slope equal to the slope of the local Mach lines. The local velocity vector is tangent to the local streamline, as shown in Figure 7.15b, and we introduce the polar representation for the velocity components

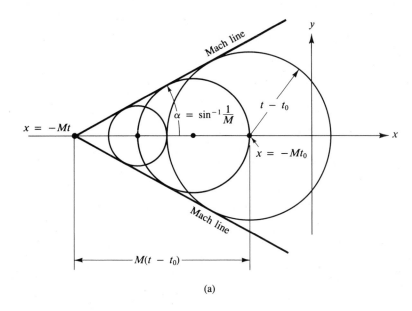

(a)

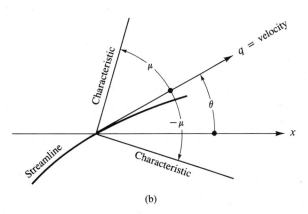

(b)

Figure 7.15

$$u = q \cos \theta \tag{238a}$$

$$v = q \sin \theta, \tag{238b}$$

where θ is the angle measured in the counterclockwise sense from the positive x direction to the velocity vector. Guided by the discussion for the case $q = M = $ constant, we denote

$$a \equiv q \sin \mu \tag{239}$$

where μ is the local Mach angle [see (237)]. If we now express u, v, and a in (236) in terms of q, θ, and μ we obtain

$$\lambda_{1,2} = \frac{q^2 \sin \theta \cos \theta \pm q \sin \mu (q^2 - q^2 \sin^2 \mu)^{1/2}}{q^2 \cos^2 \theta - q^2 \sin^2 \mu}$$

$$= \frac{\sin \theta \cos \theta \pm \sin \mu \cos \mu}{\cos^2 \theta - \sin^2 \mu} = \frac{\sin 2\theta \pm \sin 2\mu}{\cos 2\theta + \cos 2\mu} \qquad (240)$$

$$= \frac{\sin(\theta \pm \mu)\cos(\theta \mp \mu)}{\cos(\theta + \mu)\cos(\theta - \mu)} = \tan(\theta \pm \mu)$$

using trigonometric identities. Thus, the angle between the streamline and the characteristics is just the local Mach angle μ.

The Riemann Invariants

The hodograph form (89) for (233) is

$$-y_v + \frac{2uv}{u^2 - a^2} x_v - \frac{v^2 - a^2}{u^2 - a^2} x_u = 0 \qquad (241a)$$

$$y_u - x_v = 0 \qquad (241b)$$

Using (93) we compute the following characteristic slopes for (241):

$$\frac{dv}{du} = \frac{-uv \pm a(u^2 + v^2 - a^2)^{1/2}}{v^2 - a^2} \qquad (242)$$

and, as we have argued all along, the integration of (242) gives the two Riemann invariants. This integration is awkward in terms of the u, v variables, so in view of the simplification that was introduced in (240), when we used θ and μ, let us attempt to solve (242) in terms of these variables. Note that (234), (238), and (239) define a transformation of variables $(u, v) \leftrightarrow (\theta, \mu)$.

First, we compute (dv/du) using (238). This gives

$$\frac{dv}{du} = \frac{\frac{dq}{d\theta} \sin \theta + q \cos \theta}{\frac{dq}{d\theta} \cos \theta - q \sin \theta} \qquad (243a)$$

But, according to (238), (239), the right-hand side of (242) is

$$\frac{-uv \pm a(u^2 + v^2 - a^2)^{1/2}}{v^2 - a^2} = \frac{-\sin \theta \cos \theta \pm \sin \mu \cos \mu}{\sin^2 \theta - \sin^2 \mu} = -\cot(\theta \pm \mu) \qquad (243b)$$

Therefore, equating the right-hand sides of (243a)–(243b) and solving for $dq/d\theta$, we find

$$\frac{dq}{d\theta} = \mp \tan \mu \qquad (244a)$$

To express $dq/d\theta$ in terms of $(d\mu/d\theta)$ we use (234) and (239) to obtain

$$\left(\frac{dq}{d\theta}\right) = -q \frac{\sin \mu \cos \mu}{\sin^2 \mu + (\gamma - 1)/2} \frac{d\mu}{d\theta} \qquad (244b)$$

We now equate the right-hand sides of (244a)–(244b) to obtain

$$\frac{d\mu}{d\theta} = \pm\frac{\sin^2\mu + (\gamma - 1)/2}{\cos^2\mu} \tag{245}$$

This expression can be integrated explicitly to define the Riemann invariants in the form:

$$\theta + v(\mu) = \text{constant} \quad \text{on} \quad \frac{dy}{dx} = \tan(\theta + \mu) \tag{246a}$$

$$\theta - v(\mu) = \text{constant} \quad \text{on} \quad \frac{dy}{dx} = \tan(\theta - \mu) \tag{246b}$$

where v is the Prandtl-Meyer function:

$$v(\mu) \equiv \int_0^\mu \frac{\cos^2\sigma}{\sin^2\sigma + (\gamma - 1)/2} d\sigma = -\mu + \frac{\gamma + 1}{2}\int_0^{2\mu} \frac{d\sigma}{\gamma - \cos\sigma}$$

$$= \left(\frac{\gamma + 1}{\gamma - 1}\right)^{1/2} \tan^{-1}\left[\left(\frac{\gamma + 1}{\gamma - 1}\right)^{1/2}\tan\mu\right] - \mu \tag{247}$$

A solution by the method of characteristic using (246) is now easy to implement and defines θ, μ at each gridpoint. To obtain u, v there, we use (234), (238), and (239). Examples can be found in standard texts in gas dynamics, such as Chapter 1 of [3]. See also [4] for a comprehensive mathematical treatment.

Problems

1. Show that for $\Phi_x^2 + \Phi_y^2 > a^2$, the characteristics of the second-order equation (164a) of Chapter 4 are the same as those defined by (236).
2. Consider shallow-water flow over a variable bottom for two layers of fluid with constant densities ρ_1, ρ_2 for the stable problem where the lighter fluid is on top ($\rho_1 < \rho_2$). See Figure 7.16.

    We denote the average horizontal speed in each layer by $u_i(x, t)$ and the height of the layer by $h_i(x, t)$. Using appropriate dimensionless variables, the laws of mass and momentum conservation in each layer reduce to the following system of four equations:

$$\frac{\partial h_1}{\partial t} + \frac{\partial}{\partial x}(u_1 h_1) = 0 \tag{248a}$$

$$\frac{\partial u_1}{\partial t} + u_1\frac{\partial u_1}{\partial x} + \frac{\partial h_1}{\partial x} + \frac{\partial h_2}{\partial x} = -\varepsilon\frac{\partial b}{\partial x} \tag{248b}$$

$$\frac{\partial h_2}{\partial t} + \frac{\partial}{\partial x}(u_2 h_2) = 0 \tag{248c}$$

$$\frac{\partial u_2}{\partial t} + u_2\frac{\partial u_2}{\partial x} + (1 - \beta)\frac{\partial h_1}{\partial x} + \frac{\partial h_2}{\partial x} = -\varepsilon\frac{\partial b}{\partial x} \tag{248d}$$

where $\beta \equiv (\rho_2 - \rho_1)/\rho_2 > 0$, and $\varepsilon b(x, t)$ is the height of the bottom measured from some reference level. (See Problem 9 of Chapter 4.)

    Derive the normal form (32) for this problem.

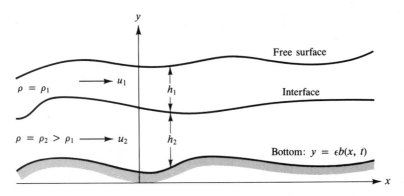

Figure 7.16

3. In this problem we wish to derive the finite difference solution along the local characteristic directions for the system (79) of two first-order equations.

To reserve subscripts for locating gridpoints, let us write equations (72) and (79) in the following forms, respectively:

$$X_\eta = \lambda^+(X, T, U, V)T_\eta \qquad (249a)$$

$$X_\xi = \lambda^-(X, T, U, V)T_\xi \qquad (249b)$$

$$AU_\eta + BV_\eta = ET_\eta \qquad (250a)$$

$$CU_\xi + DV_\xi = FT_\xi, \qquad (250b)$$

where λ^+ refers to λ_1 as given in (71a) and λ^- refers to λ_2 as given in (71b). We have also set $A = V_{11}$, $B = V_{12}$, $C = V_{21}$, $D = V_{22}$, $E = f_1 V_{11} + f_2 V_{12}$, and $F = f_1 V_{21} + f_2 V_{22}$. All eight coefficients λ^-, λ^+, A, B, C, D, E, and F are known functions of X, T, U, and V.

a. Consider the initial-value problem where Cauchy data are prescribed on the x-axis and use the subscript notation of Section 7.1.2 to locate gridpoints. Thus, the $X_{i,-i}$, $U_{i,-i}$, $V_{i,-i}$ are all given and the $T_{i,-i}$ are all equal to zero. Assume also that $\lambda_{i,-i}^- < 0$ and $\lambda_{i,-i}^+ > 0$ for all the given initial gridpoints. Derive the values of $X_{m,n}$, $T_{m,n}$, $U_{m,n}$, $V_{m,n}$ given the values of these four quantities at the two preceding gridpoints: $(m - 1, n)$ and $(m, n - 1)$.

b. Now assume that U is specified along the t-axis in addition to the Cauchy data along the x-axis. Develop the characteristic solutions for $x \geq 0$, $t \geq 0$ and indicate under what conditions you can calculate this solution.

4. Calculate u and h as functions of x and t for the piston problem discussed in Section 7.4.2 for the special case where $\varepsilon = \frac{1}{2}$ and

$$g(t) = \begin{cases} \dfrac{t}{2} + \dfrac{t^2}{4}; & 0 < t \leq 1 \\[2mm] t - \dfrac{1}{4}, & 1 \leq t \end{cases} \qquad (251)$$

5. Assume that the solution in region ③ of Figure 7.9 has been worked out and, in particular, that we know the characteristic arc BD in parameteric form

$$x = x_0(\sigma) \qquad\qquad (252a)$$

$$t = t_0(\sigma) \qquad\qquad (252b)$$

and that we also have u and h on BD in the form

$$u = u_0(\sigma) \qquad\qquad (253a)$$

$$h = h_0(\sigma) \qquad\qquad (253b)$$

a. Derive the solution in ⑥ of Figure 7.9 in parametric form.
b. Use the results in part (a) to compute the solution in ⑨.

6. At time $t = 0$ a bore traveling with speed $V_1 > 0$ is at $x = 0$, and a second bore traveling with speed $V_2 < 0$ is at $x = 1$. The water in the interval $0 < x < 1$ is at rest and has unit height. The water to the left of the bore V_1 has a constant height $h_1 > 1$ and constant speed $u_1 > 0$, whereas the water to the right of the bore V_2 has constant height $h_2 > 1$ and constant speed $u_2 < 0$. Let us specify these bores by fixing h_1 and h_2 and let us assume that $h_1 > h_2$. Thus, according to (90) of Chapter 5, we have

$$u_1 = (h_1 - 1)\left(\frac{h_1 + 1}{2h_1}\right)^{1/2} \qquad\qquad (254a)$$

$$V_1 = \left[\frac{h_1(h_1 + 1)}{2}\right]^{1/2} \qquad\qquad (254b)$$

and

$$u_2 = -(h_2 - 1)\left(\frac{h_2 + 1}{2h_2}\right)^{1/2} \qquad\qquad (255a)$$

$$V_2 = -\left[\frac{h_2(h_2 + 1)}{2}\right]^{1/2} \qquad\qquad (255b)$$

Show that for $t > 1/(V_1 - V_2) \equiv t_c$, when the two bores have interacted, the solution still consists of two bores having speeds $\bar{V}_1 > 0$, $\bar{V}_2 < 0$ bounding the interval $V_1 t_c + \bar{V}_2(t - t_c) < x < V_1 t_c + \bar{V}_1(t - t_c)$, in which the speed is a constant u_3 and the height is a constant h_3. The water to the right of the $\bar{V}_1$ bore has $u = u_2$, $h = h_2$, whereas to the left of the $\bar{V}_2$ bore, $u = u_1$, $h = h_1$. See Figure 7.17.

a. Verify that the four bore conditions governing u_3, h_3, $\bar{V}_1$, $\bar{V}_2$ in terms of the known quantities u_1, h_1, u_2, h_2 are

$$\bar{V}_1 = \frac{u_2 h_2 - u_3 h_3}{h_2 - h_3} \qquad\qquad (256a)$$

$$\bar{V}_1 = \frac{u_2^2 h_2 - u_3^2 h_3 + h_2^2/2 - h_3^2/2}{u_2 h_2 - u_3 h_3} \qquad\qquad (256b)$$

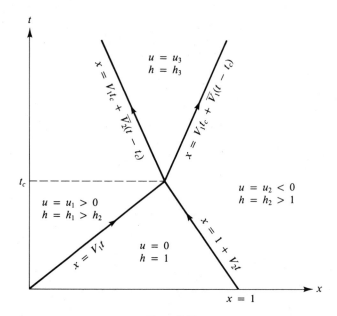

Figure 7.17

$$\bar{V}_2 = \frac{u_1 h_1 - u_3 h_3}{h_1 - h_3} \tag{257a}$$

$$\bar{V}_2 = \frac{u_1^2 h_1 - u_3^2 h_3 + h_1^2/2 - h_3^2/2}{u_1 h_1 - u_3 h_3} \tag{257b}$$

Equating the two expressions for $\bar{V}_1$ and $\bar{V}_2$, show that u_3 and h_3 are the solutions of the two equations

$$u_3 = u_2 + (h_3 - h_2)\left(\frac{h_3 + h_2}{2h_2 h_3}\right)^{1/2} \tag{258a}$$

$$u_3 = u_1 - (h_3 - h_1)\left(\frac{h_3 + h_1}{2h_1 h_3}\right)^{1/2} \tag{258b}$$

Once u_3 and h_3 have calculated from (258), we obtain $\bar{V}_1$ from either (256a) or (256b) and $\bar{V}_2$ from either (257a) or (257b).

b. Consider the special case $h_1 = 1.5, h_2 = 1.25$, for which (254) and (255) give $u_1 = 0.4564$, $V_1 = 1.369$, $u_2 = -0.2372$, and $V_2 = -1.186$. Show that (258a)–(258b) give $u_3 = 0.2186$, $h_3 = 1.804$, and (256), (257) give $\bar{V}_1 = 1.2471$, $\bar{V}_2 = -0.9546$. What numerical scheme would you use in general to calculate u_3, h_3 from (258)?

c. Compare your results with those predicted for the linear theory. See (137)–(138) of Chapter 3 and Figure 3.17.

7. Consider a uniform shock propagating to the right into a quiescent gas. Use the dimensionless variables in Section 7.5.1 so that $u = 0, \rho = 1, p = 1$ ahead of the shock and denote the pressure behind the shock by p_i. When this shock

reflects from a wall, it moves to the left into a gas with pressure p_i. Denote the pressure behind the reflected shock by p_r.

Show that

$$\frac{p_r - p_i}{p_i} = \frac{p_i - 1}{1 + (\gamma - 1)(p_i - 1)/2\gamma} \tag{259}$$

8. Consider steady supersonic flow in the corner domain *above* the boundary

$$y = \begin{cases} 0, & \text{if } x < 0 \\ x \tan \beta; \ \beta = \text{constant:} \ -\pi < \beta < 0, & \text{if } x > 0 \end{cases} \tag{260}$$

The boundary condition is that the velocity vector at the boundary be tangent to it. Given $u > 1$ and $v = 0$ at $x = -\infty$, compute the flow everywhere. Discuss what happens as β increases. In particular, for a given Mach number at $x = -\infty$, what is the angle required to have $M = \infty$ on $x > 0$?

9. For axially symmetric supersonic flow, the equations corresponding to (233) are

$$u_x + \frac{2uv}{u^2 - a^2} u_r + \frac{v^2 - a^2}{u^2 - a^2} v_r = \frac{a^2 v}{r(u^2 - a^2)} \tag{261a}$$

$$v_x - u_r = 0 \tag{261b}$$

where u and v are the axial (x) and radial (r) components of the velocity vector, respectively.

Show that the matrix $\{V_{ij}\}$ in (76c) for this case is

$$\{V_{ij}\} = \frac{1}{\lambda_2 - \lambda_1} \begin{pmatrix} \lambda_2^{-1} & 1 \\ -\lambda_1^{-1} & -1 \end{pmatrix} \tag{262}$$

where λ_1 and λ_2 are given by (236). Therefore, the system of characteristic equations (72), (79) is

$$r_\eta = \lambda_1 x_\eta \tag{263a}$$

$$r_\xi = \lambda_2 x_\xi \tag{263b}$$

$$u_\eta + \lambda_2 v_\eta = \frac{a^2 v}{r(a^2 - v^2)} x_\eta \tag{263c}$$

$$u_\xi + \lambda_1 v_\xi = \frac{a^2 v}{r(a^2 - v^2)} x_\xi \tag{263d}$$

References

1. J. D. Cole, and L. P. Cook, *Transonic Aerodynamics.* North-Holland, New York, 1986.
2. G. B. Whitham, *Linear and Nonlinear Waves.* Wiley-Interscience, New York, 1974.
3. H. W. Liepmann, and A. Roshko, *Elements of Gasdynamics.* Wiley, New York, 1957.
4. R. Courant, and K. O. Friedrichs, *Supersonic Flow and Shock Waves.* Springer, New York, 1976.

C H A P T E R 8

Perturbation Solutions

A problem that is governed by a differential eqnation (together with prescribed initial and boundary data, as appropriate) is usually amenable to solution by a perturbation procedure if one knows the solution in the limiting case: $\varepsilon \to 0$, where ε is a parameter that occurs in the dimensionless formulation of the equation or initial or boundary data. By a suitable definition of ε, we can also perturb about any given value of a dimensionless parameter.

The basic concepts that underlie perturbation methods are discussed in detail in texts devoted entirely to the subject, such as [1] and [2]. As it is not possible to cover this vast material in one chapter, we shall give only a sampling of examples in ordinary and partial differential equations to illustrate the main ideas.

8.1 Asymptotic Expansions

In our discussion so far, we have had occasion to derive certain "perturbation series" formally in powers of a small parameter ε:

$$u(x_i; \varepsilon) = u_0(x_i) + \varepsilon u_1(x_i) + \varepsilon^2 u_2(x_i) + \cdots$$

where the scalar function $u(x_i; \varepsilon)$ is the solution of a partial differential equation involving the small parameter ε and the n independent variables $x_i \equiv x_1, \ldots, x_n$. A more precise characterization of the preceding series is that it is the *asymptotic expansion* of $u(x_i; \varepsilon)$ with respect to the sequence $1, \varepsilon, \varepsilon^2, \ldots$ in the limit as $\varepsilon \to 0$.

For a comprehensive discussion of asymptotic expansions, the reader is referred to standard texts, such as [3] and [4]. In the remainder of this section, we shall review some of the basic ideas of asymptotic expansions. We shall consider functions such as $u(x_i; \varepsilon)$ and $v(x_i; \varepsilon)$ defined over some domain $\mathscr{D}$ of the independent variables x_i. For the time being, the boundary of $\mathscr{D}$ will be assumed to be independent of ε. The small parameter ε will *always* be taken nonnegative, typically in some neighborhood $I: 0 \le \varepsilon \le \varepsilon_0$ of the origin.

8.1.1 Order symbols

The statement

$$u(x_i; \varepsilon) = O(v(x_i; \varepsilon)) \qquad \text{in } \mathscr{D} \text{ as } \varepsilon \to 0 \tag{1}$$

means that for each point x_i in $\mathscr{D}$, there exists a positive $k(x_i)$ and an interval $I: 0 \le \varepsilon \le \varepsilon_0(x_i)$, where ε_0 depends in general on the choice of x_i, such that

$$|u| \le k|v| \tag{2}$$

for every ε in I. If (u/v) is defined in $\mathscr{D}$, (2) implies that $|u/v|$ is bounded above by k.

The statement (1) is said to be *uniformly valid* in $\mathscr{D}$ if k is a constant and ε_0 does not depend on the x_i. We now illustrate some of the features of this definition with the following examples.

1. Let $n = 2$; $\mathscr{D}: x_1^2 + x_2^2 < 1$; $u = (1 - x_1^2 - x_2^2 + \varepsilon)^{-1}$; $v = 1$. The statement $u = O(1)$ is correct with $k = (1 - x_1^2 - x_2^2)^{-1}$ and for any $\varepsilon_0 < 1$, but this result is *not* uniformly valid because k becomes unbounded as $x_1^2 + x_2^2 \to 1$. However, if we restrict our domain to the interior of the disc $\tilde{\mathscr{D}}: x^2 + y^2 \le a < 1$, then $u = O(1)$ uniformly in $\tilde{\mathscr{D}}$ with k equal to the constant $(1 - a)^{-1}$.

2. Let $n = 2$; $\mathscr{D}: 0 < x_1 < \infty$; $0 < x_2 < \infty$; $u = x_1 \varepsilon^\alpha$; $v = x_2 \varepsilon^\beta$, where $\alpha = $ constant, $\beta = $ constant, with $\alpha \ge \beta$. We have $u = O(v)$ because we can choose $k(x_1, x_2) = x_1/x_2$, $\varepsilon_0 < 1$ to verify that $x_1 \varepsilon^\alpha \le (x_1/x_2)x_2 \varepsilon^\beta$ holds as long as x_1 is finite and $x_2 \ne 0$. Thus, the statement that $u = O(v)$ is not uniformly valid in $\mathscr{D}$, but it is uniformly valid in any subdomain $\tilde{\mathscr{D}}: 0 < x_1 \le X_1 < \infty$; $0 < X_2 \le x_2 < \infty$, where X_1 and X_2 are positive constants, because if x_1 and x_2 are restricted to $\tilde{\mathscr{D}}$, we can pick $k = X_1/X_2 = $ constant.

This example also shows that the O symbol does not necessarily imply that u and v are of the "same order of magnitude"; for example, if $\alpha > \beta$, we see that $u < v$ for any fixed x_1, x_2 in $\mathscr{D}$ if ε is sufficiently small. (In fact, $u/v \to 0$ as $\varepsilon \to 0$.) In order to describe two functions u and v that are of the "same order of magnitude," we must say $u = O(v)$ *and* $v = O(u)$. In this case the limit of $|u/v|$, if it exists, is neither zero nor infinity. The two functions in the first example satisfy this dual requirement, but those in the second do not. In [2] the notation $u = O_s(v)$ for "strictly of the order of" is used to indicate $u = O(v)$ and $v = O(u)$; we shall adopt this notation when needed.

We now define the lower case o symbol. The statement

$$u(x_i; \varepsilon) = o(v(x_i; \varepsilon)) \qquad \text{in } \mathscr{D} \text{ as } \varepsilon \to 0 \tag{3}$$

for two functions u, v defined in $\mathscr{D}$ and $\varepsilon > 0$ means that for each point x_i in $\mathscr{D}$ and a given $\delta > 0$, there exists an interval $I: 0 \le \varepsilon \le \varepsilon_0(x_i, \delta)$, which depends in general on the choice of x_i and δ_i, such that

$$|u| \le \delta|v| \tag{4}$$

for all ε in I. Sometimes the notation $u \ll v$ is used to indicate (3). Inequality (4) states that $|u|$ becomes arbitrarily small compared to $|v|$, and if (u/v) is defined, (4) implies that $(u/v) \to 0$ as $\varepsilon \to 0$. Note that $u = o(v)$ always implies $u = O(v)$.

Again, we say that (3) holds uniformly in $\mathscr{D}$ if the right endpoint of the interval I depends only on δ and not on the choice of the point x_i in $\mathscr{D}$. The following examples illustrate these ideas.

1. $x_1 \varepsilon^\alpha = o(x_2 \varepsilon^\beta)$ in $\mathscr{D}: 0 < x_1 < \infty; 0 < x_2 < \infty$ as $\varepsilon \to 0$ for constant α, β as long as $\alpha > \beta$. To prove this we need to show that for any given δ, we can find a neighborhood of $\varepsilon = 0$ such that $(x_1/x_2)\varepsilon^{\alpha-\beta} \leq \delta$. The neighborhood of $\varepsilon = 0$ that meets our need is the interval $0 \leq \varepsilon \leq (\delta x_2/x_1)^{1/(\alpha-\beta)}$. This neighborhood shrinks to zero if either $x_2 \to 0$ or $x_1 \to \infty$. Therefore, the statement $x_1 \varepsilon^\alpha = o(x_2 \varepsilon^\beta)$ is not uniformly valid in $\mathscr{D}$. However, if we restrict attention to $\tilde{\mathscr{D}}: 0 < x_1 \leq X_1 < \infty; 0 < X_2 \leq x_2 < \infty$ for constants X_1, X_2, then the statement is uniformly valid in $\tilde{\mathscr{D}}$, and the neighborhood of $\varepsilon = 0$ that suffices is now $0 \leq \varepsilon \leq (\delta X_2/X_1)^{1/(\alpha-\beta)}$, which depends only on δ.

2. Let $\mathscr{D}$ be the triangular domain $0 < x_1 < \infty; 0 < x_2 < x_1$. For any arbitrarily large positive constant β, we have

$$e^{(x_2-x_1)/\varepsilon} = o(\varepsilon^\beta) \qquad \text{in } \mathscr{D} \text{ as } \varepsilon \to 0 \tag{5}$$

A function such as $e^{(x_2-x_1)/\varepsilon}$ satisfying (5) is said to be *transcendentally small* as $\varepsilon \to 0$. Note that $(x_2 - x_1)/\varepsilon < 0$. To prove (5), it suffices to note that $\lim_{\varepsilon \to 0}(\varepsilon^{-\beta}e^{-\alpha/\varepsilon}) = 0$ for any $\alpha > 0, \beta > 0$. Again, (5) is not uniformly valid in the triangular domain, but it is uniformly valid in the subdomain $\tilde{\mathscr{D}}: 0 < X_1 \leq x_1 < \infty; 0 < x_2 \leq x_1 - X_1$.

8.1.2 Definition of an Asymptotic Expansion

Let $\{\phi_n(\varepsilon)\}$ with $n = 1, 2, \ldots$ be a sequence of functions of ε such that

$$\phi_{n+1}(\varepsilon) = o(\phi_n(\varepsilon)) \qquad \text{as } \varepsilon \to 0 \tag{6}$$

for each $n = 1, 2, \ldots$. Such a sequence is called an *asymptotic sequence*. Thus $\{\varepsilon^{n-1}\}$ with $n = 1, 2, \ldots$ is an asymptotic sequence; so is the sequence $\log \varepsilon, 1, \varepsilon \log \varepsilon, \varepsilon, (\varepsilon \log \varepsilon)^2, \varepsilon^2 \log \varepsilon, \varepsilon^2, \ldots$.

Let $u(x_i; \varepsilon)$ be defined for all x_i in some domain $\mathscr{D}$ and all ε in some neighborhood of $\varepsilon = 0$. Let $\{\phi_n(\varepsilon)\}$ be a given asymptotic sequence. The series $\sum_{n=1}^{N} \phi_n(\varepsilon)u_n(x_i)$, where the integer N may be finite or infinite, is said to be the *asymptotic expansion* of u with respect to $\{\phi_n\}$ as $\varepsilon \to 0$, if for every $M = 1, 2, \ldots, N$

$$u(x_i; \varepsilon) - \sum_{n=1}^{M} \phi_n(\varepsilon)u_n(x_i) = o(\phi_M) \qquad \text{as } \varepsilon \to 0 \tag{7a}$$

A stronger definition, which follows from (7a), is that

$$u(x_i; \varepsilon) - \sum_{n=1}^{M} \phi_n(\varepsilon)u_n(x_i) = O(\phi_{M+1}) \qquad \text{as } \varepsilon \to 0 \tag{7b}$$

for each $M = 1, 2, \ldots, N - 1$. The asymptotic expansion is said to hold *uniformly in $\mathscr{D}$* if the order relations in (7) hold uniformly there.

We see that once u and the sequence $\{\phi_n\}$ are given, it is a straightforward matter to construct a unique asymptotic expansion by the repeated application of (7a). Thus,

$$u_1(x_i) = \lim_{\varepsilon \to 0} \frac{u(x_i; \varepsilon)}{\phi_1(\varepsilon)} \tag{8a}$$

$$u_2(x_i) = \lim_{\varepsilon \to 0} \frac{u(x_i; \varepsilon) - \phi_1(\varepsilon)u_1(x_i)}{\phi_2(\varepsilon)} \tag{8b}$$

$$u_m(x_i) = \lim_{\varepsilon \to 0} \frac{u(x_i; \varepsilon) - \sum_{n=1}^{m-1} \phi_n(\varepsilon)u_n(x_i)}{\phi_m(\varepsilon)} \tag{8c}$$

8.1.3 Asymptotic Expansion of a Given Function

To illustrate the preceding idea in its most elementary form, we assume that $u(x_i; \varepsilon)$ is given explicitly, and we wish to derive its asymptotic expansion with respect to a given sequence $\{\phi_n(\varepsilon)\}$. Let $\mathscr{D}$ be the domain $0 \le x < \infty; 0 < y < \infty$ and consider the asymptotic expansion of

$$u = \log\left(1 + \frac{\varepsilon x}{y} + \varepsilon^2 xy\right) \qquad \text{in } \mathscr{D} \text{ as } \varepsilon \to 0 \tag{9}$$

with respect to the sequence $1, \varepsilon, \varepsilon^2, \ldots$.

Since $z \equiv \varepsilon x/y + \varepsilon^2 xy$ is small for $\varepsilon \ll 1$, as long as x and y are fixed, we calculate the Taylor series of (9) (with remainder) around $z = 0$:

$$u = \sum_{n=1}^{N} (-1)^{n-1} \frac{z^n}{n} + \frac{(-1)^N z^{N+1}}{(N+1)(1 + \alpha z)^{N+1}} \tag{10}$$

where α is a constant, $0 < \alpha < 1$. Now, we expand each of the terms z^n in powers of ε and retain only terms up to $O(\varepsilon^N)$. The result gives the asymptotic expansion of (9), and our construction guarantees that the condition (7b) is satisfied. For example, for $N = 4$, we have

$$u = z - \frac{z^2}{2} + \frac{z^3}{3} - \frac{z^4}{4} + R_5(z)$$

where $|R_5| \le C_5 z^5$ and C_5 is a constant. Thus,

$$
\begin{aligned}
u &= \left(\varepsilon\frac{x}{y} + \varepsilon^2 xy\right) - \frac{1}{2}\left(\varepsilon^2 \frac{x^2}{y^2} + 2\varepsilon^3 x^2 + \varepsilon^4 x^2 y^2\right) \\
&\quad + \frac{1}{3}\left(\frac{\varepsilon^3 x^3}{y^3} + 3\varepsilon^4\frac{x^3}{y} + O(\varepsilon^5)\right) - \frac{1}{4}\left(\varepsilon^4\frac{x^4}{y^4} + O(\varepsilon^5)\right) + O(\varepsilon^5) \\
&= \varepsilon\frac{x}{y} + \varepsilon^2\left(xy - \frac{x^2}{2y^2}\right) + \varepsilon^3\left(-x^2 + \frac{x^3}{3y^3}\right) \\
&\quad + \varepsilon^4\left(-\frac{x^2 y^2}{2} + \frac{x^3}{y} - \frac{x^4}{4y^4}\right) + O(\varepsilon^5)
\end{aligned}
\tag{11}
$$

It is clear from the definition of z that (11) is not uniformly valid in $\mathscr{D}$. However, if we restrict attention to $\tilde{\mathscr{D}}: 0 \le x \le X < \infty; 0 < Y_1 \le y \le Y_2 < \infty$, where X, Y_1, and Y_2 are arbitrary positive numbers with $Y_1 < Y_2$, then the expansion is uniformly valid. The structure of z also indicates that the sequence $1, \varepsilon, \varepsilon^2$ is the "natural" one to use for this function.

8.1.4 Asymptotic Expansion of the Root of an Algebraic Equation

A less direct way to define a function $u = U(x_i; \varepsilon)$ is to have it equal to one of the roots of the algebraic equation

$$R(x_i, u; \varepsilon) = 0 \tag{12}$$

Thus, $R(x_i, U(x_i; \varepsilon); \varepsilon) \equiv 0$ for all $\varepsilon > 0$ and all x in some domain $\mathcal{D}$. We assume that the limiting value $u_1(x_i) \equiv U(x_i; 0)$ is known, and we are interested in the asymptotic expansion of the root $U(x_i; \varepsilon)$ for $\varepsilon \to 0$.

The following example illustrates the ideas. Let

$$R(x, y, u; \varepsilon) \equiv \varepsilon u^2 + 2f(x, y)u - g(x, y) = 0 \tag{13}$$

where f and g are given functions in $\mathcal{D}$: $-\infty < x < \infty$; $-\infty < y < \infty$ and $0 < \varepsilon \ll 1$. Since (13) is quadratic, the solution for the roots is readily obtained in the form

$$U^+(x, y; \varepsilon) = \frac{-f(x, y) + [f^2(x, y) + \varepsilon g(x, y)]^{1/2}}{\varepsilon} \tag{14a}$$

$$U^-(x, y; z) = \frac{-f(x, y) - [f^2(x, y) + \varepsilon g(x, y)]^{1/2}}{\varepsilon} \tag{14b}$$

and for ε sufficiently small, the two roots are real.

For a general expression R, the exact roots will not be explicitly available, as in (14). Therefore, we shall ignore these expressions for the time being and illustrate how to go about calculating their asymptotic expansions based only on (13).

Setting $\varepsilon = 0$ in (13) shows that the limiting value of one root is $u_1^+(x, y) = g(x, y)/2f(x, y)$ whenever $f \neq 0$. Let us restrict attention to the case of $f \neq 0$ and assume an asymptotic expansion for U^+ in the form

$$U^+(x, y; \varepsilon) = \frac{1}{2}\frac{g(x, y)}{f(x, y)} + \phi_2(\varepsilon)u_2^+(x, y) + \phi_3(\varepsilon)u_3^+(x, y) + o(\phi_3) \tag{15}$$

where $\phi_1 = 1, \phi_2, \phi_3, \ldots$ is an asymptotic sequence yet to be defined. If we substitute (15) into (13) and cancel out the terms of $O(1)$, we are left with

$$\varepsilon \frac{g^2}{4f^2} + \phi_2(\varepsilon)(2fu_2^+) + \varepsilon\phi_2\frac{g}{f}u_2^+ + \phi_3(2fu_3) = O(\varepsilon\phi_3) + O(\varepsilon\phi_2^2) + o(\phi_3). \tag{16}$$

The choice of ϕ_2 and ϕ_3 affects the relative importance of the various terms in the left-hand side of (16). At any rate, $\varepsilon\phi_2 \ll \phi_2$ and $\phi_3 \ll \phi_2$, so the third and fourth terms in the left-hand side of (16) are negligible in comparison with the second term proportional to ϕ_2. If we assume $\phi_2 \ll \varepsilon$, we must set $g = 0$ in order to satisfy (16) to $O(\varepsilon)$, and this is inconsistent in general. On the other hand, if we set $\varepsilon \ll \phi_2$, we must either have $f = 0$ (which we have excluded) or $u_2^+ = 0$. Having $u_2^+ = 0$ in (15) does not provide any new information about the expansion because it is always trivially possible to insert a zero term of some order ϕ_2: $\varepsilon \ll \phi_2$. The only nontrivial possibility left is that ϕ_2 be of the same order of magnitude as ε—that is,

$$\phi_2 = O_s(\varepsilon) \tag{17}$$

For simplicity, we take $\phi_2 = \varepsilon$ and conclude that

$$u_2^+(x, y) = -\frac{g^2(x, y)}{8f^3(x, y)} \tag{18}$$

in order to satisfy (16). This equation now reduces to

$$-\varepsilon^2 \frac{g^3}{8f^4} + \phi_3(2fu_3^+) = O(\varepsilon\phi_3) + O(\varepsilon^3) + o(\phi_3) \tag{19}$$

The same arguments that we used to determine ϕ_2 now imply that $\phi_3 = O_s(\varepsilon^2)$, and we choose $\phi_3 = \varepsilon^2$. In this case, we must set

$$u_3^+(x, y) = \frac{g^3(x, y)}{16f^5(x, y)} \tag{20}$$

and all the terms that we have neglected in the right-hand side of (19) are $O(\varepsilon^3)$.

Thus, we have determined the asymptotic expansion of U^+ in the form

$$U^+(x, y; \varepsilon) = \frac{1}{2}\frac{g}{f} - \varepsilon \frac{g^2}{8f^3} + \varepsilon^2 \frac{g^3}{16f^5} + O(\varepsilon^3) \tag{21}$$

It is reassuring to note that if we expand the radical in (14a) for $|\varepsilon g| < f^2$ and collect terms, we obtain (21).

The procedure that we have outlined extends to all orders, and we conclude that the sequence $\{\varepsilon^{n-1}\}$, $n = 1, 2, \ldots$ is "appropriate" for U^+. Strictly speaking, this sequence is not unique; any sequence satisfying

$$\phi_n(\varepsilon) = O_s(\varepsilon^{n-1})$$

for each $n = 1, 2, \ldots$ can be used. For example, we may choose the sequence $\{\psi_n(\varepsilon)\} = C_{n-1}\varepsilon^{n-1}$, $n = 1, 2, \ldots$, for arbitrary nonzero constants C_{n-1}. In this case, the new $u_n^+(x, y)$ that we calculate will differ from those given by (18), (20), and the like by the factors $1/C_{n-1}$, but the net result (21) will be the same. A more elaborate choice has $\psi_2(\varepsilon) = \varepsilon/(1 + \varepsilon)$, $\psi_3(\varepsilon) = \varepsilon^2(1 + \varepsilon)$, ... In this case $\varepsilon = \psi_2 + \varepsilon\psi_2 = \psi_2 + \psi_3$ and $R = \psi_2 u^2 + \psi_3 u^2 + 2fu - g = 0$. We then compute

$$U^+(x, y; \varepsilon) = \frac{1}{2}\frac{g}{f} - \frac{\varepsilon}{1 + \varepsilon}\frac{g^2}{8f^3} + \frac{\varepsilon^2}{1 + \varepsilon}\left[\frac{g^3}{16f^5} - \frac{g^2}{8f^3}\right] + o(\psi_2) \tag{22}$$

Note that if the expressions for ψ_2 and ψ_3 in (22) are expanded in powers of ε and only terms proportional to 1, ε, and ε^2 are retained, the expressions in (21) and (22) are *asymptotically equivalent* to $O(\varepsilon^2)$.

There is no advantage, a priori, in using (22) as opposed to (21). For particular choices of f, g, and ε, the three-term expansion (22) may be numerically more accurate than the three-term expansion (21). For example, if $f = g = 1$,

$\varepsilon = 0.1$, we have the exact root $U^+ = 0.4880884817$, whereas (21) gives $U^+ = 0.488125$ and (22) gives $U^+ = 0.488068$. In this case (22) is more accurate than (21), but in both cases the error is still $O(\varepsilon^3)$.

Setting $\varepsilon = 0$ in (13) shows that there is only one root U^+ that is $O(1)$. This root corresponds to the balance to $O(1)$ between the second and third terms in the expression (13) for R. The balance between the first and second terms to $O(1)$ is the only other possible one if $f \neq 0$. This means that the root U^- must be $O(\varepsilon^{-1})$, and we assume it has an asymptotic expansion in the form

$$U^-(x, y; \varepsilon) = \frac{1}{\varepsilon} u_1^-(x, y) + \lambda_2(\varepsilon) u_2^-(x, y) + \lambda_3(\varepsilon) u_3^-(x, y) + o(\lambda_3) \tag{23a}$$

Proceeding as before, we find that $1/\varepsilon$, 1, ε, ... is an appropriate sequence, and we obtain the expansion

$$U^-(x, y; \varepsilon) = \frac{1}{\varepsilon}(-2f) - \frac{1}{2}\frac{g}{f^2} + \varepsilon\left(\frac{1}{8}\frac{g^2}{f^3}\right) + O(\varepsilon^2) \tag{23b}$$

which also follows from (14b).

The two expansions (21) [or (22)] and (23b) are uniformly valid in $\mathscr{D}$: $-\infty < x < \infty$; $-\infty < y < \infty$, as long as $f \neq 0$ and g is bounded in $\mathscr{D}$. To illustrate the situation for $f = 0$, g bounded, let us assume that $f(x, y)$ vanishes along some curve $y = h(x)$ but that $f_y(x, h(x)) \neq 0$. Since our expansions fail near $y - h(x) = 0$, let us use a rescaled variable y^* defined by

$$y^* \equiv \frac{y - h(x)}{\alpha(\varepsilon)} \tag{24}$$

instead of y, and retain $x^* = x$ to see what happens to (13) for various choices of the scale function $\alpha(\varepsilon)$. We are interested in $\alpha(\varepsilon) \to 0$ as $\varepsilon \to 0$ so that holding y^* fixed in this limit implies that $y - h(x) = O(\alpha)$.

We now regard u as a function of x^* and y^* and write (13) in the form

$$\varepsilon u^2 + 2f(x^*, h(x^*) + \alpha y^*)u - g(x^*, h(x^*) + \alpha y^*) = 0$$

Assuming that f and g are sufficiently differentiable near $y = h(x)$, we can expand these functions in Taylor series. To leading order, we obtain

$$f(x^*, h(x^*) + \alpha y^*) = f_y(x^*, h(x^*))\alpha y^* + O(\alpha^2)$$

$$g(x^*, h(x^*) + \alpha y^*) = g(x^*, h(x^*)) + O(\alpha)$$

where in the expansion for f we have noted that $f(x^*, h(x^*)) \equiv 0$ for all x^*. The expression for R to leading order then becomes

$$\varepsilon u^2 + 2\alpha(\varepsilon)f^*(x^*)y^*u - g^*(x^*) = O(\alpha) \tag{25}$$

where we have denoted

$$f_y(x^*, h(x^*)) \equiv f^*(x^*); \qquad g(x^*, h(x^*)) \equiv g^*(x^*)$$

Unless $g^* \equiv 0$, the only way to achieve a dominant balance in (25) is to have u large. Suppose that we assume $u = O(\beta^{-1}(\varepsilon))$, where $\beta(\varepsilon) \ll 1$. We may implement this assumption explicitly by setting

$$u \equiv \frac{u^*}{\beta(\varepsilon)} \tag{26}$$

where $u^* = O(1)$. Equation (25) then becomes

$$\frac{\varepsilon}{\beta^2} u^{*2} + \frac{2\alpha}{\beta} f^* y^* u^* - g^* = O(\alpha) \tag{27}$$

The choice of α and β for which the most terms in (27) are in dominant balance corresponds to $\varepsilon/\beta^2 = O_s(1)$; $\alpha/\beta = O_s(1)$, or simply $\alpha = \beta = \varepsilon^{1/2}$. In this case all three terms in (27) are in dominant balance; any other choice would result in a limiting form with fewer terms. This principle for choosing the scale functions α, β is sometimes called the *principle of least degeneracy*. More often, we say that the choice $\alpha = \beta = \varepsilon^{1/2}$ leads to the "richest" limiting form of (27), and this is

$$u^{*2} + 2f^* y^* u^* - g^* = O(\varepsilon^{1/2}) \tag{28}$$

to leading order.

The solution of (28) for u^* gives the leading term of the asymptotic expansion of $U^\pm$ near $y = h(x)$ in the form:

$$\varepsilon^{1/2} U^\pm = \{-y^* f^*(x^*) \pm [y^{*2} f^{*2}(x^*) + g^*(x^*)]^{1/2}\} + O(\varepsilon^{1/2}) \tag{29}$$

Although (28) is somewhat simpler than (13), it is still a general quadratic and, strictly speaking, not any simpler to solve than (13). In a less trivial example we would, in general, not be able to solve analytically the limiting expression corresponding to (28). However, once this solution has been defined—for example, numerically—the calculation of the higher-order terms becomes straightforward. Notice again that when the u^*, x^*, y^* variables are used in (14), we obtain (29) in the limit as $\varepsilon \to 0$.

A final comment about the expansion that begins in the form (29) is that this is uniformly valid for *fixed* y^* but fails to be uniform as $|y^*| \to \infty$—that is, if $|y - h(x)| \to \infty$. But, if $|y - h(x)|$ is large, we are in the region where the expansions (21), (23b) are valid; these fail near $|y - h(x)|$ small. Therefore, we have been able to derive two expansions that are uniformly valid in "complementary" subdomains of $\mathcal{D}$ but fail to be uniform in each other's subdomain. This behavior is defined more precisely when we discuss matching of asymptotic expansions in Section 8.3.

Suppose now that $g(x, y)$ is not a bounded function as $|x|$ and/or $|y| \to \infty$. We see that (21) and (23b) fail to be uniform in this limit. The source of our difficulty is that we regarded $g(x, y) = O(1)$ in our derivations, and this statement is not uniformly valid in any domain where $|g| \to \infty$. To fix these ideas, let us choose $g(x, y) = x \sin y$. We see that $g = O(x)$ as $|x| \to \infty$. Now, if $|g|$ is large, (13) implies that both roots must be large also, and this suggests the following rescaling of variables

$$\tilde{x} \equiv \gamma(\varepsilon)x; \qquad \tilde{y} \equiv y; \qquad \tilde{u} \equiv \delta(\varepsilon)u$$

where $\gamma(\varepsilon) \ll 1$, $\delta(\varepsilon) \ll 1$ and $\tilde{x}, \tilde{y}, \tilde{u}$ are all $O(1)$. Equation (13) then becomes

$$\frac{\varepsilon}{\delta^2}\tilde{u}^2 + \frac{1}{\delta}2f\left(\frac{\tilde{x}}{\gamma},\tilde{y}\right)\tilde{u} - \frac{1}{\gamma}\tilde{x}\sin\tilde{y} = 0 \tag{30a}$$

Assuming that $f(\infty,\tilde{y})$ exists, we see that setting $\delta = \gamma = O_s(\varepsilon)$, or more simply $\delta = \gamma = \varepsilon$, results in the richest limiting expression for (30a) in which all three terms are in dominant balance—that is,

$$\tilde{u}^2 + 2f(\infty,\tilde{y})\tilde{u} - \tilde{x}\sin\tilde{y} = o(1) \tag{30b}$$

Here again, (30b) is not essentially simpler than (13) but must be solved in order to compute an expansion that remains valid for $|x|$ large.

8.1.5 Asymptotic Expansion of a Definite Integral

We have seen many applications in Chapters 1–3 in which a linear partial differential equation can be formally solved using an integral transform (for example, Laplace transform or Fourier transform). The solution is then defined by an inversion integral, which often is not possible to evaluate explicitly. A large body of work concerns the asymptotic expansion of such integrals using various methods such as *stationary phase* and *steepest descents*. We shall not discuss these methods here, and the reader is referred to standard texts (for example, [3]–[5]). To illustrate some of the ideas, we shall consider only a simple model problem based on an example first proposed by Laplace (see also the Introduction in [3]).

Consider the function $f(x,y;\varepsilon)$ defined by the real definite integral

$$f(x,y;\varepsilon) \equiv \int_0^\infty \frac{e^{-t}}{1+\varepsilon t g(x,y)}\,dt \tag{31}$$

where $0 < \varepsilon \ll 1$ and $0 < g(x,y)$.

This integral exists and defines $f(x,y;\varepsilon)$. Actually, f is a function of εg, and by changing variables of integration from t to $s = (1+\varepsilon g t)/\varepsilon g$, we can express (31) as

$$f(\varepsilon g) = \frac{e^{1/\varepsilon g}}{\varepsilon g}E_1\left(\frac{1}{\varepsilon g}\right) \tag{32a}$$

where E_1 is the exponential integral

$$E_1(z) \equiv \int_z^\infty \frac{e^{-s}}{s}\,ds; \qquad z > 0 \tag{32b}$$

The function E_1 cannot be expressed in closed form; numerical values for f are given in Table 5.1 of [6].

In order to approximate (31), suppose that we expand the integrand in series form for ε small and then integrate this series term by term. In what sense, if any, does the resulting series approximate f? We shall show that this procedure leads to a *divergent series for any fixed* ε. Nevertheless, this series is asymptotic as $\varepsilon \to 0$, and it provides a useful approximation for f if ε is a small number.

We have the following *exact* expansion for $1/(1+\varepsilon t g)$ to N terms:

$$\frac{1}{1+\varepsilon t g} = \sum_{n=0}^N (-1)^n(\varepsilon t g)^n + \frac{(-1)^{N+1}(\varepsilon t g)^{N+1}}{1+\varepsilon t g} \tag{33}$$

If we now use (33) in (31) and integrate the result, we obtain the following *exact* expression

$$f = \sum_{n=0}^{N} (-1)^n n! (\varepsilon g)^n + R_N(\varepsilon g) \tag{34a}$$

where R_N is the remainder

$$R_N(\varepsilon g) = (-1)^{N+1} (\varepsilon g)^{N+1} \int_0^\infty \frac{e^{-t} t^{N+1} \, dt}{1 + \varepsilon t g} \tag{34b}$$

and we have used the following identity for the Gamma function of $(n + 1)$, where n is a nonnegative integer

$$n! = \int_0^\infty e^{-t} t^n \, dt \equiv \Gamma(n + 1) \tag{34c}$$

We reiterate that for N finite, the expression in (34a) is exact. If, however, we ignore R_N and let $N \to \infty$, the series in (34a) diverges for any positive ε, as can be seen from the ratio test. We have

$$\left| \frac{(-1)^{n+1}(n + 1)! (\varepsilon g)^{n+1}}{(-1)^n n! (\varepsilon g)^n} \right| = n\varepsilon g$$

and $\lim_{n \to \infty} n\varepsilon g = \infty$ for any $\varepsilon g > 0$.

Nevertheless, it is easy to prove that the series

$$f_N(\varepsilon g) \equiv \sum_{n=0}^{N} (-1)^n n! (\varepsilon g)^n \tag{35}$$

is the asymptotic expansion of f to N terms (where $N = 0, 1, 2, \ldots$) with respect to the sequence $1, \varepsilon, \varepsilon^2, \ldots$. To do so, we use (34a)–(34b) to conclude that

$$|f - f_N| = (\varepsilon g)^{N+1} \int_0^\infty \frac{e^{-t} t^{N+1} dt}{1 + \varepsilon t g}$$

$$< (\varepsilon g)^{N+1} \int_0^\infty e^{-t} t^{N+1} \, dt = (N + 1)! (\varepsilon g)^{N+1}$$

Therefore,

$$|f - f_N| = O(\varepsilon^{N+1}) \qquad \text{as } \varepsilon \to 0$$

as required in order that f_N be the asymptotic expansion of f. Moreover, if g is bounded in some domain of the xy-plane, then (35) is uniformly valid there. An alternate derivation of (35) is outlined in Problem 4.

The sign of the error R_N is positive if N is odd and negative if N is even. If we denote $E_N \equiv |R_N|$, we see that for any fixed $g = g_0$ and $\varepsilon = \varepsilon_0$, the error $E_N(\varepsilon_0 g_0)$ decreases as N increases up to N equal to some integer $M(\varepsilon_0 g_0)$, which depends on the value of $\varepsilon_0 g_0$. For $N > M$, E_N increases with increasing N. Thus, for any given $\varepsilon_0 g_0$, there is a certain minimum numerical error $E_M(\varepsilon_0 g_0)$, which is achieved by retaining M terms in (35). We cannot improve the accuracy beyond this value, and, in fact, retaining more terms beyond the Mth only degrades the accuracy. We can also show that E_M decreases whereas M increases as $\varepsilon_0 g_0$ decreases.

As an illustration, we take $\varepsilon_0 = 0.1$, $g_0 = 1$ in (35) and compare our asymptotic results with the exact value $f(0.1) = 0.9156333394$ obtained from [6]. We find that $M = 9$ and $E_M = 1.7702 \times 10^{-4}$. If we take $\varepsilon_0 = 0.08$, $g_0 = 1$, $f(0.08) = 0.9304409399$. We then find $M = 11$ and $E_M = 1.544 \times 10^{-5}$.

In general, we do not need to have an exact result to determine where an asymptotic expansion begins to diverge. We need only monitor the absolute value of each successive term in the expansion; the optimal cutoff point M occurs when we reach the smallest term in absolute value.

8.2   Regular Perturbations

In this section we broaden our scope for defining a function $u(x_i; \varepsilon)$ to include the case where u is the solution of a differential equation. The small parameter ε may then occur either in the equation and/or boundary conditions. Further generalizations to integral equations, difference equations, differential-delay equations, and so on are possible but are not discussed here.

Regular perturbation problems (as opposed to singular perturbation problems discussed in Sections 8.3 and 8.4) are characterized by the property that their solution can be expressed in the entire domain of interest $\mathscr{D}$ by a single asymptotic expansion of the form (7). Of course, if the solution of the differential equation is known explicitly, the construction of its expansion is both trivial and unnecessary. We are mainly concerned with problems that have an easily calculated explicit solution if $\varepsilon = 0$ but that are either very difficult or impossible to solve explicitly if $\varepsilon \neq 0$. Often, the equation becomes linear if $\varepsilon = 0$; in other cases the boundaries or boundary conditions simplify and the problem is solvable if $\varepsilon = 0$. We shall illustrate various possibilities by means of examples as we go along.

The following fundamental assumption underlies the regular perturbation solution of a differential equation problem. We assume that the limit process (8), which defines the asymptotic expansion of the solution if it were known, produces a consistent sequence of perturbation differential equations governing the $u_i(x_i)$ when applied to the exact differential equation. Moreover, we assume that the successive solution of these perturbation equations gives the same $u_i(x_i)$ as we would obtain from the asymptotic expansion of the exact solution. We have already made a corresponding assumption implicitly in Section 8.1.4 when we expanded R as in (16). In Section 8.3, we shall extend this idea to include different expansions (associated with different limit processes) applied to the same differential equation.

8.2.1   Green's Function for an Ordinary Differential Equation

Consider the problem of a string on an elastic support under load. We discussed this problem in Section 3.1. If we restrict attention to the vertical displacement v and assume that the elastic support varies weakly in the x direction, we have the wave equation

$$v_{tt} - v_{xx} + [k_0^2 + \varepsilon k_1(x)]v = p(x, t)$$

Here k_0^2 is the predominantly constant part of the elastic support; $0 < \varepsilon \ll 1$ and

$k_1(x)$ determines the variation of the elastic support with x. The applied load is given by $p(x, t)$. For the static problem of deflection over the unit interval $0 \le x \le 1$, with $v = 0$ at the two ends, $v(x; \varepsilon)$ satisfies

$$-v'' + [k_0^2 + \varepsilon k_1(x)]v = p(x) \tag{36a}$$

$$v(0; \varepsilon) = 0 \tag{36b}$$

$$v(1; \varepsilon) = 0 \tag{36c}$$

Green's function $g(x, \xi; \varepsilon)$ for this problem satisfies

$$-g'' + [k_0^2 + \varepsilon k_1(x)]g = \delta(x - \xi) \tag{37a}$$

$$g(0, \xi; \varepsilon) = 0 \tag{37b}$$

$$g(1, \xi; \varepsilon) = 0 \tag{37c}$$

where ξ is a constant: $0 < \xi < 1$, primes denote derivatives with respect to x, and δ is the Dirac delta function.

In view of the variable coefficient $\varepsilon k_1(x)$ in (37a), an exact solution is out of reach, and we seek a perturbation expansion for g in the form

$$g(x, \xi; \varepsilon) = g_0(x, \xi) + \phi_1(\varepsilon)g_1(x; \varepsilon) + o(\phi_1) \tag{38}$$

where $\phi_1 \ll 1$ and remains to be specified. Substituting (38) into (37a) gives

$$-g_0'' + k_0^2 g_0 - \delta(x - \xi) - \phi_1 g_1'' + \varepsilon k_1 g_0 + \phi_1 k_0^2 g_1 = o(\phi_1) \tag{39a}$$

and the boundary conditions (37b)–(37c) give

$$g_0(0, \xi) + \phi_1 g_1(0, \xi) = o(\phi_1) \tag{39b}$$

$$g_0(1, \xi) + \phi_1 g_1(1, \xi) = o(\phi_1) \tag{39c}$$

Regardless of the choice of ϕ_1, as long a $\phi_1 \ll 1$, (39b)–(39c) imply the homogeneous boundary conditions

$$g_0(0, \xi) = g_0(1, \xi) = 0 \tag{40a}$$

$$g_1(0, \xi) = g_1(1, \xi) = 0 \tag{40b}$$

The leading term g_0 satisfies

$$-g_0'' + k_0^2 g_0 = \delta(x - \xi) \tag{41a}$$

whereas the perturbation term satisfies

$$-g_1'' + k_0^2 g_1 = \begin{cases} 0, & \text{if } \varepsilon \ll \phi_1 \\ -k_1(x)g_0, & \text{if } \phi_1 = O_s(\varepsilon) \end{cases} \tag{41b}$$

The choice $\varepsilon \ll \phi_1$ in (41b), combined with the boundary conditions in (40b), implies that $g_1 \equiv 0$. Therefore, we set $\phi_1 = \varepsilon$. Henceforth, we shall omit this detailed justification for the choice of gauge functions in our expansions and anticipate the result directly. The operating rule is to choose ϕ_1 so that we obtain the most general possible equation for g_1.

The solution of (41a) is calculated by noting that for $x \neq \xi$, we have a homogeneous equation that is satisfied by

$$g_0(x, \xi) = \begin{cases} A(\xi)e^{k_0 x} + B(\xi)e^{-k_0 x}, & \text{if } x < \xi \\ C(\xi)e^{k_0 x} + D(\xi)e^{-k_0 x}, & \text{if } x > \xi \end{cases}$$

where A, B, C, D are arbitrary functions of ξ. Imposing the boundary conditions at $x = 0$ and $x = 1$ determines B in terms of A and D in terms of C, and we find

$$g_0(x, \xi) = \begin{cases} A(\xi)[e^{k_0 x} - e^{-k_0 x}] \\ C(\xi)[e^{k_0 x} - e^{-k_0(2-x)}] \end{cases} \tag{42}$$

To determine A and C we impose the conditions

$$g_0(\xi^-, \xi) - g_0(\xi^+, \xi) = 0 \tag{43a}$$

$$\frac{\partial g_0}{\partial x}(\xi^-, \xi) - \frac{\partial g_0}{\partial x}(\xi^+, \xi) = 1 \tag{43b}$$

which follow from integrating (41a) with respect to x from $x = \xi^-$ to $x = \xi^+$. Using (42) into (43) gives

$$A(\xi)[e^{k_0 \xi} - e^{-k_0 \xi}] - C(\xi)[e^{k_0 \xi} - e^{k_0(2-\xi)}] = 0$$

$$A(\xi)k_0[e^{k_0 \xi} + e^{-k_0 \xi}] - C(\xi)k_0[e^{k_0 \xi} + e^{k_0(2-\xi)}] = 1$$

The solution of this linear system for A and C gives

$$A(\xi) = \frac{e^{k_0 \xi} - e^{k_0(2-\xi)}}{2k_0(1 - e^{2k_0})} \tag{44a}$$

$$C(\xi) = \frac{e^{k_0 \xi} - e^{-k_0 \xi}}{2k_0(1 - e^{2k_0})} \tag{44b}$$

and this defines $g_0(x, \xi)$. Note that for $k_0 \to 0$, this gives the piecewise linear solution

$$g_0(x, \xi) = \begin{cases} x(1 - \xi), & \text{for } x < \xi \\ \xi(1 - x), & \text{for } x > \xi \end{cases} \tag{45}$$

which is intuitively the obvious deflection profile for a unit concentrated load applied at $x = \xi$ to an unsupported string.

Now, the right-hand side of (41b) is known, and we can calculate g_1. In fact, since the homogeneous operator for g_1 is the same as that governing g_0, a convenient approach is to use the Green's function g_0, which we have just calculated. We find

$$g_1(x, \xi) = \int_0^1 g_0(x, \zeta)[-k_1(\zeta)g_0(\xi, \zeta)] \, d\zeta$$

and this defines $g_1(x, \xi)$ by quadrature once k_1 is specified. This procedure can be extended to higher orders to define subsequent terms in the perturbation expansion for g. Notice that if k_1 is a well-behaved function, our result (38) is uniformly valid in $0 \leq x \leq 1$.

Since Green's function to $O(\varepsilon)$ is given by

$$g(x, \xi; \varepsilon) = g_0(x, \varepsilon) + \varepsilon \int_0^1 g_0(x, \zeta)[-k_1(\zeta)g_0(\xi, \zeta)] \, d\zeta + O(\varepsilon^2)$$

the solution of (36) for v may be written in the form

$$v(x, \varepsilon) = \int_0^1 g(x, \xi; \varepsilon)p(\xi) \, d\xi$$

$$= \int_0^1 g_0(x, \xi)p(\xi) \, d\xi$$

$$+ \varepsilon \int_0^1 p(\xi) \left\{ \int_0^1 g_0(x, \zeta)[-k_1(\zeta)g_0(\xi, \zeta)] \, d\zeta \right\} d\xi + O(\varepsilon^2) \qquad (46)$$

This result also follows from a direct perturbation solution of (36). We assume that $v(x; \varepsilon)$ has the expansion

$$v(x; \varepsilon) = v_0(x) + \varepsilon v_1(x) + O(\varepsilon^2)$$

and substitute this into (36) to find

$$L(v_0) \equiv -v_0'' + k_0^2 v_0 = p(x) \qquad (47a)$$

$$v_0(0) = 0 \qquad (47b)$$

$$v_0(1) = 0 \qquad (47c)$$

$$L(v_1) = -k_1(x)v_0(x) \qquad (48a)$$

$$v_1(0) = 0 \qquad (48b)$$

$$v_1(1) = 0 \qquad (48c)$$

Once Green's function g_0 for the unperturbed problem is known, we can express the solutions of (47) and (48) by quadrature in terms of g_0 and the given right-hand sides. We find

$$v_0(x) = \int_0^1 g_0(x, \xi)p(\xi) \, d\xi \qquad (49a)$$

$$v_1(x) = \int_0^1 g_0(x, \xi)[-k_1(\xi)v_0(\xi)] \, d\xi$$

That is,

$$v_1(x) = \int_0^1 g_0(x, \xi) \left[-k_1(\xi) \int_0^1 g_0(x, \zeta)p(\zeta) \, d\zeta \right] d\xi \qquad (49b)$$

The result in (49a) agrees with the $O(1)$ term in (46), and the result (49b) also agrees with the $O(\varepsilon)$ term in (46) once we use the symmetry of g_0—that is,

$$g_0(\xi, \zeta) \equiv g_0(\zeta, \xi)$$

—and interchange ξ and ζ in (49b). Therefore, there is no particular advantage in using one approach in favor of the other.

8.2.2 Eigenvalues and Eigenfunctions of a Perturbed Self-Adjoint Operator

To motivate the discussion for the general case, we consider the following one-dimensional wave equation:

$$u_{tt} - u_{xx} + \varepsilon x u = 0; \qquad 0 \le x \le \pi, 0 \le t \tag{50a}$$

which may again be thought of as the equation governing the transverse displacement, now denoted by $u(x, t; \varepsilon)$, of a string on a weak ($\varepsilon \ll 1$), x-dependent elastic support ($\varepsilon x u$). We take the homogeneous boundary conditions

$$u(0, t; \varepsilon) = 0 \tag{50b}$$

$$u(\pi, t; \varepsilon) = 0 \tag{50c}$$

if $t > 0$, and general initial conditions

$$u(x, 0; \varepsilon) = f(x; \varepsilon) \tag{50d}$$

$$u_t(x, 0; \varepsilon) = g(x; \varepsilon) \tag{50e}$$

We studied the unperturbed problem ($\varepsilon = 0$) in Section 3.6 using Green's function. As indicated by the discussion in Section 3.6.2, this problem can also be solved using eigenfunction expansions. This is the approach that we shall follow here to see the effect of the perturbation term.

Let us assume that the solution has the separated form

$$u(x, t; \varepsilon) = X(x; \varepsilon) T(t; \varepsilon)$$

Substituting this into (50a) implies that X and T must satisfy

$$-\frac{\ddot{T}}{T} = -\frac{X''}{X} + \varepsilon x = \lambda > 0$$

Thus, the eigenvalue problem associated with (50a)–(50c) is

$$-\frac{d^2 X_n(x; \varepsilon)}{dx^2} + \varepsilon x X_n(x; \varepsilon) = \lambda_n X_n(x; \varepsilon) \tag{51a}$$

$$X_n(0; \varepsilon) = 0 \tag{51b}$$

$$X_n(\pi; \varepsilon) = 0 \tag{51c}$$

In this section we shall be concerned with the general eigenvalue problem for which (51) is a one-dimensional special case. Although (51) can be solved exactly, we shall concentrate on its asymptotic expansion for $\varepsilon \ll 1$ to illustrate ideas. First, we note that if $\varepsilon = 0$, the eigenvalues are $\lambda_n^{(0)} = n^2$, and the normalized eigenfunctions are

$$\xi_n^{(0)}(x) = \left(\frac{2}{\pi}\right)^{1/2} \sin nx; \qquad n = 1, 2, \dots \tag{52}$$

The eigenfunctions (52) are orthogonal in the sense that the inner product

$$\langle \xi_n^{(0)}, \xi_m^{(0)} \rangle \equiv \int_0^\pi \xi_n^{(0)}(x) \xi_m^{(0)}(x) \, dx = 0 \qquad \text{if } m \neq n \tag{53}$$

and we have normalized these eigenfunctions by the choice of multiplier $(2/\pi)^{1/2}$ in (52), so that

$$\langle \zeta_n^{(0)}, \zeta_n^{(0)} \rangle = 1 \tag{54}$$

We assume that the eigenfunctions $X_n(x; \varepsilon)$ for $\varepsilon \neq 0$ have the expansion

$$X_n(x; \varepsilon) = \left(\frac{2}{\pi}\right)^{1/2} \sin nx + \varepsilon \zeta_n^{(1)}(x) + O(\varepsilon^2) \tag{55}$$

and that the eigenvalues $\lambda_n(\varepsilon)$ in (51a) have the expansion

$$\lambda_n(\varepsilon) = n^2 + \varepsilon \lambda_n^{(1)} + O(\varepsilon^2) \tag{56}$$

Substituting (55) and (56) in (51a) gives

$$\frac{d^2 \zeta_n^{(1)}}{dx^2} + n^2 \zeta_n^{(1)} = -\lambda_n^{(1)} \left(\frac{2}{\pi}\right)^{1/2} \sin nx + \left(\frac{2}{\pi}\right)^{1/2} x \sin nx \tag{57}$$

Rather than solving (57) explicitly, which is certainly possible in this case, and then imposing the boundary conditions $\zeta_n^{(1)}(0) = \zeta_n^{(1)}(\pi) = 0$, let us express the solution $\zeta_n^{(1)}$ in the form of a series of eigenfunctions $\zeta_n^{(0)}$ (that is, a normalized Fourier sine series). Let

$$\zeta_n^{(1)}(x) = \left(\frac{2}{\pi}\right)^{1/2} \sum_{j=1}^{\infty} a_{nj}^{(1)} \sin jx \tag{58}$$

so that the boundary conditions at $x = 0$ and $x = \pi$ are automatically satisfied. Substituting this into (57) gives

$$\left(\frac{2}{\pi}\right)^{1/2} \sum_{j=1}^{\infty} (-j^2 + n^2) a_{nj}^{(1)} \sin jx = -\left(\frac{2}{\pi}\right)^{1/2} \lambda_n^{(1)} \sin nx + \left(\frac{2}{\pi}\right)^{1/2} x \sin nx \tag{59}$$

We now multiply (59) by $\zeta_k^{(0)}$, integrate the result over $(0, \pi)$, and use orthogonality to simplify the left-hand side. The result is

$$a_{nk}^{(1)}(-k^2 + n^2) = -\left(\frac{2}{\pi}\right) \lambda_n^{(1)} \int_0^{\pi} \sin nx \sin kx \, dx + \left(\frac{2}{\pi}\right) \int_0^{\pi} x \sin nx \sin kx \, dx \tag{60}$$

which is valid for arbitrary integers n and k.

If $k \neq n$, the first integral on the right-hand side of (60) vanishes, and we have

$$a_{nk} = \frac{2}{\pi(n^2 - k^2)} \int_0^{\pi} x \sin nx \sin kx \, dx \tag{61}$$

This integral can be evaluated explicitly, but the actual result is unimportant for this discussion, so we do not write down the answer.

If $k = n$, the left-hand side of (60) vanishes and the first term on the right-hand side just equals $-\lambda_n^{(1)}$, so we find

$$\lambda_n^{(1)} = \left(\frac{2}{\pi}\right) \int_0^{\pi} x \sin^2 nx \, dx = \frac{\pi}{2} \tag{62}$$

It remains to compute a_{nn}. But, recall that if X_n is an eigenfunction, any constant times X_n is also an eigenfunction. If we choose to normalize X_n, we have

$$\langle X_n, X_n \rangle = 1 = \langle \xi_n^{(0)}, \xi_n^{(0)} \rangle + 2\varepsilon \langle \xi_n^{(0)}, \xi_n^{(1)} \rangle + O(\varepsilon^2) \tag{63}$$

Since $\langle \xi_n^{(0)}, \xi_n^{(0)} \rangle = 1$, we must set

$$\langle \xi_n^{(0)}, \xi_n^{(1)} \rangle = 0 = \int_0^\pi \sin nx \left(\sum_{j=1}^\infty a_{nj} \sin jx \right) dx$$

in order to satisfy (63) to $O(\varepsilon)$, and this implies that $a_{nn} = 0$.

This procedure can be extended to any desired order in ε, and we find the successive terms in (55) and (56).

To complete the solution of the vibration problem (50), we assume $u(x, t; \varepsilon)$ in series form

$$u(x, t; \varepsilon) = \sum_{n=1}^\infty \alpha_n(t; \varepsilon) x_n(x; \varepsilon) \tag{64}$$

where α_n is the amplitude of the nth eigenfunction. Substituting (64) into (50a) and using (51a) shows that α_n obeys

$$\ddot{\alpha}_n + \lambda_n(\varepsilon) \alpha_n = 0 \tag{65}$$

so the frequency of each eigenfunction is $\lambda_n^{1/2}$ (for which an asymptotic expansion can be found) and $\alpha_n = \rho_n \sin(\lambda_n^{1/2}(\varepsilon)t + \psi_n)$, where ρ_n and ψ_n are constants. The fact that (50) is linear is reflected by the result (65) that the amplitude equation for each mode decouples from the others. In a weakly nonlinear problem, this is no longer the case. In fact, this approach is not suitable if the perturbation term is nonlinear (for example, $\varepsilon x u^2$); we can no longer separate the x-dependence of the solution into an exact eigenvalue problem as in (51). However, we can expand u itself in the form

$$u(x, t; \varepsilon) = \sum_{n=1}^\infty \beta_n(t; \varepsilon) \sin nx \tag{66}$$

and calculate *weakly coupled* oscillator equations for the β_n (see Problem 18(b) of Chapter 3). These oscillator equations can then be solved using various techniques that we discuss briefly in Sections 8.4.2 and 8.4.3 (see Section 3.5 of [1] for a detailed account).

Perturbed linear or weakly nonlinear eigenvalue problems arise in contexts other than vibration problems, and we now consider the following generalization of (51):

$$L(u_n) + \varepsilon F(x_i, u_n) = \lambda_n u_n \quad \text{in } \mathcal{D} \tag{67a}$$

Here, $\mathcal{D}$ is a given N-dimensional domain and x_i denotes $x_1, \ldots, x_N$. The nth eigenfunction is denoted by $u_n(x_i; \varepsilon)$ and the associated eigenvalue is $\lambda_n(\varepsilon)$, where ε is a small parameter. L is a self-adjoint linear differential operator, and F is a prescribed function of its arguments $(x_1, \ldots, x_N, u_n)$. We will not consider the more interesting case where instead of $F(x_i, u_n)$, the perturbation term in (67a) is

$F_n(x_i, u_i)$; that is, it is a different function for each n and may depend on all the u_i. In this case the equations (67a) for each n are coupled to $O(\varepsilon)$ with all the others.

The boundary condition is that

$$u_n(x_i; \varepsilon) = 0 \qquad \text{on the boundary of } \mathcal{D} \tag{67b}$$

The reader is referred to [7] for a discussion of self-adjoint operators and the correspondence of this theory with the eigenvalue theory for finite dimensional linear operators on a vector space.

We define the inner product between two eigenfunctions u_m and u_n by

$$\langle u_m, u_n \rangle \equiv \int \cdots \int_{\mathcal{D}} u_m(x_i; \varepsilon) u_n(x_i; \varepsilon) \, dV \tag{68}$$

where dV is the volume element $dx_1 \ldots dx_N$. Since L is self-adjoint, we know that [see (108) of Chapter 4]

$$\langle u_m^{(0)}, L(u_n^{(0)}) \rangle = \langle L(u_m^{(0)}), (u_n^{(0)}) \rangle \tag{69}$$

where $u_m^{(0)}$, $u_n^{(0)}$ are eigenfunctions of the unperturbed problem—that is, $L(u_m^{(0)}) = \lambda_m^{(0)} u_m^{(0)}$; $L(u_n^{(0)}) = \lambda_n^{(0)} u_n^{(0)}$. We shall assume that the eigenvalues associated with different eigenfunctions $u_m^{(0)}$, $u_n^{(0)}$ are distinct, in which case (69) implies the orthogonality condition

$$\langle u_m^{(0)}, u_n^{(0)} \rangle = 0 \qquad \text{if } m \neq n \tag{70a}$$

As in the example problem (51), we shall normalize the $u_i^{(0)}$ by requiring

$$\langle u_n^{(0)}, u_n^{(0)} \rangle = 1 \tag{70b}$$

Now, we expand the eigenfunctions and eigenvalues of the perturbed problem (67) with $\varepsilon \neq 0$ in the form

$$u_n(x_i; \varepsilon) = u_n^{(0)}(x_i) + \varepsilon u_n^{(1)}(x_i) + O(\varepsilon^2) \tag{71a}$$

$$\lambda_n(\varepsilon) = \lambda_n^{(0)} + \varepsilon \lambda_n^{(1)} + O(\varepsilon^2) \tag{71b}$$

For suitable F, (71a) implies that $\varepsilon F(x_i, u_n)$ has the expansion

$$\varepsilon F(x_i, u_n) = \varepsilon F_n^{(0)}(x_i) + O(\varepsilon^2) \tag{72a}$$

where

$$F_n^{(0)}(x_i) \equiv F(x_i, u_n^{(0)}(x_i)) \tag{72b}$$

Since L is linear, we have

$$L(u_n^{(0)} + \varepsilon u_n^{(1)} + O(\varepsilon^2)) = L(u_n^{(0)}) + \varepsilon L(u_n^{(1)}) + O(\varepsilon^2)$$

and therefore $u_n^{(1)}$ satisfies

$$L(u_n^{(1)}) - \lambda_n^{(0)} u_n^{(1)} = -F_n^{(0)}(x_i) + \lambda_n^{(1)} u_n^{(0)} \tag{73}$$

We assume that each $u_n^{(1)}$ may be expressed as a linear combination of the eigenfunctions $u_i^{(0)}$ in the series form [see (58)]

$$u_n^{(1)} = \sum_{j=1}^{\infty} a_{nj}^{(1)} u_j^{(0)} \tag{74a}$$

Therefore, in view of (70), the constants a_{nj} are just

$$a_{nj} = \int \cdots \int_{\mathscr{D}} u_n^{(1)}(x_i) u_j^{(0)}(x_i)\, dV \tag{74b}$$

Similarly, we assume that we may express the known functions $F_n^{(0)}(x_i)$ in series form:

$$F_n^{(0)}(x_i) = \sum_{j=1}^{\infty} f_{nj}^{(0)} u_j^{(0)} \tag{75a}$$

in which case the $f_{nj}^{(0)}$ are the known coefficients

$$f_{nj}^{(0)} = \int \cdots \int_{\mathscr{D}} F_n^{(0)}(x_i) u_j^{(0)}(x_i)\, dV \tag{75b}$$

In order to calculate the a_{nj} and $\lambda_n^{(1)}$, we proceed as in the simple example worked out earlier and multiply (73) by $u_k^{(0)}$ for some arbitrary integer k, then integrate the result over $\mathscr{D}$ to find

$$\int \cdots \int_{\mathscr{D}} [u_k^{(0)} L(u_n^{(1)}) - \lambda_n^{(0)} u_n^{(1)} u_k^{(0)}]\, dV = -f_{nk}^{(0)} + \lambda_n^{(1)} \delta_{nk} \tag{76}$$

where δ_{nk} is the Kronecker delta.

Now, multiply

$$L(u_k^{(0)}) - \lambda_k^{(0)} u_k^{(0)} = 0$$

by $u_n^{(1)}$ and integrate the result over $\mathscr{D}$ to obtain

$$\int \cdots \int_{\mathscr{D}} [u_n^{(1)} L(u_k^{(0)}) - \lambda_k^{(0)} u_k^{(0)} u_n^{(1)}]\, dV = 0 \tag{77}$$

The second term on the left-hand side of (76) and (77) is a multiple of

$$\int \cdots \int_{\mathscr{D}} u_k^{(0)} u_n^{(1)}\, dV = \int \cdots \int_{\mathscr{D}} u_k^{(0)} \sum_{j=1}^{\infty} a_{nj}^{(1)} u_j^{(0)}\, dV = a_{nk}$$

Therefore, subtracting (77) from (76) gives

$$\int \cdots \int_{\mathscr{D}} [u_k^{(0)} L(u_n^{(1)}) - u_n^{(1)} L(u_k^{(0)})]\, dV = a_{nk}^{(1)}(\lambda_n^{(0)} - \lambda_k^{(0)}) - f_{nk}^{(0)} + \lambda_n^{(1)} \delta_{nk} \tag{78a}$$

It follows from (74a), the linearity of L, and the fact it is self-adjoint, exhibited by (69), that the left-hand side of (78a) vanishes, and we are left with [see (60)]

$$a_{nk}^{(1)}(\lambda_n^{(0)} - \lambda_k^{(0)}) = f_{nk}^{(0)} - \lambda_n^{(1)} \delta_{nk} \tag{78b}$$

If $k \neq n$, $\delta_{nk} = 0$, and (78b) gives

$$a_{nk}^{(1)} = \frac{f_{nk}^{(0)}}{\lambda_n^{(0)} - \lambda_k^{(0)}} \tag{79a}$$

If $k = n$, the left-hand side of (78b) vanishes, and we have

$$\lambda_n^{(1)} = f_{nn}^{(0)} \tag{79b}$$

To complete the solution to $O(\varepsilon)$, we note that if we normalize the u_n, we must have

$$\langle u_n, u_n \rangle = 1 = \langle u_n^{(0)}, u_n^{(0)} \rangle + 2\varepsilon \langle u_n^{(0)}, u_n^{(1)} \rangle + O(\varepsilon^2)$$

that is, $\langle u_n^{(0)}, u_n^{(1)} \rangle = 0$, and this implies that $a_{nn}^{(1)} = 0$. Problems 7 and 8 illustrate these ideas.

8.2.3 A Boundary Perturbation Problem

We showed in Chapter 2 that once Green's function for Laplace's equation is known for a given domain for the case where the potential vanishes on the boundary, we could solve the Dirichlet problem for this domain. Let us study the special case where the domain has a boundary nearly equal to the unit circle; that is, we wish to solve

$$\Delta_P K(r, \theta, \rho, \phi; \varepsilon) = \delta_2(P, Q) \qquad \text{in } \mathscr{D} \tag{80a}$$

$$K(1 + \varepsilon f(\theta), \theta, \rho, \phi; \varepsilon) = 0 \tag{80b}$$

where $\mathscr{D}$ is the domain $0 \leq r \leq 1 + \varepsilon f(\theta)$ for a given 2π-periodic function $f(\theta)$. Here r and θ are the polar coordinates of the observer point P, ρ and ϕ are the polar coordinates of the source point Q, δ_2 is the two-dimensional delta function, and Δ_P denotes the Laplacian with respect to the r, θ variables.

We showed earlier (see (159) of Chapter 2), that for $\varepsilon = 0$, we have

$$K^{(0)}(r, \theta, \rho, \phi) \equiv K(r, \theta, \rho, \phi; 0) = \frac{1}{2\pi} \log \left(\frac{r^2 + \rho^2 - 2r\rho\cos(\theta - \phi)}{\rho^2 r^2 + 1 - 2r\rho\cos(\theta - \phi)} \right)^{1/2} \tag{81}$$

If we expand K in the form

$$K(r, \theta, \rho, \phi; \varepsilon) = K^{(0)}(r, \theta, \rho, \phi) + \varepsilon K^{(1)}(r, \theta, \rho, \phi) + O(\varepsilon^2) \tag{82}$$

we see that $K^{(1)}$ obeys

$$\Delta_P K^{(1)} = 0 \tag{83}$$

To compute the boundary condition for $K^{(1)}$, we first expand (80b) near $r = 1$ to obtain

$$K(1, \theta, \rho, \phi; \varepsilon) + \frac{\partial K}{\partial r}(1, \theta, \rho, \phi; \varepsilon)\varepsilon f(\theta) = O(\varepsilon^2)$$

We then substitute (82) into each term:

$$K^{(0)}(1, \theta, \rho, \phi) + \varepsilon K^{(1)}(1, \theta, \rho, \phi) + \varepsilon f(\theta)\frac{\partial K^{(0)}}{\partial r}(1, \theta, \rho, \phi) = O(\varepsilon^2)$$

Since $K^{(0)}(1, \theta, \rho, \phi) = 0$, we must have

$$K^{(1)}(1, \theta, \rho, \phi) = -f(\theta) \frac{\partial K^{(0)}}{\partial r}(1, \theta, \rho, \phi) \tag{84}$$

We use (81) to evaluate the right-hand side of (84) and obtain the following boundary condition for $K^{(1)}$:

$$K^{(1)}(1, \theta, \rho, \phi) = \frac{f(\theta)(\rho^2 - 1)}{1 + \rho^2 - 2\rho\cos(\theta - \phi)} \equiv g(\theta, \rho, \phi) \tag{85}$$

The solution of (83) subject to the boundary condition (85) is given by Poisson's formula [see (165) of Chapter 2].

$$K^{(1)}(r, \theta, \rho, \phi) = \frac{1 - r^2}{2\pi} \int_0^{2\pi} \frac{g(\theta', \rho, \phi)\, d\theta'}{1 + r^2 - 2r\cos(\theta - \theta')} \tag{86}$$

For an application, see Problem 9. This idea generalizes to any operator and domain for which Green's function is known.

8.3   Matched Asymptotic Expansions

In this section we discuss a class of problems characterized by the property that the solution $u(x_i; \varepsilon)$ has a different asymptotic expansion in certain distinguished subdomains of the x_i space. These subdomains have boundaries that depend on ε and can be established either trivially from knowledge of the exact solution or indirectly from the governing differential equation. Typically, the order of magnitude of certain terms in the governing equation will depend on the solution domain; in certain thin layers near the boundaries or along particular interior regions, a term multipled by ε in the differential equation becomes important because the term itself is large there. By considering all the possible limiting forms of the governing equation, we can usually establish the location and nature of these layers. An example of this behavior occured in Section 5.3.6, where we studied a special solution of Burgers' equation.

8.3.1 An Ordinary Differential Equation

We begin our discussion with the following linear boundary-value problem for a second-order equation of some historical interest.

$$\varepsilon u'' + u' = \tfrac{1}{2}; \qquad 0 < \varepsilon \ll 1; \qquad 0 \le x \le 1 \tag{87a}$$

$$u(0; \varepsilon) = 0; \qquad u(1; \varepsilon) = 1 \tag{87b}$$

A slightly more general version of (87) was first introduced by K.O. Friedrichs in 1942 to illustrate boundary-layer behavior. See [2] for the original reference and a more detailed discussion.

The solution of (87) is

$$u(x; \varepsilon) = \frac{1 - e^{-x/\varepsilon}}{2(1 - e^{-1/\varepsilon})} + \frac{x}{2} \tag{88}$$

and we shall examine the asymptotic behavior of this solution in order to motivate our later discussion concerning a perturbation analysis based only on (87).

Outer and Inner Limits

For any fixed x in $0 < x \le 1$, the limiting value as $\varepsilon \to 0$ of u derived from (88) is

$$\lim_{\substack{\varepsilon \to 0 \\ x \text{ fixed} \ne 0}} u(x;\varepsilon) = \frac{1+x}{2} \equiv u_0(x) \tag{89}$$

We shall refer to the limit process $\varepsilon \to 0$ with x fixed $\ne 0$ as the *outer limit process* and to $u_0(x)$ as the *outer limit* of $u(x;\varepsilon)$. We note that $u_0(x)$, which is the only term in the asymptotic expansion of $u(x;\varepsilon)$ in the outer limit, does not satisfy the boundary condition at $x = 0$. In fact, the statement

$$u(x;\varepsilon) = u_0(x) + \text{T.S.T.}$$

is not uniformly valid if $x = O(\varepsilon)$. Here T.S.T. denotes transcendentally small terms—that is, terms that are $o(\varepsilon^\alpha)$ for *any* arbitrarily large α.

The difficulty at $x = 0$ is clear; we ignored $e^{-x/\varepsilon}$ in deriving (89), but this term is important if $x = O(\varepsilon)$. Thus, in a thin *boundary layer* over the interval $0 \le x \le \varepsilon x^*$ (where x^* is an arbitrary finite positive constant), this term equals $e^{-x^*} = O(1)$. This behavior suggests that we should consider a limit process where we let $\varepsilon \to 0$ for a fixed value of the boundary-layer variable $x^* \equiv x/\varepsilon$. In this case, we replace x in (88) with εx^* and find

$$u(\varepsilon x^*;\varepsilon) \equiv u^*(x^*;\varepsilon) = \frac{1 - e^{-x^*}}{2(1 - e^{-1/\varepsilon})} + \frac{\varepsilon x^*}{2} \tag{90}$$

We define the *inner limit* of (88) to be

$$\lim_{\substack{\varepsilon \to 0 \\ x^* \text{ fixed} \ne \infty}} u(\varepsilon x^*;\varepsilon) = \frac{1 - e^{-x^*}}{2} \equiv u_0^*(x^*) \tag{91}$$

Now, we notice that $u_0^*(x^*)$ does not satisfy the boundary condition at $x = 1$—that is, at $x^* = 1/\varepsilon$. In fact, $u_0^*(x^*)$ is the leading term of the following asymptotic expansion of $u^*(x^*;\varepsilon)$ constructed in the limit $\varepsilon \to 0$ with x^* fixed and not equal to ∞.

$$u^*(x^*;\varepsilon) = \frac{1 - e^{-x^*}}{2} + \varepsilon \frac{x^*}{2} + \text{T.S.T.} \tag{92}$$

This *inner expansion*, which terminates after two terms for this example, is not uniformly valid if $x^* \to \infty$.

Extended Domains of Validity; Matching

We shall now examine more carefully in what sense the outer and inner limits approximate the function (88). The outer limit (89) is formally derived by letting $\varepsilon \to 0$ for arbitrary *fixed* x in the half-open interval $0 < x \le 1$. We shall demonstrate that (89) actually remains valid in an *extended* domain that corresponds to $\varepsilon \to 0$ *with* $x \to 0$ at some maximal rate relative to ε. To be more precise, we set $x \equiv \eta(\varepsilon)x_\eta$ for a function $\eta(\varepsilon) \ll 1$ to be specified and a *fixed* $x_\eta > 0$. Thus, as $\varepsilon \to 0$ with x_η fixed, x tends to zero "at the rate" $\eta(\varepsilon)$.

Observe now that the statement

$$\lim_{\substack{\varepsilon \to 0 \\ x_\eta \text{ fixed} \neq 0}} \{u(\eta x_\eta; \varepsilon) - u_0(\eta x_\eta)\} = 0 \tag{93}$$

holds as long as $e^{-\eta x_\eta/\varepsilon} \to 0$ and $\eta \to 0$, as postulated. Actually, $e^{-\eta x_\eta/\varepsilon}$ must be transcendentally small as $\varepsilon \to 0$, and this implies that $\varepsilon|\log \varepsilon| \ll \eta$; if $\eta = O_s(\varepsilon|\log \varepsilon|)$, then $e^{-\eta x_\eta/\varepsilon} \to O_s(\varepsilon^{x_\eta})$, which does vanish as $\varepsilon \to 0$ but is not transcendentally small. The result (93) remains true if $\eta = O_s(1)$; therefore the *extended domain of validity* of the outer limit is defined by the set of functions $\eta(\varepsilon)$ satisfying

$$\varepsilon|\log \varepsilon| \ll \eta(\varepsilon) \underset{\approx}{\ll} 1 \tag{94}$$

Here we have introduced the notation

$$\phi(\varepsilon) \underset{\approx}{\ll} \psi(\varepsilon) \quad \text{if} \quad \phi \ll \psi \quad \text{or} \quad \phi = O_s(\psi) \tag{95}$$

The shaded region in Figure 8.1a represents this domain in the $x\varepsilon$-plane. This diagram should not be interpreted literally to mean that x must be in the shaded region as $\varepsilon \to 0$, since x_η is an arbitrary positive constant. Moreover, $\eta_0(\varepsilon)$ is also arbitrary as long as $\eta_0(\varepsilon) = O_s(1)$. Roughly speaking, the left boundary of the extended domain restricts the admissible class of functions $\eta(\varepsilon)$ to those that tend to zero "more slowly" than $\varepsilon|\log \varepsilon|$.

One final point concerning this very special example is that all our statements concerning the outer limit remain true if the bracketed expression in (93) is divided by ε^n for any $n > 0$. This is because the outer expansion terminates after one term with a transcendentally small remainder.

Let us now examine the inner limit defined by (91). This limit corresponds formally to having $x^* \equiv x/\varepsilon$ fixed as $\varepsilon \to 0$. To see how far we can extend the domain of validity of this limit, we again set $x = \eta(\varepsilon)x_\eta$—that is, $x^* = \eta x_\eta/\varepsilon$. We then find that

$$\lim_{\substack{\varepsilon \to 0 \\ x_\eta \text{ fixed} \neq \infty}} \left[u^*\left(\frac{\eta x_\eta}{\varepsilon}; \varepsilon\right) - u_0^*\left(\frac{\eta x_\eta}{\varepsilon}\right) \right] = 0 \tag{96}$$

actually remains true as long as $\eta \ll 1$. Thus, the extended domain of validity of the inner limit is the class of functions $\eta(\varepsilon)$ with the property

$$\varepsilon \underset{\approx}{\ll} \eta \ll 1 \tag{97}$$

as shown in Figure 8.1b.

We note the following remarkable result for this example: The extended domains of validity of the inner and outer limits *overlap* in the sense that the intersection of these two extended domains consists of the nonempty class of functions $\eta(\varepsilon)$ with the property

$$\varepsilon|\log \varepsilon| \ll \eta \ll 1 \tag{98}$$

described qualitatively in Figure 8.1c. Any $\eta(\varepsilon)$ in the class (98) lies in the extended

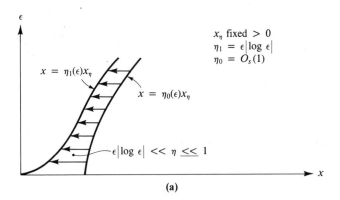

(a)

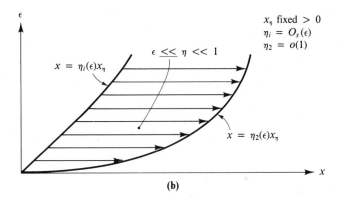

(b)

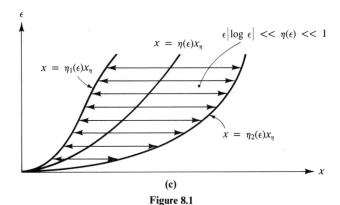

(c)

Figure 8.1

domain of validity of *both* the outer and inner limits. For such an η, (93) and (96) both hold; therefore, their difference vanishes, and we have the *direct matching condition*

$$\lim_{\substack{\varepsilon \to 0 \\ x_\eta \text{ fixed} \\ \neq 0, \neq \infty}} \left\{ u_0(\eta x_\eta) - u_0^*\left(\frac{\eta x_\eta}{\varepsilon}\right) \right\} = 0 \qquad (99)$$

for any η such that $\varepsilon|\log \varepsilon| \ll \eta \ll 1$.

The result (99) in the form

$$u_0(0) = u_0^*(\infty) \tag{100}$$

was first proposed on physical grounds by L. Prandtl in 1905 as the matching condition for the horizontal component of velocity (u_0) of an inviscid flow evaluated on the boundary of a body $(x = 0)$ with the velocity of a viscous boundary-layer flow (u_0^*) evaluated at ∞. It was later shown by S. Kaplun in 1957 that the direct matching condition (99) is a special case of a more general situation of matching each of the outer and inner limits with an *intermediate limit*. It is beyond the scope of this chapter to give an account of this theory. The reader is referred to [2] and the works cited there for a comprehensive treatment.

The result (99) (and its extension to higher order, as discussed in Section 8.3.2) turns out to be sufficient for most applications as long as we retain enough terms in each expansion. We demonstrate this in the examples we present in this section. Other examples can be found in [1] and [2].

Uniformly Valid Result to $O(1)$

We have seen that the outer limit fails when $\varepsilon \to 0$ and $x = O_s(\varepsilon)$ and that the inner limit fails as $\varepsilon \to 0$ and $x = O_s(1)$. But, each of these limits is individually valid in its extended domain. We show next that one can construct a *composite* expression that tends to the outer limit as $\varepsilon \to 0$ with x fixed $\neq 0$ and to the inner limit as $\varepsilon \to 0$ with x/ε fixed $\neq \infty$. The idea is to add the inner and outer limits and then subtract out of their sum the term that is common to both expressions in the overlap domain. In our case, this common term is $\frac{1}{2}$. Therefore, we propose

$$\tilde{u}_0(x; \varepsilon) \equiv u_0(x) + u_0^*(x^*) - \frac{1}{2} = \frac{1 + x - e^{-x/\varepsilon}}{2} \tag{101}$$

and verify that (101) satisfies the dual requirements we must meet. It is easily seen that

$$\lim_{\substack{\varepsilon \to 0 \\ x \text{ fixed} \neq 0}} \tilde{u}_0(x; \varepsilon) = \frac{1 + x}{2} = u_0(x) \tag{102a}$$

and

$$\lim_{\substack{\varepsilon \to 0 \\ x^* \text{ fixed} \neq \infty}} \tilde{u}_0(\varepsilon x^*; \varepsilon) = \frac{1 - e^{-x^*}}{2} = u_0^*(x^*) \tag{102b}$$

We have shown that $\tilde{u}$ tends to u_0 under the outer limit process and to u_0^* under the inner limit process. It then follows that $\tilde{u}_0$ approximates $u(x; \varepsilon)$ uniformly in $0 \le x \le 1$. In fact, for this example $(u - \tilde{u}_0)$ is transcendentally small as $\varepsilon \to 0$.

A sketch of the three functions $u_0(x)$, $u_0^*(x^*)$, and $\tilde{u}_0(x; \varepsilon)$ is shown in Figure 8.2. The matching condition (100) corresponds graphically to the fact the asymptotic value for u_0^* is the same as $u_0(0)$. A more important observation is that the composite expression $\tilde{u}_0$ is a smooth function of x; it is not just a "patching" of u_0^* over part of the interval $(0, 1)$ with u_0 over the remainder. In fact, for this

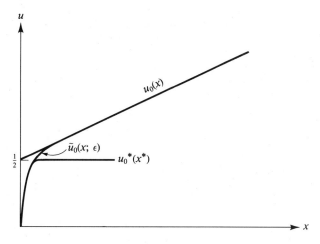

Figure 8.2

example u_0^* and u_0 do not intersect on $(0, 1)$, and it is not possible to patch these two functions at any point in $(0, 1)$.

Distinguished Limits of the Differential Equation

In the preceding discussion, we derived the outer and inner limits from a knowledge of the exact solution. A useful procedure will be possible in more general cases only if the same information can be derived directly from the governing equation in a systematic way without knowing the exact solution. We now demonstrate that this is the case for the present example; less trivial examples will be discussed in later sections.

Since the rate at which x approaches the point of nonuniformity is the key feature that distinguishes the two limits that we found, let us investigate what all the possible limit processes are for (87). We introduce the general transformation of independent variable given by

$$x_\eta \equiv \frac{x - x_0}{\eta(\varepsilon)} \tag{103}$$

to study how the various terms in (87) behave under all possible rescalings $\eta(\varepsilon)$. Here x_0 is some fixed point in $[0, 1]$ and $\eta(\varepsilon) \ll 1$ but is otherwise arbitrary. Equation (87a) transforms to

$$\frac{\varepsilon}{\eta^2} \frac{d^2 U}{dx_\eta^2} + \frac{1}{\eta} \frac{dU}{dx_\eta} = \frac{1}{2} \tag{104a}$$

or

$$\frac{d^2 U}{dx_\eta^2} + \frac{\eta}{\varepsilon} \frac{dU}{dx_\eta} = \frac{\eta^2}{2\varepsilon} \tag{104b}$$

where we are denoting $u(\eta x_\eta; \varepsilon) \equiv U(x_\eta; \varepsilon)$. Since (87) is autonomous, the transformed problem does not depend on x_0.

A fundamental observation concerning (104) is that for any $\eta(\varepsilon) = O_s(1)$, the first term in (104a) is vanishingly small in the limit $\varepsilon \to 0$ relative to the other two remaining terms. Therefore, with no loss of generality, we may set $\eta = 1$ to obtain the *outer* limiting equation

$$\eta = O_s(1): \quad \frac{dU}{dx_\eta} = \frac{1}{2} \qquad\qquad (105a)$$

Similarly, for any $\eta = O_s(\varepsilon)$, the first two terms in (104b) are in dominant balance. We set $\eta = \varepsilon$ to obtain the *inner* limiting equation

$$\eta = O_s(\varepsilon): \quad \frac{d^2U}{dx_\eta^2} + \frac{dU}{dx_\eta} = 0 \qquad\qquad (105b)$$

These are the *only two distinguished limits for* (87) in the sense that they each correspond to a function $\eta(\varepsilon)$ that is of a *specific order* in ε and for which a subset of terms in the original equation are in dominant balance. For example, the limiting equation

$$\frac{dU}{dx_\eta} = 0 \qquad\qquad (105c)$$

results for any $\eta(\varepsilon)$ with $\varepsilon \ll \eta \ll 1$; it is not a distinguished limit because it does not correspond to an $\eta(\varepsilon)$ having a specific order in ε.

We now demonstrate that knowledge of the two limiting equations (105a) and (105b), and our ability to match their solutions, uniquely determines x_0 and the correct limits $u_0(x)$ and $u_0^*(x^*)$.

First we consider (105a) and set $\eta = 1$ for simplicity. The solution for u_0 is just $u_0 = (x - x_0)/2 + C$, and with no loss of generality, we absorb the constant $-x_0/2$ in C and write this as

$$u_0 = \frac{x}{2} + C \qquad\qquad (106)$$

Let us defer the determination of C until we have established the behavior of the inner limit. With $\eta = \varepsilon$, the general solution of (105b) is

$$u_0^*(x^*) = A + Be^{-x^*} \qquad\qquad (107)$$

where A and B are constants and $x^* \equiv (x - x_0)/\varepsilon$. If $x_0 = 1$—that is, we assume a boundary layer at $x = 1$—the boundary condition $u(1; \varepsilon) = 1$ must be satisfied by (107). This means $u^*(0) = 1$—that is, $B = 1 - A$—and the inner limit has the form

$$u_0^*(x^*) = A + (1 - A)e^{-x^*} \qquad\qquad (108)$$

Now, the matching with a solution outside the boundary layer has $x < x_0$; that is, $x^* \to -\infty$ in the matching region. But e^{-x^*} grows exponentially large in this limit and must therefore be excluded by setting $A = 1$. In this case, the inner limit is simply $u_0^* = 1$, and the outer limit (106) must satisfy the boundary condition at $x = 0$, which gives $C = 0$, so $u_0 = x/2$. But now the inner and outer limits do not match. In general, a boundary-layer solution that grows exponentially in the domain of interest is inappropriate. For an exception, see Problem 10. Similarly, the choice of *any* $x_0 > 0$ will require that we set $A_0 = 1$ with the result that the inner and outer limits do not match. The only possible location of a boundary layer for this example is at $x = 0$, and we must set $x_0 = 0$. In this case, $x^* \to \infty$ in the matching region and (108) gives $u_0^* \to A$, where A is unknown. However, since we know that we can have a layer only at $x = 0$, the solution (106) must hold everywhere else. In particular, (106) must satisfy the boundary condition $u_0(1; \varepsilon) = 1$, which means $C = \frac{1}{2}$. Now we can impose the matching condition (99) or (100) to conclude the correct result that $A = \frac{1}{2}$.

8.3.2 A Second Example

A less trivial problem for which we do not attempt an exact solution is

$$\varepsilon u'' + (1 + 2x)u' + u = x \tag{109a}$$

$$u(0; \varepsilon) = 0 \tag{109b}$$

$$u(1; \varepsilon) = 1 \tag{109c}$$

The Solution to O(1)

The outer limit $u_0(x)$ obeys

$$(1 + 2x)u_0' + u_0 = x \tag{110}$$

and the general solution is

$$u_0(x) = \frac{c_0}{(1 + 2x)^{1/2}} + \frac{x - 1}{3} \tag{111}$$

where c_0 is an arbitrary constant. The choice of c_0 depends on which of the two boundary conditions derived from (109b)–(109c) is to be satisfied.

In order to answer this question, we study the inner limit of (109a) for an unspecified layer location to see where such a layer is appropriate. As in the example of the previous section, we introduce the general scaled variable x_η by

$$x_\eta \equiv \frac{x - x_0}{\eta(\varepsilon)} \tag{112}$$

with x_0 and $\eta(\varepsilon)$ still unspecified. Equation (109a) for $U(x_\eta; \varepsilon) \equiv u(x_0 + \eta x_\eta; \varepsilon)$ becomes

$$\frac{d^2 U}{dx_\eta^2} + \frac{\eta}{\varepsilon}(1 + 2x_0 + 2\eta x_\eta)\frac{dU}{dx_\eta} + \frac{\eta^2}{\varepsilon}U = \frac{\eta^2}{\varepsilon}(x_0 + \eta x_\eta) \tag{113}$$

Clearly, the terms multiplied by η^2/ε are less important than the second term for any $\eta \ll 1$. In fact, the most general limit results for $\eta = O_s(\varepsilon)$, and this is a distinguished limit. For simplicity, we choose $\eta = \varepsilon$ and denote

$$x^* \equiv \frac{x - x_0}{\varepsilon}$$

The inner limiting equation is then

$$\frac{d^2u_0^*}{dx^{*2}} + (1 + 2x_0)\frac{du_0^*}{dx^*} = 0 \tag{114}$$

with solution

$$u_0^*(x^*) = A_0 e^{-(1+2x_0)x^*} + B_0 \tag{115}$$

If $x_0 = 1$—that is, the boundary layer is at the right end—(115) must satisfy the right boundary condition $u_0^*(0) = 1$, which gives $A_0 = 1 - B_0$; (115) becomes

$$u_0^*(x^*) = (1 - B_0)e^{-3x^*} + B_0 \tag{116}$$

To prevent exponential growth in the interior of our domain, we must set $B_0 = 1$. In this case, we must choose $c_0 = \frac{1}{3}$ in order to have $u_0(x)$ satisfy the left boundary condition. Again, we see that with $c_0 = \frac{1}{3}$, (111) does not match with $u_0^* = 1$. Similarly, any choice of x_0 other than $x_0 = 0$ leads to the same difficulty.

Therefore, the boundary layer must be at $x = 0$—that is, $x_0 = 0$, $x^* \equiv x/\varepsilon$— and (115) reads

$$u_0^*(x^*) = B_0(1 - e^{x^*}) \tag{117}$$

in order to satisfy the left boundary condition. We set $c_0 = \sqrt{3}$ to have (111) satisfy the right boundary condition and determine B_0 from the matching condition to $O(1)$. This is just (99) with the preceding values of u_0 and u_0^*, and we find

$$B_0 = \frac{3\sqrt{3} - 1}{3} \tag{118}$$

Thus, matching is possible for all $\eta(\varepsilon)$ in the overlap domain

$$\varepsilon|\log \varepsilon| \ll \eta \ll 1 \tag{119}$$

By adding the inner and outer limits and subtracting the common part (B_0), we find the composite approximation

$$\tilde{u}_0(x;\varepsilon) = \frac{\sqrt{3}}{(1 + 2x)^{1/2}} + \frac{x - 1}{3} - \left(\frac{3\sqrt{3} - 1}{3}\right)e^{-x/\varepsilon} \tag{120}$$

which is uniformly valid to $O(1)$ over the entire interval. The inner limit less the common part $\left[-\left(\frac{3\sqrt{3} - 1}{3}\right)e^{-x/\varepsilon}\text{ in this case}\right]$ is often denoted as the *boundary-layer correction* because it is a term that is transcendentally small everywhere except inside the boundary layer, where it serves to satisfy the boundary condition.

The Solution to $O(\varepsilon)$

Having established the location of the boundary layer, we no longer need worry about this question and can compute the next term in each of the two limit process expansions.

As we have seen before, the choice of asymptotic sequence is dictated by our need to have the most general (richest) equations governing each successive term in our expansions. In the present case, it is easily seen that this requires choosing the next term in both the inner and outer expansions to be $O(\varepsilon)$.

We substitute the outer expansion ($\varepsilon \to 0$, x fixed $\neq 0$)

$$u(x;\varepsilon) = u_0(x) + \varepsilon u_1(x) + o(\varepsilon) \tag{121}$$

into (109a) and find that $u_1(x)$ obeys

$$(1 + 2x)u_1' + u_1 = -u_0'' = -\frac{3\sqrt{3}}{(1 + 2x)^{5/2}} \tag{122a}$$

We know now that (121) must satisfy the right boundary condition—that is,

$$u_1(1) = 0 \tag{122b}$$

The solution of (122) is easily found to be

$$u_1(x) = \frac{2 - x - x^2}{\sqrt{3}(1 + 2x)^{5/2}} \tag{123}$$

Next, we construct the inner expansion in the form ($\varepsilon \to 0$, x^* fixed $\neq \infty$)

$$u(x, \varepsilon) \equiv u^*(\varepsilon x^*; \varepsilon) = u_0^*(x^*) + \varepsilon u_1^*(x^*) + o(\varepsilon) \tag{124}$$

The change of variables $x \to x^*$ implies that $u^*(x^*; \varepsilon)$ obeys

$$\frac{d^2 u^*}{dx^{*2}} + (1 + 2\varepsilon x^*)\frac{du^*}{dx^*} + \varepsilon u^* = \varepsilon^2 x^* \tag{125a}$$

$$u^*(0; \varepsilon) = 0 \tag{125b}$$

We do not express the right boundary condition in terms of x^*, as our expansion (124) is not supposed to hold there.

Substituting (124) in (125) gives

$$\frac{d^2 u_1^*}{dx^{*2}} + \frac{du_1^*}{dx^*} = -2x^*\frac{du_0^*}{dx^*} - u_0^*$$

$$= 2B_0 x^* e^{-x^*} - B_0 + B_0 e^{-x^*} \tag{126a}$$

$$u_1^*(0) = 0 \tag{126b}$$

where B_0 is the known constant (118). The solution of (126) involving an undetermined constant B_1 is found to be

$$u_1^*(x^*) = -B_0 e^{-x^*}(x^{*2} + 3x^* + 1) - B_0(x^* - 1) + B_1(1 - e^{-x^*}) \tag{127}$$

Matching to $O(\varepsilon)$; Composite Expansion

The direct matching to $O(\varepsilon)$ of the two-term outer and two-term inner expansion requires that [see (99)]

$$\lim_{\substack{\varepsilon \to 0 \\ x_\eta \text{ fixed} \\ \neq 0, \neq \infty}} \frac{1}{\varepsilon}\left\{ u_0(\eta x_\eta) + \varepsilon u_1(\eta x_\eta) - u_0^*\left(\frac{\eta x_\eta}{\varepsilon}\right) - \varepsilon u_1^*\left(\frac{\eta x_\eta}{\varepsilon}\right) \right\} = 0 \tag{128}$$

in some overlap domain to be determined.

Note that it is not always possible to match outer and inner expansions to the same order as they are formally derived. Here, for example, we have computed each of the outer and inner expansions to order ε, and we propose to match these to order ε also. Examples given in [1] and [2] show that we may need to retain terms of order higher than the order of matching in one or the other of the two expansions. In the more general matching procedure of Kaplun, this situation does not arise. In fact, Kaplun proposed his matching principle to resolve a classical difficulty wherein the leading terms of the outer and inner expansions do not match directly (see [2] for more details).

In preparation for imposing (128), we expand u_0 and u_1 for x small and u_0^* and u_1^* for x^* large. This gives

$$u_0(x) = \sqrt{3}(1 - x) + \frac{x}{3} - \frac{1}{3} + O(x^2) \tag{129a}$$

$$u_1(x) = \frac{2}{\sqrt{3}} + O(x) \tag{129b}$$

$$u_0^*(x^*) = \frac{3\sqrt{3} - 1}{3} + \text{T.S.T.} \tag{129c}$$

$$u_1^*(x^*) = \frac{3\sqrt{3} - 1}{3} - \left(\frac{3\sqrt{3} - 1}{3}\right)x^* + B_1 + \text{T.S.T.} \tag{129d}$$

We first cancel the terms of $O(\varepsilon^{-1})$ out of (128) that have already been matched. Next, we note that the term $-\left(\frac{3\sqrt{3} - 1}{3}\right)x$ in u_0, which produces a *singular* contribution proportional to $\eta/\varepsilon \to \infty$ in (128), matches identically with a corresponding term proportional to x^* in u_1^*. This provides us with a partial check of the calculations for u_1^*. The next largest terms left are the two constants contributed by u_1 and u_1^*. Matching these gives

$$B_1 + \frac{3\sqrt{3} - 1}{3} = \frac{2}{\sqrt{3}}$$

—that is,

$$B_1 = \frac{1 - \sqrt{3}}{3} \tag{130}$$

The two terms that have not been matched to $O(\varepsilon)$ are the $O(x^2)$ remainder in (129a) and the $O(x)$ remainder in (129b). These contribute remainders of order η^2/ε and η, respectively, in (128). We need be concerned only about having $\eta^2/\varepsilon \to 0$, since $\eta \to 0$ automatically in any overlap domain. To have $\eta^2/\varepsilon \to 0$, we must restrict η such that $\eta \ll \varepsilon^{1/2}$. Since $\varepsilon|\log \varepsilon| \ll \varepsilon^{1/2}$ and we must have $\varepsilon|\log \varepsilon| \ll \eta$ in order to ignore the transcendentally small terms in (129b)–(129c), we conclude that the overlap domain is defined by

$$\varepsilon|\log \varepsilon| \ll \eta \ll \varepsilon^{1/2} \tag{131}$$

and that this is "smaller" than (119). This is typical for higher-order matching.

Even though the result (130) could have been directly deduced by comparing the various expansions in (129), it is essential for exhibiting the overlap domain (131) that we keep track of all the terms that were ignored in the matching and that we express these in terms of the matching variable. This is particularly true when higher-order terms are needed to match to a given order. For example, see Section 2.3.2 of [1].

The reader should verify that the expression

$$\tilde{u}(x;\varepsilon) = \tilde{u}_0(x;\varepsilon) + \tilde{u}_1(x;\varepsilon) + O(\varepsilon^2) \qquad (132a)$$

where $\tilde{u}_0$ is given by (120), and $\tilde{u}_1$ is defined as

$$\tilde{u}_1(x;\varepsilon) \equiv \varepsilon \left[\frac{2 - x - x^2}{\sqrt{3}(1 + 2x)^{5/2}} - \left(\frac{3\sqrt{3} - 1}{3} \right) \left(\frac{x^2}{\varepsilon^2} + \frac{3x}{\varepsilon} + 1 \right) e^{-x/\varepsilon} - \frac{1 - \sqrt{3}}{3} e^{-x/\varepsilon} \right] \qquad (132b)$$

gives the two-term outer expansion if $\varepsilon \to 0$, x fixed $\neq 0$, and it also gives the two-term inner expansion if $\varepsilon \to 0$, x^* fixed $\neq \infty$. Therefore, it is uniformly valid to $O(\varepsilon)$ in $0 \leq x \leq 1$.

Generalized Asymptotic Expansion

The result (132) is (without proof) the two-term asymptotic expansion of the exact solution and is uniformly valid in $0 \leq x \leq 1$. We note that this expansion is not in the form (7) of a series consisting of functions of ε times functions of x. In fact, it is not possible to factor out the ε-dependence in this expression. We need a more general definition of an asymptotic expansion to characterize this result.

We say that $\sum_{n=1}^{N} \tilde{u}_n(x_i;\varepsilon)$ is the *generalized* asymptotic expansion of $u(x_i;\varepsilon)$ as $\varepsilon \to 0$ with respect to the sequence $\{\phi_n(\varepsilon)\}$ if for every $M = 1, 2, \ldots, N$,

$$u(x_i,\varepsilon) - \sum_{n=1}^{M} \tilde{u}_n(x_i;\varepsilon) = o(\phi_M) \qquad \text{as } \varepsilon \to 0 \qquad (133)$$

Thus, in (132), $\phi_1 = 1$, $\phi_2 = \varepsilon, \ldots$.

Generalized asymptotic expansions arise in other contexts and do not necessarily consist of the sum of functions of x and functions of x^*. In Section 8.4, we discuss another class of problems that must be solved in terms of generalized asymptotic expansions where the dependence on the two scales x, x^* cannot be separated as in (132).

8.3.3 Interior Dirichlet Problems for Elliptic Equations

In some applications modeled by a second-order partial differential equation, the small parameter ε multiplies the second-derivative terms. In this section we study two examples for elliptic equations with Dirichlet-type boundary conditions. More details can be found in [1] and [2].

A Model Equation for a Bounded Domain

We wish to solve

$$\varepsilon(u_{xx} + u_{yy}) = u_y \qquad (134a)$$

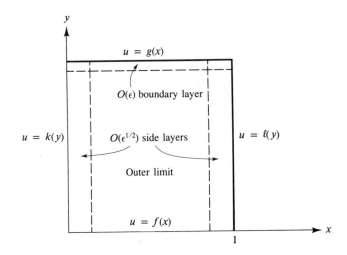

Figure 8.3

in the interior of the unit square: $0 \le x \le 1$, $0 \le y \le 1$ with Dirichlet-type boundary conditions

$$u(x, 0) = f(x) \tag{134b}$$

$$u(x, 1) = g(x) \tag{134c}$$

$$u(0, y) = k(y) \tag{134d}$$

$$u(1, y) = \ell(y) \tag{134e}$$

The choice of a square domain serves to illustrate the essential ideas without unnecessary complications (see Figure 8.3).

If we set $\varepsilon = 0$ in (134a), we conclude that the outer limit is independent of y; that is, $u = $ constant on vertical lines $x = $ constant. These vertical lines are the characteristics of the lower-order operator $\partial/\partial y = 0$. In a more general problem, this lower-order operator may contain both $\partial/\partial x$ and $\partial/\partial y$ terms; it may also involve variable coefficients, and so on. In such cases, the subcharacteristics (that is, the characteristics of the operator with $\varepsilon = 0$) define a more complicated family of curves on which u may vary. Some examples are given in [1].

For our case, we must decide which of the two boundary conditions (134b)–(134c), if any, is appropriate for the outer limit. Along any line $x = $ constant, (134a) is an ordinary differential equation in terms of the independent variable y, and we proceed just as in previous examples. We first establish the thickness and location of possible boundary layers by introducing the transformation

$$y^* \equiv \frac{y - y_0}{\eta(\varepsilon)}$$

where y_0 is a constant in the interval $0 \le y_0 \le 1$ and $\eta(\varepsilon) \ll 1$ measures the thickness of the layer. If we denote

$$u(x, y_0 + \eta y^*; \varepsilon) \equiv u^*(x, y^*; \varepsilon)$$

(134a) becomes

$$\varepsilon \frac{\partial^2 u^*}{\partial x^2} + \frac{\varepsilon}{\eta^2} \frac{\partial^2 u^*}{\partial y^{*2}} = \frac{1}{\eta} \frac{\partial u^*}{\partial y^*}$$

The distinguished inner limit corresponds to $\eta = \varepsilon$, and the limiting equation is

$$\frac{\partial^2 u_0^*}{\partial y^{*2}} = \frac{\partial u_0^*}{\partial y^*} \tag{135}$$

where $u_0^* = u^*(x, y^*; 0)$. The solution of (135) is

$$u_0^*(x, y^*) = A(x) + B(x)e^{y^*} \tag{136a}$$

Therefore, the boundary layer decays as $y^* \to -\infty$. This means that we can have a boundary layer only along $y = 1$—that is, $y_0 = 1$—and (136a) must satisfy the boundary condition at $y = 1$. We find

$$u_0^*(x, y^*) = A(x) + [g(x) - A(x)]e^{y^*} \tag{136b}$$

The solution (136b) must match with the outer limit. Since we introduced the boundary layer at $y = 1$, we except the outer limit to satisfy the lower boundary condition. In fact, in this case the outer limit is just

$$u_0(x, y) = f(x) \tag{137}$$

Now, we must match (136b) with (137) as $\varepsilon \to 0$ with x fixed and

$$y_\eta \equiv \frac{y - 1}{\eta(\varepsilon)}$$

fixed. Neither (136b) nor (137) is valid near $x = 0$ or $x = 1$, and we exclude neighborhoods of these lines for the time being. Thus, the matching condition [see (99)] in some overlap domain for η is

$$\lim_{\varepsilon \to 0} \left\{ u_0(x, 1 + \eta y_\eta) - u_0^*\left(x, \frac{\eta y_\eta}{\varepsilon}\right) \right\} \tag{138}$$

with x fixed $\neq 0$, 1 and y_η fixed $\neq 0$, ∞. Using (136b) and (137), we see that the matching requires that we set $A(x) = f(x)$. Also, in order that e^{y^*} be transcendentally small in the matching, we must have $\varepsilon|\log \varepsilon| \ll \eta \ll 1$. Thus the overlap domain for the matching to $O(1)$ is $\varepsilon|\log \varepsilon| \ll \eta \ll 1$.

The inner limit is now defined explicitly by

$$u_0^*(x, y^*) = f(x) + [g(x) - f(x)]e^{y^*} \tag{139}$$

and contains the outer limit $u_0 = f$. Therefore, (139) is the uniformly valid solution to $O(1)$ (as long as we stay away from $x = 0$ and $x = 1$).

The prescribed boundary conditions at $x = 0$, $x = 1$ can be satisfied only by introducing layers of appropriate thickness in x there.

Again, we assume a layer of unknown thickness $\gamma(\varepsilon)$ at $x = 0$ or $x = 1$:

$$\bar{x} \equiv \frac{x - x_0}{\gamma(\varepsilon)}; \qquad x_0 = 0 \text{ or } 1$$

and we denote $u(x_0 + \gamma\bar{x}, y; \varepsilon) \equiv \bar{u}(\bar{x}, y; \varepsilon)$ to obtain

$$\cdot \frac{\varepsilon}{\gamma^2} \frac{\partial^2 \bar{u}}{\partial \bar{x}^2} + \varepsilon \frac{\partial \bar{u}}{\partial y^2} = \frac{\partial \bar{u}}{\partial y} \tag{140}$$

Now, the distinguished limit corresponds to $\gamma = \varepsilon^{1/2}$, and the limiting equation is

$$\frac{\partial^2 \bar{u}_0}{\partial \bar{x}^2} = \frac{\partial \bar{x}_0}{\partial y} \tag{141}$$

This is a one-dimensional diffusion equation with y as a timelike variable (see Chapter 1).

Consider the solution at the left boundary $x_0 = 0$ (the solution at the right boundary is entirely analogous). Equation (141) must satisfy the left boundary condition—that is, $\bar{u}_0(0, y) = k(y)$—and matching the solution of (141) with the outer limit dictates that $\bar{u}_0(\infty, y) = f(0)$ ($y \neq 0, 1$). This result may be regarded as the boundary condition at $\bar{x} = \infty$ for the diffusion equation over $0 \leq \bar{x} < \infty$. In order to have a well-posed problem, we need to specify the "initial condition" for (141)—that is, $\bar{u}(\bar{x}, 0)$. But (141) is not valid if y is small. In fact, near $x = 0$, $y = 0$, we must introduce the scaling $\bar{\bar{x}} \equiv x/\varepsilon^{1/2}$, $\bar{\bar{y}} \equiv y/\varepsilon^{1/2}$, $\bar{\bar{u}}(\bar{\bar{x}}, \bar{\bar{y}}; \varepsilon) \equiv u(\varepsilon^{1/2}\bar{\bar{x}}, \varepsilon^{1/2}\bar{\bar{y}}; \varepsilon)$ and find that $\bar{\bar{u}}$ obeys

$$\frac{\partial^2 \bar{\bar{u}}}{\partial \bar{\bar{x}}^2} + \frac{\partial^2 \bar{\bar{u}}}{\partial \bar{\bar{y}}^2} = \frac{\partial \bar{\bar{u}}}{\partial \bar{\bar{y}}} \tag{142a}$$

with boundary conditions

$$\bar{\bar{u}}(0, \bar{\bar{y}}) = k(0) \tag{142b}$$

$$\bar{\bar{u}}(\bar{\bar{x}}, 0) = f(0) \tag{142c}$$

Equation (142a) is elliptic [it is the *exact* equation (134a)], and it must be solved subject to the two constant boundary conditions (142b)–(142c) and subject also to a behavior as $\bar{\bar{x}} \to \infty$, $\bar{\bar{y}} \to \infty$ that is compatible with matching with the three limiting solutions u_0, u_0^*, and $\bar{u}_0$ away from the origin. This is a complicated problem, which we do not discuss. The result that really concerns us is the behavior of $\bar{\bar{u}}$ as $\bar{\bar{y}} \to \infty$ with $\bar{\bar{x}}$ fixed. It is this behavior that dictates the initial condition $\bar{u}(\bar{x}, 0)$ once the $\bar{u}$ and $\bar{\bar{u}}$ limits are matched. We state without proof that this analysis yields

$$\bar{u}(\bar{x}, 0) = f(0) \tag{143}$$

Aside from providing this unsurprising result, the solution $\bar{\bar{u}}$ is needed only to describe what happens in a domain of order $\varepsilon^{1/2}$ in x and y near the origin. A similar solution is needed near the other three corners of the square.

The solution of (141), subject to the boundary condition $\bar{u}_0(0, y) = k(y)$ and initial condition (143) is given by combining the results (53) and (64b) of Chapter 1. We find

$$\bar{u}_0(\bar{x}, y) = f(0) \operatorname{erf}\left(\frac{\bar{x}}{2\sqrt{y}}\right) + \frac{\bar{x}}{\sqrt{4\pi}} \int_0^y \frac{k(y-s)e^{-\bar{x}^2/4s}}{s^{3/2}} ds \tag{144a}$$

In particular, if $k = k_0 = $ constant, (144a) becomes

$$\bar{u}_0(\bar{x}, y) = k_0 + (f(0) - k_0)\operatorname{erf}\left(\frac{\bar{x}}{2\sqrt{y}}\right) \tag{144b}$$

Near the edge $x = 1$, we find the limiting solution

$$\hat{u}_0(\hat{x}, y) = f(0)\operatorname{erf}\left(\frac{\hat{x}}{2\sqrt{y}}\right) + \frac{\hat{x}}{\sqrt{4\pi}}\int_0^y \frac{\ell(y - s)e^{-\hat{x}^2/4s}}{s^{3/2}}\,ds \tag{145}$$

where $\hat{x} \equiv (1 - x)/\sqrt{\varepsilon}$.

The uniformly valid solution to $O(1)$ in the interior of the square [except in $O(\varepsilon^{1/2})$ neighborhoods of the four corners] is found by adding (139), (144a), and (145) and subtracting $2f(0)$ because $f(0)$ is the common term in the matching of u_0^* with each of the limits $\bar{u}_0$, $\hat{u}_0$:

$$\tilde{u}_0(x, y; \varepsilon) = f(x) + [g(x) - f(x)]e^{y^*}$$
$$- f(0)\left[\operatorname{erfc}\left(\frac{\bar{x}}{2\sqrt{y}}\right) + \operatorname{erfc}\left(\frac{\hat{x}}{2\sqrt{y}}\right)\right]$$
$$+ \frac{\bar{x}}{\sqrt{4\pi}}\int_0^y \frac{k(y - s)e^{-\bar{x}^2/4s}}{s^{3/2}}\,ds + \frac{\hat{x}}{\sqrt{4\pi}}\int_0^y \frac{\ell(y - s)e^{-\hat{x}^2/4s}}{s^{3/2}}\,ds \tag{146a}$$

For the special case $k = k_0 = $ constant, $\ell = \ell_0 = $ constant, this reduces to

$$\tilde{u}_0(x, y; \varepsilon) = f(x) + [g(x) - f(x)]e^{y^*}$$
$$+ [k_0 - f(0)]\operatorname{erfc}\left(\frac{\bar{x}}{2\sqrt{y}}\right) + [\ell_0 - f(0)]\operatorname{erfc}\left(\frac{\hat{x}}{2\sqrt{y}}\right) \tag{146b}$$

A Heat-Transfer Problem

This problem, taken from the exercises for Section 4.1 of [1], gives a concrete illustration of an interior Dirichlet problem for an axially symmetric Laplace equation with a lower derivative term.

We have steady flow of a viscous incompressible fluid in an infinitely long circular pipe (Poiseuille flow). In dimensionless variables the horizontal flow speed is given by $(1 - r^2)$; thus the flow speed vanishes at the pipe boundary $(r = 1)$ and reaches its maximum on the axis $(r = 0)$. The steady-state temperature distribution $u(x, r; \varepsilon)$ obeys

$$\varepsilon\left(u_{rr} + \frac{1}{r}u_r + u_{xx}\right) = (1 - r^2)u_x \tag{147}$$

and we wish to consider the heat-transfer problem associated with having half the pipe wall $(x > 0)$ kept at a higher temperature than the other half. The dimensionless boundary condition that results is

$$u(x, 1; \varepsilon) = \begin{cases} 0, & \text{if } x < 0 \\ 1, & \text{if } x > 0 \end{cases} \tag{148}$$

In (147) we have normalized the radial and axial distances using the pipe radius R, the speed is normalized using its axial value U, and the temperature T

is normalized using

$$u \equiv \frac{T - T^-}{T^+ - T^-}$$

where T^+ and T^- are the constant wall temperatures for $x > 0$ and $x < 0$, respectively. The small parameter ε is [see (63c) of Chapter 3].

$$\varepsilon \equiv \frac{k/\rho c}{UR} = \frac{1}{RePr} \ll 1$$

where k is the thermal conductivity, ρ is the density, and c is the specific heat for the fluid. This problem is somewhat simpler than the one discussed in Section 8.3.3 because there are no finite boundaries in the x direction.

The outer limit has $u_x = 0$, and we conclude that we must have $u_0(x, r) = 0$. This means that the temperature is everywhere equal to its upstream $(x < 0)$ value, as would be the case if $R \to \infty$, for example; the effect of changing the boundary temperature for $x > 0$ would be felt only in a thin layer near $r = 1$.

Therefore, we introduce the trial radial scaling for $x > 0$ given by

$$r^* \equiv \frac{1 - r}{\eta(\varepsilon)} \tag{149}$$

and hold x fixed. We denote $u(x, 1 - \eta r^*; \varepsilon) \equiv u^*(x, r^*; \varepsilon)$ and write (147) in the form

$$\frac{\varepsilon}{\eta^2} \left(\frac{\partial^2 u^*}{\partial r^{*2}} - \frac{\eta}{1 - \eta r^*} \frac{\partial u^*}{\partial r^*} + \eta^2 \frac{\partial^2 u^*}{\partial x^2} \right) = \eta(2r^* - \eta r^{*2}) \frac{\partial u^*}{\partial x} \tag{150}$$

We see that we must have $\varepsilon/\eta^2 = O_s(\eta)$; that is, $\eta = O_s(\varepsilon^{1/3})$ for the distinguished boundary-layer limit. For simplicity, we set $\eta = \varepsilon^{1/3}$ in (149). Then $u_0^*(x, r^*) \equiv u^*(x, r^*; 0)$ obeys

$$2r^* \frac{\partial u_0^*}{\partial x} = \frac{\partial^2 u_0^*}{\partial r^{*2}}; \qquad x > 0; r^* > 0 \tag{151}$$

The boundary condition (at $r = 1$, $r^* = 0$) is $u_0^*(x, 0) = 0$, and matching with the outer limit requires $u_0^*(x, \infty) = u_0(x, r) = 0$. The solution of (151), subject to the preceding two boundary conditions, can be calculated using similarity arguments (see Problem 1 of Chapter 1). We find

$$u_0^*(x, r^*) = 1 - \frac{6^{1/3}}{\Gamma(1/3)} \int_0^{r^*/x^{1/3}} \exp\left(-\frac{2}{9} s^3 \right) ds \tag{152}$$

where Γ denotes the Gamma function, and this is the uniformly valid solution to $O(1)$.

8.3.4   Slender Body Theory; a Problem with a Boundary Singularity

In Section 2.4.3 we showed that the incompressible irrotational flow over a body of revolution satisfies

$$\Delta u = 0 \qquad \text{in } r \geq \varepsilon F(x) \tag{153a}$$

with the boundary condition at infinity

$$u \to x \qquad \text{as } (x^2 + r^2)^{1/2} \to \infty \tag{153b}$$

and the tangency condition on the body surface

$$\frac{u_r(x, \varepsilon F(x); \varepsilon)}{u_x(x, \varepsilon F(x); \varepsilon)} = \varepsilon F'(x) \tag{153c}$$

Here, $r = \varepsilon F(x)$ defines the body shape for a given $F(x)$ on $0 \le x \le 1$ with $F(0) = F(1) = 0$ and $0 \le F(x) \le 1$. The small parameter ε (which was not exhibited explicitly in Chapter 2) is the ratio of the maximum radius of the body to the body length.

The outer limit has $\varepsilon \to 0$ with x and r fixed at some point off the body surface. In this limit the body shrinks to a needle of zero radius, which does not disturb the free stream—that is, $u_0(x, r) = x$. We therefore construct the outer expansion in the form

$$u(x, r; \varepsilon) = x + \mu_1(\varepsilon) u_1(x, r) + o(\mu_1) \tag{154}$$

where $\mu_1(\varepsilon)$ is to be determined. Since (153a) is linear, u_1 satisfies the homogeneous Laplace equation subject to zero boundary condition at $(x^2 + r^2)^{1/2} \to \infty$

$$\Delta u_1 = 0 \tag{155a}$$

$$u_1 \to 0 \qquad \text{as } (x^2 + r^2)^{1/2} \to \infty \tag{155b}$$

If we attempt to have (154) satisfy the tangency condition on the body surface, we must have

$$\frac{\mu_1 \dfrac{\partial u_1}{\partial r}(x, \varepsilon F) + o(\mu_1)}{1 + \mu_1 \dfrac{\partial u_1}{\partial x}(x, \varepsilon F) + o(\mu_1)} = \varepsilon F' \tag{156a}$$

That is,

$$\mu_1 \frac{\partial u_1}{\partial r}(x, 0) + O(\mu_1 \varepsilon) + o(\mu_1) = \varepsilon F' \tag{156b}$$

For $\varepsilon \ll \mu_1$, we conclude from (156b) that $(\partial u_1 / \partial r)(x, 0) = 0$, which combined with (155) implies $u_1(x, r) = 0$. Thus, the right-hand side of (156b) comes into play only if $\mu_1 = O_s(\varepsilon)$, $\mu_1 = \varepsilon$, for instance. With this choice, we find

$$\frac{\partial u_1(x, 0)}{\partial r} = F'(x) \tag{157}$$

The fact that in (157) the boundary condition for u_1 is evaluated at $r = 0$ leads to difficulties, as seen from the general solution [see (88) of Chapter 2]

$$u_1(x, r) = -\frac{1}{4\pi} \int_0^1 \frac{S_1(\xi) \, d\xi}{\sqrt{(x - \xi)^2 + r^2}} \tag{158}$$

for the potential u_1 in terms of an unknown axial distribution of sources of strength/unit length S_1. It is pointed out in Problem 6 that this potential has the behavior

$$u_1(x,r) = \frac{S_1(x)}{2\pi}\log r - \frac{S_1(0)}{4\pi}\log 2x - \frac{S_1(1)}{4\pi}\log 2(1-x)$$

$$+ T_1(x) + O(r^2) \qquad \text{as } r \to 0 \tag{159a}$$

where

$$T_1(x) \equiv -\frac{1}{4\pi}\int_0^1 S_1(\xi)\,\text{sgn}(x-\xi)\log 2|x-\xi|\,d\xi \tag{159b}$$

If $S_1(x) \neq 0$, $\partial u_1/\partial r$ has a $1/r$ singularity at $r = 0$ and cannot satisfy (157). In addition, we have singularities at $x = 0$ and $x = 1$ unless we set $\lim_{x\to 0} S(x)\log x = 0$ and $\lim_{x\to 1} S(x)\log 2(1-x) = 0$. We shall discuss the implication of the singularities at $x = 0$ and $x = 1$ later on. The behavior of (159) implies that the outer expansion is not valid as $r = 0$. Note that in the exact problem, the singularity in r is still along the $r = 0$ axis, but the boundary condition (153c) is evaluated $r = \varepsilon F(x)$, which is off the axis for $0 < x < 1$. Thus, the boundary singularity in u_1 is a direct consequence of the nonuniform validity of (154) near the body. We abandon the requirement that $\partial u_1/\partial r = 0$ at $r = 0$ (which means μ_1 need not equal ε), and we look for another expansion valid near $r = 0$.

Since the body radius shrinks to zero at a rate proportional to ε, we introduce the inner variable

$$r^* \equiv \frac{r}{\varepsilon}$$

denote

$$u(x, \varepsilon r^*; \varepsilon) \equiv u^*(x, r^*; \varepsilon)$$

and consider the limit as $\varepsilon \to 0$ with r^*, x fixed.
Equation (153a) transforms to

$$\Delta u = \frac{\partial^2 u^*}{\partial r^{*2}} + \frac{1}{r^*}\frac{\partial u^*}{\partial r^*} + \varepsilon^2 \frac{\partial^2 u^*}{\partial x^2} = 0 \tag{160}$$

and the boundary condition (153c) becomes

$$\frac{\dfrac{\partial u^*}{\partial r^*}(x, F; \varepsilon)}{\dfrac{\partial u^*}{\partial x}(x, F; \varepsilon)} = \varepsilon^2 F'(x) \tag{161}$$

We construct an inner expansion in the form

$$u^*(x, r^*; \varepsilon) = u_0^*(x, r^*) + \mu_1^*(\varepsilon)u_1^*(x, r^*) + o(\mu_1^*) \tag{162}$$

Substituting (162) into (160) shows that u_0^* and u_1^* satisfy

$$\frac{\partial^2 u_0^*}{\partial r^{*2}} + \frac{1}{r^*}\frac{\partial u_0^*}{\partial r^*} = 0 \tag{163a}$$

$$\frac{\partial^2 u_1^*}{\partial r^{*2}} + \frac{1}{r^*}\frac{\partial u_1^*}{\partial r^*} = \begin{cases} 0, & \text{if } \varepsilon^2 \ll \mu_1^* \\ -\dfrac{\partial^2 u_0^*}{\partial x^2}, & \text{if } \mu_1^* = O_s(\varepsilon^2) \end{cases} \tag{163b}$$

The boundary conditions at $r = \varepsilon F$—that is, at $r^* = F$—that we derive for u_0^* and u_1^* from (161) are

$$\frac{\partial u_0^*}{\partial r^*}(x, F) = 0 \tag{164a}$$

$$\frac{\partial u_1^*}{\partial r^*}(x, F) = \begin{cases} 0, & \text{if } \varepsilon^2 \ll \mu_1^* \\ F'(x)\dfrac{\partial u_0^*}{\partial x}, & \text{if } \mu_1^* = O_s(\varepsilon^2) \end{cases} \tag{164b}$$

The general solution of (163a) is

$$u_0^*(x, r^*) = A_0(x)\log r^* + B_0(x) \tag{165a}$$

and when (164a) is imposed, we have $A_0 = 0$—that is,

$$u_0^*(x, r^*) = B_0(x) \tag{165b}$$

Matching of the inner limit (165b) with the outer limit $u_0 = x$ implies that $B_0(x) = x$—that is,

$$u_0^* = x \tag{165c}$$

Since now $\partial^2 u_0^*/\partial x^2 = 0$, u_1^* satisfies the homogeneous equation (163b) for any μ_1^*: $\varepsilon^2 \ll \mu_1^*$. Therefore,

$$u_1^*(x, r^*) = A_1(x)\log r^* + B_1(x) \tag{166}$$

The boundary condition (164b) gives $A_1 = 0$ if $\varepsilon^2 \ll \mu_1^*$ or $A_1(x) = F(x)F'(x)$ if $\mu_1^* = \varepsilon^2$. Therefore, the inner expansion has the form

$$u^*(x, r^*; \varepsilon) = x + \mu_{01}^*(\varepsilon)B_{01}(x) + \varepsilon^2[F(x)F'(x)\log r^* + B_1(x)] + o(\varepsilon^2) \tag{167}$$

where we have inserted a solution of $O(\mu_{01}^*)$ with $\varepsilon^2 \ll \mu_{01}^* \ll 1$, satisfying the homogeneous case of the boundary condition (164b), for reasons that will become clear from the matching.

As we have calculated the inner expansion to $O(\varepsilon^2)$, let us attempt a matching to this order also. We must satisfy

$$\lim_{\varepsilon \to 0} \frac{1}{\varepsilon^2}\left\{ x + \mu_1(\varepsilon)u_1(x, \eta r_\eta) - x - \mu_{01}^*(\varepsilon)B_{01}(x) \right.$$

$$\left. - \varepsilon^2\left[F(x)F'(x)\log \frac{\eta r_\eta}{\varepsilon} + B_1(x) \right] \right\} = 0 \tag{168a}$$

for x, r_η fixed and in some overlap domain for $\eta(\varepsilon)$. Here u_1 is given in (158).

We use (159) for the expansion of u_1, assume that $\lim_{x \to 0} S_1(x) \log x = 0$, $\lim_{x \to 1} S_1(x) \log(1 - x) = 0$, and obtain

$$\lim_{\varepsilon \to 0} \left\{ \frac{\mu_1(\varepsilon)}{\varepsilon^2} \left[\frac{S_1(x)}{2\pi} \log \eta r_\eta + T_1(x) \right] \right.$$
$$\left. - \frac{\mu_{01}(\varepsilon)}{\varepsilon^2} B_{01}(x) - F(x)F'(x) \log \eta r_\eta + F(x)F'(x) \log \varepsilon + B_1(x) \right\} = 0 \quad (168b)$$

In order to match the singular terms proportional to $\log \eta r_\eta$, we must set $\mu_1 = \varepsilon^2$ and $S_1(x)/2\pi = F(x)F'(x)$. In order to remove the singular term proportional to $\log \varepsilon$, we must set $\mu_{01}(\varepsilon)/\varepsilon^2 = \log \varepsilon$ and $B_{01}(x) = F(x)F'(x)$. Finally, in order to match the $O(1)$ terms, we must set $B_1(x) = T_1(x)$. The matching to $O(\varepsilon^2)$ is then implemented and defines all the unknowns

$$\mu_1(\varepsilon) = \varepsilon^2 \tag{169a}$$

$$S_1(x) = 2\pi F(x)F'(x) \tag{169b}$$

$$\mu_{01}(\varepsilon) = \varepsilon^2 \log \varepsilon \tag{169c}$$

$$B_{01}(x) = F(x)F'(x) \tag{169d}$$

$$B_1(x) = -\frac{1}{4} \int_0^1 [F^2(\xi)]'' \operatorname{sgn}(x - \xi) \log 2|x - \xi| \, d\xi \tag{169e}$$

The result (169b) relates the source strength to the rate of change of cross-sectional area of the body. In fact, with $a(x) = \pi F^2(x)$, we have $S_1(x) = a'(x)$. Our result remains uniformly valid near $x = 0$ and $x = 1$ only if the prescribed body shape satisfies the conditions $\lim_{x \to 0} a'(x) \log x = 0$, $\lim_{x \to 1} a'(x) \log(1 - x) = 0$. For example, if $F(x)$ behaves like x^α or $(1 - x)^\alpha$ near $x = 0$, $x = 1$, we must restrict α to be larger than $\frac{1}{2}$. A discussion of the behavior of the solution if $\alpha < \frac{1}{2}$ is given in Section 4.3.1b of [2]. The reader can also find there a discussion for the more general problem of a deformable body $F(x, t)$ and a calculation of the force on the body.

We note that for this example, *every* term in the inner expansion to $O(\varepsilon)$ is contained in the two-term outer expansion

$$u(x, r; \varepsilon) = x - \frac{\varepsilon^2}{2} \int_0^1 \frac{F(\xi)F'(\xi)}{[(x - \xi)^2 + r^2]^{1/2}} + o(\varepsilon)^2 \tag{170}$$

Thus, (170) is the uniformly valid solution to $O(\varepsilon^2)$; the matching here serves to determine the source strength. Also, if we are interested only in the flow near the body, as in the calculation of the pressure on the body, we do not need (170) and can use the simpler expression

$$u(x; r^*; \varepsilon) = x + (\varepsilon^2 \log \varepsilon)F(x)F'(x) + \varepsilon^2[F(x)F'(x) \log r^* + B_1(x)] + o(\varepsilon^2) \tag{171}$$

where B_1 is given by (169e).

8.3.5 Burgers' Equation for $\varepsilon \ll 1$

In Section 5.3.6 we studied a class of exact solutions for Burgers' equation corresponding to piecewise constant initial data. Here, we recalculate these results using matching of appropriate limit process equations and show that this approach is correct. We then study a class of signaling problems for which an exact solution is not known (see the discussion in Section 1.6.3) and show that a perturbation solution is also easy to compute.

Initial-Value Problem with a Shock Layer

As pointed out in Section 5.3.6, there is no loss of generality in studying the initial-value problem

$$u_t + uu_x = \varepsilon u_{xx} \tag{172a}$$

$$u(x, 0; \varepsilon) = \begin{cases} 1, & \text{if } x < 0 \\ -1, & \text{if } x > 0 \end{cases} \tag{172b}$$

if the initial condition is piecewise constant on either side of $x = x_0$ with $u(x_0^+, 0) < u(x_0^-, 0)$.

The outer expansion of (172) terminates with the outer limit $u_0(x, t) \equiv u(x, t; 0)$ given by

$$u_0(x, t) = \begin{cases} 1, & \text{if } x < 0 \\ -1, & \text{if } x > 0 \end{cases} \tag{173}$$

and the remainder is transcendentally small. Equation (173) is not uniformly valid if x is small, and we look for an appropriate inner variable

$$x^* \equiv \frac{x}{\eta(\varepsilon)} \tag{174}$$

for $u^*(x^*, t; \varepsilon) \equiv u(\eta x_\eta, t; \varepsilon)$. Substituting (174) into (172a) gives

$$\frac{\partial u^*}{\partial t} + \frac{1}{\eta} u^* \frac{\partial u^*}{\partial x^*} = \frac{\varepsilon}{\eta^2} \frac{\partial^2 u^*}{\partial x^{*2}} \tag{175a}$$

Clearly, the distinguished inner limit corresponds to $\eta = O_s(\varepsilon)$, and we set $\eta = \varepsilon$ to obtain

$$\varepsilon \frac{\partial u^*}{\partial t^*} + u^* \frac{\partial u^*}{\partial x^*} = \frac{\partial^2 u^*}{\partial x^{*2}} \tag{175b}$$

Thus, the inner limit $u_0^*(x^*, t) \equiv u^*(x^*, t; 0)$ obeys

$$u_0^* \frac{\partial u_0^*}{\partial x^*} = \frac{\partial^2 u_0^*}{\partial x^{*2}} \tag{176}$$

Integrating (176) with respect to x^* gives

$$\frac{u_0^{*2}}{2} - \frac{\partial u_0^*}{\partial x^*} = \alpha(t) \tag{177}$$

In order to match with the outer limit, we must have $u_0^* \to \pm 1$ as $x^* \to \mp\infty$ and $\partial u_0^*/\partial x^* \to 0$ as $|x^*| \to \infty$. Therefore, $\alpha(t)$ must equal $\frac{1}{2}$. Integrating (177) for $\alpha = \frac{1}{2}$ with respect to x^* gives

$$u_0^* = -\tanh\left(\frac{x^* + x_0^*}{2}\right) \tag{178a}$$

or

$$u_0^* = -\coth\left(\frac{x^* + x_0^*}{2}\right) \tag{178b}$$

where x_0^* is an arbitrary function of t. We discard the hyperbolic cotangent solution (178b) for this problem because the interior layer must hold over $-\infty < x^* < \infty$, $t > 0$, and (178b) becomes singular when $x^* = -x_0^*$. In fact, (178a) matches with (173) to all algebraic orders *for arbitrary* $x_0^*(t)$.

For this simple example, we can use symmetry arguments to conclude that $x_0^* = 0 \dots$. We note that (172) is invariant under the transformation $x \to -x$, $u \to -u$, $t \to t$. Thus, in (178a) we must have

$$-\tanh\left(\frac{x^* + x_0^*}{2}\right) = \tanh\left(\frac{-x^* + x_0^*}{2}\right)$$

for all x^*; this can hold only if $x_0^* = 0$. In general, for a curved shock, we must determine $x_0^*(t)$ by matching to $O(\varepsilon)$ (see the example discussed in [1]).

The small parameter ε is "artificial" in (172) in the sense that it can be removed from the governing equation and initial condition (for the special condition (172b)) through the transformation $u \to u$, $t \to t/\varepsilon$, $x \to x/\varepsilon$. In fact, note that the exact solution given by (128) of Chapter 5 is free of ε if we use the variables t/ε, x/ε. Thus, the limit $u_0 = -\tanh x^*/2$, which we also derived from the exact solution in Chapter 5, is the uniformly valid approximation for all x and all $t > 0$ to all algebraic orders in ε; it differs from the exact expression only by transcendentally small terms if t is not small. The exact expression is needed only if $t = O_s(\varepsilon)$ and $x = O_s(\varepsilon)$.

A limiting expression such as (178a) is easier to derive in general and it gives the *shock structure* for the solution directly even if we cannot solve the exact problem in the entire domain. A discussion of the shock structure for one-dimensional viscous heat-conducting flow, equations (56) of Chapter 3, is given in Section 6.15 of [8]. This solution describes how the flow variables actually vary in a thin shock layer. In the limit $Re \to \infty$ with x and t fixed, this thin layer shrinks to the shock of zero thickness that we calculated in Sections 5.3.4 and 5.3.5 (see also Problem 15, where the simpler problem of an isothermal shock layer is outlined).

Initial-Value Problem with a Corner Layer

If we replace (172b) with

$$u(x, 0; \varepsilon) = \begin{cases} -1, & \text{if } x < 0 \\ +1, & \text{if } x > 0 \end{cases} \tag{179}$$

the outer limit for $t > 0$ is

$$u_0(x, t) = \begin{cases} 1, & \text{if } x \geq 0 \\ \dfrac{x}{t}, & \text{if } -x \leq t \leq x \\ -1, & \text{if } x \leq -t \end{cases} \qquad (180)$$

and it is easily seen that the outer expansion again terminates with u_0. Although (180) is valid to $O(1)$ everywhere, the two partial derivatives $\partial u_0/\partial t$ and $\partial u_0/\partial x$ do not exist if $|x| = t$.

Near $x = \pm t$ we need to introduce a *corner layer* to smoothly join the solutions on either side. We look for rescaled independent variables in the form

$$x_c \equiv \frac{x \mp t}{\eta(\varepsilon)}; \qquad t_c \equiv t \qquad (181)$$

Since the layer is now near $x = \pm t$, where $u = \pm 1$ if $t > 0$, we expand u in the form

$$u(x, t; \varepsilon) = \pm 1 + \delta(\varepsilon)u_c(x_c, t_c) + o(\delta) \qquad (182)$$

We compute

$$\frac{\partial u}{\partial t} = \delta \frac{\partial u_c}{\partial t_c} \mp \frac{\delta}{\eta}\frac{\partial u_c}{\partial x_c}; \qquad \frac{\partial u}{\partial x} = \frac{\delta}{\eta}\frac{\partial u_c}{\partial x_c}; \qquad \frac{\partial^2 u}{\partial x^2} = \frac{\delta}{\eta^2}\frac{\partial^2 u}{\partial x_c^2}$$

Therefore, (172a) transforms to

$$\delta \frac{\partial u_c}{\partial t_c} + \frac{\delta^2}{\eta}u_c\frac{\partial u_c}{\partial x_c} = \frac{\delta\varepsilon}{\eta^2}\frac{\partial^2 u_c}{\partial x_c^2} + o(\delta) \qquad (183)$$

and we see that we must set $\delta = \eta = \varepsilon^{1/2}$ for the richest equation, which is the full equation

$$\frac{\partial u_c}{\partial t} + u_c\frac{\partial u_c}{\partial x_c} = \frac{\partial^2 u_c}{\partial x_c^2} \qquad (184)$$

One cannot avoid solving the exact Burgers' equation in order to calculate the corner layer. This calculation was discussed in Section 5.3.6, where we found [see (140) of Chapter 5]

$$u_c(x_c, t_c) = -\frac{2}{\sqrt{\pi t_c}}\frac{\exp(-x_c^2/4t_c)}{\operatorname{erfc}(-x_c/2t_c^{1/2})} \qquad (185)$$

It was also pointed out in Chapter 5 that using (185) in (182) gives $u \to 1$ as $x_c \to \infty, t$ fixed > 0, and $u \to x/t$ as $x_c \to -\infty, t$ fixed > 0. This is just the matching condition for the corner layer expansion (182) with the outer limit on either side. Similar results apply for the corner layer centered at $x = -t$.

In a sense, the calculation of (185) is no longer a perturbation problem. However, we need (185) only to demonstrate that u_x and u_t join smoothly across the corner layer and to find a better approximation than $u = \pm 1$ at $x = \pm t$.

Signaling Problem with Constant Initial and Boundary Data

It was pointed out in Section 1.6.3 that the signaling problem for Burgers' equation on the semi-infinite interval leads to an integral equation that cannot be solved explicitly in general. In contrast, the perturbation solution for this case is not any harder than for the initial-value problem. We illustrate ideas for the simple case where the initial and boundary values for u are constants.

We wish to compute the leading approximation for

$$u_t + uu_x = \varepsilon u_{xx}; \qquad 0 \le x < \infty; \qquad 0 \le t \le \infty \qquad (186a)$$

with initial condition

$$u(x,0) = A \qquad (186b)$$

and boundary conditions

$$u(0,t) = B \qquad (186c)$$

$$u(\infty,t) = A \qquad (186d)$$

where A and B are arbitrary constants.

This problem is discussed in [1], where it is shown that the solution has a different structure in each of the five regions of the A, B plane shown in Figure 8.4. Actually, if $A \ne 0$, the solution depends only on the ratio (B/A) because we can transform A out of (186) by setting $u \to Au$, $t \to t/A^2$, $x \to x/A$. However, in this discussion we let A and B vary independently.

First, we note that if $A = B$, the exact solution of (186) is $u = \text{constant} = A$. In the adjacent region I, we have $0 < B < A$, and the outer limit is given by

$$u_0(x,t) = \begin{cases} A, & \text{if } At \le x < \infty \\ \dfrac{x}{t}, & \text{if } Bt \le x \le At \\ B, & \text{if } 0 \le x \le Bt \end{cases} \qquad (187)$$

Therefore, we need to introduce two corner layers of the type discussed earlier along the lines $x = At$ and $x = Bt$.

In region II, we have $0 < |A| < B$, and the characteristics of the $\varepsilon = 0$ problem cross. Therefore, we must introduce a shock along $x = (A + B)t/2$, and the outer limit is

$$u_0(x,t) = \begin{cases} A, & \text{if } \dfrac{(A + B)t}{2} < x < \infty \\ B, & \text{if } 0 \le x < \dfrac{(A + B)t}{2} \end{cases} \qquad (188)$$

The shock layer that smooths the discontinuity at $x = (A + B)t/2$ is just a general version of the hyperbolic tangent layer discussed earlier. Matching with the outer limit on either side determines the constants as discussed earlier, and

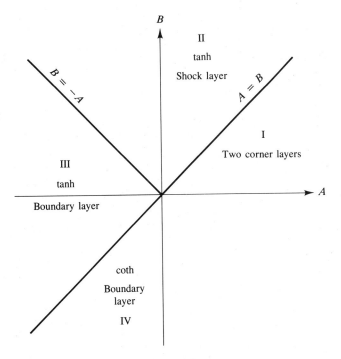

Figure 8.4

we find

$$u_0^* = \frac{A + B}{2} + \frac{A - B}{2} \tanh\left(\frac{B - A}{2}\right)x^* \tag{189a}$$

where

$$x^* = \frac{x - (A + B)t/2}{\varepsilon} \tag{189b}$$

In region III, we have $A < 0$ and $|B| < -A$. The characteristics of the $\varepsilon = 0$ problem emerge from the x-axis and point to the left as t increases. Thus, we cannot satisfy both conditions (186b)–(186c) with the outer limit. Moreover, no shock is possible for $x > 0$, $t > 0$, so we must have a boundary layer at $x = 0$. We construct this boundary layer using the general hyperbolic tangent solution of (176):

$$u_0^*(x^*, t) = -\alpha(t) \tanh \alpha(t) \frac{x^* + x_0^*(t)}{2} \tag{190}$$

The boundary condition $u(0, t) = B$ gives

$$B = -\alpha \tanh \alpha \frac{x_0^*}{2} \tag{191}$$

and the matching condition $u_0^*(\infty, t) = A$ gives

$$\alpha = -A > 0 \tag{192a}$$

With this value for α, we solve (191) for x_0^* and find

$$x_0^* = -\frac{2}{A}\tanh^{-1}\frac{B}{A} \tag{192b}$$

This result is real only if $|B/A| \leq 1$. One can show that in region IV, where $A < 0$, $B < 0$, $(B/A) > 1$, we must use a hyperbolic cotangent boundary layer at $x = 0$. For more details and a discussion of the case $A > 0$, $B < 0$, see Section 4.1.3 of [1].

8.4   Cumulative Perturbations; Solution Valid in the Far Field

In Section 3.5.4 we studied the signaling problem for a wavemaker in the limiting case of small-amplitude waves on shallow water. Using a regular perturbation expansion for the speed u and height h, we found that our results were not uniformly valid in the far field: $x = O(\varepsilon^{-1})$.

The reason for this nonuniformity is that a regular perturbation is based on the limit $\varepsilon \to 0$ with x and t fixed. Typically, small nonlinear terms in the equations introduce slow modulations, which are defined by bounded functions of εx and εt in the solution, such as $\sin \varepsilon x$ and $e^{-\varepsilon t}$. Such functions cannot be uniformly approximated by the limit process $\varepsilon \to 0$, x fixed or t fixed, if x or t are $O(\varepsilon^{-1})$.

Often the solution depends simultaneously on a "fast time" t and a "slow time" εt in the form $f(x, t, \varepsilon t)$, say. For example, we may have a damped wave: $u(x, t; \varepsilon) = \varepsilon^{-\varepsilon t}\sin(x - t)$. In such cases we cannot use a limit process expansion to express u in terms of x and $t_1 = \varepsilon t$ because $\lim_{\varepsilon \to 0}\sin t_1/\varepsilon$, with t_1 fixed $\neq 0$, does not exist.

Slowly modulated oscillations defined by systems of ordinary differential equations can be solved using *averaging* or *multiple scale* expansions. The basic ideas are discussed in Chapter 3 of [1], and a unified treatment of the two methods, including some more recent results, is given in the expository article [9]. We discuss a simple example in Sections 8.4.1–8.4.3 to illustrate ideas.

Some of these ideas carry over directly to the solution of weakly nonlinear partial differential equations. However, this theory is much less comprehensive with many open questions. In Section 8.4.4 we shall consider the perturbation solution of the shallow water equations to illustrate ideas.

8.4.1 The Oscillator with a Weak Nonlinear Damping: Regular Expansion

Consider the initial-value problem

$$\ddot{u} + u + \varepsilon\dot{u}|\dot{u}| = 0; \qquad 0 \leq t; \qquad 0 < \varepsilon \ll 1 \tag{193a}$$

$$u(0) = a > 0 \tag{193b}$$

$$\dot{u}(0) = 0 \tag{193c}$$

If we look for a regular expansion for u in the form

$$u(t; \varepsilon) = u_0(t) + \varepsilon u_1(t) + O(\varepsilon^2) \tag{194}$$

—that is, $u_0 = \lim_{\varepsilon \to 0} u(t; \varepsilon)$ with t fixed, and so on—we find that u_0 and u_1 satisfy

$$\ddot{u} + u_0 = 0 \tag{195a}$$

$$u_0(0) = a \tag{195b}$$

$$\dot{u}(0) = 0 \tag{195c}$$

$$\ddot{u}_1 + u_1 = -\dot{u}_0|\dot{u}_0| \tag{196a}$$

$$u_1(0) = 0 \tag{196b}$$

$$\dot{u}_1(0) = 0 \tag{196c}$$

Thus,

$$u_0(t) = a \cos t \tag{197}$$

and (196a) becomes

$$\ddot{u}_1 + u_1 = a^2 \sin t |\sin t| \tag{198}$$

The right-hand side of (198) is the *odd* 2π-periodic function that equals $a^2 \sin^2 t$ on $0 \le t \le \pi$ and $-a^2 \sin^2 t$ on $-\pi \le t \le 0$. Therefore, we can develop this right-hand side in the Fourier sine series

$$a^2 \sin t |\sin t| = \sum_{n=1}^{\infty} b_n \sin nt \tag{199a}$$

where

$$\begin{aligned}
b_n &= \frac{2a^2}{\pi} \int_0^{\pi} \sin^2 t \sin nt \, dt \\
&= \frac{a^2}{\pi} [(-1)^n - 1] \frac{n - n^2 + 4}{n(n^2 - 4)}, \qquad n \ne 2
\end{aligned} \tag{199b}$$

Thus, $b_n = 0$ for n even, and we can write (199a) as

$$a^2 \sin t |\sin t| = \frac{4a^2}{\pi} \sum_{n=0}^{\infty} b_{2k+1} \sin(2k + 1)t \tag{200a}$$

$$b_{2k+1} = \frac{2k^2 + k - 2}{(2k - 1)(2k + 1)(2k + 3)} \tag{200b}$$

The solution of (196) then has the form

$$\begin{aligned}
u_1 &= -\frac{4}{3\pi} a^2 t \cos t - \frac{4a^2}{\pi} \sum_{k=1}^{\infty} \frac{b_{2k+1}(2k + 1)}{4k(k + 1)} \sin(2k + 1)t \\
&\quad + \left[\frac{4a^2}{3\pi} + \frac{4a^2}{\pi} \sum_{k=1}^{\infty} \frac{b_{2k+1}(2k + 1)}{4k(k + 1)} \right] \sin t
\end{aligned} \tag{201}$$

The first term on the right-hand side of (201) is proportional to $t \cos t$. Therefore, εu_1 has a term proportional to $\varepsilon t \cos t$. This term becomes $O(1)$ when $t = O(\varepsilon^{-1})$, so our expansion is not uniformly valid for t large.

We know that the small positive damping term in (193a) must introduce a slow decay in the amplitude. Thus, instead of $u_0 = a \cos t$ with $a = $ constant, we must have a first approximation $u_0 = A \cos t$, where the amplitude A decreases slowly with time and satisfies $A(0) = a$. The term $-(4/3\pi)a^2 \varepsilon t \cos t$ is evidently the second term in a nonuniform expansion of A and shows that A must depend on εt. This reasoning prompts us to look for a solution that depends on $t_1 \equiv \varepsilon t$. But, we have seen that we cannot express a 2π-periodic function by a limit process where $\varepsilon \to 0$ with t_1 fixed. Therefore, we are led to look for a solution that depends *simultaneously* on $t_0 \equiv t$ and $t_1 \equiv \varepsilon t$, a so-called multiple scale expansion. A discussion of the various source references for this method is given in Section 3.2 of [1] and the introduction of [9].

8.4.2 The Multiple Scale Expansion

Basically, we assume that the solution of (193) has a certain structure—it involves oscillations that occur over the "fast scale" t_0, and these oscillations decay over the "slow scale" t_1. Dimensional analysis can be used to argue that the small parameter ε is the ratio of two characteristic times: the times T_0 associated with the oscillatory behavior and T_1 associated with the amplitude decay (see Section 3.2 of [1]).

At any rate, we *assume* that the solution of (193) has the form

$$u(t; \varepsilon) = U(t_0, t_1; \varepsilon) \tag{202}$$

and we expand U with respect to its ε-dependence in the form

$$U = U_0(t_0, t_1) + \varepsilon U_1(t_0, t_1) + O(\varepsilon^2) \tag{203}$$

We emphasize the fact that (203) is not a limit process expansion because it does not correspond to having either t_0 or t_1 fixed as $\varepsilon \to 0$. In fact, we regard (203) as a generalized asymptotic expansion in the sense defined by (133). We shall demonstrate that when the limit process $\varepsilon \to 0$, t_0 fixed, is applied to (203), we obtain (194) but that the limit process $\varepsilon \to 0$, t_1 fixed, does not exist.

The assumed form of (202) implies that derivatives are given by

$$\dot{u} = \frac{\partial U}{\partial t_0} + \varepsilon \frac{\partial U}{\partial t_1} = \frac{\partial U_0}{\partial t_0} + \varepsilon \left(\frac{\partial U_1}{\partial t_0} + \frac{\partial U_0}{\partial t_1} \right) + O(\varepsilon^2) \tag{204a}$$

$$\begin{aligned}
\ddot{u} &= \frac{\partial^2 U}{\partial t_0^2} + 2\varepsilon \frac{\partial^2 U}{\partial t_0 \partial t_1} + \varepsilon^2 \frac{\partial^2 U}{\partial t_1^2} \\
&= \frac{\partial^2 U_0}{\partial t_0^2} + \varepsilon \left(\frac{\partial^2 U_1}{\partial t_0^2} + 2 \frac{\partial^2 U_0}{\partial t_0 \partial t_1} \right) + O(\varepsilon^2)
\end{aligned} \tag{204b}$$

Using these expressions in (193) leads to the following equations governing U_0 and U_1:

$$\frac{\partial^2 U_0}{\partial t_0^2} + U_0 = 0 \tag{205}$$

$$\frac{\partial^2 U_1}{\partial t_0^2} + U_1 = -2 \frac{\partial^2 U_0}{\partial t_0 \partial t_1} - \frac{\partial U_0}{\partial t_0} \left| \frac{\partial U_0}{\partial t_0} \right| \tag{206}$$

The initial conditions (193b)–(193c) give

$$U_0(0,0) = a \tag{207a}$$

$$\frac{\partial U_0}{\partial t_0}(0,0) = 0 \tag{207b}$$

$$U_1(0,0) = 0 \tag{208a}$$

$$\frac{\partial U_1}{\partial t_0}(0,0) = -\frac{\partial U_0}{\partial t_1}(0,0) \tag{208b}$$

In (205), t_1 occurs as a parameter (there are no derivatives with respect to t_1), so that this equation is actually an ordinary differential equation with integration "constants" that are functions of t_1. We express the solution of (205) in the form

$$U_0(t_0, t_1) = A_0(t_1)\cos[t_0 + \phi_0(t_1)] \tag{209}$$

where A_0 and ϕ_0 are the slowly varying amplitude and phase shift, which are unknown at this stage. We know only that $A_0(0) = a$ and $\phi_0(0) = 0$, according to (207).

To calculate U_1 we use (209) to express the right-hand side of (206) explicitly. This gives [see (200) for the definition of the Fourier coefficients]

$$\frac{\partial^2 U_1}{\partial t_0^2} + U_1 = 2A_0'(t_1)\sin[(t_0 + \phi_0(t_1)] + 2A_0(t_1)\phi_0'(t_1)\cos[t_0 + \phi_0(t_1)]$$

$$+ \frac{8}{3\pi} A_0^2(t_1)\sin[t_0 + \phi_0(t_1)]$$

$$+ \frac{4A_0^2(t_1)}{\pi} \sum_{k=1}^{\infty} b_{2k+1} \sin(2k + 1)[t_0 + \phi_0(t_1)] \tag{210}$$

where a prime denotes differentiation with respect to t_1.

Again, t_1 is just a parameter as far as the integration of (210) with respect to t_0 is concerned. We know that the terms proportional to $\sin(t_0 + \phi_0)$ and $\cos(t_0 + \phi_0)$ on the right-hand side of (210) will give rise to terms proportional to $t_0\cos(t_0 + \phi_0)$ and $t_0\sin(t_0 + \phi_0)$, respectively, in the solution for U_1. These *mixed secular* terms render the solution invalid for $t = O(\varepsilon^{-1})$. The idea is to choose A_0 and ϕ_0 so that the solution for U_1 is free of mixed secular terms. In this case, it is seen that we must set

$$2A_0' + \frac{8}{3\pi} A_0^2 = 0 \tag{211a}$$

$$A_0\phi_0' = 0 \tag{211b}$$

The solution of (211a) subject to $A_0(0) = a$ is

$$A_0(t_1) = \frac{a}{4at_1/3\pi + 1} \tag{212a}$$

and the solution of (211b) with $A_0 \neq 0$ and $\phi_0(0) = 0$ is

$$\phi_0(t_1) = 0 \tag{212b}$$

Therefore, we have determined U_0 as a function of t and ε in the form

$$U_0 = \frac{a}{4a\varepsilon t/3\pi + 1} \cos t \tag{213}$$

Note, in particular, that in the limit as $\varepsilon \to 0$ with t fixed, we find

$$U_0 = a \cos t - \frac{4}{3\pi} \varepsilon a^2 t \cos t + \cdots \tag{214}$$

but the limit of (213) as $\varepsilon \to 0$ with εt fixed does not exist. Equation (214) gives precisely the mixed secular term that rendered the expansion (194) invalid for t large. For this example, the amplitude is a decreasing function of t_1, as is physically obvious. We have, in fact, shown that the amplitude behaves like $(\varepsilon t)^{-1}$ for t large if the damping is quadratic. In contrast, if the damping term in (193) is linear $(2\varepsilon\dot{u})$, the amplitude decays like $e^{-\varepsilon t}$.

Once the troublesome terms are removed from (210), we can solve this equation in the form

$$U_1(t_0, t_1) = A_1(t_0)\cos[t_0 + \phi_1(t_1)] - \frac{A_0^2(t_0)}{\pi} \sum_{k=1}^{\infty} \frac{b_{2k+1}}{k(k+1)} \sin(2k+1)t_0 \tag{215}$$

where A_1 and ϕ_1 are functions of t_1 to be determined by requiring the solution to $O(\varepsilon^2)$ to be uniformly valid. This is as far as we shall proceed with the calculation of the expansion (203). The procedure is straightforward and can be implemented, at least in principle, to higher orders. Actually, as pointed out in [1], we need to refine the assumed form (202) to depend also on $t_2 = \varepsilon^2 t, t_3 = \varepsilon^2 t$, and so on, if we wish to compute higher-order terms. Thus, the dependence of U_0 on t_2 is determined by requiring that U_1 be uniformly valid for t_1 large, and so on. If the perturbation term in (193) is of the form $\varepsilon f(u, \dot{u})$, U depends only on the "strained coordinate" $t^+ \equiv (1 + \varepsilon^2\omega_1 + \varepsilon^3\omega_2 + \cdots)t$ and εt. For more details the reader is referred to Chapter 3 of [1].

8.4.3 Near-Identity Averaging Transformations

A more systematic procedure for calculating the uniformly valid asymptotic expansion of (193) involves the use of near-identity averaging transformations, which we discuss next for this example. This procedure, often denoted as the *method of averaging*, was first proposed in rudimentary form in [10]. The point of view that we use here has evolved in the more recent literature, which is surveyed in the introduction of [9].

Reduction to Standard Form; Averaged Equations to $O(1)$

One of the crucial requirements for applying the averaging procedure is that the system of equations be in the standard form

$$\frac{dp_m}{dt} = \varepsilon f_m(p_i, q_i; \varepsilon); \qquad m = 1, \dots, M \tag{216a}$$

$$\frac{dq_n}{dt} = \omega_n(p_i) + \varepsilon g_n(p_i, q_i; \varepsilon); \qquad n = 1, \dots, N \tag{216b}$$

Here, the subscript i indicates that all components are present. The functions f_m and g_n are periodic in each of the q_i with the same period 2π. As indicated in Section 6.2.6, if $M = N$ and (216) is Hamiltonian, the p_i and q_i are normalized action and angle variables.

For our example problem, the transformation of dependent variables $(u, \dot{u}) \leftrightarrow (p, q)$ defined by

$$u = p \cos q \tag{217a}$$

$$\dot{u} = -p \sin q \tag{217b}$$

(with $p \geq 0$) is convenient and reduces (193) to standard form, as shown next.

In order that (217a) and (217b) be compatible, we must have the expression for $\dot{u}$, obtained by differentiating (217a), agree with (217b); that is,

$$\dot{p} \cos q - p\dot{q} \sin q = -p \sin q \tag{218a}$$

Next, we use the expression

$$\ddot{u} = -\dot{p} \sin q - p\dot{q} \cos q$$

obtained by differentiating (217b), to write (193a) as

$$-\dot{p} \sin q - p\dot{q} \cos q + p \cos q = \varepsilon p^2 \sin q |\sin q|$$

$$= \varepsilon \frac{4p^2}{\pi} \sum_{k=0}^{\infty} b_{2k+1} \sin(2k + 1)q \tag{218b}$$

where we defined the constants b_{2k+1} in (200b). When we solve (218) for $\dot{p}$ and $\dot{q}$ and collect equal harmonics, we find, after some algebra,

$$\dot{p} = -\frac{4\varepsilon}{3\pi} p^2 + \frac{2\varepsilon p^2}{\pi} \sum_{m=0}^{\infty} c_m \cos 2(m + 1)q \tag{219a}$$

$$\dot{q} = 1 + \frac{2\varepsilon p}{\pi} \sum_{m=0}^{\infty} d_m \sin 2(m + 1)q \tag{219b}$$

where

$$c_m = b_{2m+1} - b_{2m+3} \tag{220a}$$

$$d_m = -b_{2m+1} - b_{2m+3} \tag{220b}$$

The result (219) is in standard form. In fact, it gives us the same information that we obtained in Section 8.4.2 in the following sense. The terms under the summation signs on the right-hand sides of (219) have a *zero average* over one 2π period in q. Therefore, we argue that ignoring these terms produces no cumulative errors in the expressions for p and q that we compute from such an approximation. This gives us the *averaged system* to $O(1)$:

$$\dot{p} = -\frac{4\varepsilon}{3\pi}p^2 + O(\varepsilon) \tag{221a}$$

$$\dot{q} = 1 + O(\varepsilon) \tag{221b}$$

where the $O(\varepsilon)$ terms have a zero average.

We can actually prove rigorously (see Chapter 8 of [11]) for rather general functions f_m and g_n that the solution of the averaged system is the asymptotic approximation of the exact solution to $O(1)$ and that this solution remains uniformly valid over the interval $0 \le t \le T(\varepsilon)$, where $T(\varepsilon) = O_s(\varepsilon^{-1})$ as $\varepsilon \to 0$.

The solution of (221), subject to $p(0) = a$; $q(0) = 0$, as implied by (193b)–(193c) and (217a)–(217b), gives

$$p = \frac{a}{4a\varepsilon t/3\pi + 1} + O(\varepsilon) \tag{222a}$$

$$q = t + O(\varepsilon) \tag{222b}$$

This means that

$$u = \frac{a}{4a\varepsilon t/3\pi + 1}\cos t + O(\varepsilon) \tag{223}$$

is in agreement with (213).

Notice that it is crucial to retain the averaged term of order ε in (221a) in order to obtain the uniformly valid result (222a) for p to $O(1)$; only terms with a zero average in q may be ignored in the equations for p and q to any order. Another way to view (221a) is to note that it is actually the statement: $dp/dt_1 = -4p^2/3\pi +$ oscillatory terms of order unity with zero average.

The Averaged Equations to $O(\varepsilon)$; Near-Identity Transformations

The essential step in computing a better approximation for p and q is to find a transformation $(p, q) \to (P, Q)$ that takes (219) to the following averaged form to order ε:

$$\dot{P} = -\frac{4\varepsilon P^2}{3\pi} + O(\varepsilon^2) \tag{224a}$$

$$\dot{Q} = 1 + O(\varepsilon^2) \tag{224b}$$

where the $O(\varepsilon^2)$ terms have a zero average.

Since this transformation does not alter (221), it must reduce to the identity transformation $p = P$, $q = Q$ if $\varepsilon = 0$. Thus, we look for a "near-identity" transformation:

$$P = p + \varepsilon T(p, q) \tag{225a}$$

$$Q = q + \varepsilon L(p, q) \tag{225b}$$

where the functions T and L are to be determined by the requirement that $\dot{P}$ and $\dot{Q}$ have the form (224). The inverse transformation to $O(\varepsilon)$ is given by

$$p = P - \varepsilon T(P, Q) + O(\varepsilon^2) \tag{226a}$$

$$q = Q - \varepsilon L(P, Q) + O(\varepsilon^2) \tag{226b}$$

If we succeed in transforming (219) to (224), then we can say that $P(t; \varepsilon)$ and $Q(t; \varepsilon)$ have the form

$$P(t; \varepsilon) = \frac{P(0; \varepsilon)}{4P(0; \varepsilon)\varepsilon t/3\pi + 1} + O(\varepsilon^2) \tag{227a}$$

$$Q(t; \varepsilon) = Q(0; \varepsilon) + t + O(\varepsilon^2) \tag{227b}$$

where the initial values $P(0; \varepsilon)$, $Q(0; \varepsilon)$ follow from (225):

$$P(0; \varepsilon) = a + \varepsilon T(a, 0) \tag{228a}$$

$$Q(0; \varepsilon) = \varepsilon L(a, 0) \tag{228b}$$

Therefore, the solution for $p(t; \varepsilon) q(t; \varepsilon)$ is given by (226) in the following form explicit to $O(\varepsilon)$:

$$p(t; \varepsilon) = P(t; \varepsilon) - \varepsilon T(P(t; \varepsilon), Q(t; \varepsilon)) + O(\varepsilon^2) \tag{229a}$$

$$q(t; \varepsilon) = Q(t; \varepsilon) - \varepsilon L(P(t; \varepsilon), Q(t; \varepsilon)) + O(\varepsilon^2) \tag{229b}$$

In order for this result to be uniformly valid for $t = O(\varepsilon^{-1})$, T and L must be bounded functions of time, and the terms of order ε^2 must have a zero average. We demonstrate next that it is possible to determine functions T and L satisfying the above requirements for our example problem (219).

To derive the equations (224), we first differentiate (225) to find

$$\dot{P} = \dot{p} + \varepsilon \left[\frac{\partial T}{\partial p}(p, q)\dot{p} + \frac{\partial T}{\partial q}(p, q)\dot{q} \right] \tag{230a}$$

$$\dot{Q} = \dot{q} + \varepsilon \left[\frac{\partial L}{\partial p}(p, q)\dot{p} + \frac{\partial L}{\partial q}(p, q)\dot{q} \right] \tag{230b}$$

Now we substitute (219a)–(219b) for $\dot{p}$ and $\dot{q}$ into the right-hand sides of (230). This gives

$$\dot{P} = \varepsilon \left[-\frac{4}{3\pi}p^2 + 2\frac{p^2}{\pi} \sum_{m=0}^{\infty} c_m \cos 2(m+1)q \right] \left[1 + \varepsilon \frac{\partial T}{\partial p}(p, q) \right]$$

$$+ \varepsilon \left[1 + \frac{2\varepsilon p}{\pi} \sum_{m=0}^{\infty} d_m \sin 2(m+1)q \right] \frac{\partial T}{\partial q}(p, q) \tag{231a}$$

$$\dot{Q} = \left[1 + 2\frac{\varepsilon p}{\pi} \sum_{m=0}^{\infty} d_m \sin 2(m+1)q \right] \left[1 + \varepsilon \frac{\partial L}{\partial q}(p, q) \right]$$

$$+ \varepsilon^2 \left[-\frac{4}{3\pi}p^2 + \frac{2p^2}{\pi} \sum_{m=0}^{\infty} c_m \cos 2(m+1)q \right] \frac{\partial L}{\partial p}(p, q) \tag{231b}$$

We transform the right-hand sides of (231a)–(231b) to functions of P and Q by substituting (226a)–(226b) for p and q. The result, correct to $O(\varepsilon^2)$, is found after some algebra in the form:

$$\dot{P} = \varepsilon\left[-\frac{4P^2}{3\pi} + \frac{2P^2}{\pi}\sum_{m=0}^{\infty} c_m\cos 2(m+1)Q + T_Q\right]$$

$$+ \varepsilon^2\left\{\frac{8P}{3\pi}T - \frac{4P^2}{3\pi}T_P + \left(\frac{2P^2}{\pi}T_P - \frac{4P}{\pi}T\right)\sum_{m=0}^{\infty} c_m\cos 2(m+1)Q\right.$$

$$+ \frac{2P}{\pi}\sum_{m=0}^{\infty}[2PL(m+1)c_m + T_Q d_m]\sin 2(m+1)Q$$

$$\left.- TT_{PQ} - LT_{QQ}\right\} + O(\varepsilon^3) \tag{232a}$$

$$\dot{Q} = 1 + \varepsilon\left[\frac{2P}{\pi}\sum_{m=0}^{\infty} d_m\sin 2(m+1)Q + L_Q\right]$$

$$+ \varepsilon^2\left\{\frac{2P}{\pi}\sum_{m=0}^{\infty}[PL_P c_m - 2L(m+1)d_m]\cos 2(m+1)Q\right.$$

$$+ \frac{2}{\pi}\sum_{m=0}^{\infty}[PL_Q - T]d_m\sin 2(m+1)Q$$

$$\left.- \frac{4P^2}{3\pi}L_P - TL_{PQ} - LL_{QQ}\right\} + O(\varepsilon^3) \tag{232b}$$

where

$$T \equiv T(P,Q); \ L \equiv L(P,Q); \ T_P \equiv \frac{\partial T}{\partial p}(P,Q); \ T_Q \equiv \frac{\partial T}{\partial q}(P,Q); \ T_{PQ} \equiv \frac{\partial^2 T}{\partial p\partial q}(P,Q)$$

and so on.

The functions T and L are chosen so that the terms of order ε in (232a)–(232b) do not involve the periodic functions of Q with zero average. Thus, quadrature gives

$$T(P,Q) = -\frac{2P^2}{\pi}\sum_{m=0}^{\infty}\frac{c_m}{2(m+1)}\sin 2(m+1)Q + \underline{T}(P) \tag{233a}$$

$$L(P,Q) = \frac{2P}{\pi}\sum_{m=0}^{\infty}\frac{d_m}{2(m+1)}\cos 2(m+1)Q + \underline{L}(P) \tag{233b}$$

where $\underline{T}$ and $\underline{L}$ are arbitrary functions of P. We next choose $\underline{T}$ and $\underline{L}$ so that the terms of order ε^2 in (232) have a zero average. Using familiar identities for products of sines and cosines shows that the only terms of $O(\varepsilon^2)$ that have a nonzero average in the right-hand side of (232a) are

$$\varepsilon^2\left(\frac{8P}{3\pi}T - \frac{4P^2}{3\pi}T_P\right)$$

Therefore, we set

$$\underline{T}(P) = P^2 \tag{234a}$$

to remove these terms. Similarly, we find that the terms of order ε^2 having a nonzero average in the right-hand side of (232b) are

$$\varepsilon^2 \left(\frac{P^2}{\pi^2} \sum_{m=0}^{\infty} \frac{c_m d_m}{m+1} - \frac{2P^2}{\pi^2} \sum_{m=0}^{\infty} d_m^2 - \frac{4P^2}{3\pi} L_P \right)$$

We remove these terms by choosing

$$\underline{L}(P) = \frac{3P}{4\pi} \sum_{m=0}^{\infty} \left(\frac{c_m d_m}{m+1} - 2d_m^2 \right) \tag{234b}$$

This defines $T(P,Q)$ and $L(P,Q)$ explicitly, and the uniformly valid solution for p and q is then given by (229). If we do not intend to proceed to higher order, we do not need the explicit form of the terms of order ε^2 with zero averages in (232).

As pointed out in [9], the choice of $\underline{T}$ and $\underline{L}$ is arbitrary and does not affect the final expressions (229) as long as we keep track of the terms of $O(\varepsilon^2)$ with a nonzero average in (232). Here, we removed these terms by choosing $\underline{T}$ and $\underline{L}$. We could also, for example, set $\underline{T} = \underline{L} = 0$ at the outset. In this case, the averaged equation (224a) is the same, but the equation (224b) for $\dot{Q}$ is

$$\dot{Q} = 1 + \varepsilon^2 \left(\frac{P^2}{\pi^2} \sum_{m=0}^{\infty} \frac{c_m d_m}{m+1} - \frac{2P^2}{\pi^2} \sum_{m=0}^{\infty} d_m^2 \right) + O(\varepsilon^2)$$

When these equations for $\dot{P}$ and $\dot{Q}$ are solved subject to the initial conditions (228) and the result is used in (229), we find exactly the same solution for $p(t;\varepsilon)$ and $q(t;\varepsilon)$ as before. Either way it is necessary to derive the terms of order ε^2 with nonzero average in the transformed equations for $\dot{P}$ and $\dot{Q}$.

In [9], it is shown that the solution of the system (216) by the method of averaging is exactly the same as the solution by the method of multiple scales. In many applications—for example, in calculating the behavior of the modal amplitudes $\beta_n(t)$ in (66) for the case where (50a) is weakly nonlinear—the equations to be solved are not in the standard form. In such problems, the method of multiple scales gives a direct solution efficiently without the need for transforming to the form (216).

8.4.4   Evolution Equations for a Weakly Nonlinear Problem

In this section we give an illustration of the use of multiple scale expansions for solving weakly nonlinear partial differential equations in a form that remains uniformly valid in the far field. Developments in this area are recent and rather limited in applicability compared with the corresponding fairly comprehensive theory for systems of ordinary differential equations. A survey of the basic techniques and their limitations can be found in Section 4.4 of [1] and [12].

We study the initial-value problem in the infinite domain $-\infty < x < \infty$ and $t \geq 0$ for the Boussinesq equations for shallow-water flow:

$$h_t + (uh)_x = O(\varepsilon^4) \tag{235a}$$

$$u_t + uu_x + h_x = -\frac{\kappa^2}{3} \varepsilon h_{xtt} + O(\varepsilon^3) \tag{235b}$$

The initial conditions are expressed in the general form

$$u(x,0;\varepsilon) = \varepsilon v(x) \tag{236a}$$

$$h(x,0;\varepsilon) = 1 + \varepsilon \ell(x) \tag{236b}$$

In [1], it is shown that the pair of equations (235) provides a consistent approximation, up to the orders indicated, for shallow-water flow of an incompressible inviscid fluid. The term $-(\kappa^2/3)\varepsilon h_{xtt}$ represents the first correction term to the equations with zero right-hand sides, which we derived in Chapter 3, assuming hydrostatic balance in the vertical direction. The constant κ^2 is a similarity parameter: $\kappa^2 \equiv \delta^2/\varepsilon$ and is held fixed as $\delta \to 0$ and $\varepsilon \to 0$. Here $\delta \equiv H/L$, where H is the undisturbed water height and L is a characteristic wavelength for surface disturbances. Thus $\delta \ll 1$ corresponds to shallow water (or long waves). The second small parameter is $\varepsilon \equiv A/H$, where A is a characteristic amplitude for surface disturbances. It is shown in [1] that the choice $\delta = \kappa\varepsilon^{1/2}$ leads to the richest limiting equations. In our discussions of shallow-water flow so far in this book, we have set $\kappa \equiv 0$. This is the hydrostatic approximation and corresponds to having $\delta \ll \varepsilon^{1/2}$. We have seen in the previous two sections that we need to use the terms of order ε^{N+1} in the differential equations in order to calculate the solution to $O(\varepsilon^N)$. Thus, since $h_{xtt} = O(\varepsilon)$, the right-hand side of (235b) cannot be ignored if $\kappa^2 = O_s(1)$ in calculating u and h to $O(\varepsilon)$.

Expansion Procedure

We assume a solution in the multiple scale form

$$u(x,t;\varepsilon) = \varepsilon u_1(x,t_0,t_1) + \varepsilon^2 u_2(x,t_0,t_1) + O(\varepsilon^3) \tag{237a}$$

$$h(x,t;\varepsilon) = 1 + \varepsilon h_1(x,t_0,t_1) + \varepsilon^2 h_2(x,t_0,t_1) + O(\varepsilon^3) \tag{237b}$$

where $t_0 \equiv t$, $t_1 \equiv \varepsilon t$, and as mentioned in Section 8.4.2, we need not include a dependence on $\varepsilon^2 t$ in the expansions because we are only concerned with the solution for u_1 and h_1.

Derivatives with respect to t become

$$\frac{\partial}{\partial t} = \frac{\partial}{\partial t_0} + \varepsilon \frac{\partial}{\partial t_1} \tag{238}$$

and we find the following equations governing the terms of order ε and ε^2:

$$\frac{\partial h_1}{\partial t_0} + \frac{\partial u_1}{\partial x} = 0 \tag{239a}$$

$$\frac{\partial u_1}{\partial t_0} + \frac{\partial h_1}{\partial x} = 0 \tag{239b}$$

$$\frac{\partial h_2}{\partial t_0} + \frac{\partial u_2}{\partial x} = -\frac{\partial h_1}{\partial t_1} - u_1 \frac{\partial h_1}{\partial x} - h_1 \frac{\partial u_1}{\partial x} \tag{240a}$$

$$\frac{\partial u_2}{\partial t_0} + \frac{\partial h_2}{\partial x} = -\frac{\partial u_1}{\partial t_1} - u_1 \frac{\partial u_1}{\partial x} - \frac{\kappa^2}{3} \frac{\partial^3 h_1}{\partial x \partial t_0^2} \tag{240b}$$

The initial conditions (236) imply that u_1, h_1, u_2, h_2 must satisfy

$$u_1(x, 0, 0) = v(x) \tag{241a}$$

$$h_1(x, 0, 0) = \ell(x) \tag{241b}$$

$$u_2(x, 0, 0) = 0 \tag{241c}$$

$$h_2(x, 0, 0) = 0 \tag{241d}$$

It is convenient to introduce the characteristic dependent and independent variables associated with the homogeneous system (239) [see (124) of Chapter 3 and (114d) of Chapter 4]. We denote

$$S_i \equiv h_i + u_i \tag{242a}$$

$$R_i \equiv h_i - u_i \tag{242b}$$

for $i = 1, 2$, and regard S_i and R_i as functions of the two fast scales

$$\xi \equiv x - t \tag{243a}$$

$$\eta \equiv x + t \tag{243b}$$

and the slow scale $t_1 = \varepsilon t$.

It is easily seen that (239) and (240) transform to the following equations for the $S_i(\xi, \eta, t_1)$ and $R_i(\xi, \eta, t_1)$:

$$2 \frac{\partial S_1}{\partial \eta} = 0 \tag{244a}$$

$$-2 \frac{\partial R_1}{\partial \xi} = 0 \tag{244b}$$

$$2 \frac{\partial S_2}{\partial \eta} = -\frac{\partial S_1}{\partial t_1} - \left(\frac{S_1 - R_1}{2} \right) \left(\frac{\partial S_1}{\partial \xi} + \frac{\partial S_1}{\partial \eta} \right)$$
$$- \left(\frac{S_1 + R_1}{4} \right) \left(\frac{\partial S_1}{\partial \xi} + \frac{\partial S_1}{\partial \eta} - \frac{\partial R_1}{\partial \xi} - \frac{\partial R_1}{\partial \eta} \right) - \frac{\kappa^2}{6} [D(S_1) + D(R_1)] \tag{245a}$$

$$-2 \frac{\partial R_2}{\partial \xi} = -\frac{\partial R_1}{\partial t_1} - \left(\frac{S_1 - R_1}{2} \right) \left(\frac{\partial R_1}{\partial \xi} + \frac{\partial R_1}{\partial \eta} \right)$$
$$- \left(\frac{S_1 + R_1}{4} \right) \left(\frac{\partial S_1}{\partial \eta} + \frac{\partial S_1}{\partial \eta} - \frac{\partial R_1}{\partial \xi} - \frac{\partial R_1}{\partial \eta} \right) + \frac{\kappa^2}{6} [D(S_1) + D(R_1)] \tag{245b}$$

where D is the third-order operator

$$D \equiv \frac{\partial^3}{\partial \xi^3} - \frac{\partial^3}{\partial \xi^2 \partial \eta} - \frac{\partial^3}{\partial \xi \partial \eta^2} + \frac{\partial^3}{\partial \eta^3} \tag{246}$$

The initial conditions (241) imply that we must have

$$S_1(x, x, 0) = \ell(x) + v(x) \tag{247a}$$

$$R_1(x, x, 0) = \ell(x) - v(x) \tag{247b}$$

$$S_2(x, x, 0) = 0 \tag{248a}$$

$$R_2(x, x, 0) = 0 \tag{248b}$$

Consistency Conditions to $O(\varepsilon^2)$; Korteweg-deVries Equations

Equations (244) imply that S_1 is independent of η and R_1 is independent of ξ; that is,

$$S_1 = f_1(\xi, t_1) \tag{249a}$$

$$R_1 = g_1(\eta, t_1) \tag{249b}$$

where, according to (247), the functions f_1 and g_1 satisfy the initial conditions

$$f_1(\xi, 0) = \ell(\xi) + v(\xi) \tag{250a}$$

$$g_1(\eta, 0) = \ell(\eta) - v(\eta) \tag{250b}$$

This is as far as we can go in defining S_1 and R_1; we need to consider the solution to $O(\varepsilon^2)$ to specify f_1 and g_1.

Using the result (249) in the right-hand sides of (245) simplifies these considerably, and we find

$$
\begin{aligned}
2\frac{\partial S_2}{\partial \eta} = & -\left(\frac{\partial f_1}{\partial t_1} + \frac{3}{4}f_1\frac{\partial f_1}{\partial \xi} + \frac{\kappa^2}{6}\frac{\partial^3 f_1}{\partial \xi^3}\right) \\
& + \frac{g_1}{4}\frac{\partial f_1}{\partial \xi} + \frac{f_1 + g_1}{4}\frac{\partial g_1}{\partial \eta} - \frac{\kappa^2}{6}\frac{\partial^3 g_1}{\partial \eta^3}
\end{aligned}
\tag{251a}
$$

$$
\begin{aligned}
-2\frac{\partial R_2}{\partial \xi} = & -\left(\frac{\partial g_1}{\partial t_1} - \frac{3}{4}g_1\frac{\partial g_1}{\partial \eta} - \frac{\kappa^2}{6}\frac{\partial^3 g_1}{\partial \eta^3}\right) \\
& + \frac{f_1}{4}\frac{\partial g_1}{\partial \eta} - \frac{f_1 + g_1}{4}\frac{\partial f_1}{\partial \xi} + \frac{\kappa^2}{6}\frac{\partial^3 f_1}{\partial \xi^3}
\end{aligned}
\tag{251b}
$$

We can now integrate these expressions and find

$$
\begin{aligned}
S_2(\xi, \eta, t_1) = & -\frac{1}{2}\left(\frac{\partial f_1}{\partial t_1} + \frac{3}{4}f_1\frac{\partial f_1}{\partial \xi} + \frac{\kappa^2}{6}\frac{\partial^3 f_1}{\partial \xi^3}\right)\eta \\
& + \frac{1}{8}\frac{\partial f_1}{\partial \xi}\int^{\eta} g_1(s, t_1)\,ds + \frac{f_1}{8}g_1 + \frac{g_1^2}{16} - \frac{\kappa^2}{12}\frac{\partial^2 g_1}{\partial \eta^2} + f_2(\xi, t_1)
\end{aligned}
\tag{252a}
$$

$$
\begin{aligned}
R_2(\xi, \eta, t_1) = & \frac{1}{2}\left(\frac{\partial g_1}{\partial t_1} - \frac{3}{4}g_1\frac{\partial g_1}{\partial \eta} - \frac{\kappa^2}{6}\frac{\partial^3 g_1}{\partial \eta^3}\right)\xi \\
& - \frac{1}{8}\frac{\partial g_1}{\partial \eta}\int^{\xi} f_1(s, t_1)\,ds + \frac{f_1^2}{16}g_1 + \frac{f_1 g_1}{8} - \frac{\kappa^2}{12}\frac{\partial^2 f_1}{\partial \xi^2} + g_2(\eta, t_1)
\end{aligned}
\tag{252b}
$$

where f_2 and g_2 are functions to be determined at the next stage.

The first group of terms multipled by η in the right-hand side of (252a) must be eliminated because it contributes a component to S_2, which becomes infinite as $|\eta| \to \infty$ (that is, as $t \to \infty$ with x fixed or as $x \to \infty$ with t fixed), and this behavior implies that the expansions (237) are not uniformly valid for $x = O_s(\varepsilon^{-1})$ or $t = O_s(\varepsilon^{-1})$. The equation that results is the *evolution equation* for f_1:

$$\frac{\partial f_1}{\partial t_1} + \frac{3}{4} f_1 \frac{\partial f_1}{\partial \xi} + \frac{\kappa^2}{6} \frac{\partial^3 f_1}{\partial \xi^3} = 0 \tag{253a}$$

which must be solved subject to the initial condition (250a). Similarly, the boundedness of R_2 for $|\xi| \to \infty$ requires that we set

$$\frac{\partial g_1}{\partial t_1} - \frac{3}{4} g_1 \frac{\partial g_1}{\partial \eta} - \frac{\kappa^2}{6} \frac{\partial^3 g_1}{\partial \eta^3} = 0 \tag{253b}$$

and this is to be solved subject to the initial condition (250b).

Equations (253a)–(253b) are formally identical; the transformation $\xi \to -\eta$ takes (253a) to (253b). This equation was first derived by Korteweg and deVries in 1895, and the reader is referred to Sections 13.11–13.15 of [8] and to [13] for a survey of results. One can show that the Korteweg-deVries (KdV) equation has bounded solutions for a large class of initial conditions.

If f_1 and g_1 are bounded integrable functions of ξ and η, respectively, we see that the terms that remain in the solution (252) are also bounded, and the expansions for S and R are therefore uniformly valid in the far field. To compute S_2 and R_2 completely, we need to define f_2 and g_2, and this involves examining the solution to $O(\varepsilon^3)$, which we will not discuss.

An interesting feature of the result (253) is that the equations for f_1 and g_1 are decoupled and can be solved individually. The solution for f_1 defines a disturbance that propagates to the right (as exhibited by the dependence of f_1 on $\xi \equiv x - t$), whereas g_1 defines a disturbance that propagates to the left; both of these disturbances evolve slowly with time (as exhibited by their dependence on $t_1 \equiv \varepsilon t$). The solution for u and h to $O(\varepsilon)$ contains both components and has the form

$$u(x, t; \varepsilon) = \frac{\varepsilon}{2} [f_1(\xi, t_1) - g_1(\eta, t_1)] + O(\varepsilon^2) \tag{254a}$$

$$h(x, t; \varepsilon) = 1 + \frac{\varepsilon}{2} [f_1(\xi, t_1) + g_1(\eta, t_1)] + O(\varepsilon^2) \tag{254b}$$

The solution to $O(\varepsilon^2)$ does not split into f and g components, since S_2 and R_2 contain products of these functions.

Often, a KdV equation is derived directly from the Boussinesq equations (235) by looking for unidirectional solutions. It is emphasized here that the KdV equations (253) apply more generally. In fact, these equations describe arbitrary flows according to (254), as long as the disturbance amplitude is small ($\varepsilon \ll 1$).

For the special case of unidirectional flows (such as $g_1 \equiv 0$), our results also specialize to a KdV equation for u or h, as shown next. In this case, we have

$$u = h - 1 = \frac{\varepsilon}{2} f_1(\xi, t_1) + O(\varepsilon^2) \tag{255}$$

and the KdV equation (253a) can be expressed as an equation for u [or $(h - 1)$] as a function of x and t by noting that

$$x \equiv \xi + \frac{t_1}{\varepsilon} \tag{256a}$$

$$t \equiv \frac{t_1}{\varepsilon} \tag{256b}$$

Thus,

$$\frac{\partial}{\partial t_1} = \frac{1}{\varepsilon} \frac{\partial}{\partial x} + \frac{1}{\varepsilon} \frac{\partial}{\partial t} \tag{257a}$$

$$\frac{\partial}{\partial \xi} = \frac{\partial}{\partial x} \tag{257b}$$

$$\frac{\partial^3}{\partial \xi^3} = \frac{\partial^3}{\partial x^3} \tag{257c}$$

We solve (255) for f_1:

$$f_1 = \frac{2}{\varepsilon} u + O(\varepsilon) = \frac{2}{\varepsilon}(h - 1) + O(\varepsilon) \tag{257d}$$

We then use the transformations (257) in (253a) to obtain

$$\frac{2}{\varepsilon^2}(u_x + u_t) + \frac{3}{4}\left(\frac{2}{\varepsilon}u\right)\left(\frac{2}{\varepsilon}u_x\right) + \frac{\kappa^2}{6} \frac{2}{\varepsilon} u_{xxx} = 0$$

or

$$\frac{2}{\varepsilon^2}(h_x + h_t) + \frac{3}{4}\left[\frac{2}{\varepsilon}(h - 1)\right]\left[\frac{2}{\varepsilon}h_x\right] + \frac{\kappa^2}{6}\frac{2}{\varepsilon}h_{xxx} = 0$$

to leading order. Therefore, to leading order, u and h obey

$$u_t + u_x + \frac{3}{2}uu_x + \frac{\delta^2}{6}u_{xxx} = 0 \tag{258a}$$

$$h_t + h_x + \frac{3}{2}(h - 1)h_x + \frac{\delta^2}{6}h_{xxx} = 0 \tag{258b}$$

where $\delta^2 \equiv \kappa^2 \varepsilon$. This is the form usually given in the literature for the KdV equation governing u or $h - 1$ for unidirectional shallow-water flow. The transformation $\frac{3}{2}u + 1 \to \frac{3}{2}w$ takes (258a) to the generic form

$$w_t + \frac{3}{2}ww_x + \frac{\delta^2}{6}w_{xxx} = 0 \tag{259}$$

The exact solution of (259) can be derived for initial conditions $w(x, 0)$, which decay sufficiently fast as $|x| \to \infty$. The procedure, known as *inverse scattering theory*, was developed in 1967 [14] and has since been studied extensively in a number of applications; for example, see [15]. The result for f_1 is obtained as the solution of a certain linear integral equation and is not explicit for general initial data. A discussion of this theory is beyond the scope of this book and will not be given.

Solitary and Periodic Waves

We can compute explicit solutions of the KdV equation for special initial data. These solutions represent uniform waves [see (233) of Chapter 3] and are in the form:

$$w \equiv w_0 W(\theta) \tag{260a}$$

$$\theta \equiv kx - \omega t \tag{260b}$$

where w_0, k, and ω are constants. The amplitude scale w_0 is chosen so that the maximum value of $|W|$ is unity. We show next that it is possible to find solutions where $W \to 0$ as $|\theta| \to \infty$ (solitary waves) and solutions where W is a periodic function of θ (periodic waves).

We substitue (260) into (259) and find that W obeys ($' \equiv d/d\theta$)

$$-\omega W' + \frac{3}{2} w_0 k W W' + \frac{\delta^2}{6} k^3 W''' = 0 \tag{261a}$$

This third-order equation can be integrated twice to obtain the conservation relation

$$\frac{\delta^2 k^3}{6} W'^2 = -\frac{w_0 k}{2} W^3 + \omega W^2 + \alpha W + \beta \tag{261b}$$

where α and β are integration constants.

If we look for a solution where W and W' both vanish as $|\theta| \to \infty$, we must set $\alpha = \beta = 0$. In this case $W' = 0$ at $\pm\infty$, where W is a minimum. The slope W' also vanishes at $W = 2\omega/w_0 k \equiv W_{max}$, which is the maximal value of W. Therefore, to have $W_{max} = 1$, we choose

$$\omega = \frac{w_0 k}{2} \tag{262}$$

and (261b) becomes

$$dW = \pm \frac{(3w_0)^{1/2}}{\delta k} W(1 - W) \tag{263}$$

If we set $\theta = 0$ when $W = 1$, (263) defines the solitary wave

$$w = w_0 \operatorname{sech}^2 \left[\frac{(3w_0)^{1/2}}{2\delta} \left(x - \frac{w_0 t}{2} \right) \right] \tag{264}$$

involving the single parameter w_0. Thus, the phase speed of the solitary wave is $w_0/2$.

Consider next the case where α and β are not equal to zero. The conservation relation (261b) leads to bounded solutions for W if the cubic

$$R \equiv -\frac{w_0 k}{2} W^3 + \omega W^2 + \alpha W + \beta = 0 \tag{265}$$

has three real roots. In this case R will be positive for values of W lying between two adjacent roots, say $W = W_1$ and $W = W_2$, and the solution oscillates periodically between these two values. We do not derive the explicit form of the solution for W as a function of θ in this case. This can be worked out in terms of the Jacobi elliptic function sn; the details are given in Section 4.1 of [16]. We reiterate that solitary wave and periodic wave solutions correspond to special initial conditions. For example, we must have $w(x,0) = w_0 \, \text{sech}^2 [(3w_0)^{1/2} x / 2\delta]$ in order to generate the solution (264).

Solutions for $\kappa \equiv 0$

Explicit solutions of the evolution equations (253) can be calculated for arbitrary initial values of ℓ and v in (250) if we set $\kappa \equiv 0$—that is, $\delta \ll \varepsilon^{1/2}$.

The basic problem now satisfies the divergence relations

$$h_t + (uh)_x = 0 \tag{266a}$$

$$(uh)_t + \left(u^2 h + \frac{h^2}{2}\right)_x = 0 \tag{266b}$$

discussed in Section 5.3.3. These equations give the bore conditions

$$U[h] = [uh] \tag{267a}$$

$$U[uh] = \left[u^2 h + \frac{h^2}{2}\right] \tag{267b}$$

where $U \equiv dx/dt$ is the bore speed.

Since the evolution equations

$$\frac{\partial f_1}{\partial t_1} + \frac{3}{4} f_1 \frac{\partial f_1}{\partial \xi} = 0 \tag{268a}$$

$$\frac{\partial g_1}{\partial t_1} - \frac{3}{4} g_1 \frac{\partial g_1}{\partial \eta} = 0 \tag{268b}$$

now admit shocks, it is important to derive the correct jump conditions for (268) consistent with (267).

We denote $[u] = u^+ - u^-$, $[h] = h^+ - h^-$ and eliminate u^+ from the two bore conditions (267). This gives the quadratic expression for U [in (90b) of Chapter 5 we calculated this expression for the special case $u^- = 0, h^- = 1$]:

$$U^2 - 2Uu^- + (u^-)^2 - \frac{h^+(h^+ + h^-)}{2h^-} = 0 \tag{269}$$

For a small disturbance theory, U has the expansion

$$U = U_0 + \varepsilon U_1 + O(\varepsilon^2) \tag{270a}$$

But

$$U \equiv \frac{dx}{dt} = \begin{cases} \dfrac{d}{dt}(\xi + t) = 1 + \varepsilon \dfrac{d\xi}{dt_1}, & \text{for } f_1 \\[3mm] \dfrac{d}{dt}(\eta - t) = -1 + \varepsilon \dfrac{d\eta}{dt_1}, & \text{for } g_1 \end{cases} \tag{270b}$$

Comparing (270a)–(270b), we see that $U_0 = \pm 1$, as expected, and that we must identify

$$U_1 = \begin{cases} \dfrac{d\xi}{dt_1}, & \text{for } f_1 \\[3mm] \dfrac{d\eta}{dt_1}, & \text{for } g_1 \end{cases} \tag{271}$$

Since the f_1 and g_1 disturbances evolve independently, we can set $g_1 = 0$ to calculate the jump condition for f_1. In this case, (254) becomes

$$u^- = \frac{\varepsilon f^-}{2} + O(\varepsilon^2) \tag{272a}$$

$$h^\pm = 1 + \frac{\varepsilon f^\pm}{2} + O(\varepsilon^2) \tag{272b}$$

and $U = 1 + \varepsilon d\xi/dt_1$. Substituting these expressions into (269) gives

$$1 + 2\varepsilon \frac{d\xi}{dt_1} - 2\varepsilon\left(\frac{f_1^-}{2}\right) - \frac{(1 + \varepsilon f_1^+/2)(2 + \varepsilon f_1^+/2 + \varepsilon f_1^-/2)}{2(1 + \varepsilon f_1^-/2)} = O(\varepsilon^2)$$

and this simplifies to

$$\varepsilon\left(2\frac{d\xi}{dt_1} - \frac{3}{4}(f_1^+ + f_1^-)\right) = O(\varepsilon^2) \tag{273}$$

Therefore, the jump condition for f_1 is

$$\frac{d\xi}{dt_1} = \frac{3}{8}(f_1^+ + f_1^-) \tag{274}$$

and this implies that the correct divergence form for (268a) is

$$\frac{\partial f_1}{\partial t_1} + \frac{\partial}{\partial \xi}\left(\frac{3}{8}f_1^2\right) = 0 \tag{275}$$

A similar calculation starting with $f_1 = 0$ gives the jump condition

$$\frac{d\eta}{dt_1} = -\frac{3}{8}(g_1^+ + g_1^-) \tag{276}$$

for the evolution equation for g_1.

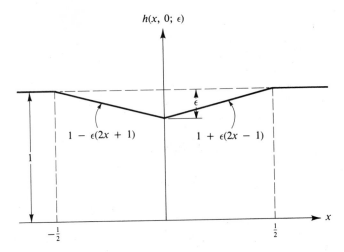

Figure 8.5

The solution of the initial-value problem (235)–(236) for $\delta \ll \varepsilon^{1/2} \ll 1$ thus reduces to the relatively simple solution of the decoupled equations (268) subject to the initial conditions (250). Whenever bores occur, these must satisfy the jump conditions (274) and (276). We shall discuss an example next.

Consider the initial-value problem for (266) with

$$u(x, 0; \varepsilon) = 0 \tag{277a}$$

$$h(x, 0; \varepsilon) = 1 + \varepsilon \ell(x) \tag{277b}$$

where

$$\ell(x) = \begin{cases} 2x - 1, & 0 \le x \le \tfrac{1}{2} \\ -1 - 2x, & -\tfrac{1}{2} < x \le 0 \\ 0, & |x| \ge \tfrac{1}{2} \end{cases} \tag{277c}$$

Thus, the initial surface has a triangular depression over the interval $-\tfrac{1}{2} \le x \le \tfrac{1}{2}$, as sketched in Figure 8.5.

The initial conditions to be satisfied by f_1 and g_1 are then

$$f_1(\xi, 0) = \ell(\xi) \tag{278a}$$

$$g_1(\eta, 0) = \ell(\eta) \tag{278b}$$

We shall consider the solution only for f_1 because the solution for g_1 follows by symmetry using

$$g_1(\eta) \equiv f_1(-\eta) \tag{279}$$

The solution of (268a) subject to (278a) follows from the characteristic equations [see (23) of Chapter 5] in the parametric form

$$\xi = \begin{cases} \frac{3}{4}(2\xi_0 - 1)t_1 + \xi_0, & 0 \le \xi_0 \le \frac{1}{2} \\ -\frac{3}{4}(2\xi_0 + 1)t_1 + \xi_0, & -\frac{1}{2} \le \xi_0 \le 0 \\ 0, & |\xi_0| > \frac{1}{2} \end{cases} \tag{280a}$$

$$f_1 = \begin{cases} 2\xi_0 - 1, & 0 \le \xi_0 \le \frac{1}{2} \\ -(2\xi_0 + 1), & -\frac{1}{2} \le \xi_0 \le 0 \\ 0, & |\xi_0| \ge \frac{1}{2} \end{cases} \tag{280b}$$

For the simple initial-value problem we are considering, we can solve (280a) for ξ_0 in terms of ξ and t_1, and then use this result in (280b) to find

$$f_1(\xi, t_1) = \begin{cases} \dfrac{4\xi - 2}{2 + 3t_1}, & -\frac{3}{4}t_1 \le \xi \le \frac{1}{2}, \ t_1 < \frac{2}{3} \\ -\dfrac{4\xi + 2}{2 - 3t_1}, & -\frac{1}{2} \le \xi \le -\frac{3}{4}t_1, \ t_1 < \frac{2}{3} \\ 0, & |\xi| \ge \frac{1}{2} \end{cases} \tag{281}$$

As shown in Figure 8.6a, the straight characteristics emerging from $-\frac{1}{2} \le \xi_0 \le 0$ intersect at $\xi = -\frac{1}{2}$, $t_1 = \frac{2}{3}$, and we need to introduce a bore for $t_1 > \frac{2}{3}$. The bore condition (274) for our case has $f_1^- = 0$, and $f_1^+ = (4\xi - 2)/(2 + 3t_1)$. Solving this equation defines the bore trajectory

$$\xi = \frac{1}{2} - \frac{1}{2}(2 + 3t_1)^{1/2} \equiv b(t_1) \tag{282}$$

Note that $d\xi/dt_1 \to 0$ as $t_1 \to \infty$. The solution of f_1 for $t > \frac{2}{3}$ is then given by $f_1 = 0$ to the left of $\xi = b(t_1)$, by f_1^+ to the right of the bore up to $\xi = \frac{1}{2}$, and by $f_1 = 0$ for $\xi \ge \frac{1}{2}$ (see Figure 8.6b).

For a more general initial curve $\ell(x)$, such as $\ell(x) = 4x^2 - 1$, the characteristics emerging from the interval $-\frac{1}{2} \le \xi_0 \le 0$ do not intersect at a point. In this case, the arc AB of the bore trajectory is defined using the solution f_1^+ along the characteristics emerging from $-\frac{1}{2} \le \xi_0 \le 0$, and the arc BC (where $C \to \infty$) is defined using the solution along the characteristics emerging from $0 \le \xi_0 \le \frac{1}{2}$. The point B is an inflection point, and again the bore speed tends to zero as $t_1 \to \infty$ (see Figure 8.6c).

If we keep in mind that $\xi \equiv x - t$ and $t_1 \equiv \varepsilon t$, the qualitative behavior of the f_1 disturbance is easy to discern. For short times $t = O(1)$, $t_1 = O(\varepsilon)$, and the initial profile in f_1 propagates to the right with speed $dx/dt \approx 1$ and remains essentially unchanged. Similarly, the initial profile in g_1 propagates to the left with speed $dx/dt \approx -1$. The values of u and h are given by (254). In particular, we note that the initial depression in the free surface splits up into two equal components, which propagate to the right (f_1) and left (g_1). This is just the result given by linear theory. However, as t_1 becomes $O(1)$—that is, $t = O_s(\varepsilon^{-1})$—the cumulative effect of the weak nonlinearities causes the front parts of the f_1 and g_1 waves to steepen, and eventually the bores defined, respectively, by $\xi = b(t_1)$ and $\eta = -b(t_1)$ are formed.

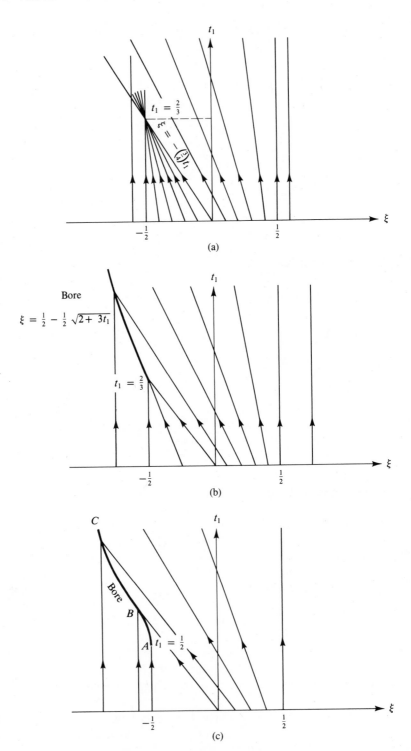

(a)

(b)

(c)

Figure 8.6

In the far field, the location of these bores is defined to $O(1)$ only, and for our example problem we have

$$x = t + \frac{1}{2} - \frac{1}{2}(2 + 3\varepsilon t)^{1/2} + O(\varepsilon); \qquad t > \frac{2}{3\varepsilon} \tag{283a}$$

$$x = -t - \frac{1}{2} + \frac{1}{2}(2 + 3\varepsilon t)^{1/2} + O(\varepsilon); \qquad t > \frac{2}{3\varepsilon} \tag{283b}$$

As t_1 increases, the f_1 and g_1 disturbances are confined over an interval that gradually widens, and the amplitude of these disturbances decays.

It is tempting to argue that the solution of the evolution equation (268a) or (268b) gives the outer limit (as $k \to 0$) of the solution of (253a) or (253b), and that appropriate shock layers can be constructed to smooth out the discontinuities. This is not true! Replacing the second-derivative term in Burgers' equation by the third-derivative term of the KdV equation dramatically alters the behavior of the solution near a discontinuity of the reduced equation—that is, the equation with $\kappa \equiv 0$. We can show this by considering an initial-value problem such as (278a), where ℓ is an isolated disturbance defined by a smooth function on $-\frac{1}{2} \le x \le \frac{1}{2}$ with $\ell = 0$ if $|x| > \frac{1}{2}$. Then, the reduced problem (268a) for f_1 will always have a discontinuity along some curve $\xi = b(t_1)$. If we now solve (253a) numerically for the same initial condition and for a sequence of successively smaller values of $\kappa > 0$, we find that the solution for the $\kappa \ne 0$ problem does not tend to the solution of (268a) near $\xi = b(t_1)$; as long as $\kappa \ne 0$, the solution has high-frequency oscillations over an interval in $|\xi - b(t_1)|$, which is not small (see [17]).

A rigorous explanation of the behavior of exact solutions of (253) as $\kappa \to 0$ is rather complicated. Superficially, we can see that an interior layer approximation for (253a) does not make sense because we cannot match such a solution with a solution of the reduced problem. To show this, we look for an interior layer of (253a) near $\xi = b(t_1)$ and introduce the inner variables (for some $\mu(\kappa) \ll 1$)

$$\xi^* = \frac{\xi - b(t_1)}{\mu(\kappa)} \tag{284a}$$

$$t_1^* = t_1 \tag{284b}$$

and denote

$$f_1(b(t_1^*) + \mu\xi^*, t_1^*) = f_1^*(\xi^*, t_1^*; \varepsilon) \tag{285}$$

We see that f_1^* satisfies

$$\frac{\partial f_1^*}{\partial t_1^*} - \frac{b'}{\mu}\frac{\partial f_1^*}{\partial \xi^*} + \frac{3}{4\mu}f_1^*\frac{\partial f_1^*}{\partial \xi^*} + \frac{\kappa^2}{6\mu^3}\frac{\partial^3 f_1^*}{\partial \xi^{*3}} = 0 \tag{286a}$$

Thus, $\mu = \kappa$ corresponds to a distinguished limit, for which $f \equiv f_1^*(\xi^*, t_1^*; 0)$ obeys

$$\frac{1}{6}\frac{\partial^3 f}{\partial \xi^{*2}} + \left[\frac{3}{4}f - b'(t_1^*)\right]\frac{\partial f}{\partial \xi^*} = 0 \tag{286b}$$

But this is essentially the same equation (261a) that we derived for uniformly propagating solutions. We showed that bounded solutions of (286b) for $|\xi^*| \to \infty$ must be either solitary or periodic waves. A solitary wave can match only with a zero solution on either side and is therefore inappropriate as a candidate for a shock layer. A periodic wave does not have a limit as $|\xi^*| \to \infty$ and is also not a suitable interior layer. Therefore, solutions of the distinguished limit (286b) cannot be used as interior layer approximations of (253a) for $\kappa \ll 1$.

In the physical description of shallow-water flow, the model (235) with $\kappa \neq 0$ is more accurate than the approximation (266). However, neither of these descriptions take viscous effects into account. A physically consistent description of the flow near a bore of the system (266) must also take dissipative effects into account. This problem has not yet been solved systematically. An "artificial" viscosity term proportional to h_{xx} may be inserted into (258b), for example, to see what effect it has on solutions. It is shown in Section 13.15 of [8] that such a term leads to a bore structure that looks qualitatively correct.

Problems

1. Verify the following order relations in $\mathscr{D}: 0 < x < 1$.
   a. $\sin \varepsilon x = O(1)$ uniformly in $\mathscr{D}$ as $\varepsilon \to 0$ (287a)
   b. $\sin \varepsilon x = O(2\varepsilon x/\pi)$ uniformly in $\mathscr{D}$ as $\varepsilon \to 0$ (287b)
   c. $\sin(x/\varepsilon) = O(x)$ as $\varepsilon \to 0$, but not uniformly in $\mathscr{D}$ (287c)
   d. $\sin(x/\varepsilon) = O(1)$ uniformly in $\mathscr{D}$ as $\varepsilon \to 0$ (287d)

2. a. What is the largest domain in the xy-plane over which

$$u(x, y; \varepsilon) \equiv e^{-x/\varepsilon}\sin y + \frac{\sin \sqrt{1 - \varepsilon^2}\, y}{1 + \varepsilon x} = \sin y + O(\varepsilon) \qquad (288)$$

   uniformly as $\varepsilon \to 0$?

   b. Derive the three-term asymptotic expansion of (288) in the form

$$u(x, y; \varepsilon) = \sin y + \varepsilon u_1(x, y) + \varepsilon^2 u_2(x, y) + O(\varepsilon^3) \qquad (289)$$

   as $\varepsilon \to 0$ in the domain that you found in part (a). Indicate where (289) fails to be valid and why.

   c. Now introduce the new variable $x^* \equiv x/\varepsilon$ and write $u(x, y; \varepsilon)$ as

$$u^*(x^*, y; \varepsilon) \equiv e^{-x^*}\sin y + \frac{\sin \sqrt{1 - \varepsilon^2}\, y}{1 + \varepsilon^2 x^*} \qquad (290)$$

   What is the largest domain in the x^*y-plane in which

$$u^*(x^*, y; \varepsilon) = (e^{-x^*} + 1)\sin y + O(\varepsilon) \qquad (291)$$

   uniformly as $\varepsilon \to 0$?

   d. Derive the three-term asymptotic expansion of (290) in the form

$$u^*(x^*, y; \varepsilon) = (e^{-x^*} + 1)\sin y + \varepsilon u_1^*(x^*, y) + \varepsilon^2 u_2^*(x^*, y) + O(\varepsilon^3) \quad (292)$$

   as $\varepsilon \to 0$ in the domain you found in part (c). Indicate where (292) fails to be valid and why.

3. Consider the cubic

$$R(x, y, u; \varepsilon) \equiv \varepsilon u^3 + u^2 - f^2(x, y) = 0 \tag{293}$$

where $f(x, y)$ is a prescribed function in $\mathscr{D}$: $-\infty < x < \infty$, $-\infty < y < \infty$.

a. Show that if $f^2 < 4/27\varepsilon^2$, (293) has three real roots: $U^+(x, y; \varepsilon) > 0$, $U^-(x, y; \varepsilon) < 0$, and $U^{(3)}(x, y; \varepsilon) < U^-(x, y; \varepsilon) < 0$. We wish to compute the asymptotic expansions of these roots.

b. Show that U^+ and U^- have the expansions

$$U^{\pm}(x, y; \varepsilon) = \pm f + \varepsilon(-\tfrac{1}{2}f^2) + \varepsilon^2(\pm\tfrac{5}{8}f^3) + O(\varepsilon^3) \tag{294}$$

and that

$$U^{(3)}(x, y; \varepsilon) = -\frac{1}{\varepsilon} + \varepsilon f^2 + \varepsilon^3 2f^4 + O(\varepsilon^5) \tag{295}$$

c. For the case $f = x \tanh y$, show that (294) and (295) fail to be uniformly valid as $|x| \to \infty$. What is the correct rescaling for x, y, u in (293) to calculate uniformly valid results?

4. We note from (32) that the asymptotic expansion (35) for $\varepsilon g \to 0$ corresponds to the asymptotic expansion of $ze^z E_1(z)$ as $z \to \infty$. One approach to calculating the asymptotic expansion of $E_1(z)$ as $z \to \infty$ is to use repeated integration by parts. Let

$$E_n(z) \equiv \int_z^\infty \frac{e^{-s} ds}{s^n}; \qquad z > 0; \qquad n = 1, 2, \ldots \tag{296}$$

Integrating this expression by parts gives

$$E_n(z) = \frac{e^{-z}}{z^n} - nE_{n+1}(z); \qquad n = 1, 2, \ldots \tag{297}$$

a. Use (297) as a recursion relation to show that

$$E_1(z) = \frac{1}{ze^z} \sum_{n=0}^N (-1)^n \frac{n!}{z^n} + (-1)^{N+1} E_{N+1}(z) \tag{298}$$

which is equivalent to (34).

b. Use repeated integration by parts to show that the error function

$$\operatorname{erf}(z) \equiv 1 - \frac{1}{\sqrt{\pi}} \int_{z^2}^\infty \frac{e^{-t}}{t^{1/2}} dt \tag{299}$$

has the asymptotic expansion

$$\operatorname{erf}(z) = 1 - \frac{e^{-z^2}}{\sqrt{\pi}} \left[\sum_{n=1}^N \frac{(-1)^{n-1} 1 \cdot 3 \cdot 5 \cdots (2n-3)}{2^{n-1} z^{2n-1}} + O(z^{-2N-1}) \right] \qquad \text{as } z \to \infty \tag{300}$$

5. In many applications, we need to calculate the singular behavior of a function defined in integral form. This requires some care in deriving the asymp-

totic expansion. As an example, consider the function

$$f(t) \equiv \int_0^t \frac{\sin(t - \tau) \, d\tau}{[1 - \sin\tau + \tau\cos\tau]^2} \tag{301}$$

which is singular as $t \to \pi/2$.

To compute the asymptotic expansion of $f(t)$ as $t \to \pi/2$, we first introduce the new independent variable $s \equiv \pi/2 - t$ and change the variable of integration in (301) from τ to $\sigma \equiv \pi/2 - \tau$.

a. Show that (301) may be written in the form

$$\tilde{f}(s) \equiv f\left(\frac{\pi}{2} - s\right) = \cos s \int_s^{\pi/2} \frac{\sin\sigma}{D(\sigma)} \, d\sigma - \sin s \int_s^{\pi/2} \frac{\cos\sigma}{D(\sigma)} \, d\sigma \tag{302}$$

where

$$D(\sigma) \equiv \left[1 - \cos\sigma + \left(\frac{\pi}{2} - \sigma\right)\sin\sigma\right]^2 \tag{303}$$

b. Next, develop the integrands of (302) near $\sigma = 0$ to find

$$\frac{\sin\sigma}{D(\sigma)} = \frac{4}{\pi^2\sigma} + O(1) \qquad \text{as } \sigma \to 0 \tag{304a}$$

$$\frac{\cos\sigma}{D(\sigma)} = \frac{4}{\pi^2\sigma^2} + \frac{8}{\pi^3\sigma} + O(1) \qquad \text{as } \sigma \to 0 \tag{304b}$$

Therefore, (302) may be written as

$$\begin{aligned}
\tilde{f}(s) = \cos s \int_s^{\pi/2} \left[\frac{\sin\sigma}{D(\sigma)} - \frac{4}{\pi^2\sigma}\right] d\sigma + \frac{4}{\pi^2}\cos s \int_s^{\pi/2} \frac{d\sigma}{\sigma} \\
- \sin s \int_s^{\pi/2} \left[\frac{\cos\sigma}{D(\sigma)} - \frac{4}{\pi^2\sigma^2} - \frac{8}{\pi^3\sigma}\right] d\sigma \\
- \sin s \int_s^{\pi/2} \left[\frac{4}{\pi^2\sigma^2} + \frac{8}{\pi^3\sigma}\right] d\sigma
\end{aligned} \tag{305}$$

where we have subtracted out the terms that become singular as $s \to 0$ from the integrands in (302); then we have added these back to obtain an identity. In fact, (305), which is exact, has the following explicit form:

$$\begin{aligned}
\tilde{f}(s) = \cos s \int_s^{\pi/2} F(\sigma) \, d\sigma + \frac{4}{\pi^2}\left(\log\frac{\pi}{2}\right)\cos s - \frac{4}{\pi^2}(\cos s)\log s \\
- \sin s \int_s^{\pi/2} G(\sigma) \, d\sigma + \frac{8}{\pi^3}\sin s - \frac{4}{\pi^2 s}\sin s \\
- \frac{8}{\pi^3}\left(\log\frac{\pi}{2}\right)\sin s + \frac{8}{\pi^3}(\log s)\sin s
\end{aligned} \tag{306}$$

where

$$F(\sigma) \equiv \frac{\sin \sigma}{D(\sigma)} - \frac{4}{\pi^2 \sigma} = O(1) \qquad \text{as } \sigma \to 0 \tag{307a}$$

$$G(\sigma) \equiv \frac{\cos \sigma}{D(\sigma)} - \frac{4}{\pi^2 \sigma^2} - \frac{8}{\pi^3 \sigma} = O(1) \qquad \text{as } \sigma \to 0 \tag{307b}$$

The functions $F(\sigma)$ and $G(\sigma)$ are regular in $(0, \pi/2)$ and the two integrals in (306) are well behaved as $s \to 0$.

c. Now all the singular terms in $\tilde{f}(s)$ have been isolated, and we can let $s \to 0$ in the other terms. Show that we obtain the asymptotic expansion

$$\tilde{f}(s) = C_1 \log s + C_2 + C_3 s \log s + O(s) \qquad \text{as } s \to 0 \tag{308}$$

where

$$C_1 \equiv -\frac{4}{\pi^2} \tag{309}$$

$$C_2 \equiv \int_0^{\pi/2} F(\sigma) \, d\sigma + \frac{4}{\pi^2}\left(\log \frac{\pi}{2} - 1\right) \tag{310}$$

$$C_3 \equiv \frac{8}{\pi^3} \tag{311}$$

d. Calculate the term of $O(s)$ in (308).

6. In Chapter 2 we showed that the potential due to distribution of sources of strength $S(\xi)$/unit length along the x-axis is given by [see (88) of Chapter 2]

$$u(x, r) = -\frac{1}{4\pi} \int_0^1 \frac{S(\xi) \, d\xi}{[(x - \xi)^2 + r^2]^{1/2}} \tag{312}$$

Use the fact that

$$[(x-\xi)^2 + r^2]^{-1/2} = -\frac{\partial}{\partial \xi} \log\{x - \xi + [(x-\xi)^2 + r^2]^{1/2}\} \quad \text{if } \xi \le x \tag{313a}$$

$$[(x-\xi)^2 + r^2]^{-1/2} = -\frac{\partial}{\partial \xi} \log\{\xi - x + [(x-\xi)^2 + r^2]^{1/2}\} \quad \text{if } x \le \xi \tag{313b}$$

and integration by parts to show that for $x \ne 0$, $x \ne 1$,

$$u(x, r) = \frac{1}{2\pi} S(x) \log r - \frac{1}{4\pi} S(0) \log 2x - \frac{1}{4\pi} S(1) \log 2(1 - x)$$

$$-\frac{1}{4\pi} \int_0^1 S'(\xi) \operatorname{sgn}(x - \xi) \log 2|x - \xi| \, d\xi + O(r^2) \qquad \text{as } r \to 0 \tag{314}$$

7. Consider the weakly nonlinear eigenvalue problem

$$u_n'' + \lambda_n u_n - \varepsilon x u_n^2 = 0; \qquad 0 < \varepsilon \ll 1 \tag{315a}$$

$$u_n(0; \varepsilon) = 0 \tag{315b}$$

$$u_n(\pi; \varepsilon) = 0 \tag{315c}$$

Assume that $u_n(x; \varepsilon)$ and $\lambda_n(\varepsilon)$ have expansions

$$u_n(x; \varepsilon) = \left(\frac{2}{\pi}\right)^{1/2} \sin nx + \varepsilon u_n^{(1)}(x) + O(\varepsilon^2) \tag{316a}$$

$$\lambda_n(\varepsilon) = n^2 + \varepsilon \lambda_n^{(1)} + O(\varepsilon^2) \tag{316b}$$

a. Show that if we expand $u_n^{(1)}$ as in (74a) we obtain

$$a_{nk}^{(1)} = \begin{cases} \dfrac{(-1)^k}{n^2 - k^2} \left(\dfrac{2}{\pi}\right)^{1/2} \left[\dfrac{4n^2}{k(k^2 - 4^2n)}\right], & \text{if } k \neq n; \ k \neq 2n \\[2ex] 0 & \text{if } k = n \\[2ex] \dfrac{1}{8n^3} \left(\dfrac{2}{\pi}\right)^{1/2}, & \text{if } k = 2n \end{cases} \tag{317a}$$

$$\lambda_n^{(1)} = \frac{4}{3n} \left(\frac{2}{\pi}\right)^{1/2} (-1)^n \tag{317b}$$

b. For $n = 2$, calculate the expansion of $x_0(\varepsilon)$ to two terms, where x_0 is defined by $u_2(x_0; \varepsilon) = 0$.

8. Calculate the eigenfunctions corresponding to the first two eigenvalues λ_1 and λ_2 (where $0 < \lambda_1^{(0)} < \lambda_2^{(0)}$) for the following linear problem inside the unit circle

$$\frac{\partial^2 u_n}{\partial r^2} + \frac{1}{r} \frac{\partial u_n}{\partial r} + \frac{1}{r^2} \frac{\partial^2 u_n}{\partial \theta^2} + (\lambda_n - \varepsilon r^2 \sin 2\theta) u_n = 0 \tag{318a}$$

$$u_n(1, \theta) = 0 \tag{318b}$$

$$u_n(r, \theta) = \text{finite as } r \to 0 \tag{318c}$$

9. We wish to solve

$$\Delta u = 0 \qquad \text{in } \mathscr{D} \tag{319a}$$

$$u(1 + \varepsilon f(\theta), \theta) = h(\theta) \tag{319b}$$

where $\mathscr{D}$ is the domain $0 \leq r \leq 1 + \varepsilon f(\theta)$, $0 \leq \theta \leq 2\pi$, and f and h are both prescribed 2π periodic functions of θ.

a. Calculate the solution directly to $O(\varepsilon)$ in integral form, assuming that u has the expansion

$$u(r, \theta; \varepsilon) = u_0(r, \theta) + \varepsilon u_1(r, \theta) + O(\varepsilon^2) \tag{320}$$

Thus, u_0 is given by [see (165) of Chapter 2]

$$u_0(r, \theta) = \frac{1 - r^2}{2\pi} \int_0^{2\pi} \frac{h(\theta') \, d\theta'}{1 + r^2 - 2r \cos(\theta - \theta')} \tag{321a}$$

Show that u_1 is given by

$$u_1(r, \theta) = -\frac{1 - r^2}{2\pi} \int_0^{2\pi} \frac{f(\theta') \dfrac{\partial u_0}{\partial r}(1, \theta')}{1 + r^2 - 2r \cos(\theta - \theta')} \tag{321b}$$

and evaluate $(\partial u_0 / \partial r)(1, \theta)$ using (321a).

b. Use Green's function $K^{(0)} + \varepsilon K^{(1)}$ calculated in (81) and (86) in the generalized Poisson formula (143) of Chapter 2 to calculate the solution of (319) to $O(\varepsilon)$. Show that this agrees with the result you found in part (a).

10. Calculate the exact solution of

$$\varepsilon u'' + xu' + u = 0; \qquad -1 \le x \le 1; \qquad 0 < \varepsilon \ll 1 \tag{322a}$$

$$u(-1; \varepsilon) = 1 \tag{322b}$$

$$u(1; \varepsilon) = 2 \tag{322c}$$

in the form

$$u(x; \varepsilon) = e^{(1-x^2)/2\varepsilon} R(x; \varepsilon) \tag{323a}$$

where

$$R(x; \varepsilon) \equiv \frac{3}{2} + \frac{1}{2} \frac{\displaystyle\int_0^x e^{t^2/2\varepsilon}\, dt}{\displaystyle\int_0^1 e^{t^2/2\varepsilon}\, dt} \tag{323b}$$

Thus, $R(-1; \varepsilon) = 1$, $R(1; \varepsilon) = 2$, and $1 \le R \le 2$ for all x in $-1 \le x \le 1$. Equation (323a) then implies that the two boundary layers at $x = \pm 1$ grow exponentially toward the interior of the interval. In fact, the solution reaches the large value $u(0; \varepsilon) = \frac{3}{2} e^{1/2\varepsilon}$.

11. Calculate and match the outer and inner expansions to $O(\varepsilon)$ for

$$\varepsilon u'' + \frac{u'}{\sqrt{x}} - u = 0; \qquad 0 \le x \le 1; \qquad 0 < \varepsilon \ll 1 \tag{324a}$$

$$u(0; \varepsilon) = 0 \tag{324b}$$

$$u(1; \varepsilon) = e^{2/3} \tag{324c}$$

In particular, show that the overlap domain for the matching to $O(\varepsilon)$ is

$$\varepsilon^2 (\log \varepsilon)^2 \ll \eta(\varepsilon) \ll \varepsilon^{2/3} \tag{325}$$

and the uniformly valid solution to $O(\varepsilon)$ on $0 \le x \le 1$ is

$$\tilde{u}(x; \varepsilon) = \exp\left(\frac{2}{3} x^{3/2}\right) - (2x^{*1/2} + 1)\exp(-2x^{*1/2})$$

$$+ \varepsilon \left\{ \left[\frac{9}{10} - \frac{x}{2} + \frac{2}{5} x^{5/2} \exp\left(\frac{2}{3} x^{3/2}\right)\right] - \frac{9}{10}(2x^{*1/2} + 1)\exp(-2x^{*1/2}) \right\} \tag{326}$$

where $x^* \equiv x/\varepsilon^2$.

12. Calculate the next term in the uniformly valid solution of each of the following.

a. The boundary-value problem (134) for the case $k = k_0 = $ constant, $\ell = \ell_0 = $ constant

b. The heat-transfer problem (147)–(148)

13. The following is a mathematical model to illustrate the asymptotic behavior of certain "collision" trajectories that occur in celestial mechanics. We wish to calculate $u(t; \varepsilon)$ over the interval $0 \leq t \leq T(\varepsilon)$, where $T(\varepsilon)$ is defined by $u(T(\varepsilon); \varepsilon) = 1$, for the initial-value problem

$$\ddot{u} + u - \frac{\varepsilon}{(u-1)^2} = 2\sin t \tag{327a}$$

$$u(0; \varepsilon) = 0 \tag{327b}$$

$$\dot{u}(0; \varepsilon) = 0 \tag{327c}$$

Thus, if $\varepsilon = 0$, u describes the secular behavior

$$u_0(t) = \sin t - t \cos t \tag{327d}$$

which has $u_0(\pi/2) = 1$. But if $u = 1$, the term in (327a) that is multiplied by ε becomes infinite. Thus, the outer expansion of (327) cannot be uniformly valid near $t = \pi/2$.

a. Show that this outer expansion is

$$u(t; \varepsilon) = \sin t - t \cos t + \varepsilon \int_0^t \frac{\sin(t-\tau)}{(1 - \sin\tau + \tau\cos\tau)^2} d\tau + O(\varepsilon^2) \tag{328}$$

and refer to Problem 5 for the singular behavior of (328) as $t \to \pi/2$.

b. In the neighborhood of the collision time $t = T(\varepsilon)$, define the new dependent variable u^* by

$$u \equiv 1 - \varepsilon u^*(t^*; \varepsilon) \tag{329a}$$

and independent variable t^* by

$$t^* \equiv \frac{t - T(\varepsilon)}{\varepsilon} \tag{329b}$$

Expand $u^*(t^*; \varepsilon)$ in the form

$$u^*(t^*; \varepsilon) = u_0^*(t^*) + O(\varepsilon) \tag{330}$$

and show that u_0^* satisfies

$$\frac{d^2 u_0^*}{dt^{*2}} + \frac{1}{u_0^{*2}} = 0 \tag{331}$$

c. Match the outer and inner expansions for du/dt to show that the appropriate integral of (331) is

$$\frac{1}{2}\left(\frac{du_0^*}{dt^*}\right)^2 + \frac{1}{u_0^*} = -\frac{\pi^2}{8} \tag{332}$$

d. Use the condition $u = 1$ at $t = T$ to express the solution of (332) as the inverse of

$$t^* = -\frac{2}{\pi} u_0^*\left(1 + \frac{8}{\pi^2 u_0^*}\right) + \frac{16}{\pi^3}\log\left\{\left[u_0^*\left(1 + \frac{8}{\pi^2 u_0^*}\right)\right]^{1/2} + (u_0^*)^{1/2}\right\}$$

$$+ \frac{8}{\pi^3}\log\frac{\pi^2}{8} \tag{333}$$

e. Use (333) to calculate the asymptotic behavior of u_0^* as $t^* \to -\infty$ in the form

$$u_0^* = -\frac{\pi}{2}t^* + \frac{4}{\pi^2}\log(-t^*) + \frac{4}{\pi^2}\left(\log\frac{\pi^2}{2} - 1\right)$$

$$+ O\left[\frac{1}{t^*}\log(-t^*)\right] \qquad \text{as } t^* \to -\infty \qquad (334)$$

f. Match the outer and inner expansions of u to $O(\varepsilon)$ to show that

$$T(\varepsilon) = \frac{\pi}{2} + \frac{8}{\pi^3}\varepsilon\log\varepsilon - \left[\frac{2C_2}{\pi} + \frac{8}{\pi^3}\log\frac{\pi^2}{2} - 1\right]\varepsilon + o(\varepsilon) \qquad (335)$$

where C_2 is the constant defined in (310) of Problem 5.

14. Consider the following signaling problem for a pair of nonlinear first-order equations

$$\ell(v)v_t = u \qquad (336a)$$

$$\ell(v)u_x = -u \qquad (336b)$$

We are interested in the solution of (336) for $u(x, t; \varepsilon)$ and $v(x, t; \varepsilon)$ on $0 \le x < \infty$, subject to the initial condition

$$v(x, 0; \varepsilon) = \varepsilon = \text{constant} > 0 \qquad (337)$$

and the boundary condition

$$u(0, t; \varepsilon) = f(t) = \text{prescribed if } t > 0 \qquad (338)$$

Here $\ell(v)$ is a prescribed function of v.

a. Show that we must have

$$\frac{\partial}{\partial t}\left[\ell(v)\frac{\partial v}{\partial x} + v\right] = 0 \qquad (339)$$

and use this result to express the solution for v in the form

$$x = -\int_{v(0,t;\varepsilon)}^{v(x,t;\varepsilon)} \frac{\ell(s)\,ds}{s - \varepsilon} \qquad (340)$$

where $v(0, t; \varepsilon)$ is obtained from the solution of

$$\int_{\varepsilon}^{v(0,t;\varepsilon)} \ell(s)\,ds = \int_0^t f(\tau)\,d\tau \qquad (341)$$

Therefore, (340)–(341) define $v(x, t; \varepsilon)$, and (336a) gives $u(x, t; \varepsilon)$ once v is known.

b. Specialize your results to the case

$$\ell(v) = v^{3/2} \qquad (342a)$$

$$f(t) = 1 \qquad (342b)$$

c. Derive a uniformly valid expression for $v(x, t; \varepsilon)$ to $O(1)$ for the case $0 < \varepsilon \ll 1$, and sketch curves of v as a function of x for fixed values of t.

This problem is an idealized model for the heating (temperature $= v$) of a semi-infinite plasma column by shining a laser at one end.

15. In the isothermal one-dimensional flow of a gas, we assume that the pressure is proportional to the density $\bar{\rho}$. The flow is then defined in terms of $\bar{\rho}$ and the speed $\bar{u}$ by the laws of mass and momentum conservation:

$$\bar{\rho}_{\bar{t}} + (\bar{\rho}\bar{u})_{\bar{x}} = 0; \qquad \text{mass} \tag{343a}$$

$$(\bar{\rho}\bar{u})_{\bar{t}} + (\bar{\rho}\bar{u}^2 + \bar{C}_0^2\bar{\rho} - \mu\bar{u}_{\bar{x}})_{\bar{x}} = 0; \qquad \text{momentum} \tag{343b}$$

where $\bar{C}_0$ is the ambient speed of sound, a constant, and μ is the coefficient of viscosity, also a constant.

We wish to study the "piston" problem analogous to the one described in Section 5.3.4 (page 295). A piston is impulsively set in motion with constant speed $\bar{v} > 0$ into gas at rest with ambient properties $\bar{\rho}_0 = $ constant, $\bar{C}_0 = $ constant. The initial conditions are

$$\bar{\rho}(\bar{x},0) = \bar{\rho}_0 \tag{344a}$$

$$\bar{u}(\bar{x},0) = 0 \tag{344b}$$

for $\bar{x} > 0$, and the boundary condition at the piston is

$$\bar{u}(\bar{v}\bar{t},\bar{t}) = \bar{v} \tag{344c}$$

a. Introduce the dimensionless variables

$$x \equiv \frac{\bar{x}}{L}; \qquad t \equiv \frac{\bar{t}\bar{C}_0}{L}; \qquad u \equiv \frac{\bar{u}}{\bar{C}_0}; \qquad v \equiv \frac{\bar{v}}{\bar{C}_0}; \qquad \rho \equiv \frac{\bar{\rho}}{\bar{\rho}_0}$$

where L is a characteristic length. Show that (343a)–(344c) become

$$\rho_t + (\rho u)_x = 0 \tag{345a}$$

$$(\rho u)_t + (\rho u^2 + \rho - \varepsilon u_x)_x = 0 \tag{345b}$$

$$\rho(x,0) = 1 \tag{346a}$$

$$u(x,0) = 0 \tag{346b}$$

$$u(vt,t) = v, \qquad t > 0 \tag{346c}$$

where

$$\varepsilon \equiv \frac{\mu}{\rho_0\bar{C}_0 L} \tag{347}$$

Thus, ε is an artificial small parameter because there is no length scale in the problem formulated except the scale $\mu/\rho_0\bar{C}_0$; choosing $L \equiv \mu/\rho_0\bar{C}_0$ gives $\varepsilon = 1$.

b. For $\varepsilon = 0$, solve the piston problem and derive the shock speed and density behind the shock.

c. For $0 < \varepsilon \ll 1$, calculate the shock structure; then verify that ε is indeed an artificial small parameter. Introduce a coordinate system moving with the shock and show that the shock structure is the solution of the pair of ordinary differential equations that result from (345) in this frame.

16. Reconsider Problem 9 of Chapter 4; that is,

$$h_t + (uh)_x = 0 \tag{348a}$$

$$u_t + uu_x + (h + \varepsilon B)_x = 0 \tag{348b}$$

for the case

$$B(x) = \begin{cases} 1 - 4x^2, & \text{if } -\tfrac{1}{2} \le x \le \tfrac{1}{2} \\ 0, & \text{if } |x| \ge \tfrac{1}{2} \end{cases} \tag{349}$$

and the initial condition

$$u(x, 0; \varepsilon) = F = \text{constant} \ne 1 \tag{350a}$$

$$h(x, 0; \varepsilon) = 1 - \varepsilon B(x) \tag{350b}$$

The bore conditions for (348) are the same as (267) for a flat bottom. (Why?)
a. Look for a solution in the form

$$u(x, t; \varepsilon) = F + \varepsilon u_1(x, t, t_1) + O(\varepsilon^2) \tag{351a}$$

$$h(x, t; \varepsilon) = 1 + \varepsilon h_1(x, t, t_1) + O(\varepsilon^2) \tag{351b}$$

and show that S_1 and R_1, defined in terms of u_1 and h_1 by (242), are now given by

$$S_1 = -\frac{B(x)}{F + 1} + f_1(\xi, t_1) \tag{352a}$$

$$R_1 = \frac{B(x)}{F - 1} + g_1(\eta, t_1) \tag{352b}$$

where

$$\xi \equiv x - (F + 1)t \tag{353a}$$

$$\eta \equiv x - (F - 1)t \tag{353b}$$

b. Show that the evolution equations governing f_1 and g_1 are still (268).
c. Solve the initial-value problem for f_1 to obtain

$$f_1(\xi, t_1) = \frac{1 + 8\xi ct_1 - [1 + 16t_1(\xi + ct_1)]^{1/2}}{6ct_1^2} \tag{354a}$$

where

$$c \equiv \frac{3F}{4(F + 1)} \tag{354b}$$

d. Show that a bore starts at $\xi = -\tfrac{1}{2}$, $t_1 = \tfrac{1}{4}c$ and satisfies

$$\frac{d\xi}{dt_1} = \frac{1}{2} f_1(\xi, t_1) \tag{355}$$

where f_1 is given by (354). Verify that the bore has the qualitative behavior given in Figure 8.6c and that $\xi \sim t_1^{1/2}$ as $t_1 \to \infty$ along the bore.

e. For $F \approx 1$, assume

$$F = 1 + \varepsilon^\lambda F^* \tag{356}$$

where $\lambda > 0$ is to be determined and F^* is a fixed constant independent of ε. Look for a multiple-scale expansion for u and h in the form

$$u(x, t; \varepsilon) = 1 + \varepsilon^\lambda F^* + \varepsilon^\beta u_1^*(x, t, t^*) + \varepsilon^{2\beta} u_2^*(x, t, t^*) + O(\varepsilon^{3\beta}) \tag{357a}$$

$$h = 1 + \varepsilon^\beta h_1^*(x, t, t^*) + \varepsilon^{2\beta} h_2^*(x, t, t^*) + O(\varepsilon^{3\beta}) \tag{357b}$$

where $\beta > 0$ is to be determined and $t^* = \varepsilon^\beta t$.

Show that the richest approximation corresponds to $\lambda = \beta = \frac{1}{2}$ and that u_1^* and h_1^* have the form

$$u_1^* = F^* - \tfrac{1}{2}g_1^*(x, t^*) \tag{358a}$$

$$h_1^* = \tfrac{1}{2}g_1^*(x, t^*) \tag{358b}$$

By requiring the terms of order ε in (357) to be bounded, derive the following evolution equation for g_1^*:

$$\frac{\partial g_1^*}{\partial t^*} + (F^* - \tfrac{3}{4}g_1^*)\frac{\partial g_1^*}{\partial x} = B'(x) \tag{359}$$

and initial condition $g_1^*(x^*, 0) = 0$. The solution of (359) and other results are given in [17].

References

1. J. Kevorkian, and J. D. Cole, *Perturbation Methods in Applied Mathematics*, Springer, New York, 1981.
2. P. A. Lagerstrom, *Matched Asymptotic Expansions—Ideas and Techniques*, Springer, New York, 1988.
3. A. Erdelyi, *Asymptotic Expansions*, Dover, New York, 1956.
4. J. D. Murray, *Asymptotic Analysis*, Springer, New York, 1984.
5. G. F. Carrier, M. Krook, and C. E. Pearson, *Functions of a Complex Variable, Theory and Technique*, McGraw-Hill, New York, 1966.
6. M. Abramowitz, and I. A. Stegun, *Handbook of Mathematical Functions*, National Bureau of Standards, 1964.
7. B. Friedman, *Principles and Techniques of Applied Mathematics*, Wiley, New York, 1956.
8. G. B. Whitham, *Linear and Nonlinear Waves*, Wiley-Interscience, New York, 1974.
9. J. Kevorkian, "Perturbation techniques for oscillatory systems with slowly varying coefficients," *S.I.A.M. Rev.* 29; (1987): 391–461.
10. N. M. Krylov, and N. N. Bogoliubov, *Introduction to Nonlinear Mechanics*, Academy of Sciences Ukrainian SSR, 1937. Translated by S. Lefsehetz, Princeton University Press, Princeton, N.J., 1947.
11. Y. A. Mitropolski, *Problèmes de la Théorie Asymptotique des Oscillations Non Stationnaires*, Gauthier-Villars, Paris, 1966. Translated from the Russian. See also the English translation published by Israel Program for Scientific Translations, Jerusalem, 1965.

12. C. L. Frenzen and J. Kevorkian, "A review of multiple scale and reductive perturbation methods for deriving uncoupled evolution equations" *Wave Motion*, 7, (1985): 25–42.

13. R. M. Miura, "The Korteweg-deVries equation: A survey of results," *S.I.A.M. Rev.* 18, (1976): 412–59.

14. C. S. Gardner, J. M. Greene, M. D. Kruskal, and R. M. Miura, "Method for solving the Korteweg-deVries equation," *Phys. Rev. Lett.* 19, (1967): 1095–1097.

15. M. J. Ablowitz, and H. Segur, *Solitons and the Inverse Scattering Method*, S.I.A.M. Philadelphia, 1981.

16. R. M. Miura, and M. D. Kruskal, "Application of a nonlinear WKB method to the Korteweg-deVries equation," *S.I.A.M. Journal on Applied Mathematics* 26 (1974): 376–95.

17. J. Kevorkian and J. Yu, "Passage through the critical Froude number for shallow water waves over a variable bottom," *J. of Fluid Mechanics*, 204 (July 1989), 31–56.

INDEX